WOMEN SCIENTISTS IN AMERICA

Women Scientists in America

Struggles and Strategies to 1940

MARGARET W. ROSSITER

The Johns Hopkins University Press
Baltimore and London

This book has been brought to publication with the
generous assistance of the National Endowment for the Humanities.

Originally published, 1982
Second printing, 1983
Third printing, 1984

Johns Hopkins Paperbacks edition, 1984

The Johns Hopkins University Press, Baltimore, Maryland 21218
The Johns Hopkins Press Ltd., London

Library of Congress Cataloging in Publication Data

Rossiter, Margaret W.
Women scientists in America.

Bibliography: p. 399
Includes index.
1. Women scientists—United States.
2. Women in science—United States. I. Title
Q130.R68 331.4′815′0973 81-20902
ISBN 0-8018-2443-5 AACR2
ISBN 0-8018-2509-1

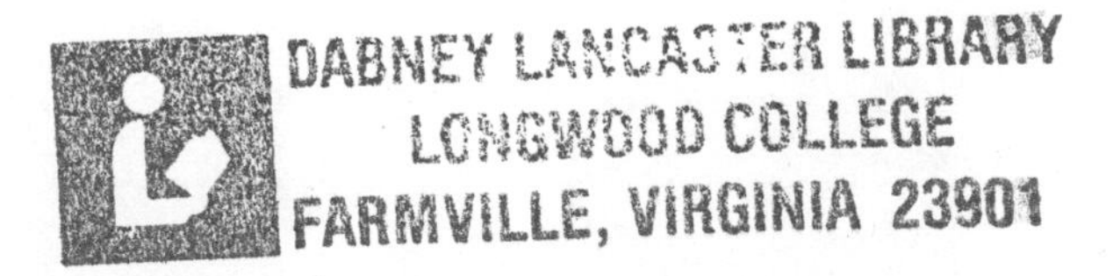

CONTENTS

ILLUSTRATIONS

PLATES

FIGURES

TABLES

PREFACE

This book is an analysis of women's participation in American science from the "beginning" to the recent past. It seeks to identify major and minor figures, to analyze significant education and employment patterns, to note striking achievements, and to examine the types of recognition accorded (or withheld from) women scientists. It has grown out of my realization that, although close to nothing was known, until recently, of the history of women in American science, women have been an integral part of the scientific community for well over a century. I can still recall my astonishment when I discovered in 1972 some women's entries in the old *American Men* [*sic*] *of Science* directories, and when I read biographies of several scientists in the then-new *Notable American Women*. Here were people who had been present at many of the familiar places and events, but who were totally unknown even to those of us well versed in the history of American science. I felt like a modern Alice who had fallen down a rabbit hole into a wonderland of the history of science that was familiar in some respects but distorted and alien in many others. Learning more about these women and bringing their stories into closer connection with the rest of the history of this period became a compelling and absorbing intellectual task.

The initial stumbling block was locating material, since most of the women scientists bordered, for a variety of reasons, on the "invisible." Overcoming this obstacle required several years of detective work in published and unpublished sources, especially old directories, bibliographies, and manuscript collections. In time the material threatened to grow beyond my control. Thus my initial problem of locating enough material on a host of hitherto obscure women became, over the years, the more usual one of trying to reduce an overwhelming amount of documentation to a coherent narrative. This meant, alas, that many topics and issues not part of the main story had to be relegated to the notes for other interested researchers.

My accumulated material not only greatly expanded the number of women and the range of topics that could be examined, but also deepened the level of analysis possible. In particular, the manuscripts exposed the motivation, attitudes, and behavior behind the statistics, and contemporary vocational guidance supplemented the experiences of the highly publicized but atypical "firsts" and "exceptions" enough to help establish the more usual patterns and normal expectations. Eventually these materials transformed the project from a kind of collective biography of women scientists (hardly any of whose names are household ones) into a history of an occupational group whose status had risen and fallen over time as the women's role responded to external events and pressures. My findings raised additional questions about the nature and history of scientific work, especially why some kinds had been more valued and honored than others. Thus the project stretched beyond the historical sociology of science and onto a kind of historical labor economics of science. In the process the manuscript grew so long that it seemed best to break the whole into two volumes: to finish off the first on the formative years to 1940 before continuing with the subsequent major developments: the involvement of the federal government in science, education, and employment.

I hope that this part of the story will be of interest to a broad audience—to those interested in science and how it has worked in the past and to those concerned with women's experience in the professions (including here academia and the government). In a sense the women's experience, hitherto so obscure, demonstrates that there were very definite limits to the supposed openness and rationality of the scientific community in the years before 1940.

ACKNOWLEDGMENTS

It is a great pleasure at the end of a long project to look back and recall the many persons and institutions that were so helpful along the way. I think first of the many librarians at the University of California at Berkeley, especially the almost infallible Marcia Goodman of Interlibrary Loan and Cathy Gordon, first in Documents and later in Education/Psychology, whom no reference, however obscure, could deter. The number of helpful librarians and archivists I encountered at other institutions is so large that even this list is too short: Patricia Ballou, Susan Boone, Margaret Campbell, Sandra Chaff, Deborah Cozort, William Deiss, Clark Elliott, Jim Glenn, Frances Goudy, J. William Hess, Kathleen Jacklin, Mary Jordan, Dolores Lahrman, Paul McClure, Marion MacPherson, Wyndham Miles, Eva Moseley, Mary-Elizabeth Murdock, Roxanne Nilan, Mary Jo Pugh, Judith Schiff, Warren Seamans, Mary Shane, Wilma Slaight, Murphy Smith, Patricia Bodak Stark, Mark Stauter, Elaine Trehub, Mary S. Trott, and Diane Windham. The nation is better served by these guardians of its heritage than most people realize.

I wish to thank also the many relatives and friends, especially those situated conveniently near to the many libraries and archives between Boston and Washington, who made space, shared their hospitality, and put up with the disruption of a visiting scholar for varying amounts of time (and often more than once): my parents, twin brother Charles, Michele and Mark Aldrich, Mary Ellen Bowden, Sharon Gibbs, Sally and David Kohlstedt, and Mary Ann Osley. Ann and Arthur Norberg also cheered me on when the future looked bleakest.

I am also grateful to the Charles Warren Center for Studies in American History at Harvard University for a fellowship in 1972/73, to the AAUW Educational Foundation for a research grant in 1975, and subsequently to the History and Philosophy of Science Program at the National Science Foundation and its manager, Ronald Overmann, for grants #SOC 77–22159 and #SOC 79–07562, which made it possible to persist with this project, even as it grew and grew; to Alexandra Oleson, John Voss, and their staff at the American Academy of Arts and Sciences in Cambridge, who administered the grants with great efficiency; and to the staff of the Office for the History of Science and Technology at the University of California at Berkeley for their professional courtesies over several years. Although it is a pleasure to acknowledge all this support, nevertheless the ideas presented here are my responsibility and not necessarily the official views or policies of any of these persons or institutions.

But I owe most gratitude to my fellow scholars in the history of science, American history, and economic history, particularly Sally Gregory Kohlstedt, who from the beginning was sure that the project could be done, had helpful comments and criticisms on almost every page of two full drafts, shared her own researches in an unpublished form, and politicked for the project at key stages. I'll always wonder whether it could have been done without her help. Other scholars who took time from their own work to read the manuscript in various stages and offer useful criticisms were Gayle Gullett Escobar, Jeffrey Escoffier, Peggy Aldrich Kidwell, Rosalind Rosenberg, Deborah Warner, Mary Brownlee, and Samuel Haber. The numerous others who had more specific suggestions are thanked in the relevant notes.

Sylvia Turner has been an excellent typist throughout. At The Johns Hopkins University Press, acquisitions editor Henry Y. K. Tom signed me up on the basis of an earlier, imperfect draft; copy editor Jackie Wehmueller fought heroically to bring order to the text and notes; and production editor Wendy Harris has turned the whole unwieldy mass into its final book form.

Permission to quote unpublished material has been graciously granted by: American Philosophical Society; The Bancroft Library; Barnard College Archives; Bentley Historical Library; California Institute of Technology Archives; Special Collections, The Joseph Regenstein Library, University of Chicago; Rare Book and Manuscripts Library, Columbia University; Department of Manuscripts and University Archives, Cornell University Libraries; Duke University Archives; Clark Elliott; Special Collections Department, Robert W. Woodruff Library, Emory University; W. A. Ferguson; Harvard University Archives; The Alan Mason Chesney Medical Archives of The Johns Hopkins Medical Institutions; Manuscript Division of The Library of Congress; National Academy of Sciences Archives; National Anthropological Archives, Smithsonian Institution; New York Botanic Garden Archives; Princeton University Library; University of Rochester Library; Rockefeller Archive Center; Schlesinger Library, Radcliffe College; College Archives, Smith College; Sophia Smith Collection (Women's History Archive), Smith College; Stanford University Archives; Barker Texas History Center, Archives and Manuscripts, University of Texas; Vassar College Library; Wellesley College Archives; University of Wisconsin Archives Oral History Project; Western History Research Center, University of Wyoming; and Yale University Library.

INTRODUCTION

Although the research on this project necessarily proceeded gradually upward from the specific to the overarching themes, it is only fair to prepare the reader for the wealth of material that follows by introducing some general concepts first. It is important to note early that women's historically subordinate "place" in science (and thus their invisibility to even experienced historians of science) was not a coincidence and was not due to any lack of merit on their part; it was due to the camouflage intentionally placed over their presence in science in the late nineteenth century. This arrangement, worked out by both sexes, was the result of the partial convergence of two major, though essentially independent, trends in American history between about 1820 and 1920. On the one hand there was the rise of higher education and expanded employment for middle-class women; on the other there was the growth, bureaucratization, and "professionalization" of science and technology in America.

Although occurring simultaneously, the two movements long seemed unrelated, largely because of various mental and cultural constructs of the time. Even as women's educational level rose and their role outside the home expanded, they were seen as doing only a narrow range of "womanly" activities, a stereotype that linked and limited them to soft, delicate, emotional, noncompetitive, and nurturing kinds of feelings and behavior. At the same time, the stereotype of "science" was seen rhetorically as almost the opposite: tough, rigorous, rational, impersonal, masculine, competitive, and unemotional. In terms, therefore, of nineteenth-century stereotypes or rhetorical idealizations, a woman scientist was a contradiction in terms—such a person was unlikely to exist, and if she did (and more and more of them were coming into existence), she had to be "unnatural" in some way. Women scientists were thus caught between two almost mutually exclusive stereotypes: as scientists they were atypical women; as women they were unusual scientists. Coming to grips with such "exceptions" long proved a problem to American society, including its historians of science. Moreover, this conceptual element meant that much of the history of women in science would be worked out not simply in the realm of objective reality, of what specific women could or did do, but covertly, in the psychic land of images and sexual stereotypes, which had a logic all its own. This psycho-political element was of overriding importance, for it affected much of women's and men's behavior about women's entrance into science, especially their strategies for advancement. In particular, women seeking greater opportunities in science could either capitalize upon these pervasive stereotypes, using them for such short-term gains as sex-typed employment, or try to defy them when they proved too restrictive after all. But they could not ignore these stereotypes—appropriate behavior and roles were too closely linked with one's gender for that.

The convergence of these two trends, the emergence of women and of science, falls into three main periods in the United States: before 1880, 1880 to 1910, and after 1910. Of these the second is both the most interesting and the most important, for it was then that acceptable conditions for women's presence in science were worked out. It is also worth noting that this chronology corresponds with few others in either American history, which relies heavily on such events as wars, economic depressions, and presidential administrations, or in the history of science, which has to date emphasized internal intellectual events such as scientific breakthroughs or "revolutions." What seems to have been more important for the women, and what has barely been touched upon in previous research, was a less dramatic, almost silent, process of economic, social, and demographic differentiation in the late nineteenth century whereby new roles and opportunities were unfolding at the same time that new persons were becoming available to fill them. By 1910, however, this period of great fluidity and innovation had ended, and a new rigidity set in. Despite much protest by feminists of both sexes, women's subsequent experience in science was more one of containment within previously demarcated limits than expansion into newer and greater opportunities beyond them.

Besides chronology, another set of distinctions necessary to emphasize at the outset are those among education, employment, and recognition. In the case of women they are not as closely linked as most writing on the sociology of science (which is largely based on the careers of successful men) would indicate. For women, higher degrees in science did not necessarily lead to desirable jobs—often they led only to unemployment; nor did good work and major publications frequently lead to advancement or better working conditions, as they often do for upwardly mobile men. Outstanding women frequently held lowly titles and were recognized only belatedly, as in their obituaries, decades after their achievements. The women's experience thus reflects a basic inconsistency in America in these years: American society, and especially its university faculties, became far more willing to educate women in science than to employ them, and were almost adamantly opposed to advancing or promoting any but the most extraordinary.

This impasse was the result of rationales that had been formulated earlier about women's position in American society. The initial justification for educating women had been that their expected roles of wife and mother would benefit. Without enlarged expectations for those prospective roles in the early nineteenth century, there would have been little reason for education beyond needlework and cookery. Hardly anyone expected middle-class women to, or wanted them to, hold jobs outside the home—or to vote. Raising and teaching sons who would work and vote, however, were deemed to be such overwhelmingly important full-time tasks that it was felt that mothers must be educated through the secondary and, later, college levels. How such well-educated women were to spend their leisure time received little attention. Before 1880 this rationale of "Republican Motherhood" worked rather well, and women devoted their free time to such acceptable areas as religious works, social philanthropy, and local reading circles, including a few scientific clubs.

Increasingly in the 1870s and thereafter, educated women joined scientific organizations and sought work in museums and observatories, thus seeming to most men to be encroaching upon what had formerly been exclusively masculine territory. Such incursions brought on a crisis of impending feminization, and a series of skirmishes in the 1880s and 1890s resulted in the women's almost total ouster from major or even visible positions in science. Although still allowed to enter most areas of science, they could hold only subordinate, close to invisible, and specifically designated positions and memberships. In this context at least part of the so-called "professionalization" of science in the 1880s and 1890s begins to look more like a deliberate reaction, conscious or not, by men against the increasing feminization of American culture, including science, at the end of the century. Ejecting women in the name of "higher standards" was one way to reassert strongly the male dominance over the burgeoning feminine presence. Thus, even though women could claim by 1920 that they had "opened the doors" of science, it was quite clear that they would be limited to positions just inside the entryway.

Henceforth when better-educated and more qualified women tried to move beyond this territorial demarcation or up the hierarchy, they were met with strong resistance. Rather than broadening the subsequent opportunities open to these highly educated women, employers and advisers of various sorts instead tried to dissuade them from aspiring to full equality or advancement through a pattern of "restrictive logic." This consisted of a series of limited stereotypes, double binds, resistant barriers (as in the phrase "no precedent"), and other no-win situations and fallacies that limited the subsequent role of almost all women scientists, regardless of their interests or abilities, to a narrow band of acceptable, because stereotypically "feminine," jobs.

In their attempts to get around these artificial barriers and inconsistencies, early women scientists developed a great many strategies. These tended to be of two sorts. One was the idealistic, liberal-to-radical, and often confrontational strategy of demanding that society reject all stereotypes and work for the feminist goal of full equality. This involved writing angry letters and otherwise documenting the "unfairness" of the unequal opportunities open to men and women. The most prominent and successful strategist of this school was Christine Ladd-Franklin, a Vassar College graduate of the 1860s, would-be physicist turned mathematician, psychologist, and logician, who for fifty years worked shrewdly and tirelessly for educated women. Her greatest triumph was in opening graduate schools to women in the 1890s and thus allowing women to earn the same doctorates as men.

The alternative strategy was the less strident and more conservative and "realistic" tactic of accepting the prevailing inequality and sexual stereotypes but using them for short-term gains such as establishing areas of "women's work" for women. Strategists for this approach emphasized that women had "unique skills" and "special talents" that justified reserving certain kinds of work for them. Ellen Swallow Richards, another early Vassar College graduate, typified this approach. Appropriately enough, her greatest achievement was the establishment of the field of home economics in the 1890s.

Although political opposites, these two strategies were together able to exploit all the available niches for women within a short period of time. Wherever the idealistic arguments that men and women were equal failed, as in the workplace, the realists could still make some partial gains by stressing the women's "special skills." For example, in the 1920s, when, after a decade of agitation, the idealists had again failed to produce any greater equality in scientific work, the conservatives came forward to stress the less threatening goals of building up and striving for excellence within women's separate realms, as at the women's colleges, in the schools of home economics, and in several separate women's clubs in science. They advocated such virtues as patience, self-discipline, and stoicism, which, rather than the liberals' protest and confrontation, remained the women scientists' dominant values until fairly recent times.

Even the limited success that women scientists had attained by 1940 had required the best efforts of a host of talented women, who, seeing how both science and women's roles were changing around them, took steps to carve out a legitimate place for themselves in the new order. If success can be judged in numbers, women scientists had done very well indeed, for by 1940 there were thousands of such women working in a variety of fields and institutions, whereas sixty or seventy years earlier there were about ten at a few early women's colleges. This great growth, however, had occurred at the price of accepting a pattern of segregated employment and underrecognition, which, try as they might, most women could not escape. Since moving beyond such patterns would still be difficult for later generations armed with federal legislation and executive orders, one can only marvel at what these earlier women, who had far less leverage, were able to accomplish in the century before 1940.

WOMEN SCIENTISTS IN AMERICA

1

WOMEN'S COLLEGES: THE ENTERING WEDGE

One of the major developments of the nineteenth century that is now often taken for granted was the rise of higher education for women. Early American society had been so antifeminist and its leading figures so fearful of "learned ladies" (whom they ridiculed) that women's education was restricted to the dame-school level until the 1820s. Starting then, however, and continuing into the next several decades, secondary and then higher education for women spread so rapidly in the United States that by mid-century this nation led the rest of the world in the amount of public and private education available to its women. This remarkable turnabout was basic for women's subsequent entrance into the scientific profession for two reasons: it allowed far more women than ever before to study science systematically, and it greatly increased the number of jobs and even professorships open to future generations of women in science. By the late nineteenth century there were enough such science professorships at the major women's colleges to form the nucleus of a female enclave within the American scientific community, the starting point from which most later developments flowed.

Ironically, such an outcome was far from, was almost the opposite of, what the advocates of higher education for women had claimed in the early nineteenth century. They were not professing to open new careers to women, but they had expected that their higher education would produce better wives and mothers for the American republic; thus their scheme fit neatly within the prevailing concept that historian Linda Kerber has termed "Republican Motherhood."[1] According to this political ideology, women could and should be educated, even to fairly high levels, if they were careful to use all this learning only within a clearly "womanly" sphere, as in raising moral and patriotic sons. To go beyond this restricted role was to jeopardize the precarious support for women's education and reawaken much conservative opposition to this innovation. Yet by working carefully within these constraints, a series of nineteenth-century educators were as free as financial resources would permit to expand and upgrade their institutions' offerings. By the end of the nineteenth century a great transformation had occurred in women's education in the United States—informal learning had given way to academies, and they in turn to colleges, some of which had faculties of more than fifty women and offered a mental training as rigorous as that available to men. Collectively these coveted positions constituted the women's "entering wedge" into the increasingly professionalized scientific world of the late nineteenth century.

Although some leading citizens of Philadelphia were sufficiently liberal toward the education of women to support a female academy as early as the 1780s,[2] most Americans of the 1790s and early 1800s considered educated women a threat rather than an asset to society. Still overwrought from the excesses of the French Revolution, citizens feared, it appears from the attitudes that can be documented, that educated women would develop deviant social or political behavior. Such women might, for example, take on undemocratic "airs" and refuse to do housework or even to obey their husbands, validating Molière's satires of Parisian women in his comedies written over a century earlier, *L'Ecole des femmes* (1662) and *Les Femmes Savantes* (1672). Others feared that educated women might become politically radicalized, as, it seemed to them, had happened to the (barely educated) Mary Wollstonecroft of England. She had pled for more education and jobs for women in her feminist manifesto, *A Vindication of the Rights of Women* (1792), but its message was discredited and ridiculed when it was revealed after her death in 1797 that she had had several lovers and had borne an illegitimate child. A third common fear, at the other extreme, was that educated women might become "masculinized," expect to be included in men's activities, and try to take over men's jobs. If these feelings were not enough, everyone "knew" that educating women would also undermine their health and that of their future children as well as violate God's will and the Bible. In order to avoid any of these anticipated social calamities or deviations from the desired American norm, conservative public opinion long recommended keeping women at home and out of the public eye. Such, for example, were the recommendations of the anonymous *Letters of Shahcoolen* (1802), a tract attributed to Benjamin Silliman, a young, strait-laced chemist from Yale College, who denounced in it both Mary Wollstonecroft and the flirtatious young women he was meeting as a student in Philadelphia.[3]

Even the lack of formal education beyond the elementary level had not, however, entirely prevented some early American women from studying on their own; a very few had even become scientists. These few were instructed at home by their educated menfolk, especially fathers, and encouraged to read books and develop their minds. The best known of these early women in science was Jane Colden, the botanist daughter of Cadwallader Colden of Newburgh, New York, a botanist and government leader who taught his daughter at home and who utilized her as his assistant for several years. Jane Colden classified over 300 species of plants from the lower Hudson River Valley in the 1750s, but she is best known for her identification and description of the gardenia, which she was the first to identify. Since Colden also helped her father with his large botanical correspondence, which included exchanges with the famed Swedish botanist Carolus Linnaeus, she was near the center of the renowned international "natural history circle" of the eighteenth century. Although Colden's botanical career was brief and her publications few, both ending with her marriage in 1759, she was America's pioneer (and only) woman scientist for almost ninety years.[4]

By the early nineteenth century some women were beginning to use the new means of informal learning that were coming into existence. Attending public lectures and visiting museums both quickly became accepted as properly "womanly" behavior in American cities and towns, first in Philadelphia but increasingly elsewhere as well. Charles Willson Peale of Philadelphia seems to have been the first person to invite women to his museum and lectures on natural history, starting in 1799 and continuing for several years thereafter. In the early 1800s the practice spread to other urban centers, including Albany, New York City, and Wilmington, and by the 1830s women's participation in rural lyceums had become an accepted practice. A strong stimulus toward liberality in these ventures was the usually precarious financial condition of such organizations. Under these circumstances even women's admission fees could be a welcome addition to an impoverished lecturer's income.[5] Before long other scientists would become aware of the economic importance of the female market for their scientific productions as well, and this subtle power of the purse would help to broaden the learning that was made available to women.

Probably the most widespread (and easily documented) aid to women's informal education in science in the early to mid-nineteenth century were the popular books and textbooks designed for female readers. Their prevalence in America in both foreign and domestic editions is another sign of the growing economic importance of women readers for scientific authors. These early popularizations were typically on botany and of English origin or derivation. The earliest of these seems to have been Jean-Jacques Rousseau's posthumous *Letters on the Elements of Botany Addressed to a Lady* (London, 1785), which went through eight English editions before 1815. This was followed by Priscilla Bell Wakefield's *Introduction to Botany, in a Series of Familiar Letters* (London, 1796), which had at least nine English editions by 1841 and several other American ones based upon them.[6]

As the market expanded greatly in the 1830s (with the advent of female academies), older works were republished. One was Leonhard Euler's eighteenth-century classic, *Lettres à une Princesse d'Allemagne sur divers sujets de physique et de philosophie* (1768–74), which went through five American editions in the 1830s and 1840s based upon an Edinburgh edition by David Brewster. The most widely read of these foreign scientific books for women were those of the well-known English popularizer Jane Marcet. Her *Conversations on Chemistry* (1806) went through more than fifteen editions in the United States before 1860, her *Conversations on Natural Philosophy* (1819) at least fourteen, and her *Conversations on Vegetable Physiology, Comprehending the Elements of Botany* (1829) at least seven.[7] By the 1830s and 1840s this vigorous popularization of science, characteristic of these decades in both England and the United

States, had managed to make the elements of several sciences available to all women who could read, and thus legitimized their having some elementary knowledge of science. (They were, of course, also free to read other textbooks that did not specify female readers in their titles.) Before long many American authors would also be tapping this burgeoning textbook market.

By 1820 the prevailing anti-intellectual and antifeminist attitude was proving inadequate to the needs of the republic. The traditional belief of Silliman and others that denying women an education would assure their virtue and make them good mothers had not proven correct—all too many uneducated women had evidently failed to deserve the high esteem held out for their sex and had proven to be quite foolish and frivolous. Not all women were "naturally" good mothers. Nor had the informal education that some women were pursuing on their own unsexed or radicalized them, as feared earlier. Ladies' textbooks, matinée lectures, and "ladies' day" at the museum had helped to soften the earlier incongruity between science or learning and "womanliness."

The new solution, therefore, was to systematize this new trend and to provide women with a more extensive but also properly moral education. It was Emma Hart Willard, more than anyone else, who wrought the basic revolution in the nation's attitude toward the education of women between 1819 and the 1830s. She was the first of several pioneers who, despite their own lack of education, did much to increase the opportunities and raise the level of education available to the next generation of American women. In 1819 she formulated a highly effective argument for the secondary education of women as part of a daring appeal to the New York state legislature for government support of her proposed school for girls.

In accordance with earlier views on female education, Willard linked the education of young women with their anticipated future sex roles, but in such an ingenious way as to reassure rather than alarm the citizenry. Thus she did not say that education would train women to think independently or prepare them to hold jobs. Nor did she mention preparing them for citizenship or the vote, since both were far in the future, and she never openly supported women's suffrage. Instead she stressed the conservative idea that education would make women better mothers. To buttress her case, Willard pulled together many prevailing sexist and nationalistic notions about motherhood and women's role into a potent argument for women's education. Her love of plane geometry is evident in her logic, which historian Anne Firor Scott has paraphrased in a short syllogism: "1. It is the duty of government to provide for the present and future prosperity of the nation. 2. This prosperity depends on the character of the citizens. 3. Character is formed by mothers. 4. Only thoroughly educated mothers are equipped to form characters of the quality necessary to insure the future of the republic."[8] Willard left it to the reader to supply: "5. Therefore state aid for women's education is necessary. Q.E.D."

Willard's argument was astute, especially when one considers how weak her leverage was. Basically she turned the increasing rhetorical tendency to idealize motherhood into an opportunity for women: since some American women were not becoming good mothers naturally, they all needed an education to do this. She could not cite any actual examples in 1819 of well-educated women who were model mothers, but she could indicate from her own teaching of history that many republics that had failed in the past (Greece and Rome, particularly) had also failed to educate their women. If America was to live up to its ideals and last more than a few generations before giving way to the corruption and decay that had befallen its predecessors, it must educate its women.

Willard's rationale had the further advantage that once women's education had been presented as necessary for motherhood, it was difficult for opponents to criticize it on any ground other than cost. For this too Willard had a solution: the more frequent employment of women schoolteachers in the expanding public school systems of the nation. She pointed out that, not only were more women available for this task than men, but they would teach school (as she herself had done) for a fraction of men's salaries. (This financial advantage did not escape the tight-fisted schoolboards of the time; when the first systematic data became available in New England in the late 1830s, they showed that almost one-half of all public school teachers were women and that they were being paid just 40 percent of the men's wages.)[9] Thus Willard's persuasive arguments hastened the spread of high schools for women in the 1820s and especially in the 1830s. There were apparently enough unhappy families in early America (whose problems were blamed on negligent mothers) for townfathers to take that heretofore "wasteful" step of spending money to educate women through the high school level.

Although the New York legislature did not give Willard state funds for her school, it did provide a charter, and before long the city of Troy, New York, offered to subsidize the academy. By all accounts this Troy Female Seminary prospered, and by the time Willard retired in 1838 it had trained several thousand students. Historians have now begun to analyze the inconsistencies between Willard's public rhetoric and her school's actual circumstances. Anne Firor Scott has suggested that Willard had remarkable political sagacity and skill as well as a complex personality. For example, she successfully applied camouflage to the sensitive issue of her pupils' subsequent vocations. Although most of her students did become wives and mothers, quantitative evidence now available indicates that they married less often and had fewer children than did women of the time who had not attended such academies. Had this been known in Willard's time, it is not clear what would have happened. She herself reportedly brushed off any suggestions that she was overeducating her students for marriage with the comment that young men were fighting so hard to find wives among them that she couldn't keep enough of them in teaching. In fact, so many of Willard's students took up teaching either for a few years before marriage or for most of their lives that some historians have characterized the Troy Female Seminary as one of the nation's first normal schools and Emma Willard's signature on a letter of recommendation as the first teacher accreditation in America.[10] Only in the late 1830s, at least fifteen years after Willard opened her Troy Seminary, were the first state-supported normal schools established.

Another ambiguity in Willard's venture was in the curriculum, which was far more rigorous than most "womanly" instruction of the time. Yet Willard's changes even here were carefully disguised. Thus although she openly scrapped embroidery, the previous core of the female curriculum (which she considered a waste of educated women's time), she featured such domestic subjects as household management and even pastry-cooking, insisting that if a woman failed at such womanly arts she had failed at everything. But alongside these "feminine" subjects, Willard introduced such demanding subjects as mathematics (through trigonometry) and science (including her favorites, physiology and natural philosophy or physics), which females had hitherto been thought incapable of understanding. She even considered naming her school a "college," but shrewdly and cautiously refrained, lest the notion of women's attending a "college" provoke undue ridicule.[11]

Thus Willard, who advocated many of the same reforms that Mary Wollstonecroft had suggested in England in the 1790s, had also learned the lesson of that Englishwoman's unfortunate experience (as should any would-be reformer): if one wishes to introduce seemingly radical reforms, it helps to be personally discreet and socially conservative. For most of her life Willard maintained a highly respectable lifestyle as a wife, mother, and, later, widow, and took pains to understate her ambitions and camouflage them with conservative rhetoric. She thus earned a credibility and respectability that permitted her to do what were in reality rather radical things, such as upgrading women's education by introducing some of the elements of a "masculine" curriculum and expanding their employment opportunities by training some of the nation's first "career women." Yet such an interpretation raises the questions of how aware Willard was of her own radicalness and how deliberate her dissimulation was. The answers remain tantalizingly ambiguous since her own statements on women's role are contradictory. Yet whether she realized it or not, Willard, her school, and others like it provided the essential starting point for women in science and the professions.[12]

Meanwhile Willard's sister Almira Hart Lincoln also worked hard, both at the Troy Female Seminary and later at her own academy in Maryland, to inject more science into the women's curriculum. Growing up in western Massachusetts, she attended various local academies and taught at a few others before her first marriage in 1817. Finding herself six years later widowed, destitute, and the mother of two, she resumed teaching and later became vice-principal of her sister's flourishing Troy Female Seminary. Already interested in science, Lincoln took advantage of her location in Troy to attend (as had her sister before her) the lectures of Amos Eaton, professor of science at the nearby Rensselaer School. An early self-taught geologist, chemist, and botanist, Eaton was particularly adept at the teaching of science to popular audiences through lectures and textbooks. Because he also held advanced ideas on the education of women, he helped several early women to further their education in science and to train other women in it as well.[13] Inspired by her sister's success in the textbook market, Eaton's encouragement, her own financial needs, and her students' evident intellectual wants, Lincoln began to write textbooks in the late 1820s. Her first book, the *Familiar Lectures on Botany* (1829), was also her most successful, going through at least seventeen editions and selling over 275,000 copies by 1872. This and her other textbooks on chemistry and natural philosophy eventually made her a wealthy woman. Thus the indefatigable Lincoln (later Phelps, a mother of four and stepmother of six) helped to kindle an interest in science among thousands of men and women in America from the 1830s through the 1870s and helped to improve the science teaching that was then available. She was also a pioneer in the generally invisible role of textbook or science writer, a role that is now taken for granted but one that, since it requires little institutional support, has continued to attract women scientists ever since.[14]

Since Lincoln's books were so widely purchased, it is worthwhile to examine their contents and note the reasons she gave for women's studying science. Though Lincoln's books were considered elementary at the time (necessarily so, since there would not have been such a large market for a more advanced text), they now seem quite rigorous in their stress on nomenclature and taxonomy, which she was at pains to explain in detail. Lincoln's books also urged the women of the time to get outdoors, look at the plants, form their own collections, and in general busy themselves with subjects of greater importance than the "trivial objects" to which women's attention was usually directed. Although she also stated that the discipline of studying science would be good training for their minds, her stress on science's value in increasing their respect for nature and bringing them closer to God probably accounts for her books' wide popularity. This was a common belief in the 1830s through 1850s and one that was often stressed at the female academies, which probably also used Lincoln's text. Thus Lincoln, cautious like Willard, did not advise women to plan to become scientists themselves (she never published a scientific article and apparently did not correspond with the leading male botanists of the day) or even to teach science to others, as many of the future schoolteachers in her class would soon be doing. Instead she urged them to use science to supplement their traditional family and domestic roles with a stronger religious faith. The study of science was thus not to threaten the established social order by taking women out of the home but to enrich and elevate their domestic life. She made her conservative philosophy explicit in a series of addresses to her students which she published in 1872: a woman's only safeguard in this world was a strong religious faith (and, therefore, not the vote).[15]

Once Willard had, with Lincoln's assistance, cleared the way, other women and men started many other "female academies," "seminaries," and even a few "female colleges" in the 1830s through the 1850s. One of the most important of these for the sciences was Mary Lyon, the founder of Mount Holyoke Seminary (later College). Growing up in western Massachusetts, she attended local academies and began to teach in a series of similar schools. One particularly fortunate assignment was in Amherst, Massachusetts, where she boarded with the Reverend and Mrs. Edward Hitchcock, he a geologist at Amherst College and she an illustrator of his publications. While staying with them, Lyon, like Willard and Lincoln, was able to attend the chemical lectures of Amos Eaton, who visited the college in 1824. Eaton was so impressed with Lyon's interest in chemistry that he invited her to spend her summer vacation at his house in Troy, where he could teach her the elements of chemical experimentation. It was the main scientific training of her career. Later, growing dissatisfied with the academies at which she was teaching, Mary Lyon visited Emma Willard and determined in 1833 to start her own school, the Mount Holyoke Seminary in western Massachusetts, as soon as she could raise the money. Here, though burdened by administrative duties, she continued to teach chemistry, her favorite subject, and she gave the sciences such a strong place in the college's curriculum that her vision of their importance has stood as an inspiration to her successors there ever since. Her stress on science stemmed partly from a belief, akin to that of Lincoln, that an understanding of the natural world would increase the religious fervor of the students, many of whom did become missionaries. Apparently uninterested in or unable to pursue research herself, Mary Lyon published no scientific articles; nor does she seem to have belonged to any scientific societies, although she was in contact with Hitchcock and other scientists at Amherst and Williams colleges. Most of her driving energy went into building up Mount Holyoke Seminary, over 80 percent of whose graduates in her day (1837–1850) taught school for an average of six years apiece.[16]

Mount Holyoke was just one of many such institutions coming into being at this time. The limited evidence available about their course offerings indicates that, despite slender budgets, they offered considerable amounts of science. Deborah Warner has examined the catalogs and advertisements of many of these schools, such as the Albany Female Academy and the Packer Collegiate Institution of Brooklyn, New York, and found many signs of scientific instruction and apparatus. Thomas Woody, the pioneer in the history of women's education, has indicated that many of these academies and seminaries stressed science because they were church related or church supported and, as in Lincoln's textbooks, they saw science or a science-oriented curriculum as a good way to reassure parents and students of the high moral character and religious values taught there. (Literature, especially novels, were by contrast quite suspect.) An 1871 survey of these "institutions for the superior instruction of females" by the U.S. commissioner of education found that about one-half of the 182 schools reporting claimed to have a chemical laboratory, a natural history museum, and a philosophical cabinet, all standard apparatus of the day. Eighteen even had an astronomical observatory, and 9 others a telescope. The academies' religious atmosphere was also still strong, since 84 (46.2 percent) were headed by clergymen or Roman Catholic nuns.[17]

Nevertheless, most of this instruction may have been quite elementary. Another survey taken by the commissioner of education a year later found that of the 24,663 women attending the 223 institutions reporting in 1873, one-quarter

(6,319) were only "preparatory" or subcollegiate students. He also found that most of these institutions were in a highly precarious financial situation: most had no endowments; only 9 of the 223 reported endowments of over $10,000, and only one over $40,000 (the Robinson Female Seminary of Exeter, New Hampshire, with a reported $200,000).

Another characteristic of these institutions, perhaps as much the result of their poverty as any overt feminism, was the high percentage of women on their faculties. Of the 2,124 persons employed on the faculties of these 223 institutions for women in 1873, 1,553 were female (73.1 percent). If one estimates that one-third to one-quarter of these women taught science, though probably not specializing in any one branch, then there were at least 400 women science instructors in the United States by 1873.[18] Little is known about these early women science teachers; the best qualified ones were themselves probably just recent graduates of these same academies. Within a few decades, therefore, the popularization of science and the growth of the academies had led to several hundred jobs for women in science teaching. Thus, whatever the quality of the actual instruction at these schools, the very existence of so much science teaching at so many institutions of at least secondary grade indicates that the movement to give women some kind of scientifically oriented higher education had already come a long way in the fifty years since Emma Willard had skirted public ridicule and opened her pioneering female seminary.

Although these academies and seminaries would seem to have been well suited to their time and place, and although they taught thousands of women all that they would need to know for several decades to come, later educators criticized their instruction for its obvious inferiority to that available to men at colleges. Thus the next step in women's higher education would be to upgrade these institutions and the instruction they offered. This would be difficult, since, for the reasons discussed earlier, little could be said about the necessity of improving education so that American women could be trained for employment (other than teaching) or for citizenship or for the vote. Most of the advocacy for "raising standards" would instead stress the feminist ideal that women should have and could benefit from the same (not just an "equal") college curriculum as that of men—men were, all agreed, headed for very different lives. What these women might eventually do with all this education, especially if they did not marry, was not clear at this time. Nor was it evident at the outset that "raising standards" for women could have its drawbacks as well as its advantages. In time, however, the desire for a full "collegiate" education for women would have important consequences for professional women.

Although Oberlin College had been coeducational since its founding in 1833 and several, primarily southern, women's institutions had called themselves "colleges" before the Civil War, the real impetus toward the full collegiate education of women came with the opening of Vassar College in Poughkeepsie, New York, in 1865. By 1870 many of the state universities, especially the new land-grant institutions, were also accepting their first women students; the most important of these coeducational institutions for women scientists were Cornell

PLATE 1. A physics lecture at the coeducational University of Michigan in the late 1880s or early 1890s was racially integrated but sexually segregated. (Courtesy of Michigan Historical Collections, Bentley Historical Library, University of Michigan.)

University and the University of Michigan (see plate 1).[19] Of even greater importance for women in science, as shown in tables 1.1 through 1.4, because they not only educated women in science but also employed sizeable numbers of them on their faculties, was the establishment of several more independent women's colleges in the 1870s and 1880s: Smith College (1871), Wellesley College (1875), and both Bryn Mawr College and the Baltimore College for Women (later renamed Goucher College) in 1885. Besides these, Mount Holyoke Seminary became a degree-granting college in 1888 and Columbia University established Barnard College as its coordinate institution for women in 1889.[20] Thus within a few decades after the Civil War the number of opportunities for women scientists at the collegiate level expanded greatly.

The chief advantage of these new elite colleges seem to have been: (1) their sizeable endowments or, in the case of the coordinate colleges like Barnard, their limited access to full university resources; (2) their nondenominational rather than sectarian administrations, which attracted a more national and urban clientele; and, perhaps most importantly, (3) their almost feminist commitment to excellence in women's higher education. These advantages enabled the new colleges to build impressive campuses (Wellesley's was valued at $2 million in

TABLE 1.1. Baccalaureate Origins of Female Scientists before 1920

Institution	Total	Botany	Zoology	Psychology	Medical Sciences	Mathematics	Chemistry	Home Economics	Geology	Physics	Astronomy	Anthropology
Wellesley	36	6	2	4	7	8	3	0	1	4	1	0
Vassar	34	1	6	6	2	3	4	4	2	1	5	0
Smith	29	5	5	5	2	5	3	1	0	0	3	0
Mount Holyoke	26	1	8	4	2	3	4	1	1	2	0	0
Cornell	23	4	5	3	3	2	1	1	2	2	0	0
Bryn Mawr	18	0	4	1	3	1	3	0	4	2	0	0
Chicago	18	5	3	3	2	0	2	2	1	0	0	0
Barnard	17	2	1	5	1	1	1	2	1	1	0	2
Michigan	17	4	5	3	0	0	2	2	0	0	1	0
Goucher	13	1	5	0	2	1	3	0	0	1	0	0
Pennsylvania	11	3	4	1	1	0	2	0	0	0	0	0
Nebraska	10	4	0	5	0	0	1	0	0	0	0	0
Wisconsin	9	3	2	1	0	1	1	0	1	0	0	0
Stanford	9	3	2	0	2	0	0	2	0	0	0	0
Oberlin	9	4	1	1	0	2	0	0	0	1	0	0
Kansas	8	0	5	1	1	0	0	0	1	0	0	0
Ohio State	8	4	1	0	0	0	0	0	3	0	0	0
Indiana	7	2	1	1	0	2	0	0	1	0	0	0
Radcliffe	7	1	2	1	0	1	0	0	0	0	2	0
MIT	6	0	1	0	0	0	0	1	2	2	0	0
78 other U.S. institutions	124	27	17	22	17	11	5	9	3	7	6	0
Total	439	80	80	67	45	41	35	25	23	23	18	2

SOURCES: First three editions of *American Men of Science* (1906, 1910, and 1921).

the 1880s), to increase their scientific apparatus (Vassar's led all others at $54,000 in 1887), and to upgrade their faculties (Bryn Mawr was the first to stress doctorates).[21] In addition to all these tangible improvements over the earlier academies, these new colleges seem also to have unleashed an enthusiastic ésprit de corps among the early faculties and students, who responded energetically to the challenging opportunity before them.

The most prominent woman professor on these early college faculties and the first real "scientist" since Jane Colden in the 1750s,[22] in that she published at least seven scientific articles, was the astronomer Maria Mitchell (plate 2), certainly the most important woman scientist in America in the nineteenth century. Her election to a professorship at Vassar College opened a new era for women in American science. Like Lyon and Lincoln, she attended local academies, but like Colden of the previous century, she apparently learned most by assisting her father, a banker and astronomical observer, at home on Nantucket Island, off the Massachusetts coast. In 1847 Maria Mitchell not only calculated the position of a new comet but also observed it across the sky and thereby won (after some controversy) the gold medal that the king of Denmark had promised to the discoverer. This made the twenty-eight-year-old Mitchell a celebrity. She was soon elected the first woman member of the American Academy of Arts and Sciences (over the protests of Asa Gray), the American Association for the Advancement of Science (supported by Louis Agassiz), and eventually (1869) one of the first American women in the American Philosophical Society of Philadelphia. But her fame and recognition did not lead to any immediate improvement in her employment prospects, for she continued to work as a part-time librarian at the Nantucket Athenaeum, assist her father at his observatory, and perform "computer" work for the U.S. Coast and Geodetic Survey on the side. She was thus highly qualified and available in 1862, when Matthew Vassar, who was anxious to have a prominent woman scientist on his new faculty, offered her the job. She accepted and became one of the nation's first women professors of science. At Vassar College Mitchell lived with her father in the observatory built for her and was a strong force on the faculty, always favoring high educational standards for the students, and fighting at one point (with Alida C. Avery, the college physician) for female salaries equal to those of the male members of the faculty. Like other women professors later, Mitchell took students on her field trips and included them in her observational schedules, and, although not liked by all of them (even protégée Mary Whitney later recalled her as having been "brusque but brilliant"), she nevertheless inspired several to become scientists.[23] At least three of these became leaders in the growing movement to increase women's opportunities in science: Mary Whitney, '68; Christine Ladd-Franklin, '69; and Ellen Swallow Richards, '70. Although in different fields and with varied political styles (no two could have been more different than Ladd-Franklin and Richards), they carried Mitchell's determination and vision of what women could become in American science into the twentieth century. Thus Mitchell, herself a product of the older family and

PLATE 2. Maria Mitchell, shown here in the E. T. Billings portrait of the 1880s, was America's best-known woman scientist of the nineteenth century. Her appointment to a professorship of astronomy at the newly opened Vassar College in 1865 was a pioneering step for all women in science. (Courtesy of Helen Wright and the Nantucket Atheneum.)

academy traditions in women's education, started at Vassar a new collegiate one that has continued down into the recent past.[24]

Although Vassar's opening in the 1860s had been accompanied by high expectations and much public endorsement, by the 1870s it had become a target of a backlash against women's increasing participation in medicine, science, and higher education. In 1873 Dr. Edward Clarke of Boston published a scurrilous attack on women's higher education, *Sex in Education*, which asserted (with examples allegedly from Vassar College) that women's health was being ruined by intensive study. Although quickly refuted by the Vassar physician, the charge struck such a responsive chord of public opinion that it put the advocates of women's higher education on the defensive for at least a decade, though with no obvious impact on college enrollments.[25] In this continuing debate the standard argument against women's participation in science stressed their supposedly delicate physical health, small and light brains, and greater "variability" in all measureable physical and mental traits. This "evidence" purported to show either that women were not physically able to undertake the study of science, or, if they did, that their bodily and mental functions would be so seriously impaired as to endanger the future of the entire race. To these somewhat hysterical medical arguments the proponents of a greater role for women in science responded by scanning the history of women scientists to cite examples of women who had contributed to science without endangering either their health or their femininity. In this Mitchell, so prominently placed at Vassar College, was a frequent example, as were her own heroines, the British astronomer Caroline Herschel and mathematician Mary Somerville. Neither side ever convinced the other, however, and the argument has persisted in various guises and with many refinements into recent times,[26] confirming perhaps what sociologists already know—that mere examples, even telling ones like these, never quite puncture stereotypes. Rather than disproving or modifying incorrect presuppositions, as they should logically, such counterexamples are instead easily neutralized as "exceptions" and the stereotypes allowed to persist unchanged.

At this time, Mitchell, whose outspoken political style contrasted sharply with that of the circumspect Emma Willard several decades before, took the offensive and helped to found the Association for the Advancement of Women (AAW) in New York City in 1873. This group, recently examined by Sally Gregory Kohlstedt, sought to unite those women who were working to increase their opportunities outside the home. Its chief activity was an annual "Congress of Women" at which members gave speeches, issued reports, and otherwise discussed such budding controversial issues as temperance, woman's suffrage, domestic science, higher education for women, and openings for women in the professions, especially medicine and science.[27] Although staid by comparison with later suffragist activities, these meetings attracted wide and largely favorable press coverage in the 1870s and 1880s. Some clippings of these articles can still be found in private scrapbooks where young women of the time who were groping toward some larger professional role carefully pasted them.

Mitchell held a prominent postion in the AAW. First as president, from 1874 through 1876, and later as chairperson of its committee on science (1876 through 1888), Mitchell reported annually to the AAW on women's advances and setbacks in science. Here her major concerns were their attempts to find scientific employment other than teaching, their varied reception by the major scientific societies, and their participation in such related activities as the nation's centennial exposition in Philadelphia in 1876. Realizing that most of what was thought to be true of women scientists was in the realm of stereotypes and conjectures, Mitchell also took steps to determine how many women scientists there were in the United States and how they were faring. Twice (in 1876 and 1880) she attempted to survey the women scientists of the nation in order to determine their interests and the obstacles they faced, but her efforts were hampered by the women's geographical dispersion and the limitations of her informants, who were usually Vassar alumnae. Even when progress seemed slow, Maria Mitchell did not despair of women's eventual contribution to science; instead, the more she saw of the foibles of the men in the field, the more urgent she felt the arrival of women; as she reported in her presidential address in 1875: "In my younger days, when I was pained by the half-educated, loose, and inaccurate ways which we [women] all had, I used to say, 'How much women need exact science,' but since I have known some workers in science who were not always true to the teachings of nature, who have loved self more than science, I have now said,

'How much science needs women.' "[28] It may have been such stirring rhetoric that inspired her students and women elsewhere to take up science and push harder for full careers within it.

Yet even as Mitchell spoke, the demand for women trained in the sciences was increasing in one quarter at least—in the faculties of the many women's colleges, academies, and normal schools. Although one would like to discuss in detail the most important women scientists at the women's colleges in these years, they were so numerous and material about them is so scarce that it is difficult to obtain more than passing glimpses of their personalities. Since, however, their obituaries show that most of them led similar lives, these women perhaps can be studied best collectively, in a composite biography.[29]

Because other women scientists were not as well known as Maria Mitchell, the president or trustees of most new colleges seeking to recruit women for their first faculties had to write to the prominent male scientists of the time, such as botanist Asa Gray or mathematician Benjamin Peirce, both of Harvard University, in hopes that these men might be able to recommend some appropriate candidates. Such professors might have corresponded with or met a woman scientist or allowed her to attend their Harvard lectures (unofficially, of course) or otherwise have been familiar with and been able to evaluate women who were either, like Mary Whitney, at home, or, like Susan Hallowell in Maine, teaching in an academy. Thus the first generation of women faculty at these women's colleges had an uneven educational background—most had attended a female seminary or academy, some were already graduates of other early colleges, but others had little more formal training than Mary Lyon had had in the 1830s. This variation would disappear in the 1890s and early 1900s as doctorates came to be required at the better women's colleges.

It also went without saying that according to the mores of the time, all candidates had to be of good Christian character and not only single but in no danger of marrying. Married women were not even considered for employment at the early women's colleges, even, it seems, when they were clearly the best candidate available. Thus Wellesley College, established in 1875 and committed to hiring women—twenty-one of its first faculty of twenty-two were women; only the music professor was a man—could hardly have done better than hire Ellen Swallow Richards, the most prominent woman chemist in New England at the time. Yet there is no evidence that founder Henry Durant ever considered her for Wellesley's professorship of chemistry, perhaps because she was married. Similarly, any women already on the college faculty who did marry, as happened on occasion, resigned immediately.[30]

Because resignation upon marriage was a matter of course, the rationale for the practice was not articulated until one brave soul refused to conform. Then, as happened in a well-documented case at Barnard in 1906, both sides were forced to state their position and thus reveal the illogic as well as the impasse between them. The crisis at Barnard started when Harriet Brooks, an instructor in physics, wrote Dean Laura Gill of her engagement and willingness to resign if her new duties interfered with her teaching. But because she did not expect them to, she planned to continue teaching: "I think also it is a duty I owe to my profession and to my sex to show that a woman has a right to the practice of her profession and cannot be condemned to abandon it merely because she marries. I cannot conceive how women's colleges, inviting and encouraging women to enter professions, can be justly founded or maintained denying such a principle."[31]

The dean (not one of Barnard's best) responded icily that the trustees expected a married woman to "dignify her home-making into a profession, and not assume that she can carry on two full professions at a time," adding threateningly that they would be displeased by her bringing on "detrimental publicity and unpleasantness by a protest against the enforcement of the rules."[32] The dean was so convinced of her course that even the plea made by Margaret Maltby, the longtime chairman of Barnard's physics department, to let Brooks stay because she was uniquely qualified for the job and "not a usual type of woman"[33] fell on deaf ears. (Brooks had already published two major articles on radioactivity with Ernest Rutherford at McGill University and had been the first [American?] woman to study at the famed Cavendish Laboratory in Cambridge, England. Both Rutherford and Nobel laureate J. J. Thomson thought she was on the verge of a great career in physics.)[34] The thoroughly nettled dean cited to Maltby the trustees' rule on the employment of married women. It put them in a classic double-bind, blaming married women who chose to work for one reason if not for another: " 'The College cannot afford to have women on the staff to whom the college work is secondary; [yet] the College is not willing to stamp with approval a woman to whom self-elected home duties can be secondary;' this was the phrasing given to the matter by one of our most influential men."[35] Thus if the woman thought that she could do both duties successfully, as Brooks kept insisting, then her character was defective! Eventually the dean won out and Brooks resigned, though, as events turned out, more because of the awkwardness of a broken engagement to a fellow Columbia graduate student than because she, still single after all, had in the end violated any Barnard rules.[36] Yet in the larger sense, both Barnard and physics clearly lost by her resignation: since Brooks's replacement, Grace Langford, never did finish her Columbia doctorate,[37] physics failed to prosper at Barnard College, and Brooks herself never did any more physics.

Male faculty at the women's colleges, on the other hand, were expected to be married. Only Bryn Mawr College hired unmarried men, several of whom eventually found wives on the campus: geneticist Thomas Hunt Morgan married student and assistant Lillian V. Sampson in 1904, and Clarence Ferree married colleague Gertrude Rand in 1918.[38] Though such a marriage prevented these talented women from holding more than a "research associateship," if that, ever again (see chapter 7), they were not always totally lost to science. Many continued to work in someone's laboratory, often as their husband's assistant, and others tried to help younger women either from behind the scenes or by influencing their husbands, thus offering a partial explanation as to why some male professors were kinder to the women than were others.

From the single woman's point of view, a call to the new women's colleges was the turning point of her life. She could never have expected such an opportunity and had probably resigned herself to spending the rest of her life with her family, perhaps tending the ubiquitous sick relatives, living with a sibling's family as a spinster aunt, or taking up teaching at an academy and boarding with a series of local families, as Mary Lyon had done. The move to the women's college opened to her a new life of greater intellectual challenge and usefulness than the old and also gave her a new freedom and a new "family" with which to share her experiences. Having had little experience with money and few, if any, other job prospects, these early women faculty members (unlike Mitchell) rarely protested the low salaries they received and the stipulation that they live among the students on campus.[39] Some independently wealthy ones, like Emily Gregory, professor of botany at Barnard College, who, despite her doctorate from Zurich, had despaired of ever finding suitable academic employment, not only went unpaid, but were only too glad to devote their services and their remaining fortune to building up an academic department.[40] They were devoted to the college, found their closest friends on the faculty, and planned to spend the rest of their lives there, as most of the early women in fact did.

These early women faculty sometimes had chivalrous professional contacts with male professors at nearby institutions. Asa Gray, Benjamin Peirce, and especially E. N. Horsford (formerly of Harvard) were all very helpful to the women at Wellesley College in the 1870s and 1880s, as Amos Eaton had been to the Troy women earlier. Charles A. Young, a Dartmouth and later Princeton astrophysicist, advised his niece Annie Young, also an astronomer, and the other women at Mount Holyoke. These men enjoyed playing a helpful "big brother" role, escorting the women to meetings, helping them select equipment, advising them on professional matters, and, in Horsford's case, even offering legal advice and devising a pension plan.[41] Women faculty members also corresponded with some of the friendlier men in their field and on occasion invited them to give a lecture at the college or come have tea with a visitor. In fact, the women's ties to these supportive men in the area were often stronger than those with the women at the other women's colleges. Strangely, it was not until decades later that they had much contact with these other women, and then certain friendly rivalries developed. Margaret Ferguson of Wellesley, for example, was pleased to note at one point that more of her students had attended a meeting than had students from Mount Holyoke's botany department.[42] Thus the first generation found its energies focused and consumed by myriad college activities in an almost entirely local but apparently satisfying and rewarding existence.

For the most part these first-generation women science professors published little (Cornelia Clapp of Mount Holyoke and Maria Mitchell were exceptions), since they were occupied with heavy teaching loads, often in more than one science, and the great burden of planning the college's first science buildings. This task could take years of worry and effort, with the selection of the laboratory apparatus a particularly troublesome problem, since it involved dealing with instrument makers and designing the college's curriculum for decades to come. However, the results could be impressive, as they were for the meticulous plan-

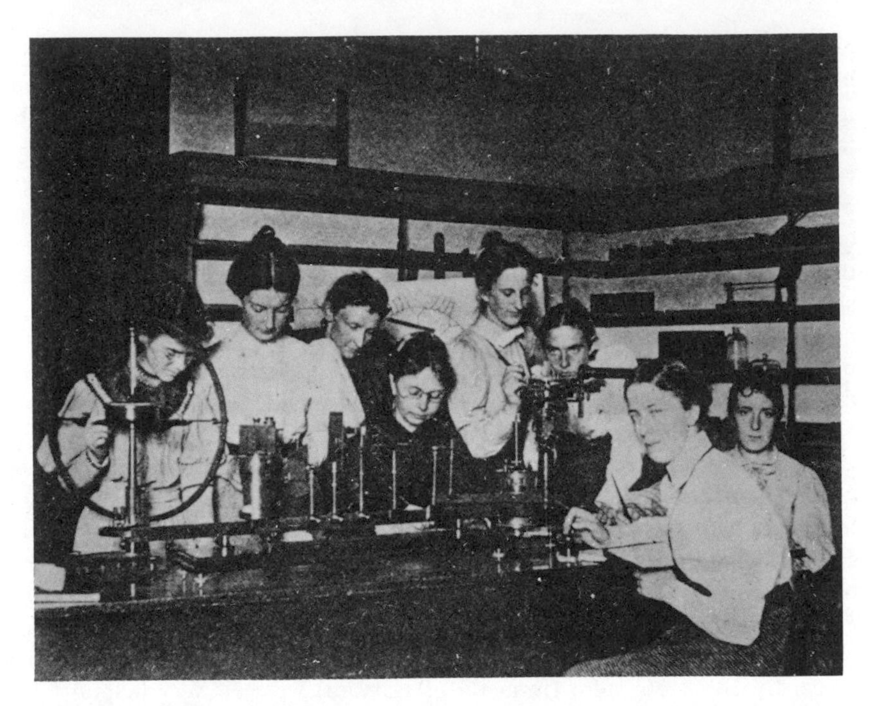

PLATE 3. Physicist and astronomer Sarah Whiting sought to have the latest equipment and laboratories for her students at Wellesley College, shown here in the academic year 1895–1896. Annie Jump Cannon, later an important astrophysicist at the Harvard College Observatory, is a postgraduate assistant, third from the left. (*Popular Astronomy* 35 [1927]:541.)

ner Sarah Whiting of Wellesley, shown in plate 3, who developed one of the earliest and best college physics laboratories in the country in the 1870s.[43]

Another important activity for some of this first generation of women scientists was the selection of bright protégées to succeed them on the faculty, since apparently most women preferred to train their own successors rather than have them hired from elsewhere. For some professors this selection was not urgent, but for others, who either were not in good health or had not been young when they had started, the need was pressing. When the professor found the right student, she encouraged the girl greatly, and a close relationship would result (often, as their correspondence reveals, they would have pet names for each other). Then the older woman would supervise the girl's selection of a graduate school, keep close contact with her progress, and later, when she was almost finished, arrange to have the college hire her back. The student was, of course, thrilled by the prospect of joining and succeeding her professor, whom she had long admired, and she considered such a job the greatest career possible for her, which, since she had few other prospects before World War I, oftentimes it was. For some women professors, the inability to find a protégée became a serious worry in later years, as an obituary of Charlotte Haywood of Mount Holyoke reveals, for having no successor made her feel that she had been a failure and let the school down.[44]

Some of these protégée chains could go on for several generations and develop over the years into long traditions of which the whole college could be proud. Table 1.2 presents the more notable of these traditions in ten sciences at seven major women's colleges before 1940. The two longest and most inbred of these protégée chains were at Mount Holyoke, in chemistry, where Emma Perry Carr, in 1937 the first woman to win an award from the American Chemical

TABLE 1.2. Notable Female Scientists at Seven Major Women's Colleges to 1940

Field	Barnard	Bryn Mawr	Goucher	Mount Holyoke	Smith	Vassar	Wellesley
Anthropology	Reichard	Swindler* Goldman*		---------------	Hawes*	Beckwith†	--------------
Astronomy	-----------			Young Farnsworth		Mitchell Whitney Furness Makemson	Whiting
Botany	Gregory	---------------	(Langdon)	Shattuck	(Snow)		Hallowell Ferguson Creighton
Chemistry	Reimer		Kelley	Lyon (Shattuck) (Goldthwaite) Carr Hahn			
Geology	Ogilvie	Bascom Wyckoff		Talbot Lochmann			(Fisher)
Mathematics		Scott Pell-Wheeler Noether	(Bacon) (Lewis) (Torrey)			Gentry	(Shafer) (Pendleton)
Nutrition	-----------	---------------		---------------	-----------	Wheeler	--------------
Physics	Maltby		(Gates) (Barton)	Laird	Anslow	Carter Wick	Whiting McDowell
Psychology		Rand†				Washburn	Calkins Gamble
Zoology			Welsh King	(Bowen) Clapp Adams Morgan Turner (Haywood)	(Wilder) Sampson		

SOURCES: Obituaries in college archives; *Notable American Women*; and *Notable American Women: The Modern Period*.

NOTES: ----------- indicates that the subject apparently was not taught at that institution. ( ) indicates important teachers who were not researchers.

*Classical archaeology rather than anthropology per se.

†Research only.

Society, was proud to trace her professional "lineage" back two generations to Mary Lyon herself and poured her energies into developing what is still perhaps the best undergraduate chemistry program in the country, and in zoology, where Susan Bowen and Cornelia Clapp (shown in plates 4a and 4b) started a line of zoologists which continued from about 1870 until at least 1961, when Charlotte Haywood retired, apparently, as mentioned, without leaving an heir. Similarly, Wellesley had distinguished traditions in botany and physics, and Bryn Mawr in mathematics, geology, and classical archaeology. At Vassar, the professors of astronomy were all students and grandstudents of Maria Mitchell until 1932, when Maud Makemson, the first outsider (a University of California graduate and doctorate), succeeded Caroline Furness. The female line was not broken until 1958, when Vassar's trustees elected Henry Albers of Butler University to the Maria Mitchell Alumnae Professorship.[45] To a certain extent these protégée chains or areas of excellence account for the pattern of baccalaureate origins of later women scientists, as shown in tables 1.1 and 6.5.

Once assured that the field was safely in the hands of her protégée, the pioneer woman scientist, now getting on in years, was freer to relax somewhat or take up new activities. She might devote these later years to building up the school's collections (an endless task) or rebuilding them after "the fire" (an unfortunate but all too common occurrence on most campuses),[46] or in any of the other innumerable tasks of running the college. A few of the early women scientists took on administrative duties as deans or even presidents of their colleges: Helen Shafer, professor of mathematics at Wellesley College, was its third president (1887–1894), and her protégée, the beloved Ellen Fitz Pendleton, who spent most of her term rebuilding the college after a disastrous fire in 1914, its sixth (1911–1936). If the older woman scientist was still interested in missionary work (which had often been a strong motivation in the founding of the early colleges), she could find new challenges in going abroad for a few years to such schools as the American College for Girls in Istanbul or the Women's Christian College in Madras, India, or to other such institutions in Japan, China, Spain or Latin America.[47] If she were feeling a little behind the times in her field, the woman professor could also take a leave of absence and go to graduate school herself, as the energetic Cornelia Clapp did with some of her own students in the 1890s.

And then, after thirty or forty years of teaching, this hardy pioneer would retire, often to a cottage near the campus, where she would live out her remaining years with a sister or retired colleague. When she did pass away, she would be greatly mourned and given certain final honors. The laboratory building for which she had worked so hard would, now that it was finally finished, be named in her honor (as was Clapp Hall at Mount Holyoke), or funds would be raised for an endowed professorship in her honor (such as the Susan M. Hallowell Professorship of Botany at Wellesley). Then the protégée, by now chairman of the department, would write the major obituary (usually for the alumnae bulletin, less often for a professional journal or *Science*), in which she would defend the woman's lack of publications and praise her selfless devotion and beloved per-

PLATE 4a. Zoologist Cornelia Clapp (at right), shown here in the late 1880s, was a dynamic and enthusiastic faculty member at Mount Holyoke College for forty-four years. (Courtesy of Mount Holyoke College Library/Archives.)

PLATE 4b. She also earned two doctorates and spent forty summers at Woods Hole, where she was one of the few women trustees of the Marine Biological Laboratory. She is shown here in 1934, the year of her death. (Courtesy of the Library, Marine Biological Laboratory, Woods Hole, Massachusetts.)

sonality. Later it would be revealed that she had left a bequest of several thousand (or more) dollars to the college.

Since such sentiment and stately procedures were the norm, the exceptions become interesting. Although human failings are rarely mentioned in these obituaries, a few women professors were apparently misunderstood or even disliked by their colleagues. Ellen Hayes of Wellesley, for example, had such unpopular political opinions, such limited training (Oberlin, 1878) and such a generally difficult personality that the college created a separate department of applied mathematics for her alone in 1897. In another case Emma Byrd, the first professor of astronomy at Smith College, mystified and embarrassed her colleagues when in 1906 she suddenly resigned in protest over the school's acceptance of grants from the Rockefeller Foundation and Carnegie Institution. From these critics' viewpoint, however, the women's colleges had failed their earlier idealistic vision. Higher education for women had, they thought, promised individuality, radical thought, and a chance to change the world dramatically. Yet all that they saw indicated that the women's colleges were taking on and being "fettered" by conservative capitalistic values of the world about them. Hayes battled on until her death to reach a few women's minds, and Byrd threw it all off in disgust. They thus serve as reminders that, however proud these colleges are of their history and achievements, many compromises were made.[48]

Meanwhile the second generation of women professors at these colleges (as in other fledgling coed schools like them) had, of course, greater opportunities than their predecessors had had and were often more oriented toward research and keeping up with their field. These younger women were better trained (often holding a doctorate) and, as the second (or third) person in the department, could hope to teach their specialty to the advanced students or majors. They might also be able to do research and publish, especially if the laboratory had been finished, and they could design a project that utilized the limited skills of the advanced undergraduates and minimized the college's lack of graduate students, specialized colleagues, and expensive equipment and other facilities. Thus, even though they might not be at the center of their field institutionally (they might be if they had chosen their topic well), they could develop a series of publications such as the "Vassar College Studies in Astronomy" or the "Wellesley College Studies in Psychology." Usually these articles were descriptions of observations or experiments in a quiet corner of the field, but the *Bryn Mawr College Monographs,* which were published from 1901 to 1926, presented the pioneering work of Thomas Hunt Morgan and his many graduate and undergraduate students (including Nettie Stevens, Helen D. King, Lillian V. Sampson, Alice M. Boring, and Florence Peebles), many papers by the eminent Charlotte Scott and her students in mathematics and physics, and several volumes of work by Gertrude Rand and her husband, C. E. Ferree, on psychology and physiological optics—all landmarks in their specialities. Less often, student papers were published in standard professional journals.[49]

But research was expensive and there was never enough money, even at the wealthiest of the women's colleges, for everyone to be able to do it. Because of a shortage of funds and also a certain official ambivalence toward research, few women's colleges did much to encourage individuals to continue with their work. Although these colleges recruited their second-generation and later faculty on the basis of their dissertations and based their promotions quite often on further scholarship, they did little to support research or make it easier, as they might have with paid sabbaticals, released time, research assistance, or necessary equipment. Even Bryn Mawr College, which maintained a small graduate school and awarded some fellowships to foreign and American graduate students, apparently provided no funds or released time for those professors who were engaged in research. In fact, its attention to pomp and ceremony even restrained some women professors who were eager to get out and do fieldwork; there were many petty restrictions, as when President Thomas insisted that Florence Bascom, professor of geology, remain on campus during the commencement week, even though as a geologist she preferred to spend that precious time in the field.[50] An occasional solution was the establishment of a research professorship for a specific woman by her friends or relatives. Thus Martha Beckwith was officially a professor at Vassar College, but, impatient with the students there, spent most of her time in Hawaii and Jamaica working on folklore, often with those very friends who, unknown to her, were supporting her "professorship." Similarly, the trustees of Bryn Mawr College established a research professorship for Nettie Stevens, their great geneticist, who had been supported by the Carnegie Institution, but she died before it could be set up.[51] But such special circumstances were rare.

Thus though most of the women on these college faculties, after the first generation of pioneers, came with the best credentials of their time—often with doctorates from Chicago, Columbia, or the German universities plus, quite often, postdoctoral research and fellowships—they tended to retire thirty to forty years later having made few or no further contributions to science. Since the administrators of these early colleges knew that their faculties were being watched and their accomplishments and productivity compared to those of the best men of the time, it does seem that they should have done more to increase the research opportunities for their faculties. Yet, as William Henry Welch of The Johns Hopkins Medical School put it in 1922, the problem was bigger than any one college or administration. Sex discrimination was so widespread in academic hiring that there was no incentive for the women at these colleges to do any more research. They already held the best jobs open to women, and even with outstanding research accomplishments they were not going to be called to a major graduate school, as the better men at these colleges often were. Although some of the stronger women, as described earlier, did do research, such pervasive discrimination and institutional poverty must have diminished the energies and motivation of even these most favored women of their time and made them less productive than they might have been.[52]

Moreover, what research was accomplished at these colleges was often done at the expense of other, less demanding, departments. With every prize and award that she won, Margaret Ferguson of Wellesley (the first woman president of the Botanical Society of America) was able to justify claiming an even larger share of the college's precious research funds for her experiments (on the cytogenetics of the petunia). But this indulgence came at a time when the zoology department, which historically had been weak at Wellesley and languished further when a fire destroyed its building in 1914, was forced to go still another few years without a proper laboratory.[53] Over the years such differences among departments could lead to major imbalances. A professor in one persistently weak department who found herself appealing to the alumnae for funds in the 1920s might feel more like the pioneers of forty or fifty years before than she did like her contemporaries in those other fields which already had good facilities and were now demanding (and often getting) even more.

As a result these colleges, as table 1.2 also shows, had certain notable curricular or faculty weaknesses that counteracted their strengths. None was strong in all the sciences. Bryn Mawr, for example, apparently dropped botany from its curriculum after some initially unsuccessful experiences. Geology was quite weak at most women's colleges; even Florence Bascom grew discouraged at Bryn Mawr, when, despite her pioneering work and her graduate students (she trained most of the important women geologists in the nation), President Thomas threatened in 1911 to terminate her department.[54] Similarly, anthropology was generally not taught at the women's colleges until after World War II, although Gladys Reichard, who started teaching it at Barnard in 1921, had an outstanding group of students.[55] Mathematics was distinguished at Bryn Mawr, but of little note elsewhere, despite devoted teachers on most campuses.[56] Few of these schools wished to have anything to do with the fields of engineering or nutrition, although several of their chemistry graduates became home economists and, for reasons explained in chapters 3 and 7, some of the women's colleges' later presidents, such as Katherine Blunt of Connecticut College for Women and Sarah Gibson Blanding of Vassar College, were former home economists.[57]

Thus these seven women's colleges had by 1940 or even 1920 evolved a division of excellence in the sciences. These complementary imbalances seem not to have developed intentionally, as the result of a cooperative agreement or consortium among them, but rather were the outcome of many historical events and accidents—a strong personality here, an early death there, a large benefaction here, a tragic fire there. The overall effect of these circumstances was that (almost) all the sciences were represented by a distinguished program someplace. This was fortunate, since it is unlikely that any one of them could have afforded all the laboratory space and all the faculty (had they been available) that would have been necessary for excellence and distinction in all of them. It also meant that each of these elite colleges (except Smith?) had an area of strength and female excellence, which was important not only for the faculty's morale but also for the students' training and career decisions. When a freshman interested in "science" arrived at one of these colleges, she not only would be exposed to excellence and enthusiasm in at least one field of science, she would be actively recruited into it. This experience would to a certain extent affect her decision about a major field, and if she went on for a doctorate or a career in science, would later be reflected in data on the "baccalaureate origins" of women scientists, presented in tables 1.1 and 6.5.[58]

In the early twentieth century quantitative data became available which showed the central place of the women's colleges in the employment of women scientists in the United States. In 1902 James McKeen Cattell, a psychologist-turned-sociologist at Columbia University, decided to construct a biographical listing of all the persons in the nation who were "contributing to science." He made a systematic search of *Who's Who in America,* the membership of fifty scientific societies, and the faculty rolls of seventy colleges and universities (including the major women's colleges). He also published open letters inviting entrants in *Science,* the *Nation,* and the *Popular Science Monthly*. From the responses he published a directory of 4,131 persons, including women, but misleadingly entitled *American Men of Science* (*AMS*), in 1906.[59] With its later editions it still constitutes the chief source of data, numerical and biographical, on American scientists before 1950.

An analysis of the women entrants in the first three editions of *AMS* (1906, 1910, and 1921) reveals that most of the 149, 204, and 450 women scientists listed there, respectively, worked in academic institutions. Of those who held the faculty rank of assistant professor or higher, over two-thirds (36 of 52, or 69.2 percent) were, in 1906, at fourteen women's colleges, as shown in table 1.3. In the next fifteen years the number of women scientists holding such faculty rank more than tripled nationwide. Even though the largest increase was in the number holding positions at the coeducational institutions, over one-half were still at the women's colleges (99 of 178, or 55.6 percent) in 1921. Their increase was the result of both expansion at some of the older colleges for women and the establishment of (or attainment of collegiate status by) several newer ones, most notably Sweet Briar and Connecticut College for Women.

The importance of the jobs at the women's colleges varied greatly among the different sciences. Although some colleges employed, as shown in table 1.3, a total of ten or more women scientists in 1921, only a few of these women would be in any one field. In some sciences, even these few jobs could be almost the only openings available to women at the time. In physics, for example, the women's dependence on these colleges reached an extreme, as shown in table 1.4, which lists *all* the women physicists in the first three editions of the *AMS* by their employer in 1920. Nineteen of these twenty-four were employed at ten women's colleges, and three of the five who were working elsewhere in 1920 had been at such colleges earlier. Furthermore, nine of these nineteen, or almost one-half, were at just two of the colleges, Wellesley and Vassar, and Margaret Maltby, given a star in the *AMS* and thus considered one of the most eminent physicists of the time, was at Barnard.[60] In the larger sciences the women's colleges played a proportionately smaller role, but even there these professorships were quite prestigious and were usually held by the top women in the field, such as psychologists Margaret Washburn at Vassar and Mary Calkins at Welles-

TABLE 1.3. Academic Institutions Employing Female Scientists, 1906, 1910, and 1921

1906		1910		1921			
Wellesley	14	Wellesley	12	Wellesley	22	Columbia Teachers	4
Mount Holyoke	8	Mount Holyoke	8	Vassar	15	Connecticut College	4
Bryn Mawr	5	Vassar	6	Mount Holyoke	13	Ohio State	4
Vassar	4	Chicago	6	Smith	13	Sweet Briar	4
Barnard	3	Bryn Mawr	5	California at Berkeley	11	Harvard	4
Smith	3	Smith	5	Chicago	10	Goucher	3
Chicago	3	Barnard	4	Barnard	8	Illinois	3
Goucher	2	Illinois	4	Bryn Mawr	7	Nebraska	3
Sophie Newcomb	2	Goucher	3	Minnesota	7	Stanford	3
Johns Hopkins	2	Johns Hopkins	3	Johns Hopkins	6	Rhode Island College	3
Minnesota	2	Harvard	3	Wisconsin	5	Kansas State	3
Stanford	2	Nebraska	3	Columbia	5	California at Los Angeles	3
Rockford	2	Western College	3	Sophie Newcomb	4		
New York State Normal	2			Kansas	4	Iowa State	3
				Cornell	4	Carnegie Tech.	3
						American College for Girls in Istanbul	3
(+21 others with one each)		(+7 others with 2 each +28 others with 1 each)		(+18 others with 2 each + 43 others with 1 each)			
Total 75		107		263			

SOURCES: First three editions of *American Men of Science* (1906, 1910, and 1921).

TABLE 1.4. Employers of Female Physicists, 1921

Employer	Rank	Name	Highest Degree (Year)	
Wellesley	Professor Emeritus	Sarah Whiting	AB	Ingham College (n.d.)
	Professor	Louise McDowell	PhD	Cornell (1909)
	Associate Professor	Grace Davis	AM	Wellesley (1905)
	Assistant Professor	Frances Lowater	PhD	Bryn Mawr (1906)
	Assistant Professor	Lucy Wilson	PhD	Johns Hopkins (1917)
Vassar	Professor	Edna Carter	PhD	Wurzburg (1906)
	Associate Professor	Frances Wick	PhD	Cornell (1908)
	Instructor	Laura Brant	AM	Brown (1909)
	Instructor	Helen Gilroy	AM	Bryn Mawr (1912)
Mount Holyoke	Professor	Elizabeth Laird	PhD	Bryn Mawr (1901)
	Associate Professor	Margaret Shields	PhD	Chicago (1917)
Barnard	Associate Professor	*Margaret Maltby	PhD	Gottingen (1895)
	Instructor	Grace Langford	BS	MIT (1900)
Connecticut College	Assistant Professor	Lillian Rosanoff	PhD	Clark (1914)
Goucher	Instructor	Lulu Joslin	AM	Brown (1906)
Sweet Briar	Professor	Isabelle Stone	PhD	Chicago (1897)
Lake Erie	Instructor, retired	Mary C. Noyes	PhD	Cornell (1896)
Newcomb-Tulane	Instructor	Eleanor Reames	PhD	Tulane (1913)
American College for Girls in Istanbul	Professor	Eleanor Burns	AB	Cornell (1904)
Illinois	Dean of Women (former head of department at Goucher)	Fanny Gates	PhD	Pennsylvania (1909)
National Bureau of Standards	Assistant Physicist	Mabel Frehafer	PhD	Johns Hopkins (1919)
H. C. Keith Company	Assistant (former head of department at Mount Holyoke)	Marcia Keith	BS	Mount Holyoke (1892)
Philadelphia High School	Director of Science	Bertha Clark	PhD	Pennsylvania (1907)
Pipestone, Minnesota, High School	Teacher	Maude Stewart	AM	Columbia (1913)

SOURCES: First three editions of *American Men of Science* (1906, 1910, and 1921).

NOTE: Maltby was the only woman physicist "starred" (*) in these editions of *American Men of Science*.

ley and botanist Margaret Ferguson, also at Wellesley, all of whom were also starred in the *AMS* (see chapter 10).

Thus by the early twentieth century just a few women's colleges were supporting the careers of a large proportion of those women "contributing to science" in the United States. These colleges and these jobs were all the accomplishments of the previous few decades. The purpose of the women's academies and seminaries had been to train women students to be good wives and mothers, and this had justified the hiring of many women to teach them (just as the needs of female patients had justified the training of women physicians in the nineteenth century). In time these institutions had been upgraded into full-scale colleges which sought to make women's education equal to that of men. This "share of the market" rationale had opened careers as science professors to many, especially single, women, several of whom were by 1900 achieving distinction. Collectively these few departments had created a first plateau or "entering wedge" for women who sought full careers within the scientific profession.

Yet one could take a less congratulatory and more cautious view of the achievement of these women's colleges for women in science and wonder what would become of all their well-educated graduates. Whether single or married, they (as well as the alumnae of coeducational colleges) might be in for quite a bit of frustration, since they were overeducated for any attainable kind of employment (besides teaching). Not even the growing number of women's colleges would be able to hire all their unmarried graduates. Many would be unemployed and constitute such an obvious waste of talent that some broadening of "women's sphere" within science (beyond that of the separate women's college) would develop in the next few decades. But even as they sought careers elsewhere, many of the graduates of the women's colleges often retained a fond, nostalgic note for their spiritual, if no longer physical, home back there.[61]

Simultaneously with this impending need to expand women's role in science, another kind of pressure mounted to replace the women on these early faculties with men. Science was changing so fast in the 1880s and 1890s, especially in the realm of higher educational "standards" and "professionalization," that many women hired earlier were no longer considered suitable faculty members at the better women's colleges. The coming of the doctoral degree to higher education in the 1880s created a certain dilemma for the presidents of women's colleges: which was the best way to give their students a top quality education—by hiring the best women available, as some of the older colleges had done, or by giving these precious jobs to persons with doctorates, when they would, because of the discriminatory practices at the major American and German (but not Swiss) graduate schools, almost always be men? M. Carey Thomas of Bryn Mawr, who had fought hard at several institutions for her own doctorate and was one of the first college presidents of either sex to hold this degree, stressed them in her hiring, even to the point that her faculty had a far lower percentage of women (33 percent) than had the older academies (75 percent) and the other elite colleges (over 90 percent at Wellesley and Mount Holyoke).[62] In time all these colleges would begin to hire doctorates, but before many of them could be women, a major reform had to take place at the graduate schools—and in the 1890s it did.

2

DOCTORATES: INFILTRATION AND CREATIVE PHILANTHROPY

Although the lofty rhetoric accompanying the founding of the major American graduate schools seemed to indicate that they would be open to both sexes, the first women applicants quickly discovered that this was usually not the case.[1] Since most American graduate schools were patterned after the German universities, which had introduced the Doctor of Philosophy degree in the eighteenth century and had never admitted women, American deans also rejected coeducation at the graduate level for several decades. Overcoming this resistance and gaining first, admission, and, later, degrees from these graduate schools thus became an important next step for women seeking to enter the learned professions in the late nineteenth century. Their eventual success, which came as early as the 1870s at Boston University and as late as the 1960s at Princeton University, can be seen as the result of a long campaign spread over four decades and characterized by important skirmishes at several European as well as American universities, at least two major strategies, a large cast of participants, and the active support (financial and otherwise) of a few women's groups, especially the young Association of Collegiate Alumnae. Although a complex and far-flung campaign, it progressed generally through three basic stages: a long latent phase before 1890, when women were admitted only as "special students" and not given degrees; a short period, 1890 to 1892, when they were suddenly and officially admitted to at least six major graduate schools; and a final, highly embattled stage, 1893 to 1907 and after, during which several diehard institutions in the United States and Germany were induced to accept their first women and to award them doctorates. This last struggle called forth such vigorous efforts and created such a heady atmosphere in the 1890s that many academic feminists of the time later considered the movement to have been the high point of their lives. Although continuing discrimination in employment greatly circumscribed the magnitude of the women's victory (and the acceptance of such segregation may have made the leniency on degrees possible), opening the graduate schools nevertheless ranks as one of women scientists' greatest triumphs.

If the women's colleges had had the wealth and personnel to expand indefinitely into all areas of knowledge, it is conceivable that they might have grown into women's universities with graduate schools, each of which might have had leading programs in a few fields. To a certain extent this did happen at Bryn Mawr College, where by dint of great efforts its dean and later president M. Carey Thomas and her successors were able to maintain small doctoral programs of excellence in several fields: zoology, mathematics, geology, English, history, and classics (including classical archaeology).[2] The obstacles to such a separate women's graduate school were so immense, however, that aside from the legal fiction of the Radcliffe Graduate School and the master's programs in some fields at Mount Holyoke, Wellesley, and Smith colleges, few doctoral programs were maintained at women's colleges. A major factor was money; even the men's graduate schools were usually in financial straits. Since women made up only a small percentage of the persons in a given specialty, the cost of providing graduate facilities for these few was very high, especially in the sciences.

Yet even if the money and the faculty had been available, it is debatable whether such institutions as women's graduate schools would have become widespread, since few persons, not even all the women professors at the women's colleges, favored such a segregated experience at the graduate level. M. Carey Thomas was unusual in thinking that every woman's college should exert itself to have a full-fledged (and possibly even coeducational) graduate school. Wellesley College psychologist and philosopher Mary Whiton Calkins may have been closer to the norm in advocating just enough graduate work to keep the faculty on its mettle. Vassar psychologist Margaret Washburn would not go even that far, however. She argued that the heart of a graduate education was communication, interaction, and criticism from the authorities and fellow students in one's field, most of whom were men already at the major graduate schools. Besides, Vassar was so poor that it could not even subscribe to the major professional journals in her field. In 1910 she wrote to an assistant professor at Harvard, "I wouldn't have a graduate student under any circumstances," and noted to a Vassar alumna in 1914 that, though the college had formerly had a fellowship for a graduate student, it now wisely gave it every year to an alumna to use elsewhere.[3]

When the first women did apply to the major graduate schools after the Civil War, the standard reply was that there was "no precedent" for admitting a woman, regardless of her qualifications. Nor were many schools willing to step forward and set one. Thus Ellen Swallow (later Richards), who applied to MIT for a graduate degree in chemistry in 1870, was turned down solely on the grounds (according to her husband, a professor there in another department) that the chemistry department did not want its first graduate degree to go to a woman.[4] Later women found that even when they were not the first applicant in a program, but the *n*th, the argument was the same—there was "no precedent" for admitting them or giving them a degree—a logic that seemed to close the issue for all time, for until a woman was admitted such a precedent would not exist.

Yet the situation was both more favorable and more complex than this straightforward rejection would imply, since most universities were willing to make "exceptions" in particular cases. Interestingly, they then went to considerable trouble to hide this liberality and make sure that such action set no legal precedents. Thus MIT did not reject Ellen Swallow (as did all the other universities to which she applied), but admitted her as a candidate for a second bachelor's degree (she already had one from Vassar) and as a "special student" who did not have to pay tuition. They had thus taken the liberal step of admitting her, their first woman, but had so camouflaged their action that, if anyone complained, they could deny she was officially enrolled. (She thought the tuition had been waived because she was poor and later insisted that, had she known the real reason, she would have protested.)[5] Similarly in 1878 The Johns Hopkins University admitted the brilliant "C[hristine] Ladd" as a "special student" whose name would not appear in the catalog and whose case would not set any precedents for future action, as numerous other applicants to Hopkins were later informed.[6] Gaining admission, however, was only the first stage of a long process, and these two women, Richards and especially Ladd-Franklin, would be the leaders of the movement to get doctorates for women in the 1890s.

What these early women learned from their "special student" status in the 1870s was that the process of earning a degree was actually a two-tiered one. The first step, gaining access to graduate education, was relatively easy, informal, and decentralized; permission could be granted by almost any friendly professor. But the second step, being awarded a degree, was much more "political" and involved far more than just meeting the stated requirements. It was a formal, almost legal, proceeding, that involved not only the professors and the department but also the president and board of trustees, many of whom long refused to award higher degrees to women or even to acknowledge their presence on campus. To gain doctorates from resistant universities the women needed some way to create pressure on particular schools. Individually they were powerless and had to await others' decisions; collectively, and with a strategy, they might be more effective.

When all the attempts by women to gain higher degrees at universities in the United States and Germany over three decades (1870 to 1900) are viewed together, they can be seen as a process of infiltration, a kind of educational "guerilla warfare" or slow "war of attrition" against the universities. Under this almost military strategy, individual women sought to test the repressive system on as many fronts (departments and universities) as possible, probing for weak points and using what friends they had to help them evade the rules informally, and, when enough "exceptional" women had been admitted in this way and had surpassed their fellow students without the imagined disruption, to push for a change in policy, which then could be seen as harmless, "only fair," and long overdue, and could be enacted quietly. Thus over several decades a series of women eventually accomplished their objective, but at great human cost.

Like other graduate students, women applicants were usually attracted to a university by the presence of a particular professor in their chosen field, and they (or a patron) consulted directly with him about their attending his lectures. Some of these early professors were quite helpful to the early women graduate students, as was Benjamin Peirce of Harvard University when astronomer Maria Mitchell of Vassar College asked to attend his lectures on quaternions in 1868. When she then suggested in 1869 that he let her former student Mary Whitney attend them also, he again obliged, invited the rather shy Miss Whitney, and chivalrously met her at the college gate to escort her to a seat at the back of his classroom. Similarly in 1876 Henry Durant, the founder of Wellesley College, made arrangements for Sarah Whiting to attend the MIT physics classes of Edward C. Pickering, a liberal supporter of women in science who had pioneered the new laboratory method of physics instruction. Also in 1876 Frederick A. Genth of the Towne Scientific School at the University of Pennsylvania admitted two women to his chemical laboratory.[7]

Dealing directly with friendly professors was generally more effective than seeking permission from less cooperative administrators, as can be seen in two cases that occurred at the newly opened Johns Hopkins University in 1878 and 1879. In 1878 President Daniel Coit Gilman, whom Hugh Hawkins has termed a "social conservative" and who thought that women did not belong in graduate school with men, refused to let M. Carey Thomas attend Basil Gildersleeve's lectures on the classics. (She could study with him privately, however.) One year later Professor J. J. Sylvester not only allowed Christine Ladd to attend his lectures on mathematics but also had his department award her a $500 fellowship, an unheard of honor for even brilliant women in the 1870s. Yet not even he could get her a degree when she submitted her dissertation in 1882.[8]

Although such informal procedures continued into the 1880s (and the 1890s at some schools), both the women students and their professors were growing increasingly dissatisfied with these arrangements, which were full of uncertainities and did not lead to degrees. Before long many women turned instead to other, less highly regarded, institutions that would give them degrees. Starting in 1877 and increasingly through the 1880s, several institutions began to award doctorates to women. Walter Crosby Eells has searched university catalogs and found that by 1889, ten colleges and universities had granted twenty-five such degrees to women.[9] Syracuse University led the way with seven, and Boston University and the then University of Wooster followed with four each.[10] Smith College,[11] the University of Michigan, and Cornell University each awarded two in these years, and four other institutions gave one apiece. Only eight of the twenty-five degrees were in the sciences, of which two were in what would later be called the "social sciences" (sociology and law) and six in zoology, botany, and mathematics, three of which were from Syracuse. Some of these early woman doctorates already held or would later hold influential positions in women's colleges and coeducational universities: Julia Gulliver became the president of Rockford College, Henrietta Hooker and Cornelia Clapp were both professors at Mount Holyoke College, and Martha Foote (later Crow) was the "lady principal" at what is now Grinnell College and later was the dean of women at Northwestern University. Some other women married influential academics and scientists. Helen Magill, the nation's first woman doctorate (Boston University,

1877) married Andrew D. White, president of Cornell University and later an ambassador to Russia; May Preston, an early Cornell doctorate, married E. E. Slosson, founder and longtime president of the Science Service; and Alice Carter married O. E. Cook, a fellow student at Syracuse who spent many years in Africa, first as an agent for the New York Colonization Society and later as a plant explorer for the U.S. Department of Agriculture.[12] Some of the pioneers, like Syracuse doctorates Jane Bancroft (later Robinson) and Martha Foote Crow, were active in contemporary battles to open more universities to women and to find suitable employment for them.[13]

Little is known about the circumstances and debates surrounding the awarding of these early degrees to women. Apparently the innovation raised far less controversy at such modest "universities" as Boston, Syracuse, or Wooster in the 1870s and 1880s than it would at some more status-conscious institutions later. Graduate programs at such schools in the 1870s and 1880s tended to be the personal enthusiasm of an ambitious chancellor (such as Alexander Winchell at Syracuse), president (Archibald A. E. Taylor at Wooster), or dean (M. Carey Thomas at Bryn Mawr) who, especially if anxious for students, could welcome women candidates eagerly. Graduate work was also still small and informal in these years, and standards flexible. (Wooster, for example, had no residency requirement for its doctoral degree, an attraction for many employed ministers and professors.)[14] Although later attempts to raise standards and accredit graduate schools were designed to upgrade such loose postgraduate programs or drive them out of existence altogether, these innovations may have served to liberalize later doctoral programs at other more prestigious institutions, which as late as 1889 were still at the stage of enrolling "special students" and avoiding precedents.

Public interest in women's admission to male-dominated graduate schools increased greatly in the years after 1889, as articles and books began to appear surveying the situation at home and abroad. Events were moving so rapidly, in fact, that one journal, the *Nation,* became a regular clearinghouse in the years 1889 through 1897 of information on how the women were faring at the various universities. The appearance in 1890 of German feminist Helene Lange's *Frauenbildung* (translated as *The Higher Education of Women in Europe*), which surveyed current practices in several countries, may have been the catalyst that fused many isolated cases into an international movement. Despite a hostile review in the *Popular Science Monthly,* Lange's book became basic reading for the academic feminists of the 1890s, who were beginning to formulate an organized program to secure women's admission to European as well as American universities.[15]

In the midst of all this heightened interest and discussion, six major American graduate schools decided in rapid-fire order in the early 1890s to admit women on an equal basis with men and to award them Doctors of Philosophy degrees when their work merited it. The decision was closely entwined with the far more controversial issue of undergraduate coeducation. Two universities (Yale and Pennsylvania) linked their admission of women to graduate work with a continued refusal to allow them into the undergraduate college; two others (Columbia and Brown) tied their decision to admit them to the graduate school to the formation of a coordinate college for women undergraduates; and two new institutions in the west (Stanford and the University of Chicago) were the most liberal of all and announced in 1891 that they were instituting full coeducation at both the graduate and undergraduate levels.[16]

Since the Yale decision was seen at the time as the major turning point for women seeking higher degrees in this country, the little that is known about it may be of some interest. Both President Timothy Dwight and Arthur Hadley, dean of the graduate school, defended the decision on the grounds that though it might seem radical "on the surface," it really was not. Since Yale had for years been offering a series of special provisions to women, whose work had always been good, Hadley wrote, "it seemed only fair to give such students official recognition as members of the university; to encourage them instead of barely tolerating them; to award them the degree of Doctor of Philosophy if their work was such to deserve it."[17] Dwight also reasoned that because the degree was increasingly necessary for an academic career and because many of these women could be teaching at the women's colleges, it was Yale's duty to make its unrivalled facilities available to them, or, as he put it chivalrously in his annual report for 1891,

> The privilege of coming under the instruction of the best and ablest professors in a large university, of using its libraries, and of enjoying the many facilities for study under the highest advantages which it furnishes, must be a privilege of inestimable value. The University becomes by the offering of this privilege, not a rival or opponent of the colleges for women, but an ally and helper to them. It offers its graduate courses and its degree of Doctor of Philosophy to their graduates, and thus presents its gift to these graduates as an addition to the gift which they have already received from their own institutions.[18]

On these grounds and on the strict stipulation that the undergraduate college would remain unchanged, the Yale faculty had approved the move in a vote that was, as Hadley reported, "all but unanimous." Beyond this official explanation, both Dwight and Hadley, like officials at Syracuse and Wooster earlier, had a personal interest in the matter. Dwight came from a long family of supporters of women's education. His mother had been highly intelligent and convinced him that women deserved a full education, and his grandfather (of the same name and also a president of Yale) had been so sure of it that he had shocked his contemporaries in the 1790s when he taught girls as well as boys at his Connecticut academy. Hadley's wife, Helen (Vassar, 1883), was also so concerned about women at Yale that years later a graduate dormitory for women was named for her.[19]

Interested women and men exulted and spread the news joyfully: "How good is the news from Yale!" Marion Talbot wrote Millicent Shinn in California, who would be Berkeley's first woman doctorate in 1898, and in April 1893 Professor William Gardener Hall of the University of Chicago addressed the Chicago branch of the Association of Collegiate Alumnae on "The Significance of the Recent Opening of Graduate Courses of Study to Women at Yale University, The University of Pennsylvania, and Brown University." After decades of evasion and procrastination the pace of change had become almost dizzying. Before long, Marion Talbot predicted, graduate education would be as freely available to women as undergraduate instruction already was. "Indeed," she added, "it is this tremendous onrush of the movement that makes me tremble."[20] In 1893 twenty-three women enrolled at Yale, and among the first class of graduate students at the new University of Chicago, the largest contingent from any one school was fourteen from Wellesley College (encouraged, no doubt, by their beloved president, Alice Freeman Palmer, who was also going to Chicago as an undergraduate dean).[21] Off to such a vigorous start, Yale and Chicago rapidly overcame their predecessors and quickly became the largest "producers" of women doctorates in the nation. This influx of women into these newly opened institutions confirmed Timothy Dwight's prediction that they desired a more rigorous training than had been available to them before 1892. One Syracuse doctorate of the 1880s (Cornelia Clapp, Ph.D. 1889) even felt it necessary to take a three-year leave of absence from her job at Mount Holyoke College to earn a second doctorate at Chicago in 1896.[22]

Eells's data also shows how swiftly the total number of doctorates awarded to women increased in the 1890s. Compared to the 25 granted before 1890, 204 were awarded through 1900—an eight-fold increase. (The number of doctorates awarded to men also increased sharply, though not as dramatically as for women, in the 1890s—tripling from 731 in the 1880s to 2,372 in the 1890s.)[23] As earlier, a few institutions dominated. Over one-half of the doctorates awarded to women from 1877 to 1900 were given by just four universities: Yale (36), Chicago (29), Cornell (28), and New York University (20). When this data is broken down further by field or department, as shown in table 2.1, one can see that, despite the official admission of women to these universities by the 1890s, they still tended to be clustered in certain departments within them. This distribution may indicate that women were receiving a friendly welcome in some of the major programs in the field. In table 2.1, where those doctorates accounting for more than 25 percent of the women in a field are italicized, one can see the influence of a few relatively strong and liberal departments. Thus Yale was already by 1900 quite dominant in English, history, mathematics, and chemistry (even before its famed department of physiological chemistry had begun to train numerous women biochemists and home economists [plate 5]). The University of Chicago was very influential in eight fields before 1900: zoology, political science, sociology, physiology, French, physics, religion, and comparative philology (later linguistics), where it stood alone. Cornell was still strong in philosophy (which had given two women doctorates in the 1880s), psychology (where it would soon be eclipsed by Columbia University), botany, and physics.[24] New York University's importance rested on just one department, that of education or pedagogy, which

TABLE 2.1. Institutions Awarding Doctorates to Women before 1900
(By Field)

Institution	Total	English	Greek/Latin	Education	History	Philosophy	Chemistry	German	Mathematics	Psychology	Botany	Zoology	Sociology	Political science	Physiology	Economics	French	Astronomy	Physics	Law	Religion	Comparative philology	Art	Geology	No field given
Yale	36	*11*	4	0	*4*	3	*4*	1	*4*	1	0	0	1	1	0	1	1	0	0	0	0	0	0	0	0
Chicago	29	5	3	1	0	1	1	1	0	1	0	*3*	*2*	*3*	*2*	1	*2*	0	*1*	0	*1*	*1*	0	0	0
Cornell	28	2	1	0	3	*5*	1	2	2	*4*	*2*	0	1	0	0	1	0	0	*1*	0	0	0	0	0	3
New York University	20	2	0	*18*	0	0	0	0	0	0	0	0	0	0	0	0	0	0	0	0	0	0	0	0	0
University of Pennsylvania	19	4	2	0	1	1	*4*	0	0	2	*3*	1	0	0	0	0	1	0	0	0	0	0	0	0	0
Bryn Mawr	15	4	3	0	2	0	0	1	2	0	0	1	0	0	*2*	0	0	0	0	0	0	0	0	0	0
Syracuse	13	5	0	0	3	1	0	0	0	0	*2*	1	0	0	0	0	0	0	0	0	0	0	*1*	0	0
University of Wooster	9	0	1	1	1	1	1	*3*	0	0	0	0	1	0	0	0	0	0	0	0	0	0	0	0	0
Boston University	9	1	4	0	0	1	0	0	0	0	0	0	0	0	0	0	0	0	0	*2*	*1*	0	0	0	0
Michigan	9	3	2	0	0	3	0	0	0	0	0	1	0	0	0	0	0	0	0	0	0	0	0	0	0
19 other institutions*	34	4	3	1	4	1	2	3	1	1	1	0	1	1	0	2	1	*2*	*1*	0	0	0	0	*1*	4
4 unidentified institutions	7	0	0	0	0	0	0	0	0	0	0	0	0	0	0	0	0	0	0	0	0	0	0	0	7
Total	228	41	23	21	18	17	13	11	9	9	8	7	6	5	4	5	5	2	3	2	2	1	1	1	14

SOURCE: Walter Crosby Eells, "Earned Doctorates for Women in the Nineteenth Century," *Bulletin of the American Association of University Professors* 42 (1956): 647 and 649–51.

NOTE: Italicized numbers indicate those doctorates accounting for more than 25 percent of the women in the field.

*These were: Brown (1), California (Berkeley) (2), Carleton (1), Columbia (7), De Pauw (1), George Washington (1), Gettysburg (1), Johns Hopkins (1), McKendree (3), Minnesota (3), Northwestern (1), Ohio State (1), Smith (2), Stanford (2), Taylor (1), Tulane (1), Washington (St. Louis) (1), Western Reserve (1), and Wisconsin (3).

PLATE 5. Physiological chemist Lafayette B. Mendel (seated third from right) was known for his many women graduate students at Yale University, including here in 1908 Edna Ferry (left) and Mary Swartz (right). Swartz later married Anton Rose (middle row, second from left). Stanley Rossiter Benedict (seated, right) later married Ruth Benedict (plate 24). (Reprinted by permission of the publisher from Juanith Archibald Eagles, Orrea Florence Pye, and Clara Mae Taylor, *Mary Swartz Rose, 1874–1941: Pioneer in Nutrition*. [New York: Teachers College Press, Copyright © 1979 by Teachers College, Columbia University. All rights reserved.])

reportedly not only was the nation's first graduate program in that field (established in 1890) but also, according to Eells's data, almost the only one to give women a doctorate in the highly feminized field of education.[25]

It is more difficult to interpret the smaller numbers in table 2.1 (as it is a lack of evidence generally). They may mean either that there was no doctoral program in that field at that university, or that there was a program, even a large one, which did not receive women. A similar caution would also apply to the many universities not listed in table 2.1. Thus one generally needs more information, as from Festschrifts or memorial volumes to major professors in these departments, to determine which were, like many of the Yale and Chicago programs, friendly and encouraging to women and which were not.

Despite these important steps taken in the years 1890, 1891, and 1892, other major American graduate schools of the 1890s—Harvard University, The Johns Hopkins University, and possibly Clark University[26]—as well as all the German universities still excluded women. Their continuing refusal, even to admit women as "special students" in some German institutions, was now, after so many successes elsewhere, increasingly intolerable to the women applicants and their supporters. At the same time the victories, especially that at Yale, gave the women a new eagerness and momentum in their struggle to win over the remaining diehard institutions. Partial success also meant that they could modify their strategy to focus more intensively on the few remaining targets, and by playing one institution off against another, prod them into opening their doors also. The movement thus shifted in the early 1890s into a higher gear as the women banded together, raised funds, and worked to force the issue at the remaining universities at home and abroad. Their strategy, still basically that of infiltration, now had a new militancy or determination that bordered on confrontation in a few instances.

Since in the 1890s most American graduate schools still emulated their German counterparts, which continued to offer some of the world's best professors, laboratories, and research training, especially in the sciences, one key battle to open the American programs to women necessarily took place abroad, at the German universities. These institutions had been admitting and granting degrees to large numbers of American men since the 1850s, and were only slowly opening to women students. In the late 1870s there were stories in the *Atlantic Monthly* about a few American women who had been allowed to enter certain language classes at Leipzig, where M. Carey Thomas had also gone after her unsatisfactory stay at The John Hopkins University. But even the relatively receptive Leipzig would not grant degrees to women. For these precious documents the women had to go to Zurich, across the border in more liberal Switzerland. In the 1880s several other American women visited and studied at the German universities, but if Helen Abbott Michael's account of her grand tour of 1887 is any guide, they met only discouragement when they asked to attend classes at any German university. Occasionally a friendly professor would show her his collections or let her watch an experiment in progress, but, as she put it, for degrees "all women are shoved to Zurich." Even the relatively liberal August Kekulé, a chemist known for his discovery of the benzene ring, told Michael that he refused to take any more women students—he had had two Russian women already, but neither had been very impressive. One had spent all her time reading novels, and the other had committed suicide. He thus felt justified in excluding all women forever.[27]

The continuing refusal of the German universities to admit women bothered Christine Ladd-Franklin and Ellen Richards, who, as recounted earlier, had themselves been denied graduate degrees. Now employed in subordinate positions at largely or all-male institutions (Ladd-Franklin as a temporary lecturer in psychology first at The Johns Hopkins and later at Columbia, and Richards as an instructor in sanitary chemistry at MIT), they were aware of the large differences in postgraduate opportunities open to men and women. They were anxious to take steps to correct this inequity but were unsure how to proceed until 1888, when Ladd-Franklin (plate 6) had the idea of establishing a graduate fellowship for a woman who wished to study abroad. When she suggested the idea to the governing board of the fledgling Association of Collegiate Alumnae (ACA), its

PLATE 6. Christine Ladd-Franklin, an early graduate of Vassar College and mathematician-turned-psychologist, fought for several decades to open both the graduate schools and full academic careers to women. (Courtesy of Vassar College Library.)

members took up the project eagerly. They raised the money ($500), formed the selection committee, and awarded the first fellowship in 1890.[28]

The Association of Collegiate Alumnae had been formed in Boston in 1882 by eighteen young college graduates (Ellen Richards, at forty, was the oldest) who sought to defend "college women" from the hostility and suspicions that greeted them in many quarters in the 1880s. The ACA was thus an early "pressure group" on behalf of educated women. Its typical projects were the collection of statistics to show that college graduates were as healthy as other women (though they married less often) and that some were holding jobs other than schoolteaching. Thus a fellowship to send a woman abroad for advanced study was a suitable and important project for the ACA. It was feasible and laden with political overtones: if its fellows were successful in opening foreign universities to women, this could be an important "entering wedge" (one of the group's favorite phrases) in broadening opportunities for women everywhere. Within just a few years, and largely through Christine Ladd-Franklin's efforts, the ACA's European fellowship would prove to be a resoundingly successful kind of "creative" (or coercive, depending on one's point of view) philanthropy indeed.[29]

Shortly after Ladd-Franklin had called the ACA members' attention to the need for a European graduate fellowship, public interest began to shift from

England, which had led the way in graduate education for women in the 1880s, to Germany, whose universities did not yet admit women in any capacity. The American women's victory at Yale may have had some role in this change. Officials at English universities had been proud as late as 1891 that they had been giving women students larger and larger graduate opportunities (though not yet degrees), and thus were offended at righteous American insinuations that they were lagging. One such official responded to an American's inquiry as to why women could attend certain lectures but not enter the laboratories or museums with the testy retort, "You must not reproach us till you have converted Harvard, Johns Hopkins, and Yale. Have any of these done as much for women as Oxford or Cambridge?"[30] Though the Yale victory a year later might have seemed to entitle the Americans to prod the British universities to greater efforts, by then the German universities seemed a greater challenge, or, as Bessie Bradwell Helmer, chairman of the ACA fellowship committee wrote Phoebe Apperson Hearst in 1894, "The battle in our own land has been almost won." The ACA was now doing "some missionary work" in the "prejudiced land" of Germany. In particular, it had "sent brilliant young women to storm the coveted citadels of learning" and open those "sacred precincts" to the feminine gender.[31]

The first ACA fellows to go to Germany were no more successful than their predecessors in the 1880s had been. Prussian authorities allowed Ruth Gentry, Michigan, 1890, and later associate professor of mathematics at Vassar College, to be the first woman to attend lectures at Berlin, but would not let her enroll for a degree. Julia Snow, a graduate of Cornell and later a professor of botany at Smith College, settled for a doctorate at Zurich in 1893, and Alice Walton, who already held a doctorate from Cornell in classics, used her ACA fellowship to spend a postdoctoral year at Leipzig.[32]

Meanwhile Ladd-Franklin herself may have helped turn the tide at Göttingen, at least, when she spent the academic year 1891–1892 in Berlin and Göttingen with her husband, Fabian Franklin, a professor of mathematics at The Johns Hopkins and later a newspaperman in New York City. (He too championed liberal causes, including that of advanced degrees for women.) She did experimental work on her new color theory in the laboratory of vision expert G. E. Muller in hopes of earning there the doctorate that Johns Hopkins had withheld from her in 1882. At first she was optimistic (and wrote William James so), but in the end she was only partially successful. Muller and mathematician Felix Klein of Göttingen were more than sympathetic, but the courts ruled that the latter could do no more than admit her to his lectures as an auditor ("Gastzuhörerinnen"). Klein assured her that, though this would not please her, it was only the beginning ("nur als Anfang") for women at Göttingen.[33] In this he was correct, since in just three years Göttingen would be the first German university to award women doctorates.

By 1893 several more ACA fellows were determined to go to Germany to fight for the cause. Three of them would be successful in earning the coveted degree in 1895 and 1896, including two from Göttingen. Margaret Maltby, an instructor at Wellesley College, studied physics with Waldemar Voigt and was

PLATE 7. Physiologist Ida Hyde, shown here at Woods Hole in the 1890s, was proud of her German doctorate (Heidelberg, 1896) and her many professional "firsts." (Courtesy of the Library, Marine Biological Laboratory, Woods Hole, Massachusetts.)

allowed to take her degree at Göttingen in July 1895, the first American woman to do so. Preceding her by just a few weeks was the Englishwoman Grace Chisholm (later Young), with a degree in mathematics from Felix Klein. Also living at the same pension with Maltby and Chisholm while they worked on their Göttingen degrees was another American, Mary Winston (later Newson), who recently had been a graduate student at the University of Chicago. She had met the friendly Professor Klein at the World's Fair in that city in 1893 and had received some modest encouragement (though no guarantees) from him. When Christine Ladd-Franklin heard about this, she sent Winston $500 of her own money to speed her on her way to Göttingen. Winston later became an ACA fellow and in 1896 became the third foreign woman (second American) to earn the coveted Göttingen doctorate.[34]

Meanwhile physiologist Ida Hyde, another ACA fellow (shown in plate 7), struggled on alone with her professors, first at Strasbourg, and then, when its officials proved unwilling to give her a degree, at Heidelberg. She was apparently unaware of events at Göttingen and assumed the whole time that she would be, if successful, the first woman with a German doctorate. Although she regretted for the rest of her life that she had not been the first (she did not get her degree until 1896), the belief that she was a pioneer sustained her through many trying circumstances. The problems she encountered (and later described in a humorous account "Before Women Were Human Beings . . .") illustrate well the frustrations facing such a pioneer in Germany in the mid-1890s.[35] She recounted her many visits to individual professors, her numerous requests for official permissions of various sorts, and her frequent appeals to local university and government officials (including at one point a ministry in Berlin). Working hard on her own, borrowing the assistant's lecture notes, and capitalizing on all the occasional courtesies and off-hand remarks of the more well-disposed professors, the determined Hyde astounded her professors by piecing together the studies and doing the work necessary for the official examinations without attending any classes. A crisis arose when the many delays required that she spend an extra year in Germany in order to get the degree. Yet the ACA fellowship committee proved equal to the emergency. Its chairman wrote two frantic letters to Phoebe Apperson Hearst in California, pointing out to her the importance to the cause of Hyde's being able to continue. Hearst, who was already supporting several women students at the University of California, responded with the needed $400.[36] With help of this kind, Hyde persevered and eventually outlasted the Germans' delaying tactics. She long remembered how physiologist Willy Kühne, who had originally laughed at her desire for a degree, had required six faculty meetings before he would agree to serve on the committee, had repeatedly postponed her examination and then prolonged it an extra hour, and finally had refused to grant her the degree summa cum laude. (She had to settle for a magna cum laude, which Winston also got at Göttingen that year.) Yet one suspects that all these difficulties made Hyde and her supporters savor her eventual triumph all the more. An easy victory would not have been so heroic or required such an indomitable personality. The authorities had done their best (short of total refusal) to thwart her, and she had met every demand.

It would be unfair to claim, however, that the ACA fellows alone opened the doors to German universities to women in the mid-1890s, since there were already a great many other women knocking on them. What the fellowship did accomplish in a very few years was to recruit and support for the duration those strong and venturesome personalities that were best able to withstand the repeated frustrations and refusals involved in earning German degrees at the time. These were the very women whom Christine Ladd-Franklin and Ellen Richards, now middle-aged, wished to reach and encourage. ACA fellows knew that the hopes and future prospects of many other women depended on their efforts. They must not fail—they must jump whatever hurdles German officialdom could erect and emerge victorious, whatever the cost. Against such uncompromising determination at least one German university's resistance slowly withered and eventually collapsed.

Yet one suspects that for these victories to have occurred in 1895 and 1896, the Germans' will to resist must already have begun to falter. Felix Klein had apparently been busy convincing his colleagues at Göttingen that there was no harm in giving degrees to foreign women, and the officials at Heidelberg, who could have refused Hyde as firmly as did those at Strasbourg, unsuspectingly underestimated her determination and relied only on stiff procedural barriers: a woman could have a degree, laughed Willy Kühne, *if* she could meet his standards. But then having agreed to set the terms (however hypothetically), he was obliged to award the degree to a woman when she did meet them.

Upon their return to the United States, some of these first American women holders of German doctorates took steps to smooth the way for their followers. Feeling that the ACA's official sponsorship would help reassure the friendlier German professors and encourage them to take other American women seriously, the ACA started in 1896 a Council for Foreign University Work with Ida Hyde, Ph.D., and Margaret Maltby, Ph.D., as members. Its purpose was to "consider and pass upon the qualifications of women to pursue advanced work in European universities." But the need for such a council never materialized. Relatively few women availed themselves of its services, and soon the German government moved to ease applications through its own consular offices.[37] Meanwhile Isabel Maddison, a "wrangler" from Cambridge University, England, who had come to Bryn Mawr College for a doctorate in mathematics, greatly eased foreign study with her *Handbook of British, Continental and Canadian Universities, with Special Mention of the Courses Open to Women* (1896). She issued a supplement in 1897 and a second edition in 1899,[38] by which time, however, the appeal of foreign doctorates had begun to wane for Americans of both sexes. It is somewhat ironic and anticlimatic that so soon after being opened to women at such great emotional cost, the German universities began to lose their competitive edge over American ones and never did play the important role in the careers of American women scholars that Ladd-Franklin and her cohorts must have expected.

(In fact, the chief beneficiaries of this heroic "entering wedge" at the German universities seem to have been the German women, who were not directly involved in the struggle of the 1890s, but who ten years later reaped the benefits of its consequences. It was German women the German professors were most adamantly opposed to admitting in the 1890s. Foreign women were far less of a threat, since they would return home and not expect to teach in Germany. Nevertheless the acceptance of foreign women [like the "infiltration" earlier at the American graduate schools] helped set a precedent upon which German feminists capitalized later. Starting as early as 1902 at some institutions and as late as 1908 at others, the German universities admitted their own countrywomen and awarded them degrees. After about 1904 one begins to hear of such great German women scientists as Lise Meitner in physics and Emmy Noether in mathematics. Some of the first German graduates were faculty daughters who had been watching the American women in the 1890s and waiting enviously for their own chance to attend the universities and earn degrees there too.)[39]

Yet Ladd-Franklin had never intended to stop her campaign with the opening of the German universities to American women. All along she had planned to use the publicity about the women's success abroad to create additional pressure on lagging institutions back home, or, as she wrote Marion Talbot in 1896, "I had in mind much less the aid and comfort of the individual women than the making it well-known that some women were engaged in studying in foreign universities, and with some attendant distinction."[40] Harvard and The Johns Hopkins were particularly susceptible targets for this tactic since they prided themselves on their similarities to the German universities. As early as 1892, Ladd-Franklin had alerted William James, professor of psychology at Harvard, that Göttingen would soon be admitting women students. He understood the implied comparison and responded, "Of course we are going to have women in Harvard soon. Göttingen mustn't be allowed to get ahead of us."[41]

By 1892 several professors in Harvard's philosophy (later divided into psychology and philosophy) department, including James, Hugo Münsterberg, Josiah Royce, and George Herbert Palmer (husband of Alice Freeman Palmer, former president of Wellesley College), had been welcoming women graduate students. By the 1890s they had admitted them not only to their lectures but even to their laboratories, a privilege hitherto withheld from them. Before long the evident brilliance of some of the early women and the unfairness of university policy toward them spurred this relatively liberal department to risk a faculty fight to get the women Harvard degrees. The first test case was Mary Whiton Calkins, who passed all her examinations brilliantly in 1895, but who was denied a degree.[42] Then, in 1898, Ethel Puffer (later Howes) also passed all her examinations with such distinction that she was deemed by a committee of eight full professors to be "unusually well qualified" for a Harvard Ph.D., but the corporation again refused to grant it.[43] Nevertheless the department appointed her an "assistant" on the faculty, undoubtedly a "first," but not one of which the university was proud. Because the authorities did not include Puffer's appointment in the official catalog, the only record of it is an unidentified clipping in her papers at Smith College:

Woman in Harvard Faculty

Miss Ethel Puffer, assistant in psychology at Harvard, is a recognized member of the University faculty, but her name is not printed in the catalog for fear that it would create a dangerous precedent. Miss Puffer was the younger member of her class at Smith College and after a few years of teaching in her alma mater, went abroad and studied psychology under Professor Munsterberg, now at the head of the department of Psychology at Harvard but then lecturer at Freiburg, Germany. He is reported as saying that Miss Puffer is his most brilliant and most thorough disciple.[44]

A few years later, in 1902, when the Harvard Corporation had resolved the issue by forming the Radcliffe Graduate School to grant the university's degree to women, both Puffer and Calkins were offered Radcliffe doctorates. Puffer accepted, but Calkins refused, despite the strong urging of Hugo Münsterberg, and insisted on a Harvard degree or none at all. Since a later appeal by her friends was unsuccessful (see following text) and she died long before the policy was changed in 1963, Calkins never did get a doctorate. Apparently the loss did not hurt her career, however, because she became a full professor at Wellesley College and not only was the first woman elected president of both the American Psychological Association (1905) and the American Philosophical Association (1918), but was (with William James and John Dewey) one of just three persons of either sex ever to hold both positions.[45]

The chief obstacle to the admission of women to the new The Johns Hopkins University, proclaimed as the first full-scale graduate school in America when it opened in 1876, was apparently the attitude of its trustees and president Daniel C. Gilman, who was not sympathetic to the women and felt he had enough problems without taking on the thorny issue of coeducation. In 1877 he offered instead to let the women buy their way into the new institution by forming their own coordinate college nearby, but he refused to let them take classes with the men, as M. Carey Thomas learned and as a thick file of rejection letters at The Hopkins reveals. Despite this official policy of exclusion, a few women were, as explained earlier, admitted as "special students" in the late 1870s and 1880s, but none was awarded a doctorate, however brilliant her work.[46]

By the early 1890s, however, changes elsewhere and at its own medical school (see what follows) brought some softening at The Hopkins, and in 1893 the university awarded its first doctorate to a woman, not retroactively to Ladd-Franklin, however, but to the less controversial Florence Bascom, in geology. Admitted in 1891 as a "special student," Bascom performed so well and proved so helpful to her professor, who used her fieldwork in his reports to the Maryland Geological Survey, that in 1892 he recommended her admission to candidacy. Apparently her case was greatly aided not only by her having all the "ability, energy and enthusiasm that could be expected of any man," as her professor put it, but also by the support of another, older professor, who had been a college classmate of her father. (John Bascom, then the president of Williams College and the former president of the University of Wisconsin, may also have known Gilman personally.) Besides citing these strong personal assets, Bascom's supporters could also inform the trustees that she would probably be hired in a year or two by Bryn Mawr College, whose dean M. Carey Thomas was at the time involved in a major fund-raising campaign for the new The Johns Hopkins Medical School. Thus if the Hopkins trustees were ever to make an exception to their prohibition on degrees for women, Bascom was as good and as safe a candidate as they were likely to get. This argument worked, and the trustees made the exception, though they emphasized that her case was a special one that did not set any precedents for others. This proviso, however, did not deter many Baltimore, New York, and other journalists from confidently predicting that The Hopkins would soon be admitting women to all its postgraduate programs.[47]

Yet the anticipated change did not come for fourteen more years. In 1896 Ladd-Franklin, who was by then a lecturer in psychology at The Hopkins, was still jubilantly expecting an official change in policy to be announced soon. This was premature, however, since a strong backlash against it developed among the rest of the faculty, including especially physicist Henry A. Rowland, who wrote an angry letter to the editor of the *Nation* denouncing women applicants and defending Gilman's decision to postpone their admission. It was not until 1907, thirty-one years after the university had opened, that Gilman's successor, President (and by then suffragist) Ira Remsen could finally admit women officially. Remsen defended the new policy, as Hadley had his at Yale fifteen years earlier, as "a simple act of justice" that merely recognized what was fast becoming standard university practice anyhow. He also softened the blow for coeducation's diehard opponents by reserving them the right to refuse women students if they wished. Women were admitted to all classes provided there was "no objection on the part of the instructor concerned." Even so, old practices seem to have lingered on in some departments, since in 1913 philosopher Arthur O. Lovejoy was pointing out to the administration the inconsistencies and potential embarrassment to the university if it were discovered that it was still not putting women in the catalog. Finally, in 1926, at its fiftieth anniversary celebration, The Johns Hopkins University awarded a long overdue doctorate to one of its most talented graduates, Christine Ladd-Franklin, who, now a sprightly seventy-nine-year-old, made it a point to attend the ceremonies and collect her degree forty-four years late.[48] (A year later, a petition signed by thirteen Harvard men mostly in psychology and presented to President Abbott Lawrence Lowell of Harvard University requested that the Harvard Corporation follow suit and award Mary Calkins her long-overdue doctorate. This request received the curt reply that the corporation felt that there was no adequate reason for granting her a Harvard degree.)[49]

It should be noted that the official acceptance of women graduate students by The Johns Hopkins Graduate School in 1907 came almost fifteen years after the same university had grudgingly admitted them to its medical school in 1893. The fact that women gained entry to The Johns Hopkins Medical School so much earlier than to its graduate school points up the relative weakness of the strategy of slow "infiltration" and the comparative strength of a second approach, that of "coercive (or creative) philanthropy"—the offering of large gifts with key strings attached. (This approach also had its limitations.) In 1890 the Hopkins trustees launched a $500,000 fund drive for a medical school designed, in the tradition of the Hopkins graduate school, to be the best in the nation. In due course the trustees approached Mary Garrett, heiress to the Baltimore and Ohio Railroad fortune, and close friend of M. Carey Thomas, dean and soon to be president of Bryn Mawr College, for a contribution. She offered $60,000 on the conditions that (1) baccalaureate degrees be required for admission, a novelty at the time, and most importantly here, (2) that women be admitted and given the "same" (not just an "equal" and so probably separate) education at the new medical school. At first the trustees were unwilling to accept these conditions, but when their fund drive faltered and Garrett increased her offer to $307,000, President Gilman, Dr. William Welch, dean of the as yet unborn medical school, and the trustees anxious to complete the drive finally relented and decided to accept the conditions in 1893. The school quickly became known for its long line of distinguished women graduates in medicine and medical research, with the greatly honored Florence Sabin, M.D. 1900, one of its first.[50]

Yet, lest one think that money always "talks" in this effective way, he or she should be reminded of two other cases where benefactors' attempts to buy women's way into major universities in these years were unsuccessful. In 1898 the University of Michigan learned that a Dr. Elizabeth Bates of New York had left them the bulk of her estate ($133,000) for a professorship of gynecology and pediatrics provided that they would expand their offerings, especially in clinical instruction, to women medical students (whom they had been admitting since 1871) to make their education fully equal to that of its men students. The university gladly accepted the gift but never used it for this purpose. Instead the money was deflected into underwriting the expenses of the obstetrics department and adding a children's wing to the university hospital. Apparently Dr. Bates's executors were unwilling or unable to enforce the conditions of her gift,[51] thus demonstrating further Mary Garrett's shrewdness in making her gift while she was still able to oversee its full implementation.

Another such gift that not only was deliberately misallocated but actually almost backfired on the women's attempts to enter the graduate schools was given to the University of Pennsylvania between 1887 and 1892. The university had, before many others, quietly become coeducational at the graduate level in the mid-1880s when two of its first four graduate students in 1885 were women. Then in 1887 Joseph Bennett, a Philadelphia businessman, initiated a series of gifts to the university, eventually valued at over $400,000, to be used for an undergraduate "College for Women." The trustees accepted the properties and money, but since the president, like Dwight at Yale, opposed the introduction of coeducation into the undergraduate college, he used the buildings to house a separate "Graduate School for Women" established in 1892. Rather than restricting graduate women at Pennsylvania, however, this "school" apparently just regularized their status, for the women at Pennsylvania continued to attend coed graduate classes, and a steady stream earned doctorates after 1894.[52] Thus these two cases illustrate that "coercive philanthropy" was not always successful in advancing the higher education of women in previously all-male universities—administrators would accept the gift but often neglect its conditions and deliberately deflect it to other uses more in line with their own wishes. Only a very astute and vigilant donor (like Mary Garrett) would enforce unpopular conditions to a gift. "Coercive philanthropy" could be a highly successful strategy for changing institutional policies, but it could also fail totally.

Some of these same "creative philanthropists" utilized their strategy again later in the 1890s when it became clear that large discrepancies persisted in the postdoctoral opportunities open to women and men. For example, although Harvard University, The Johns Hopkins University, the Smithsonian Institution, and some other organizations were supporting tables for Americans at the famed Naples Zoological Station, where many German researchers did their experimental work, none provided space for women investigators. Ida Hyde learned of this in 1896, when, after earning her doctorate at Heidelberg, she was invited to continue her researches at one of the German tables at the Naples station. Greatly enjoying the experience and realizing how invaluable it was for the training of any future biologist, Hyde formed a committee in 1897 to subsidize a table there for an American woman investigator ($500 annually). M. Carey Thomas, whom Hyde had known when at Bryn Mawr in the early 1890s and who had been calling in the ACA's publications for some kind of encouragement for American women scholars, was enthusiastic about the project and immediately headed a new group, which in 1898 called itself the "Naples Table Association for Promoting Laboratory Research by Women" and which was to be closely affiliated with the ACA. Thomas had little difficulty in convincing several women's colleges and wealthy individuals to contribute $50 each to a fund to underwrite this highly coveted research experience for a series of American women biologists. Most of the distinguished recipients of this award were or would soon be professors at or otherwise affiliated with the women's colleges, as were Mary Willcox of Wellesley and later Vassar, Florence Peebles of Bryn Mawr and later Goucher, Emily Ray Gregory* of Wells College and later the American College for Girls in Constantinople, Cornelia Clapp of Mount Holyoke, and Nettie Stevens of Stanford University and Bryn Mawr College. In 1901 the Naples Table Association felt sufficiently wealthy to support a second "table" for a woman, this time at the recently reorganized marine biological station at Woods Hole, Massachusetts. Over the years this too was held by a series of highly regarded women.[53]

The enthusiastic Naples Table Association did not stop even there. Finding itself in the early 1900s with more money than it needed for its two zoological "tables" and realizing, as its leaders put it, that "the possibilities of wider work had opened to them," the NTA decided to enlarge its philanthropy to cover the more general task of "the promotion of scientific research by women." They voted to establish another award, this time a prize for "the best thesis written by a woman (of any nationality) on a scientific subject embodying new observations and new conclusions." The cash award was to be $1,000, or almost a year's salary. (In 1911 it was renamed the Ellen Richards Research Prize after the death of one of the committee's more active members.) Despite certain difficulties in selecting winners, the NTA awarded its prize six times before 1924: to Florence Sabin of The Johns Hopkins Medical School in 1903; Nettie Stevens of Bryn Mawr College in 1905; Florence Buchanan of University College (London) in 1910; Ida Smedley MacLean, also of University College, in 1915; Eleanor E. Carothers of the University of Pennsylvania in 1921; and Mary E. Laing of the University of Bristol (England) in 1924.[54] (For more on this later, more troubled, history of the Naples Table Association, see chapter 11.)

A third and even more significant postdoctoral award for women of this period, including those in the sciences, and a particularly interesting case of

*Not to be confused with Emily L. Gregory (1840–1897), the Bryn Mawr and Barnard botanist.

"creative or coercive philanthropy" in its own right, was the Sarah Berliner Research or Lecture Fellowship. Probably the first moveable postdoctoral fellowship for a scientist of either sex, it was long the personal project of Christine Ladd-Franklin, and it bore the mark of her particular style. As early as 1898 she expressed concern that so many women doctorates in general and ACA fellows in particular were failing to find satisfactory employment upon completion of their studies. Some found jobs in women's colleges, a few had positions in coeducational institutions, but many apparently were underemployed or unemployed and forced to return either to schoolteaching or to staying at home with their families. In order to do something about this continuing problem, she advocated in 1905 the establishment of some endowed "peripatetic professorships" for women in order to take a few of the top young women of the country and "tide them over the years that must elapse between their becoming mature enough and distinguished enough to be full professors; to prevent them from sinking into plain schoolteachers."[55]

As earlier, with her use of the women's success in earning doctorates from Göttingen as leverage against Harvard, Ladd-Franklin had a larger, ulterior motive for her proposed professorships: "But, by far the most important [purpose] of all [is], to create a few first-class women college professors who . . . would . . . make a distinct contribution toward the furthering of the rights and privileges of the sex in general."[56] Her plan was for the ACA to pay a woman's salary for a year while she did research and taught a light schedule at a college of her choice, preferably a coeducational university. In return for her (free) services, the school would list her name in the catalog as a faculty member, an arrangement Ladd-Franklin thought few schools could refuse. The results of this exchange were to be not only a profitable experience for the woman and a strong boost to her career, but also public evidence that women could teach successfully at coeducational universities, thus opening the doors still further for other women. If the coeducational universities would not hire women, even "exceptional" ones, on their own, the ACA would entice them into it by offering the first ones free.

Although Ladd-Franklin did not succeed in raising the endowment necessary for a full professorship, she did induce Emile Berliner, the wealthy inventor of the phonograph record, to endow (with $25,000) a postdoctoral fellowship for a woman scientist in honor of his recently deceased mother, Sarah Berliner. In 1909 he and Ladd-Franklin offered this fellowship with the moderate sums of $1,200 (for research) or $1,500 (for research and lectures). It immediately attracted a series of distinguished candidates, many of whom had already won other ACA awards. Most either were or would later be professors or affiliates of the women's colleges, as Edna Carter (physics at Vassar), Gertrude Rand (psychology at Bryn Mawr), Elizabeth Laird (physics at Mount Holyoke), Ethel Browne Harvey (independent research at Princeton and Woods Hole), Janet Howell Clark (physiology at The Johns Hopkins), and Carlotta Maury (independent paleontologist and petroleum consultant). Although occasionally criticized by other women for both her seemingly high-handed and autocratic methods and the whole idea of forcing coeducational institutions into "hiring" women with such blatant blackmail, Ladd-Franklin was undeterred and administered the fellowship personally until 1919, when, at the age of seventy-two, she passed the endowment over to the ACA, which still awards the fellowship annually. Because most of the Berliner fellows remained at the women's colleges, the project failed in its second and larger purpose of inducing the major universities to hire prominent women scientists and scholars. Nevertheless, it did succeed in its first purpose of rescuing highly trained women from "sinking into plain schoolteaching" by giving them an opportunity to pursue their important researches even farther.[57]

By the early twentieth century academic feminists could rejoice that their efforts had played some role in inducing most of the American and German graduate schools to open their highest degrees to women. Some of this progress had occurred quite readily, but converting the remaining institutions had been a long and difficult struggle that had required the best energies of many persons of vision, determination, and wealth. In the process of working toward their goal, several women scientists had helped to develop a variety of distinctive strategies and pressure tactics that were useful in also creating new opportunities for women beyond the doctorate, as gaining access to certain laboratories, offering new postdoctoral research opportunities, and subsidizing up to a year of suitable employment at a coeducational institution. Since such postdoctoral opportunities were increasingly necessary for successful careers in science and remained far from equal for the best men and women of the time, it was becoming clear that women needed special supplements such as fellowships, awards, and even subsidized jobs if they were not to fall even farther behind. The ACA and especially the NTA were pioneers in the creation of new honors and forms of recognition that, they sensed correctly, were becoming essential features of science in the twentieth century.[58]

Nevertheless these reforms did not lead to the full academic careers for women that the women activists must have expected. On the contrary, it was during these same years, when the more liberal universities were agreeing that it was "only fair" to award women the same degrees as men, that they were also rapidly institutionalizing a separate and decidedly unequal employment policy for women, if they hired them at all. In fact, it may have been the consolidation of such segregated employment in the 1890s that lay behind the graduate schools' sudden liberalism in allowing women to enroll and take degrees there. In any case the women now had to face the basic underlying inconsistency that those very institutions that would educate them and award them doctorates would not hire them for their faculties. Until the women could find a way around that dichotomy, their positions in science and academia would remain marginal, despite their hard-won doctorates and fellowships.

3

"WOMEN'S WORK" IN SCIENCE

If by 1910 women had succeeded in being allowed to earn degrees from almost all German and American universities granting them to men, they were far less successful in these years in gaining equal treatment in the world of employment.* The resistance to women's holding the same jobs as men was much stronger than it was to women's earning the same degrees as men, largely because most jobs were traditionally labelled or "sex typed." Thus when women sought those jobs to which their hard-won degrees seemed to entitle them, they found themselves the victims of powerful social forces and traditions that channelled men and women into separate and decidedly unequal forms of employment. Systematically refused "men's jobs" but in need of some kind of scientific employment, most women scientists of the 1880s and 1890s abandoned the agressive tactic of "infiltration" that they were using so successfully at the graduate schools and instead advocated (publicly at least) the more moderate goal of creating separate, specifically "feminine," jobs for women in science.

Although it has often been asserted that the practice of science was open equally to both sexes (or, to use sociological terms, "universalistic" or "sex blind"), in fact a separate labor market for women emerged in the sciences in the 1880s and 1890s, when they began to seek scientific employment in significant numbers, and it was firmly established in several fields by 1910. Although the practice of such "sex segregation" was usually justified with the essentially conservative rhetoric that women had "special skills" or "unique talents" for certain fields or kinds of work, the phenomenon basically seems to have been an economic one whose origin and perpetration were the result of three forces: (1) the rise of a new supply of women seeking employment in science, including the first female college graduates; (2) strong resistance to this female work force's entering traditional kinds of scientific employment (such as university teaching or government employment); and (3) the changing structure of scientific work in the 1880s and after which provided new roles and fields for these entrants. Thus the way in which these first women were incorporated into the world of scientific

*This chapter appeared in slightly different form in *Isis* 71 (1980):381–98.

employment between 1880 and 1910 gives some clues as to how scientific work was expanding and research strategies changing in these years.

When the movement to give women a higher education had begun to take hold in the United States in the 1870s and 1880s, little thought had been given to the careers such graduates might take up. Because of the prevailing notion of "separate spheres" for the two sexes, it was assumed that most women were seeking personal fulfillment and were planning to become better wives and mothers. Advocates of their study of science saw it as offering a rigorous and satisfying intellectual experience to women who led essentially "aimless lives." Even such accomplished scientists as entomologist Mary Murtfeldt of St. Louis, astronomer Maria Mitchell of Vassar College, ornithologist Graceanna Lewis of Philadelphia, and physicist Edward C. Pickering of the Massachusetts Institute of Technology expected that the women would participate in science only as amateurs. There were still so few women scientists in the United States in the 1870s that there was barely a hint of sexual stereotyping or "women's work"—all fields were presumed to be open to women, who were assumed to be equally adept at all of them. Accordingly, the Reverend Phebe (*sic*) Hanaford opened her chapter on the women scientists of the 1870s with the words "Science knows no sex."[1]

As the numbers of college women increased, however, in the 1880s, expectations began to rise on all sides that their training lead somewhere. If it had been wasteful in the 1870s for women to sit idly home,[2] it was much more intolerable for college graduates to lack useful and respectable work. Accordingly the advocates of higher education for women began in the 1880s to talk hesitantly of improved job prospects as well as personal fulfillment for college graduates.[3] In her 1882 article "Scientific Study and Work for Women," Mary Whitney, Mitchell's student and successor in astronomy at Vassar, went beyond the standard view that science would help to develop a woman's mind and "introduce a more definite purpose into her life" to proclaim that it also laid the basis for "useful, and I hope, in the future, remunerative labor" for her. Unfortunately this was not yet a real possibility for most women, since, as Whitney put it, "we cannot say the present offers many examples." Some women had been successful as physicians and others as professors, but Whitney also thought there would soon be opportunities for women in those other areas for which they were particularly well suited and in which some women (no names were given) had already been active, fields such as practical chemistry, architecture, dentistry, and agriculture. To her credit Whitney resisted the temptation to minimize the problems this pioneer generation would face. She warned them that they would have difficulty in finding any kind of employment, since even at the women's colleges "the chances are largely in favor of the man." But after urging young women to prepare themselves for such careers, Whitney showed her own ambivalence by concluding that in any case scientific training would make them excellent mothers, which after all was "the highest profession the world has to offer."[4]

Fortunately for the women seeking to enter science in the 1880s and after, at least three forces were shaping scientific work at that time which would provide new roles and opportunities for them: (1) the rise of "big science" or large budgets, which could support staffs of assistants at a few research centers; (2) a new concern for the nation's growing social problems which created the need for several new hybrid or "service" professions (or semiprofessions) designed to solve them; and (3) the need for new faculty and other personnel at the coeducational land-grant agricultural colleges. Since some of these new jobs offered women a chance to use their special "feminine" skills in ways that did not threaten men directly (and even enhanced their dominant role), the 1880s and 1890s saw much explicit channelling of the newly available women into certain "appropriate" callings.

Advocates of such "women's work" had no trouble developing a rationale for separate kinds of jobs for women. They had merely to urge women to capitalize on the relatively warm welcome that they were already receiving in the marketplace for two kinds of jobs: those that were so low paying or low ranking that competent men would not take them (and which often required great docility or painstaking attention to detail) and those that involved social service, such as working in the home or with women or children (and which were often poorly paid as well). The literature on "women's work" glorified these positions, considered them very suitable for women in science, and advocated more of them.[5] This message was quickly communicated to an eager audience by a new social mechanism, the middle-class magazines, whose contributors seized upon any work found suitable for such women and with evident relief advised others to enter it as well. Like the more explicit "vocational guidance" of the twentieth century, these magazines needed only a hint of a success story to unleash a torrent of articles (many written by women) extolling the new opportunities awaiting women in the newest area of "women's work."

Events were occurring so rapidly in the early 1880s that even as Mary Whitney was writing her transitional and ambivalent article in 1882, the first kinds of "women's work" in science were appearing. Although a few women (including Maria Mitchell herself) had previously worked at home as "computers" for others' astronomical projects, "women's work" in astronomy was just entering a new phase. The change apparently grew out of a fortunate but not unusual set of circumstances at the Harvard College Observatory in 1881. In that year, as the story goes, Edward Pickering, the advocate of advanced study for women cited earlier and the newly elected director of the observatory (the first astrophysicist to attain the position), became so exasperated with his male assistant's inefficiency that he declared even his maid could do a better job of copying and computing. He promptly put Williamina P. Fleming, age twenty-four, Scottish immigrant, public school graduate, divorcée, and mother, to the test, and she did so well that he kept her on for the next thirty years. She not only became one of the best-known astronomers of her generation, but she also showed such good executive ability and "energy, perseverance and loyalty," as one obituary put it, that Pickering put her in charge of hiring a staff of other women assistants (see plate 8), whom he paid the modest sum of twenty-five to thirty-five cents per hour to sort photographs of stellar spectra. Between 1885 and 1900 she hired twenty such assistants, including the college graduates Antonia C. Maury, Vassar 1887; Henrietta P. Leavitt, Radcliffe 1892; and Annie Jump Cannon, Wellesley 1884 (who had been at home in Delaware for a decade). Several of them made such prodigious contributions not only to the observatory's main project in those years, the Henry Draper Star Catalog of stellar spectra, but to other areas of astrophysics, that they became highly regarded in their own right.[6]

Within a few years the fame of this novel employment practice began to spread, and women astronomers became the subject of several favorable magazine articles.[7] In addition, Fleming openly propagandized for the Harvard arrangement in an address on "A Field for Woman's Work in Astronomy" at the World Columbian Exposition ("World's Fair") in Chicago in 1893. Like other contemporary accounts, hers praised Pickering's progressive attitude in hiring the women and talked of their many contributions, but also moved beyond this into some sex stereotyping of the skills involved. Thus Fleming was on safe ground when she urged other observatory directors to hire female assistants, since such women "if granted similar opportunities would undoubtedly devote themselves to the work with the same untiring zeal." But she was on more precarious ground when she tried to describe what this case illustrated about the comparative skills and abilities of the two sexes, and in fact it is not clear what she did mean by her rather confused conclusion: "While we cannot maintain that in everything woman is man's equal, yet in many things her patience, perseverance and method made her his superior. Therefore, let us hope that in astronomy, which now affords a large field for woman's work and skills, she may, as has been the case in several other sciences, at least prove herself his equal!"[8] Apparently she agreed with the prevailing idea that women were generally inferior to men, but she felt that by overachieving in the bottom ranks (or far outstripping what persons at their level were expected to do), they might prove themselves "equal" to men who had far greater opportunities.

This pattern of segregation/sex typing proved popular and spread to most other major observatories in the United States in the 1890s and after. Of the 20 women astronomers listed in the third edition of the *American Men of Science* (1921), 8 worked as assistants at major observatories (and Fleming, who died in 1911, had been listed earlier), one sign that this work attracted a sizeable proportion of the women astronomers in the country. (Seven others taught at women's colleges, as had three retirees.) Yet this count only skims the surface of the phenomenon, for Pamela Mack has collected a list of 164 women, mostly high school graduates, who worked at various observatories for a year or more between 1875 and 1920. (Of the few college graduates in her list, those trained at Vassar by Mitchell, Whitney, and later Caroline Furness were in particular demand and were sought after by many observatory directors.) Almost every large observatory hired these women; Mack found 24 at the Dudley Observatory in Albany, New York; 12 at Yerkes Observatory in Wisconsin; 12 at Mount

PLATE 8. The Harvard College Observatory was known in the 1890s for the many women it employed to classify stellar spectra. Among those shown here in 1892 are Henrietta S. Leavitt (third from left), Williamina P. Fleming (standing), and Annie Jump Cannon (far right). (*New England Magazine*, n.s. 6 [1892]:166.)

Wilson in southern California; 6 at the U.S. Naval Observatory in Washington, D.C.; and smaller numbers at most of the other observatories, such as Columbia, Allegheny, Lick, and Yale, where talented doctorate Margaretta Palmer was especially notable.[9]

Such a wide acceptance of female assistants in astronomy in these years would seem to have been the result of something more pervasive than Pickering's personality and his practice at Harvard. More likely it grew out of certain competitive forces within the field of astronomy itself in the 1880s and after. Although still a small field, astronomy was apparently growing rapidly in the 1880s and 1890s, when several new observatories were built, and the whole new field of astrophysics was just appearing. This rapid expansion created two problems for the older observatories: maintaining a large staff of good assistants, which would be difficult because the more experienced (male) assistants might be offered better positions elsewhere at any moment, and keeping up in observational astronomy, which would be almost impossible because the newer observatories often had larger telescopes and were located in areas with better viewing conditions. Thus, as Pamela Mack has explained it, most of the newer observatories used their advantages to concentrate on the traditional observational astronomy, and they offered men from the older institutions exciting new opportunities in this field. Although these new observatories also hired some women, they restricted them to the tedious and laborious "computer" work women had long done for male astronomers. For the women the better opportunities in this period of change were not in observational astronomy at all but in the newer specialities, where they got some of the work (but not the actual jobs) that the more mobile young men had left behind. They fared much better at Harvard, for example, than they did in the West, because in order to compete with his new rivals, Pickering moved away from observational astronomy and into another specialty, the new field of photographic astrophysics. His adoption of this more advanced technology of cameras and spectroscopes had great implications for women in science, since it required a different labor force: Pickering needed fewer observers ("men's work") and many more assistants ("women's work") to classify as cheaply as possible the thousands of photographic plates his equipment was generating.[10] The term *proletariatinization* has been applied to this phenomenon of downgrading the job and then allowing it to be feminized.[11]

If Pickering (and some other observatory directors) were progressive in greatly expanding women's employment in astronomy in the 1880s and 1890s, they were not so far ahead of their time as to promote them to "men's work" or pay them men's wages, even for admittedly important or outstanding work. Most female assistants remained at the same level for decades, and thus had no alternative but to make a whole career out of a job that should have been just a stepping stone to more challenging and prestigious roles. In a sense science benefited from this practice, since these women could complete many long-term projects. One example is the massive Henry Draper Star Catalog, compiled by women when Pickering became too involved in administration and fund raising to do much astronomy himself. Knowing their "place" and having few if any options, women graciously (and for the most part gladly)[12] accepted what was offered to them and stayed as long as they were wanted. Thus the new "women's work" in astronomy trapped many into low-level jobs; Fleming, for example, appears to have had strong executive abilities, and might, had she been given the chance, have made a good director of one of the many new observatories opening up in these decades. Nevertheless, the only female directors of American observatories to date have been those at the women's colleges and at the small Maria Mitchell Observatory on Nantucket Island, a feminist memorial and outpost in the Atlantic Ocean.[13]

The most advancement these women could hope for was to a position like Mrs. Fleming's in directing the work of other women. Even she, however, seldom got a raise, a circumstance that had begun to bother her greatly by 1900—after almost twenty years of devoted services. For two months of that year she kept a private diary about her job as part of a university-wide historical project to collect and preserve for posterity descriptions of the work of the Harvard "officers," among whom she was the only female. In her account, recently uncovered in the Harvard University Archives, she describes her daily activities for the month of March and part of April. Most of the diary simply recounts her duties directing the work of eleven women and assisting the director at the observatory, but at times she records some of her own reactions to her job. There one learns that personally she would rather have spent her days making her own astronomical discoveries, but the director, whom she greatly respected, thought it was better for the observatory for her to spend most of her time preparing others' work for publication, a task she found "very trying" at times and "generally extremely difficult," especially when the would-be authors were the less articulate members of the staff. But it was her low salary that caused her greatest private complaints. On 12 March 1900 Fleming erupted with a lengthy and forthright analysis of the injustice of her salary of just $1,500 per year:

> During the morning's work on correspondence &c. I had some conversation with the Director regarding women's salaries. He seems to think that no work is too much or too hard for me, no matter what the responsibility or how long the hours. But let me raise the question of salary and I am immediately told that I receive an excellent salary as women's salaries stand. If he would only take some step to find out how much he is mistaken in regard to this he would learn a few facts that would open his eyes and set him thinking. Sometimes I feel tempted to give up and let him try some one else, or some of the men to do my work, in order to have him find out what he is getting for $1500 a year from me, compared with $2500 from some of the other assistants. Does he ever think that I have a home to keep and a family to take care of as well as the men? But I suppose a woman has no claim to such comforts. And this is considered an enlightened age! I cannot make my salary meet my present expenses with Edward in the Institute [MIT] and still another year there ahead of him. The Director expects me to work from 9 A.M. until 6 P.M., although my time called for is 7 hours a day, and I feel almost on the verge of breaking down. There is a great pressure of work certainly, but why throw so much of it on me, and pay me in such small proportion to the others, who come and go, and take things easy?
>
> The [rest of the] day was occupied with the usual work. . . .[14]

A month later, upon reviewing her previous entries, Fleming sought to clarify her mixed feelings of personal respect for Pickering but continuing impatience with her low salary:

> I find that on March 12 I have written at considerable length regarding my salary. I do not intend this to reflect on the Director's judgment, but feel that it is due to his lack of knowledge regarding the salaries received by women in responsible positions elsewhere. I am told that my services are very valuable to the Observatory, but when I compare the compensation with that received by women elsewhere, I feel that my work cannot be of much account.[15]

A similar bureaucratization resulting in sex-segregated employment was also underway in the expanding world of natural history museums in the late nineteenth century. Although there are instances of women assistants on museum staffs as early as the late 1860s, more were added in the next few decades to help museums cope with the explosive growth in collections brought back from the increasing number of expeditions to far-off places. The whole scale of natural history was changing. A naturalist, for example, who might formerly have devoted his career to the classification of the flora or fauna of a given region would now be overwhelmed by the vast quantities of specimens that were piling up. Entire museums were necessary to house them, and staffs of bright (but poorly paid) assistants would be needed to catalog and classify them and to publish taxonomic descriptions. Although women were accepted in this new kind of work, and by 1873 twelve were listed as working at Harvard's Museum of

Comparative Zoology, they received far less publicity than did the women astronomers.[16]

By the 1880s some of these museum assistants were being supported partially at least by the U.S. Fish Commission, which underwrote much zoological research in these years. Two of these were selfless spinsters from scientific families who rejoiced in even this modest opportunity to become productive marine biologists. In 1879 Addison Emery Verrill, a zoologist at Yale's Peabody Museum of Natural History, hired Katharine Jeannette Bush (plate 9) as an assistant. Although not a college graduate, Bush later enrolled in Yale's Sheffield Scientific School as a "special student," and in 1901 she became the first woman to earn a Yale doctorate in zoology. But there is no sign that the degree made much difference in her job situation. Although she worked at the Peabody Museum until 1913, she may have been paid for only twelve of these thirty-four years, and then by the federal government: by the Fish Commission in the 1880s, when she reported on the various mollusks dredged by the commission's vessels, and by the U.S. National Museum in the 1890s, when she reported on some of its collections. In time she published at least nineteen articles on marine invertebrates and was the best known woman in Yale's zoological circle at the turn of the century. Also there were two of her sisters: one, a librarian at the Peabody Museum, worked as an assistant to paleontologist O. C. Marsh, and the other was married to Wesley Coe, professor of zoology.[17]

Similarly, Mary Jane Rathbun (also pictured in plate 9) started her long scientific career in the summer of 1881, when as a high school graduate she accompanied her older brother Richard to Woods Hole, where he was an employee of the Fish Commission, and volunteered to assist his boss, Spencer Baird. She returned each summer thereafter with her brother, and in 1884 became a paid clerk. Then in 1886 Baird transferred her to a job as a copyist at the U.S. National Museum in Washington, D.C., and she began to work on the classification of crabs. Although her job titles always remained subordinate (scientific assistant, second assistant curator, and finally assistant curator), she took over much of the general supervision of the museum's invertebrate zoology division, which her brother headed in the 1890s. Unlike Katharine Bush, Rathbun had a government salary that rose steadily from $960 per year in 1894 to $1,680 per year in 1910. In 1914, however, she returned her pay to the museum so it could be reallocated to a younger curator with a large family. She then worked unpaid for almost thirty years. By the time she died in 1943, she was one of the world's experts on crabs, having published 158 articles on them.[18]

Likewise, in the anthropological museums of this period, women began in the 1880s to hold marginal positions, from which they often made sizeable contributions to science. Women anthropologists in these years were often loosely affiliated, almost "free-lance," field workers who were only rarely paid a salary but who were allowed to publish in the museum's proceedings and to have some official museum identification, which aided them in the field. Among the most notable of these women were Alice C. Fletcher, Erminnie A. P. Smith, and Zelia Nuttall, who were affiliated with the Peabody Museum of Archeology and

PLATE 9. Marine biologists Katharine J. Bush (second from left) of Yale University and Mary Jane Rathbun (far left) of the U.S. National Museum are shown here at the Marine Biological Laboratory at Woods Hole, possibly in the 1890s. (Smithsonian Institution Archives, Mary Jane Rathbun Papers, box 8.)

Ethnology at Harvard, the University Museum of the University of Pennsylvania, and the museum at the University of California at Berkeley. A few other, wealthier, women also took up the matronly role of financial patron and power-behind-the-director at late-nineteenth-century anthropology museums. Two of these powerful women were Sara Yorke Stevenson at the University Museum of the University of Pennsylvania in the 1890s and her protégée, Phoebe Apperson Hearst, who started her own museum at the University of California in 1901.[19]

Although most of the women employed by botanical gardens in the 1880s and 1890s tended to be illustrators retained on a piece-work basis, San Francisco physician Mary Brandegee became curator of botany at the California Academy of Sciences in 1883 and retained her position for eleven years, despite many controversies with others over the proper nomenclature for the region's diverse flora. She was succeeded in 1894 by Alice Eastwood, her assistant, who gained instant fame in 1906, when, after the city's earthquake, she had the courage and presence of mind to run back into the academy's building and save the "type specimens" before the oncoming fire could destroy them. Despite this heroism, Eastwood's position was long in jeopardy because of various factions among the academy's officers.[20]

All these women and their various positions are considered here together, since they typify and show, with rare exceptions, the limited range of women's opportunities in the scientific research institutions of the 1880s: they were chiefly in the marginal, subordinate positions that can be termed a hierarchical kind of "women's work." It should be noted therefore that most of the discussion of these jobs has been largely in economic terms.[21] Women were willing to do the often tedious and difficult tasks required for far lower salaries than would satisfy competent men. The women had so few other opportunities that they grabbed these low-ranking jobs and often did superbly well with little support. In addition, they were often relatives of the men in charge.

One should also pay some attention to the arguments that were not being used to restrict women's place in these scientific institutions in the 1880s and 1890s. Although many persons seem to have thought that women "naturally" made good assistants, apparently no one went a step further to use the second standard argument for sex stereotyping of occupations, claiming that the subject matter (here, the stars or the crustacea) was somehow inherently "feminine" or that the science's tools or facilities (observatories or museums) were really "homes" that women could tend better (or more "naturally") than men, a commonly held perception about libraries and public schools at the time. Thus sex stereotyping in astronomy and zoology remained "hierarchical" and did not take on the additional rhetorical and psychological trappings that "territorial" kinds of women's work—child psychology,[22] librarianship,[23] social work, or "home economics"—did at the time. Women in these fields also received low pay, but, in addition, generations of persons believed (and advised others) that women "belonged" in them because there was something uniquely feminine about their subject matter.

The development of scientific employment within the federal government in the 1880s and 1890s provides another vantage point on the sex typing and sex segregation that did and did not quite occur in several other fields. These cases also help to clarify the relatively rare circumstances necessary for the emergence and spread of "women's work" within a science. First, the field had to have a shortage of available men, since there was a general reluctance to appoint women scientists to any job for which men were in good supply. Yet in practice the situation was more complex than this, since the actual supply of suitable workers depended on the job description, which was often in flux in these years. Thus many scientific jobs could be upgraded (masculinized) or downgraded (feminized) over time or as the budget required, manipulating the kind of workers desired. Second (and this is closely related to the first), the employer had to want to hire women, for the appointment process required him (officially under federal Civil Service rules, unofficially elsewhere) to specify beforehand which sex he wished to appoint. Generally only those employers with strong feelings or economic incentives would request a woman. Otherwise inertia and prevailing stereotypes meant that most appointments would go to men. Third, women had to be alerted that this was work for which they would be hired. This receptivity was usually communicated by participants (like Mrs. Fleming) or enthusiastic journalists who described the job's duties almost exclusively in terms of prevailing sexual stereotypes. (Anthropology, for example, could be described as a field in which women could make unique contributions because they could study women and children better than men could.) This publicity was a kind of "market signal" to both potential workers and future employers as to what type of person would be hired.[24] The result of these processes was that what was accepted as "men's" and what was "women's work" was oftentimes not particularly logical or even consistent, but rather the result of a series of employer preferences and economic incentives. In general, however, the women got the less powerful, less prestigious, and lower-paying jobs. The rest were reserved for the men.

The prime example of sex segregation in the federal government occurred in the field of botany, or, more precisely, in the new specialty of plant pathology, in the 1880s. One sign that the signal had by this time already been passed to the public that botany was a "feminine" science was an article reprinted in *Science* in 1887 entitled "Is Botany a Suitable Study for Young Men?" Although none of its four arguments (mental discipline, outdoor exercise, practicality, and lifelong happiness) seems to be very sex linked today, popularizers had already propagandized botany's suitability for young ladies so effectively that some persons thought this protest necessary.[25] Perhaps this feminine image was one reason why several divisions of the U.S. Department of Agriculture that were growing rapidly and would soon become its Bureau of Plant Industry started in 1887 to make it a regular practice to hire women as "scientific assistants." Erwin Frink Smith, the USDA's plant pathologist-in-charge, made it a point to hire a woman (Effie Southworth, later Spalding), and notified the Civil Service ahead of time, as its rules required. Smith was quite proud of this practice and continued it until his retirement in the 1920s. Over the years he hired more than twenty women assistants, including such talented ones as Nellie Brown, Clara Hasse, Charlotte Elliott, Agnes Quirk (see plate 10), Della Watkins, and Mary Bryan, who earned modest fame for their outstanding work on such agricultural problems as crown galls, citrus canker disease, and corn and chestnut blight.[26]

This employment practice was unusual, and the motives behind it are not at all clear. There may have been economic incentives, or Smith and others at the BPI may have been taking advantage of certain highly discriminatory restrictions at the time on women's taking Civil Service examinations. Though sources on this point differ, women were presumably prohibited from taking the examinations for the standard entry positions of "junior botanist" or "junior plant pathologist" before 1919. Until then women, even those with masters' degrees in botany, could take only the exam for "scientific assistant," a lower category. Under

PLATE 10. The Bureau of Plant Industry of the U.S. Department of Agriculture hired many women botanists to study plant diseases. Shown here in 1905 are (left to right) Dean Swingle, Agnes Quirk, Florence Hedges, and Alice Haskins. (Courtesy of National Archives and Records Service.)

these conditions, a shrewd laboratory director might have been able to hire highly qualified personnel at bargain rates. For their part the women might have been very glad to work at projects suited to their skills (and several praised Smith for this) rather than at the more tedious ones to which the Civil Service rules limited them. Then, too, any training given such women botanists would be a good investment, since they would not be promoted to an administrative position or leave for a better position in a state university or an experiment station, all of which were also expanding rapidly in these same years. In addition to these economic motives for preferring women assistants, there may also have been a psychological one—Smith and the other male bosses at the USDA may have (like Pickering at the Harvard Observatory) liked the "harem effect" of being surrounded by a bevy of competent female subordinates who would not be as threatening as an equal number of bright young men. Thus employing women might have been an effective way to limit the turnover and competition in a period of great opportunity and rapid growth, when maintaining a good staff might otherwise have been very difficult. In any case (and there is a 650-page biography of Smith which gives no clues as to why he preferred women assistants), the Bureau of Plant Industry was the only agency of the federal government to become so highly feminized before World War I.[27]

Yet if there were demonstrable economic and psychological advantages to hiring women subordinates in government agencies, one would expect that the practice would have become as widespread as the employment of women astronomical assistants apparently did. Since the economic potential for sex stereotyping was there, perhaps in all fields, it becomes interesting to observe where it did and did not take place in other federal agencies before 1910. Several other agencies appointed one or more women to their staffs, usually on a temporary basis before 1900, but this was apparently not enough to set a precedent, label the field, and so reserve its jobs for other women. Although no other agency became as feminized as the Bureau of Plant Industry, there were two other places (besides the Naval Observatory) where one suspects it might have happened: the Patent Office in the 1850s and the Bureau of American Ethnology in the 1880s. The appointment of three women clerk-copyists, apparently the first in the federal government and including Clara Barton, later the founder of the American Red Cross, in the Patent Office in the early 1850s might have led to a series of women patent examiners doing such detailed, painstaking, and indoor ("feminine"?) work as the job required (perhaps only on inventions submitted by women), but in 1855 a new secretary of the interior rejected even this possibility and put a stop to the "impropriety" of mixing the sexes in a government office by assigning the women elsewhere.[28]

At the Bureau of American Ethnology, Director John Wesley Powell financed the expeditions of at least two women anthropologists, Erminnie Smith and Matilda Stevenson, in the 1880s. Since there were also a great many other women anthropologists around Washington at the time (including Alice Fletcher and patron Phoebe Apperson Hearst), one suspects that female fieldwork was on the verge of becoming a regular feature of the bureau's projects, perhaps even justifying a separate women's division within it. But instead the opposite happened: there were so many women anthropologists in Washington, D.C., in the 1880s that a backlash developed. Not only did the BAE not start a women's branch, but the all-male Anthropological Society of Washington sternly refused to admit women members in 1885, and by 1901 Franz Boas was writing Phoebe Hearst that the way to upgrade the field was by training "a small number of young men."[29]

Meanwhile at least three other kinds of "women's work" were developing at coeducational colleges and universities in the 1890s.[30] They chiefly demonstrate the second kind of occupational sex segregation: besides the hierarchical form, such as in astronomy, where women were employed as assistants to higher-ranking persons, there was a territorial kind, where women did all the work in a specific, highly sex-typed, field or location. These jobs could be faculty positions in a "womanly" subject or staff jobs concerned with the women students' "special problems."

Although one can list a series of historical "firsts" to document the trickle of women onto the faculties of coeducational schools from the 1850s on, the opposition to even these "exceptions" was usually intense. Maria Mitchell had been aware as early as the 1870s that the coeducational schools were not hiring many women and had mentioned it in an address to the Association for the Advancement of Women (AAW) in 1875. She recounted the tale of a relatively liberal president of a coeducational college who said that he would hire a woman scientist if she was as good as Mary Somerville, the renowned British mathematician. Mitchell pointed out that he was creating a double standard and requiring more of the women than he was of the men, for, as she put it, "If he applied the same standard to his choice of gentlemen professors, his chairs must be vacant today."[31]

By the 1890s the topic was no longer humorous, and other women were calling attention to the small proportion and systematic exclusion of women from the faculties of coeducational schools. In what was probably the first of that genre later known as "reports on the status of women," Octavia Bates reported to the AAW in 1891 that though women were now attending coeducational colleges and universities in large numbers, few of the faculty at these institutions were women.[32] The idea was surfacing that the proportion of women on the faculty should bear some correlation to their representation in the student body. In a way this "share-of-the-market" argument did lead to the reserving of a certain percentage of faculty jobs for women, but in a way that was very typical of the separate and subordinate world of "women's work" in the 1890s. Those women who were allowed on the faculty found themselves restricted to the segregated fields of "home economics" and "hygiene" and to the subfaculty position of "dean of women." That they were lucky to get even this much at the coeducational schools is clear from the experience at Cornell University in these years.

The opposition to having women on the faculty at Cornell University was so intense in the 1890s and after as to politicize greatly the few appointments of women that were made. Women's status and rank were deliberately lowered, thus setting precedents and limits for future appointments at that school. Cornell had opened in 1867, and after much hesitation had admitted its first women students in 1872, but had not allowed any women on the faculty until the late 1890s. Even then two women appointees were allowed only in the bottom ranks. When Liberty Hyde Bailey, Cornell's beloved and energetic professor of horticulture and later dean of its college of agriculture, urged the appointment of entomologist Anna Botsford Comstock as "assistant professor" of nature study, the trustees insisted that she be only a "lecturer." Nor would they allow Agnes Claypole to rank higher than a mere "assistant," despite her doctorate in zoology. It was thus to be expected that the appointment of the first women full professors at Cornell in 1911 would call forth a pitched battle. Faculty arguments against their promotion included those that the university would lose status by appointing women to professorships, that the women did not have families to support and did not need the money (a common fallacy), that these women were not as well trained as most men (agricultural faculties generally lacked doctorates), and that there was no need to bring the women into competition with the men. The debate was finally settled in favor of the women, but only because they were in the new department of home economics or, as the historian of Cornell has put it, "After a long and acrimonious argument, the faculty voted (18 October 1911) that 'while not favoring in general the appointment of women to professorships, it would interpose no objection to their appointment in the Department of Home Economics.' "[33] To the Cornell faculty, home economics was a field of such low status (because of its feminization or territorial segregation) that full professorships there were tolerable. But such high rank was not acceptable elsewhere, in the more masculine parts of the faculty. There were no women full professors in Cornell's College of Arts and Sciences until 1960, and not even an assistant professor there until 1947.[34] At Cornell hierarchical segregation was more rigid than the territorial kind.

The increasing numbers and percentage of women students at coeducational institutions in the 1890s played a major role in staking out an area of "women's work" for the women faculty, but in a way that now seems to have hurt them as much as it helped. When there had been only a few female students they had been tolerated without any special problem, but as their numbers grew their presence had become so visible as to disturb the men and create much pressure for segregation, both in the curriculum, with special women's subjects (the humanities plus home economics), and in student housing, with special women's dormitories. Soon this duplication would require new personnel, whose status was unclear, to take care of the women's "special problems." And who was better suited to worry about these matters than the young woman on the faculty, who would never be promoted anyway? She knew the school, was of high moral character, and had always taken a special interest in the women students. Though single, she might even have the appropriate "maternal instincts" for the job. In a flash a woman chemist could become a home economist, a physiologist an instructor in "hygiene," or an assistant professor a dean of women—almost whether she wanted to or not.

Probably the largest area of scientific "women's work" in academia was the field of "home economics." Several factors contributed to its rise in the 1890s and its rapid institutionalization as an academic field for women after 1910. The subject seems, in brief (though no history of it has been written), to have been the product of two long-range trends which merged in the 1890s and after: one of nutrition research, which was creating a large supply of new information, and another of popularization, which fed a strong and increasing demand for practical advice. The feminization of the field, which had become pronounced by 1900, was the result of the men's aversion (and inability?) to advise women on domestic matters and their willingness to let the women do it instead. Had the men chosen to take over the field and make it into a profession like medicine or religion or even the law, where men commonly advise female patients or clients, it is hard to see how the women could have stopped them. Apparently the field was already too "feminine" by the 1890s, since few men seemed to have tried to enter it.

The field of nutrition research grew out of the nineteenth-century sciences of analytical chemistry and biochemistry that received great impetus from the works of the great German chemist Justus Liebig, which were published in the 1840s and after. He and his many German and American followers greatly stimulated the scientific study of foods and human metabolism, which by the 1880s were being studied by workers at the young American experiment stations, especially by W. O. Atwater at Connecticut. Other scientists, such as Graham Lusk, Russell Chittenden, Lafayette B. Mendel, and Francis Benedict, were also studying these subjects, which grew up independently of medical research, at various universities and private institutes. There were no women in this classic research tradition until 1910, when Mary Swartz Rose earned her doctorate at Yale and started her own research program at Teachers College, Columbia University.[35]

There were many women, however, in the second long-term root of "home economics," the tradition of "advice literature" for the family and home, of which a classic in the 1840s and several decades thereafter had been the *Treatise on Domestic Economy for the Use of Young Ladies at Home and at School* (1841) by nonscientist Catherine Beecher. This sort of literature sought to popularize and circulate among persons who would later be called "consumers" the latest scientific advice on how best to run their lives, their homes, and their families. Here chemistry, bacteriology, and psychology would all in due course make their contributions to a primitive form of what would later be partially institutionalized and subsidized by the federal government in the Department of Agriculture's "extension service." One particularly receptive audience or market for this literature was the female one of wives and mothers whose needs and willingness to listen created a demand for a series of lay (female) advisors.[36]

Perhaps it was inevitable that the researchers or experts would be dissatisfied with much of the advice that these popularizers circulated and would want to have some hand in upgrading it. But in addition to this, two other nonscientific factors began to bring the researchers and the popularizers closer together in the 1890s and to create the need for new well-trained hybrids or "home economists" who could better fit the women's needs. These new forces were the rise of the agricultural college and the massive immigration to eastern cities which seemed to many reformers of the time to require numerous social services (including libraries, schools, settlement houses, hospitals, and social welfare agencies) to train, "Americanize," and generally homogenize and upgrade these unwashed hordes into respectable middle-class citizens. For many reasons, women, especially the new college graduates, seemed best able to take on this overwhelming social task. They were presumed to have the female's traditional interest in the home (which, because of their earlier church and charity work, could now be extended to include the neighborhood and even the whole city), to be more venturesome than their stay-at-home sisters who had not gone to college, and to be available and willing to work at low pay on these herculean social problems. Thus, like the schoolteachers, social workers, librarians, and settlement house workers, the women home economists could act as missionaries trying to save society and its victims through better nutrition and home life. There were enough diseases and other public health problems, especially among city children (who accounted for one-half of all deaths in the 1890s), to create a real demand (in the eyes of the middle-class reformers, if not among the immigrants themselves) for new and better methods of hygiene and diet. It would be easier for such self-appointed ministers to the unfortunate, however, if they had some authority or expertise other than that they thought they knew best. A scientific background and thus some claim to the role of "expert" might give these women the authority to tell others how to live. In time, of course, they, like the other women in similar subprofessional roles, would need to be upgraded with master's degrees (in education, librarianship, or social work), but in the 1890s women college graduates who "meant well" felt equal to the task.[37]

Besides this large but unorganized (and perhaps manufactured) urban demand for home economics, there was also a very strong rural interest in it, one that in the long run created more jobs for women at the university level. As a result of the Morrill Land-Grant Act of 1862 many colleges had been created in rural areas of the country to teach the "agricultural and mechanic arts" to local youths of both sexes. Since agriculture was still at that time as much an art as a science, the content of the new colleges' curriculum was somewhat problematic. On the one hand there was much antielitist sentiment for such colleges to offer "practical" instruction; on the other hand their offerings had to be more rigorous or worthwhile than the information their students could pick up at home on the farm. Some fields, like "economic entomology" or "soil chemistry," straddled this problem successfully and were both "practical" and "scientific" at the same time.

But what were the young women at such schools to study? Most chose to enroll for courses on "cookery," "sewing," and the "household arts." What sort of faculty was best for this? It apparently had to be female—the association with "women's work" was too strong for any man to teach such subjects in the late nineteenth century, though one can imagine that the agricultural colleges could have imported male (French) chefs for the purpose if they had wished. There is no evidence that men ever contested the issue. Then, too, what methods were suitable for teaching such domestic arts? Could one lecture on "cookery" or assign readings on sewing, or were demonstrations enough? The fact that the students often had little or no scientific background (and might, like the immigrants, balk at having to learn science just to cook) limited the depth of the material that could be presented. One could easily go too far, however, in pleasing the students and their parents and end up scorned by the rest of the college faculty (as well as the university administration) for not having a German doctorate and for teaching a subject that lacked intellectual rigor. Thus there was pressure on the early faculties of "domestic science" or "home economics" at the land-grant colleges to upgrade their curriculum, to make it seem "scientific" and demanding, and to hire women with doctorates to teach the subject as rigorously as the traffic would bear. These teachers should also do research in some aspect of the field and work to upgrade it into an almost (but not quite) regular academic science. Treading the narrow path between these two cultures and meeting the pressures for both prestige and practicality would be a continuing and dominant theme in the field's history.[38]

The founder of "home economics," and one whose leadership and character touched her contemporaries deeply, was Ellen Swallow Richards, Vassar 1870 and MIT 1873 (plate 11). Between 1880 and 1910 she almost single-handedly created the field—she propagandized for it, ran demonstration projects, raised money, performed many chemical analyses, wrote several handbooks, trained and inspired her coworkers, and organized its main activities and professional associations. In a sense she had been preparing for this role all her life. As a student at MIT she had learned how to make a place for herself (and other women) by capitalizing on woman's traditional role or, as she put it in 1871, "Perhaps the fact that I am not a Radical or a believer in the all powerful ballot for women to right her wrongs and that I do not scorn womanly duties, but claim it as a privilege to clean up and sort of supervise the room and sew things, etc., is winning me stronger allies than anything else."[39] After marrying Robert Richards, an engineering professor at MIT (who proposed, appropriately enough, in the chemistry laboratory), she volunteered her services (and about $1,000 annually) to the "Woman's Laboratory" there, which she induced philanthropic Bostonians to support from 1876 until 1883. Her curriculum there was not initially sex typed—her students were mostly schoolteachers whose normal school training had lacked laboratory work and who now wished to perform chemical experiments and learn mineralogy, one of her own specialities. In 1880, however, Richards started stressing chemistry's value to the homemaker, asserting in an address to the AAW that "laboratory work, rightly carried out, makes women better housekeepers, better cooks, better wives, and mothers more fitted to care for the versatile American youth."[40] This change in emphasis was probably the result of both the increasing interest in what would later be called "pure food and drug" issues and her own precarious position at MIT.

After MIT began to admit women directly in 1878 and the need for her separate "Woman's Laboratory" lessened (it was closed in 1883), Richards lacked a position. But when, a year later, MIT set up a new laboratory to study sanitation, apparently the first of its kind in the nation, she was appointed an instructor in sanitary chemistry, a position she held until her death in 1911. There she helped MIT professors associated with the laboratory analyze the state's water samples and also developed her interests in the composition of food and other groceries, safe drinking water, and low-cost diets for the poor. She prepared many popular works, and in 1889 she also helped several college women in Boston start their "New England Kitchen," where they prepared nutritious soups for the city's poor. It was this experience that convinced Richards that the field of nutrition education (as it is now called) offered great opportunities to women college graduates who should, she felt, be using their trained minds and special talents to understand and help solve the social problems around them. In 1890 she presented a paper entitled "The Relation of College Women to Progress in Domestic Science" to the Association of Collegiate Alumnae, of which she had been a founder (and, as Marion Talbot later put it, "like an elder sister" to the other early members). In the speech Richards stressed how challenging efficient housework could be and how much it could be improved by the application of

PLATE 11. Chemist Ellen Swallow Richards (shown here late in life), who was a student of Maria Mitchell, made a career for herself and many others by founding the field of "home economics" specifically for women. (Permission of M.I.T. Special Collections.)

scientific principles. She thought the new subject of "domestic science" should be taught at all the women's colleges. It would not only help college women lead more efficient home lives, but would also bring them into touch with pressing local social problems. When few liberal arts colleges seemed to follow this lead, she taught "domestic science" herself to the more vocationally oriented women students at Simmons College in Boston.[41]

By 1893 Ellen Richards was (like everyone else) at the World's Columbian Exposition in Chicago, the greatest showpiece of women's achievements since the Centennial Exposition in Philadelphia in 1876. She ran a "Rumford Kitchen" that offered nutritious and scientifically cooked lunches to visitors for thirty-two cents. Six years later Richards and Melvil Dewey, director of the New York State Library, author of the "Dewey Decimal System" of book classification, and advocate of "women's work" in librarianship, called the first of the ten Lake Placid Conferences on Home Economics. These annual meetings brought together the diverse elements within the movement—the urban cooking-school leaders, the public school supervisors, and the increasingly strong contingent of faculty members from the agricultural and teachers colleges—to discuss the field and its problems, especially its terminology and objectives, and to formulate model curricula. In 1908, at its tenth conference, the group formed the American Home Economics Association and elected Ellen Richards its first president.[42]

Yet even before her death in 1911, the movement was already moving beyond Richards's vision of it and on to a deliberately more academic phase, as it saw its future tied less to urban cooking schools and demonstration kitchens and more to the growing agricultural colleges of the Midwest and West. By 1911 many of these colleges had already formed programs and even departments of home economics, and others were eager to do so. Some of the great leaders in the field in the next few decades were getting their start around 1910: Isabel Bevier at the University of Illinois, Mary Swartz Rose at Teachers College, Columbia University, Agnes Fay Morgan at the University of California, Abby Marlatt at the University of Wisconsin, and Flora Rose and Martha Van Rensselaer at Cornell University. Yet the very success of this kind of "women's work" on major campuses helped to harden the sexual segregation for future generations still further. Rather than being accepted for other scientific employment once the pioneers had shown women could handle this employment, the women found themselves *more* restricted to "women's work" than ever. Since women were finding such good opportunities in this field, many persons (including the first vocational guidance counselors, a new specialty around 1910) urged ambitious young women interested in science to head for home economics. It was the only field where a woman scientist could hope to be a full professor, department chairman, or even a dean in the 1920s and 1930s.[43]

Another new and highly feminized field, which looked for a while as if it might attract women physiologists the way home economics absorbed women chemists, was that of "hygiene" or "hygiene and physical education." Started by a number of male physicians in the late nineteenth century, this field sought to understand the scientific bases of both personal and public health, especially in relation to physical exercise. About the same time, several of the early women's colleges and coeducational universities employed women doctors to teach hygiene as well as to be the college physician, since administrators were anxious to minimize the physical ailments that were suspected at the time to accompany mental exertion in females. Some women doctors apparently found this position a congenial one, as did for example the sisters Dr. Clelia Mosher, appointed at Stanford University in 1893, and Dr. Eliza Mosher, hired by the University of Michigan in 1896. Another, Dr. Lillian Welsh of Goucher College (appointed in 1894), not only taught hygiene for thirty years but also gradually developed the only full department of physiology and hygiene at a woman's college into a strong premedical program that was famed for the number of its graduates who went on to the nearby The Johns Hopkins Medical School. By the 1920s, however, and perhaps even earlier, for reasons that were then and still are unclear, the subject of "hygiene" did not flourish like "home economics" but floundered and was rarely taught in American colleges and universities. Possibly it had been downgraded; it was quickly replaced by the more popular subject of physical education, which by then had its own kinds of specialists trained in the normal schools and accredited by its own professional societies, though they often taught "health" as well in the public schools. Thus though "hygiene" might have followed the path of home economics, and many universities hired women with doctorates in physiology to do research and train high school teachers of "physical education," the field developed quite differently.[44]

A third kind of academic hybrid of "women's work" which arose at the coeducational colleges in the 1890s was that of the "dean of women." As the number of women students increased, and especially as they were required to live in dormitories on campus, the need arose for some sort of supervision. Although the duties of the office were not at first clear, most administrators considered almost all faculty women capable of the job. Accordingly the earliest office-holders included several highly trained scientists and physicians, such as Marion Talbot, assistant professor of sanitary science at the new University of Chicago, who was appointed dean of women in 1892; Eliza Mosher, M.D., became a dean of women as well as professor of hygiene at the University of Michigan in 1896; psychologist Margaret Washburn was appointed warden of Sage College (the women's dormitory) at Cornell in 1900; Mary Bidwell Breed, assistant professor of chemistry at Indiana University, was appointed its dean of women in 1901; Lucy Sprague (later Mitchell), who was later active in the child study movement, replaced a woman physician as dean of women at the University of California in 1906; and Fanny Cook Gates, formerly professor of physics at Goucher and Grinnell colleges, became the dean of women at the University of Illinois in 1916. Despite these early appointments, it soon became clear that not all women were suited to such a position of moral authority over the students, for personality and temperament were of paramount importance. Thus the research-oriented Margaret Washburn, who disliked her job at Cornell intensely, was only too glad to give it up after two years, and Fanny Gates left Illinois after two years when gossip hinted that she may have been addicted to drugs. (The president of the University of Illinois vowed never to make the mistake of hiring a Ph.D. in science for that job again.)[45] President William Faunce of Brown University had, however, sensed this nicety as early as 1900, when he described the position in these terms:

> It is a position of peculiar responsibility and opportunity; not because of the large number of students, as yet—we have about one hundred and fifty—but because things are in plastic shape, and the whole future of women's education in this region can be molded by the one who occupies this position.
>
> We want a woman who can teach, in order to emphasize the intellectual life of the College; but we want, quite as much, one who can create and maintain the right social atmosphere and keep before the young women womanly ideals.
>
> I feel that we ought to have a woman in whose nature the religious element is not lacking, and who could sympathize as well as instruct.[46]

Another problem facing these early deans of women was the declining status of their position. Although the first such deans tended to have full faculty status and taught a few classes in their specialty, the position was in constant danger of being downgraded to a staff or administrative appointment. This had apparently happened by 1911, when a report on the duties and status of deans of women at fifty-five colleges and universities across the nation revealed that, though the deans were usually required to have doctorates, they were only barely tolerated on the institutions' often nearly all-male faculties.[47]

Yet attitudes were beginning to change somewhat by 1910. The revival of the women's rights movement (which led eventually to the ratification of the suffrage amendment in 1920) emboldened some women scientists, including the archetype of "women's work," Marion Talbot, professor of household science and dean of women at the University of Chicago. She and others were beginning to realize around 1910 that sex-typed employment had not proven the opening wedge to broader opportunities that it might have. It had brought them jobs in science and academia, but it was now clear that many of these were marginal or subordinate positions, easily downgraded and rarely accorded recognition (such as a star in the *American Men of Science*).[48] Even the jobs in home economics, which were some of the best positions for women outside the women's colleges, were both separate and unequal (deliberately so, as in the Cornell faculty's vote), and thus had not brought women that much closer to the final and, now for the first time, visible goal of full equality. "Women's work" no longer seemed as progressive a step as it had in the 1880s and 1890s.

Unfortunately, however, such segregation suited so many other needs and constituencies so well that later generations would find it difficult to move beyond "women's work" and into the mainstream of scientific employment. What had formerly been a fairly nonhierarchical collection of independent investigators had become, in some fields at least, highly bureaucratized "big science" with all the gradations in status and role that this implied: henceforth, some persons would have to be "hired hands" on projects directed by others. Government science was also expanding rapidly, as were several applied or service-oriented fields that academics scorned and tried to keep at a distance. It was a tense time of jockeying for position and status, as some jobs and roles were downgraded, others created, and still others expanded and promoted. The presence of women created new opportunities for the more liberal "empire-builders" of the time, but it worried other, more vulnerable, men, whose scientific standing it seemed to threaten. Thus a segregated, low-status, almost invisible kind of "women's work" offered a harmonious method of incorporating the newcomers into the scientific labor force: women were introduced in ways that divided the ever-expanding labor, but withheld most of the ever more precious recognition. Later women unable to change this pattern of segregated employment, would devote much energy and ingenuity to creating new forms of awards and recognition that would help to compensate for their structural invisibility in the scientific work force.

4

A MANLY PROFESSION

In the same decades when women were seeking their first scientific employment (and finding themselves restricted to the least powerful positions or "women's work"), they were also beginning to enter many of the scientific societies that are now taken for granted but which were often just being formed at the end of the nineteenth century. These organizations rarely welcomed women very warmly, however, for they were already in the midst of upgrading themselves into nationwide "professional" societies by excluding or diminishing the influence of persons perceived pejoratively as "mere amateurs." An influx of women who rarely held important positions in science, if any at all, usually was seen in these years as a threat to a group's precarious "prestige" and triggered an intense discussion of the need to "raise standards" for membership. Since the concepts of prestige, status, and professionalism were at the time closely intertwined with that of masculinity, the new membership requirements that were introduced in the 1870s through the 1890s were often deliberately harder on women than on men. Although most organizations eventually admitted women, they frequently, for example, restricted them to lower levels of membership, demanded of them more degrees for admission (as a doctorate) than they did of the male members, rejected them outright, or ousted those who were already members through reorganization. Other practices, such as featuring certain kinds of social entertainments at the annual meeting, also discouraged women's participation more than men's. The result of these antifeminist practices was a drastic defeminization of the higher ranks of those branches of science that women were then entering in appreciable numbers: natural history, anthropology, botany, and marine biology. Even this was only the most noticeable part of the phenomenon, for in most other scientific fields participation in a "professional" society presupposed visible employment in the field, a criterion that, because of prevailing practices, excluded all but the women professors from membership. One thus has a situation whereby, although there were more women "in science" in the 1870s and 1880s than ever before, only a few were considered "professionals." "Prestige" and "professionalism" were thus concepts that would within a few decades reshape "science" to make it seem even more masculine than it really was.

The women's only recourse in the face of these new barriers was to strengthen their existing but also largely invisible subculture of women's clubs in science. Yet these largely self-help groups could at best be only partial substitutes for women's broader participation in the main professional organizations of the time. Most were just local societies, a factor that greatly limited the degree of specialization possible. In a few fields there were enough women to support a separate scientific society, but its members were still cut off from the men in the specialty. To a large extent, therefore, it was the most qualified women, those who had the most to contribute to the field and the least to learn from the other women, who were cut off from fellowship with suitable colleagues and who bore the brunt of the men's fight for professional status and prestige at the end of the nineteenth century. By the early twentieth century, the structure of women's presence in science had been restricted to fit the desired stereotype: science looked to the public like a manly profession, even though there were many talented women in it too.

American society, as explained in chapter 1, had long considered it impossible (or at least "unnatural" or "unfeminine") for a woman to be interested in serious learning or professional pursuits. To many antebellum minds it was almost a contradiction in terms for a woman, however bright or accomplished in appropriately feminine pursuits, to be at all "learned." The prospect confounded normal logic, as this unidentified quotation reveals:

> Culture in young women should never develop into learning; for then it ceases to be delicate feminine culture. A young woman cannot and ought not to plunge with the obstinate and persevering strength of a man into scientific pursuits, so as to become forgetful of everything else. Only an entirely unwomanly young woman could try to become so thoroughly learned, in a man's sense of the term; and she would try in vain, for she has not the mental faculties of a man. In opposition to these sentiments I may be directed to learned ladies, a second-rate article which, thank God, is extremely rare.[1]

Emma Willard's advocacy of seminaries for American women had put a modest dent in this pervasive antifeminism, and by the 1830s increasing numbers of future wives, mothers, and schoolteachers had begun to receive an often highly scientific academy education. Yet the stated goal of this training had not been to make women learned or scholarly or worthy contributors to science but to make them more religious and self-sacrificing mothers. If, however, an unusual woman did overcome this social pressure to remain a dilettante and did attempt to make scientific discoveries and then communicate them to others, the prevailing propriety required that she proceed cautiously and demurely. It was unseemly for a woman to expose her talents outside the home, especially to a male audience, or to acknowledge, let alone enjoy, any compliments or recognition she might receive. Thus one finds among early women scientists an almost painful modesty, as in a Mrs. Percival's letter to botanist John Torrey about her rather large collection of plants, where she referred to herself only in the third person,[2] and in this depiction of Mary Treat, a Philadelphia botanist and entomologist who was praised by contemporaries for so nearly achieving this feminine ideal: "Her most prominent characteristic is a modesty so shrinking as to make any public recognition of her services painful to her, while her joyous enthusiasm for her life-work is so great and so contagious that her home is always the center of attraction, where are welcomed all who come to learn even the alphabet of her beloved book of nature, and where she dispenses the bounty of her gifts and attainments with a modest lavishness and an unwearied patience, which appears to be to her their own reward."[3]

Starting in the 1840s, however, some women began to create their own scientific organizations that would successfully combine their growing interest in science with their continuing personal modesty. These were the numerous local all-women scientific clubs, many of whose records Sally Gregory Kohlstedt has found and brought to light: the Female Botanical Society of Wilmington, Delaware, in the 1840s; the Dana Natural History Society of Albany, New York, in the 1860s; and some years later the Syracuse Botanical Club and the Botanical Society of Philadelphia. Other such clubs existed in Jersey City, New Jersey, in the 1870s and 1880s and in Boston in the 1890s and early 1900s, and still others may reappear when more work has been done in local manuscript collections.[4] The members of such groups seem to have been both single women and married women whose domestic tasks did not consume all their energies. Having studied science, especially botany, either at school or independently from the textbooks of the time, and seeking a congenial (and perhaps "feminine") hobby outside the home, many women found such private science clubs an appropriate activity. These groups, as far as can be determined, seem to have been less interested in publishing new findings and seeking recognition or prestige, which would have been "unwomanly" in any case, than in private learning, mutual growth, and contributions to local education and conservation. These clubs were very effective in building bonds among women outside the women's colleges in the decades before such women could join other scientific organizations and then again in the 1890s, when many had been ousted by these same groups. Over the years these clubs thus served women at levels of science both as a substitute and as a supplement to their broader participation in other, more central, scientific organizations.

Because antebellum ideas of feminine modesty required that women be unobtrusive in public and take pains to camouflage any special talents, particularly intellectual ones, it is hardly surprising that few of these many women who were interested in science before the Civil War took the daring step of joining a scientific society. The first woman to join an American scientific society seems to have been Lucy Way Say, the widow and former assistant of entomologist Thomas Say. She was elected an associate member of the Academy of Natural Science in Philadelphia in 1841, possibly, it has been conjectured, since she did not participate in the meetings, in gratitude for contributing her husband's collections to the academy's museum.[5]

Several years later, in 1848, astronomer Maria Mitchell became the first woman elected to the American Academy of Arts and Sciences in Boston, and in 1850 both she and entomologist Margaretta Morris of Germantown, Pennsylvania, were accepted by the American Association for the Advancement of Science (AAAS), as was Almira H. L. Phelps, the botanical educator, in 1859. Despite these elections, however, the topic of the admission of women remained a recurring one at the AAAS's annual business meetings throughout the 1850s. Although the members were willing to elect individual women whose work was known to them, they resisted opening their doors to all interested women, an invitation they extended freely to men. They justified this double standard on the Victorian grounds that they had to protect the two sexes from each other: members of the stereotypically delicate female sex might either be embarrassed at the scientific discussion of biological facts or divert stouthearted men from the serious pursuit of science. It was inconceivable to these men that some women, at least, might have something to contribute to science or could profit by attending. The result of this rather inhospitable attitude was that although three women did join the association in the 1850s, they conformed to the appropriate sex-role expectations and neither presented papers nor held elective office. (One nonmember, the unknown Eunice Foote of New York City, did read a paper in 1857, however.)[6] In this context the invitation in 1853 by the newly formed California Academy of Sciences in San Francisco to women to join was extended only because of the highly unusual dearth of educated persons of either sex on the West Coast at the time. Even so, the academy still had no women members ten years later.[7]

So many social taboos on women's participation in public groups weakened during the Civil War that shortly thereafter a steady stream of women sought to enter the major scientific associations of the time. By the 1870s and 1880s the flow was strong enough to cause certain policy changes or restrictions in some groups, such as the AAAS in the 1870s. As soon as the AAAS resumed its meetings after the Civil War, several women (including Emma Willard, that bellwether of feminine propriety) came and joined. At first there were fewer than three new women per year, but then in 1870, when the AAAS met in Troy, New York (and even held some sessions at the famed Troy Female Seminary), ten women joined, followed by eleven in 1871 and fifteen in 1872. This influx may have worried the leaders, since the next year, when there were forty-six women members (6.8 percent of the total), the AAAS executive committee decided, for reasons that were unspecified, to amend the constitution to introduce a second, higher, level of membership, that of "fellows," for those members "as are professionally engaged in science, or have by their labors aided in advancing science." Those persons already on the rolls could choose to be listed as fellows, but any additions would have to be nominated by a standing committee of the association and elected by a majority of the members and fellows present at the business meeting. The result of this new distinction was to create a sexual division of membership levels, for, though women continued to join the AAAS at a rate of four to eight per year through the 1870s, only two (the unknown entomologist Sophie Herrick of Baltimore and the well-known chemist Ellen Swallow Richards of MIT) became fellows in the 1870s. The discrepancy became even greater in the early 1880s as more and more women became members (a record-breaking forty-seven joined in Boston in 1880 and fifty-one in Philadelphia in 1884), but only five became fellows in those same five years (Erminnie Smith in 1880, Cornelia Clapp, Sarah Whiting, and Alice Fletcher in 1883, and Cora Clarke in 1884).[8] Starting in the 1870s, therefore, and increasingly thereafter, men and women scientists seem to have struck a compromise about the acceptable level of women's participation in scientific organizations that was very similar to that reached over their relegation to "women's work." Women could now join formerly masculine scientific organizations, but few would be raised to the recently introduced higher levels or rewarded for contributing "professionally" to science. In fact the very word *professional* was in some contexts a synonym for an all-masculine and so high-status organization.

The intrepid Ellen Swallow played an even stronger role in opening the Boston Society of Natural History to women in the 1870s, since it was her three papers (read to the members on her behalf by their president) in 1875 that apparently precipitated the whole question of female membership. The result was that after a year of debates, the members decided to follow the path of the AAAS. At their annual meeting in May 1876 they voted to divide their previous category of "resident members" in two: all current "resident members" would henceforth be known as "corporate members," but all newcomers could only be "associate members," though they might later be raised to full "corporate" status. The sexual overtones of this decision became clear a few months later when the members elected twenty-one women and one man to the new "associate" status. Sally Gregory Kohlstedt has analyzed the unpublished minutes and correspondence of this particularly revealing episode in the society's history and found that, like the men in the AAAS in the 1850s, the male Bostonians of the 1870s still professed the Victorian fear that women members would be offended by some of the indelicate remarks that might accompany biological arguments and that their presence might deflect men from serious discussions. (The one woman consulted thought that those women attending would be able to take care of themselves.) Yet the compromise of restricting the women to partial membership did not address this problem. In fact it specifically allowed the women to attend the meetings and even join in the discussion, though it barred them from voting on any matters and holding any office. Kohlstedt thus suggests that by the 1870s the problem was no longer the potential embarrassment of a few women's attendance, which could have been handled in some other way,* but a fear of loss of control and prestige that was raised by the presence of many women. The Boston naturalists were, she concludes, so "worried about maintaining credibility with the professionals" that they had to devise a way to keep the women from

*They could have allowed the women to vote but restricted their attendance to balconies or erected a screen to keep them out of the view of men, as in Islamic practice ("purdah") and as a geographer later suggested another society do.

lowering their already endangered prestige still further.[9] Their practice of admitting women but restricting them to a visibly subordinate level was so effective that it soon became standard procedure in natural history groups. When, for example, in 1885 the all-male American Ornithological Union, established two years before, accepted its first woman, Florence Merriam Bailey of Washington, D.C., it was as an "associate member."[10]

Meanwhile the few women chemists in the country in the 1870s and 1880s (including once again Ellen Swallow Richards) were unable to get even this standard compromise of secondary membership in the new American Chemical Society, despite an encouraging beginning in 1874. In that year, when discussion in the *American Chemist* centered on how best to celebrate the upcoming centennial of the discovery of oxygen, Rachel Bodley, a graduate of the Wesleyan Female Seminary in Cincinnati and professor of chemistry and dean of the Women's Medical College of Philadelphia, wrote the journal to suggest that the chemists of America commemorate the occasion by meeting at Joseph Priestley's former home in Northumberland, Pennsylvania. Once assembled at that remote spot, the group voted to establish an American Chemical Society as soon as possible. The participants even showed their gratitude to Bodley for her suggestion by electing her one of the interim committee's thirteen honorary vice-presidents (the only woman so honored by the ACS before the 1970s). This recognition is all the more remarkable since Bodley did not bother to attend the meeting herself, preferring to spend the summer botanizing around Denver with friends. Several other women chemists did attend, however, including Ellen Swallow of MIT, Lydia Shattuck of Mount Holyoke Seminary, and Bessie Capen of the Girls' High School in Boston and later Wellesley College. But it was a danger signal in 1874 indicative of their marginal status in American chemistry that none of these three women was included in the famous photograph taken of the major participants at the Northumberland meeting (see plate 12). They were instead placed decorously off to the side, probably with the male scientists' wives and children.[11]

Despite these three early participants and the honor afforded Bodley, no other women joined the ACS before the 1890s. One reason may have been the early society's preoccupation with industrial chemistry; another may have been its generally "clubby" atmosphere, which included stag dinners like that hosted by member Henry Morton, president of the Stevens Institute of Technology, at the annual meeting of the AAAS in Boston in August 1880. What makes this occasion different from many others like it was that its ribald proceedings were taken down by a male stenographer, privately printed, and distributed to members as *The Misogynist Dinner of the American Chemical Society*. After this even Bodley, the lone woman member, dropped out, and there were no women members until 1891, when Rachel Lloyd, assistant professor of analytical chemistry at the University of Nebraska, was elected. She was followed into the society shortly thereafter by Mary Engle Pennington, a recent doctorate from the University of Pennsylvania. By then, however, most women chemists were taking the men's hint and following Ellen Richards, who must surely have known of the mi-

PLATE 12. Although several women chemists, including Ellen Swallow, attended this 1874 Priestley Centennial, which led to the formation of the American Chemical Society, this official photo did not include them, probably because such "professional" societies wished to project a totally masculine image. (Samuel A. Goldschmidt, "The Priestley Centennial," *Journal of the American Chemical Society* 48 [1926]:4.)

sogynists' dinner in Boston in 1880, into the more womanly and receptive field of "home economics."[12]

Even these restrictions, intimidating as they must have been in certain fields, did not stop women from participating in and contributing to science in the 1880s and 1890s more than ever before. Data collected by historian of science Clark Elliott and presented in table 4.1 show that the women of the post-Civil War decades were not idle amateurs by any means, but were publishing scientific papers in record numbers as well as joining societies in force. Elliott has counted the number of American women who published articles in those scientific journals of the nineteenth century indexed in the *Royal Society Catalog*. His results show that the increase after the 1860s was quite dramatic, and that by the mid-1880s women's participation was not only visible but also must have been either impressive or alarming, depending upon one's point of view.[13]

One source of many of these papers by women scientists was the AAAS *Proceedings*, where, despite their status as "members" rather than "fellows," several women were presenting papers by the 1880s. Although a few women had been presenting papers at AAAS meetings since 1869, the first to publish in the *Proceedings* were Ellen Richards and Alice W. Palmer, both of the Woman's Laboratory at MIT in 1878 ("Notes on Antimony Tannate"). The next year Erminnie A. P. Smith of Jersey City, New Jersey, published one on jade, and in

TABLE 4.1. Publications by Women in American Scientific Journals, 1800–1900

Years	Number of Women Authors	Frequency	Rate of Increase
1860–63	3	.05/year	
1864–73	5	.50/year	10-fold
1874–83	36	3.6/year	7-fold
1884–1900	about 400	about 25/year	7-fold

SOURCE: Clark Elliot, unpublished data.

1880 the outspoken Ellen Hardin Walworth of Saratoga Springs, New York, published "Field Work by Amateurs," in which she pleaded for more popular science, especially among women and in geology. Yet the bulk of the papers given by women in the early 1880s were by just two remarkable individuals, both middle-aged, and in just one field, that of anthropology: Erminnie Smith and Alice Fletcher, both protégées of Frederic W. Putnam, director of Harvard's Peabody Museum of American Archaeology and general secretary of the AAAS. Together these two women accounted for eleven of the fourteen papers published by women in the *Proceedings* from 1881 to 1884 and ten of the sixteen printed "by title only." In 1885 Smith was rewarded by being elected secretary of section H (anthropology), the first woman to hold any office in the AAAS, an event that is depicted in plate 13. Only her sudden death a year later eliminated the opportunity for what might well have been even greater accomplishments. Fletcher was chosen chairman of section H in 1896, another feminine "first." Thus personal encouragement by a prominent scientist and an official in the AAAS was enough to help two energetic women scientists make major contributions to their field and thus to overcome the AAAS's otherwise low expectations for them.[14]

Putnam's influence and importance appear all the greater when one compares the women's performance in section H of the AAAS with their reception by the all-male Anthropological Society of Washington in 1885. There, despite (or perhaps, perversely, because of) the women's strong showing in the AAAS at the time, the ASW members rejected Matilda Coxe Stevenson's application for membership because of her gender and presumed lack of field training. (It is unclear what training the men in the group had had at this time. In the days before there were doctoral programs in the subject, most "anthropologists" either were self-taught or had other miscellaneous backgrounds. The members' occupations were listed as government official, physician, banker, engineer, "antiquarian," and "none.") Since Stevenson had published an article on the Zuñi Indians and assisted her ailing geologist husband with his own anthropological field work in New Mexico, she determined not to accede to the group's rejection but to fight back. Encouraged by Edward Tylor, the British anthropologist, who had accompanied her and her husband in the field and who believed that women had a special contribution to make to anthropology, Stevenson met with ten other

PLATE 13. Anthropologist Erminnie A. P. Smith was the first woman officer of the American Association for the Advancement of Science, serving as secretary of Section H in 1885. Frederic W. Putnam, the permanent secretary of the AAAS, who encouraged Smith and several other women anthropologists, is seated at the far left. (Courtesy of the National Anthropological Archives, Smithsonian Institution.)

PLATE 14. Another Putnam protegée, Alice Fletcher, shown here with Winnebago Indians Meepe and Martha between 1887 and 1889, was, with Ellen Richards, Cornelia Clapp, and Elizabeth Britton, one of the few women scientists of her time to be considered a "professional" scientist. (Courtesy of the National Anthropological Archives, Smithsonian Institution.)

women to form the separate Women's Anthropological Society of America in Washington, D.C., in 1885. Although the group soon grew to fifty-six members, including as corresponding members such friends of the cause as Christine Ladd-Franklin, Graceanna Lewis, Zelia Nuttall, and Ellen Richards, its most distinguished members were (besides Stevenson) Erminnie Smith, briefly, and Alice Fletcher, who, though affiliated with Harvard and often in the field (see plate 14), was frequently in Washington on special assignments as President Grover Cleveland's consultant on Indian problems. For fifteen years the group met twice monthly to present papers to each other as a means of mutual encouragement and to hear such visitors as Alfred Russel Wallace on "The Great Problems of Anthropology" in 1887 and Frederic W. Putnam on an unspecified topic in 1890.[15]

Although the men's group relented as early as 1891 and elected Stevenson to membership and the two groups held several joint sessions in the 1890s, they did not merge officially until 1899, and then only because of the fear of a rival organization in New York City. In 1903 the combined Washington groups elected Fletcher president over the opposition of some of the men, but by then the honor may have been somewhat diminished, since the ASW had been eclipsed by the founding of the new American Anthropological Association in 1902. Although Fletcher had been one of the founders of this group (which grew out of section H of the AAAS) and its membership was about 10 percent female in 1903, she was never chosen its president. Nor was any woman so honored until the rather late date of 1940, when Elsie Clews Parsons of New York City, a wealthy benefactor of many of Franz Boas's projects, was accorded the privilege. Meanwhile Fletcher, whose executive skills are evident in her correspondence, was chosen president of the young American Folklore Society in 1905.[16]

One response to women's desire to join scientific societies had been to relegate them to a secondary level of membership, and a second had been to exclude them for an extended period of time; a third was to retreat to the higher ground of a more exclusive, more highly "professional" (and so almost totally masculine) society. The first of this genre was the American Society of Naturalists, established in 1883 after Samuel Clarke, Ph.D., of Williams College and twelve other men issued a call for an open meeting of "professional naturalists." Those who came discussed the need to upgrade work in natural history and to improve upon the current AAAS, which often met in far-flung places and which had "a very large and varied membership." They preferred to have a select membership meet in the East, for, as they put it, "If a few men, who are thoroughly in earnest, meet together at an appointed place and time, for a single definite purpose, the chances are all in favor of their accomplishing something worth the doing."[17] They felt so strongly about the characteristics of desirable members that they specified in both their new constitution and their by-laws that they be "professionally engaged."

> Membership in this society shall be limited to persons professionally engaged in some branch of Natural History as Instructors in Natural History, Officers in Museums and other Scientific Institutions, Physicians, and others.[18]

> The following persons shall be considered professionally engaged in natural history within the meaning of Article II, Section 1:—Only those who regularly devote a considerable portion of their time to the advancement of natural history; *first,* those who have published investigations in pure science of acknowledged merit; *second,* teachers of natural history, officers of museums of natural history, physicians, and others who have essentially promoted the natural-history sciences by original contributions of any kind.[19]

Enough other persons felt the same way for the society to be able to boast of a membership of 124 men by the end of the year. There were no women members in this select organization until 1886, when Emily Gregory, then an associate in botany at Bryn Mawr College, was elected. She was followed six years later by Ida Keller, another botanist temporarily at Bryn Mawr. Since they were just two of the several women naturalists in the country at the time who would have met the ASN's stated membership criteria of publishing research, what made them acceptable and the others not was apparently their foreign doctorates (both from Zurich); they were among the few women to hold doctorates this early. Both were elected shortly after their return to the United States.[20]

At least one other fledgling professional society of the 1880s required a doctorate only of its women members, though this was never stated in its regulations. Thus when the Geological Society of America, established in 1888, elected as its first two women Mary E. Holmes in 1889 and Florence Bascom in 1894, it could hardly have been a coincidence that they were the first two women in the nation to earn this degree in the earth sciences. Since only about one-quarter of the male members who elected them held such a degree, the GSA was clearly requiring a far higher standard for female applicants than it was for men.[21]

Even if women were expected to hold a doctorate in order to be accepted as "professionals" in the 1880s and early 1890s (and not even all of the growing number of women doctorates were so accepted), at least one highly productive woman botanist without the degree was accepted into a highly selective "professional" group in 1893. Her experience in two clubs provides much circumstantial evidence of the internal workings of the simultaneously upgrading/defeminizing process. Although there were no women among the thirty-one founders of the Torrey Botanical Club of New York City in 1870, one, Elizabeth Knight (later Britton), shown in plate 15, was admitted in 1879. After this the club's policy changed considerably, and so many women joined in the 1880s that by 1890 they comprised 40 percent of the club's membership. Though several of these women were professors at the women's colleges (Emily Gregory, Ph.D., of Barnard was an active member, while Lydia Shattuck of Mount Holyoke and Susan Hallowell and two others from Wellesley were corresponding members), most were instead local, seemingly affluent, perhaps even "society women" whose family responsibilities did not require all of their time. Some may have been college graduates, who, unlike their predecessors in the all-female botanical clubs, preferred such

PLATE 15. Botanist Elizabeth (Knight) Britton of the New York Botanic Garden, shown here in 1902, was an expert on mosses, and was the only woman charter member of the Botanical Society of America. (*Journal of the New York Botanical Garden* 35 [1934]:99.)

mixed clubs for social as well as scientific reasons. They could associate with other like-minded persons of both sexes, go on field trips, and exchange with and admire each other's collections. The high point for such collectors might be an occasional publication or a compliment from one of the staff members at the Columbia College Herbarium, where the group met originally, or later at the New York Botanical Garden.[22]

In these ways lay persons could feel a part of science, even though they held no advanced degrees and were not employed in a professional capacity. Some women belonged to these local clubs for decades, working hard on particular projects and building up strong loyalties, and a few became mainstays of their local societies and even exerted considerable local leadership, as did Annie Morrill Smith of the William Sullivant Moss Society (later the American Bryological Society, whose "principal founder" was Elizabeth Knight Britton), and Mary Louise Duncan Putnam of the Davenport (Iowa) Academy of Sciences.

These women felt that, whatever the opinion of the so-called professionals who were beginning to scorn such groups, they were contributing as much to science as anyone.[23]

Yet a crisis was coming. The new academic, professional, and increasingly well-funded male academics, who had formerly been content to lead such groups and exhort their members to greater efforts, were by the late 1880s becoming acutely embarrassed by the nonprestigious amateurs among them.[24] Thus shortly after the women's percentage in the Torrey Botanical Club (TBC) passed 40 percent, some of its leaders became charter members of a new national organization, the Botanical Society of America, whose membership was at first restricted to those botanists who had published several articles or otherwise made important "professional" contributions to the science. Accordingly, when this group was formed in 1893, there was just one woman among its twenty-five charter members (or 4 percent, just one-tenth of the female composition of the TBC). Its lone woman was once again Elizabeth Knight Britton, who with Ellen Richards in home economics, Alice Fletcher in anthropology, and Cornelia Clapp in zoology was one of the most prominent women scientists of the 1890s. Though not a professor or even a holder of a doctoral degree (she was a graduate of the New York, later Hunter, Normal School), she was nevertheless a highly productive botanist who published over 300 articles on the mosses in her lifetime. She was also by 1893 the wife of Nathaniel Lord Britton, the new director of the New York Botanical Garden (whose establishment had been her idea). Yet if a later letter to a friend is any indication, Britton's position within the BSA had its socially awkward moments. When Annie Morrill Smith, the avid amateur, inquired about attending an upcoming meeting of the Botanical Society of America in New York City, Britton's response reminded her not only of the group's unwritten prohibition on female attendance, but also of her acquiescence to such practices: "I think it will not be a meeting suitable for women to attend, as the discussions will be rather embarrassing and the men will probably prefer to dine alone."[25] Though Britton had done the research, had read her papers to the members and listened to theirs, the possibility of her discussing botany over dinner was considered so potentially "embarrassing" as to justify her exclusion. Nor was her experience in such restrictions at all unique.

Thus the highly feminized field of botany had by the early 1890s put forward the image of an almost totally masculine professional science. This was done by stressing publications as necessary for membership, a requirement that generally (but not exclusively) limited membership to academics at the richer universities and omitted those at the women's colleges, who rarely published at this time. As a result the prominent women botany professors Susan Hallowell and Emily Gregory, who had been in the TBC, were not invited to join, nor were any of the several women plant pathologists at the U.S. Department of Agriculture (not even their director, E. F. Smith, was included), whose publications were chiefly in government bulletins. Only Elizabeth Britton was accepted, and she, though professionally very visible with her numerous publications, was conditioned enough to know her place and to be voluntarily invisible at "professional" banquets. The anxiety evident in the 1887 article in *Science* that had asked rhetorically "Is Botany a Suitable Study for Young Men?" had now been relieved, and the question had been answered in the affirmative.[26] No one need now think that botany was any less rigorous, scientific, or "professional" than other seemingly more "masculine" fields of science. Even the physicists would have two women among their charter members a few years later.

Another case, and one that telescopes the whole three-decade-long process of, first, the feminization of natural history in the 1870s, and then its enforced, and here rather brutal, defeminization under the guise of higher standards in the 1890s, occurred at the Marine Biological Laboratory at Woods Hole, Massachusetts. Women, especially science teachers, had been conspicuous in the early days of marine biology in this country, notably in 1873 and 1874, when Elizabeth and Louis Agassiz (and after his death, his son, Alexander) had run a summer school on Penikese Island off the coast of Massachusetts. Despite some initial protest by the male students, the Agassizes admitted fifteen women to their first class of fifty students (or, by another report, sixteen in the final forty-three) and increased their number to twenty in 1874. Among these early women at the "Anderson School of Natural History" were several from the women's colleges, such as Lydia Shattuck, Susan Bowen, and Cornelia Clapp from Mount Holyoke Seminary and Susan Hallowell from Wellesley College. Historian Joan Burstyn has traced the subsequent careers of these and other Penikese women and found that this one summer school, which lasted only two years, played an important role in improving science teaching in the United States. The Anderson School not only offered these early women teachers and professors, who had few other opportunities for advanced training in science or biological research in the 1870s, the rare chance to study nature directly, but it also stimulated them to a higher level of ambition and achievement. In addition, the school had a certain impact on the young male instructors, some of whom would later be important figures at major universities. David Starr Jordan, for example, not only later married student Susan Bowen, but he also invited other women to join them on walking tours of the United States and Europe, and when in the 1890s he became the first president of Stanford University, he hired several women faculty members. The whole Penikese experience, then, brief as it was, had an important place in the history of science teaching in the United States and in the entrance of some women into the scientific profession.[27]

Although the Anderson School lasted only two years, its success inspired many followers. The most notable was that at Annisquam, Massachusetts, run by another Agassiz student, Alpheus S. Hyatt, and supported by the influential Women's Education Association (WEA) of Boston, a group of wealthy and socially prominent Boston women that had opened many doors for women in that city in the 1870s and 1880s. Dissatisfied, however, with the makeshift facilities at Annisquam, the WEA joined other interested groups, including the Boston Society of Natural History, in starting the Marine Biological Laboratory at

PLATE 16. Summer classes at Woods Hole, Massachusetts, in the 1890s attracted many women. This botany class in 1893 included (at right) the Claypole twins, Agnes (later Moody) and Edith, and, to their left, Ida Ogilvie, later a professor of geology at Barnard College. (Courtesy of the Library, Marine Biological Laboratory, Woods Hole, Massachusetts.)

Woods Hole, Massachusetts, in 1888 as a permanent seaside summer school for teachers and researchers (see plate 16). Of the eight original trustees of the new MBL, three were women from Boston: Anna Phillips Williams; Susan Minns, a longtime "associate member" of the BSNH; and Florence M. Cushing, a former student of Maria Mitchell at Vassar College (1874), an influential alumnae trustee, and founder of both the Association of Collegiate Alumnae and the Naples Table Association. Thus by the 1880s women scientists and trustees had had a long and important relationship with a series of marine laboratories around Greater Boston.[28]

Despite the women's long association with the laboratory and its predecessors, the clientele of the laboratory began to change in the 1890s as the new breed of young professionals from the universities found it an increasingly desirable place to do their researches and spend their summers. They, however, soon began to feel constricted by the local, cautious, and amateur Board of Trustees with its limited vision of Woods Hole's future. Several of the new professionals chafed at the trustees' attitudes, since, like the laboratory's director, Charles O.

Whitman, professor at the new University of Chicago, they knew millionaires and were optimistic they could raise the funds necessary for a major expansion and overhaul of the MBL. Then in 1897 the storm that had been gathering for several years finally broke, and the aggressive professionals managed to wrest control from the Boston trustees, set up their own board, and embark on a vigorous new campaign to upgrade the laboratory. Of course this changeover would not necessarily have hurt the women biologists at Woods Hole, a few of whom were by 1897 as professionally qualified as the men and just as eager for improved facilities. But the new leaders had little use for these women, and though many women continued to go there after 1897 and be elected "members of the corporation," as practically all investigators at Woods Hole were, only two, Cornelia Clapp (with her two doctorates) and much later Ethel Browne Harvey, were among the one hundred trustees elected in the next fifty years.[29] Thus by the late 1890s the scale and power of science had shifted from what could be labeled small-time philanthropy (under $10,000 per year), where wealthy women like Florence Cushing had been able to play a fostering role, to the realm of professionals, millionaires, big business, and, soon, the big foundations, which few women ever penetrated. Rather than sitting on the board of trustees and running the institution from within, as they had for a decade, the women at Woods Hole were after 1897 back on the outside. About all they could do for women thereafter was repeat the strategy of "creative philanthropy" that they were already using at the Naples Zoological Station: raise a modest sum of money to buy an opportunity (a "table") for a few aspiring women zoologists to work with the male professionals.[30]

By the 1890s such defeminized national professional organizations had become the norm not only in those fields formerly designated as "natural history," but in most other sciences as well. Female participation in these groups was at best restricted to a fraction of those relatively few women who already had the best credentials (American or European doctorates), the best positions (usually professorships at the women's colleges), most publications, or, in lieu of these, as in engineering, strong personal contacts with prominent men in the field. Thus, though it was never stated that only "exceptional" women, if any at all, were welcome in such organizations, it is their names that recur on any list of examples. By the early 1900s a handful of the most eminent of these even held an occasional elective office (usually a vice-presidency). For example, although there were no women present at the founding of the American Psychological Association in 1892, Margaret Washburn, then at Wells College, and Christine Ladd-Franklin, a lecturer at The Johns Hopkins University, joined two years later. Alice Hamlin (later Hinman) of Mount Holyoke College was elected to its council in 1897, and Mary Calkins of Wellesley College became its first woman president in 1905.[31] Similarly, in mathematics, Charlotte Scott, a Cambridge University "wrangler" on the faculty of Bryn Mawr College, and Ruth Gentry of Vassar, who had been the first woman to attend lectures at the University of Berlin, were the only female charter members of the American Mathematical Society, which was formed in New York City in 1894. Scott was later elected to its council and served as vice-president in 1906.[32] The two women present at the founding of the American Physical Society in 1899 were once again from the women's colleges: Marcia Keith, chairman of the department at Mount Holyoke College, and Isabelle Stone of Vassar, the first woman doctorate in physics from the University of Chicago, who had spent a postdoctoral year in Berlin.[33] (Margaret Maltby, formerly of Wellesley and soon to go to Barnard, might also have attended had she not been in Germany at the time.) In fact so great was the dependence of women physicists on the women's colleges for employment at this time, as demonstrated in table 1.4, that almost any woman physicist present at an APS meeting necessarily would have come from such a school.

Meanwhile the medical sciences of anatomy and physiology were facing certain additional pressures and tensions as they sought to professionalize in the 1880s and 1890s, since their devotees had also to separate themselves from the rank and file of ordinary physicians, who, usually holding M.D.s, were already "professionals" in another sense. Since women were often excluded from the regular medical schools, especially the most prestigious ones, they faced additional difficulties in gaining admission to the new professional societies in the medical sciences, and there were very long delays before any women were admitted to these groups. The Association of American Anatomists, for example, was formed in 1886, but it had no women members until 1894, when it took the unusual step of electing the fifty-seven-year-old Mary Blair Moody, M.D. (University of Buffalo, 1876), of Connecticut as well as her son Robert, an instructor at the Yale Medical School (and later husband of Agnes Claypoole Moody, a zoologist at Cornell). After this it was another eight years before the association elected any women, but then in 1902 and 1903, for reasons that are unclear, it suddenly elected seven more women, including Lydia DeWitt, longtime instructor at the University of Michigan; the brilliant but unemployed Susanna Phelps Gage of Ithaca, New York; Florence Sabin, one of the first women to earn an M.D. at The John Hopkins Medical School and the protégée of its distinguished anatomist, Franklin P. Mall; and several current and former "demonstrators" at the women's medical colleges. Sabin became a vice-president of the association in 1909.[34]

The founders of the American Physiological Society were even more restrictive, however, and not only held the line against physician-mothers of young professionals but also seemed to require foreign training (preferably a German Ph.D.) as well as an M.D. for membership. This society was founded in 1887, when, according to its self-serving official history, there "were no women physiologists in the country," although this was a matter of definition. Eight years later, Frances E. White, M.D., and long a professor of physiology at the Woman's Medical College in Philadelphia, was refused admission on the grounds that the quality of her publications was not high enough. Apparently this rejection deterred the many other women professors of hygiene and physiology in the 1890s (see chapter 3) from applying, since it was not until 1902 that another woman made the attempt. In that year the veteran Ida Hyde, extremely well qualified with her German Ph.D., her postdoctoral research at Naples, her further work at the Harvard Medical School (where she was reportedly the first woman to do research), her presentations at the society's own meetings and publications in its journals, and most recently her associate professorship in physiology at the University of Kansas, applied for membership. A "full discussion" ensued, and she was finally elected, but she never held an office and was not followed by another woman until 1913.[35] One sees in all this a pattern: consciously or unconsciously, some societies required such a high standard of documented achievement and qualification from their applicants, such as a German doctorate, that they were in effect excluding women, who had long been barred from holding such a credential. In fact physiology would, for reasons that are unclear, long remain one of the fields offering the least amount of recognition to its women (see chapter 10).

Similarly, geography was a field so dominated by European influence that its young American professional society required foreign training of its women members but not of all its men. Established in 1903, before any American university had awarded a doctorate in the field, the Association of American Geographers had forty-eight original members, only two of whom were women (4.2 percent). One of these was Martha Krug Genthe, the only original member to hold a doctorate in the subject (from Heidelberg; two men held doctorates in economics); the other was Ellen Semple, who had studied with Friedrich Ratzel at Leipzig in 1891 and 1892 but who had not been allowed to take a degree. Neither woman ever held a regular appointment in any of the new American departments that were springing up at the time (Chicago should have hired Semple in 1902), but their reactions to this exclusion varied. Semple continued to publish, and became one of the major geographers of her generation (she was, in 1921, the only woman ever to serve as president of the AAG). Genthe, by contrast, taught in a private girls' school in Hartford, published only one major article, and was one of the first members to resign from the association.[36]

Requiring membership credentials that were not generally available to women, such as German or American doctorates before the late 1890s or professorships at major universities, might now be considered a kind of second-order or de facto kind of discrimination in which barriers in one realm are used to justify the continued exclusion of women from another. Such linkages or credentialism put an additional burden on any would-be pioneers in certain fields and organizations, since they had to jump several barriers in succession. For this reason, the engineering societies long presented the greatest difficulties to women aspirants, and "firsts" and "exceptions" there tended to have either highly unusual backgrounds or strong personal contacts with persons already in the field that enabled them to override or evade the stated regulations.

When engineers began to "professionalize" in the late nineteenth and early twentieth centuries, they deliberately created distinctions and barriers between the engineers with college degrees and relevant professional experience and those other "engineers" who had learned their jobs by experience and lacked such credentials. Most of the engineering societies established in the 1880s had a strict hierarchy of requirements necessary for each of several levels of membership. Women's exclusion from most of the nation's engineering schools and from most jobs thus proved a fairly effective means of keeping them out of the societies and so the "profession" as well. Nevertheless, some women did manage to obtain the necessary training when a few schools, like MIT and Cornell's Sibley College of Engineering, started admitting women ("Sibley Sues") in 1878 and 1884, respectively; some others obtained the required experience in family businesses. Thus Ellen Richards, who became the first woman member of any engineering society when she was elected a full member of the American Institute of Mining Engineers in 1879, was aided by her MIT degree and her publications on mineral chemistry. But it can hardly have been irrelevant or a coincidence that her husband was a vice-president of the organization at the time. It was fifteen years before the next woman was admitted to an engineering society: Lena Allen Stoiber, who owned a silver mine in Nevada, became an "associate member" of the American Society of Mining and Metallurgical Engineers in 1894. In 1909 Nora Blatch de Forest, a "Sibley Sue" in the top five of her class, became a "junior member" of the American Society of Civil Engineers (ASCE), but she was unable to advance any higher. When her "junior membership" expired in 1916 the ASCE refused to promote her to full membership, despite her having met the stated requirements, and dropped her from its rolls instead. In response, this fiery engineer, ex-wife of Lee de Forest, the inventor of the radio tube, and a prominent suffragist, brought a lawsuit against the society. Unfortunately, she lost her case. No woman attained the rank of "full member" of the ASCE until the rather late date of 1927, even though Kate Gleason, another Cornell graduate who had assisted her father in his Rochester, New York, gear works, was the first female elected to that rank in the American Society of Mechanical Engineers in 1914. Engineering would long remain one of the slowest fields to accept women.[37]

Yet the highly qualified women who did belong to these professional scientific societies, including even those who held office in them, often felt ill at ease and unsure of their welcome. They often did not know, for example, how to react to official invitations to the society's social functions and informal get-togethers. Had the host *really* meant to include them, or was the invitation merely a pro forma one that should be ignored or politely declined? Usually professors at the women's colleges, these women members were sufficiently well mannered (or "ladylike") to be conscious of the potentiality of intruding on others' territory and, rather than protesting their exclusion, they either did not go or made cautious inquiries beforehand lest their presence disturb the men unduly.

Although banquets at professional meetings (like the chemists' "misogynists' dinner" of 1880) had long excluded women, the ban began to seem a little less intimidating around 1900, when several women scientists began in their own quiet way to challenge some of these age-old restrictions. Thus, for example, Mary Whitney of Vassar College, who had attended the founding meeting of the American Astronomical Society at Yerkes Observatory in Wisconsin in 1899 with her protégée and successor, Caroline Furness, was still not sure whether they would be welcome at the society's banquet in Washington, D.C., in 1902. President Simon Newcomb noticed her unease and wrote to assure her that they were indeed expected to attend: "I am much disappointed to notice that although you hope to be here at our meeting, you do not propose to join in the dinner. Possibly you may be under a misapprehension, supposing that the dinner is only for the men of the society. Permit me, therefore to assure you that all members are equal, and that we should like very much to have our lady members with us."[38] Newcomb's encouragement induced these women to go, and thereby set a precedent for later meetings. The next generation of women astronomers would feel confident enough of their welcome to invite the society to their campuses for meetings, as did the Smith and Mount Holyoke departments jointly in 1920, the one at Vassar in 1923, and that at Wellesley in 1940.[39]

Smoking and organized "smokers," apparently an important and often deliberately intimidating part of the male professional culture at the turn of the century, greatly complicated the women's comprehension of whether or not they were actually welcome. Properly bred women apparently did not smoke or enter rooms where men were smoking before the 1920s, when the cigarette industry, acutely sensitive to appropriate sex-role behavior, started a publicity campaign to induce them to take up the habit.[40] Thus to include smoking on the program or to allow men to smoke after dinner was, whether the men realized it or not, long an overt social message to the women not to attend. Most women would not attend a smoker unless the man in charge specifically requested their presence. Physicist Sarah Whiting, for example, later reported that though she had avoided the banquet and the male smoker when she first attended meetings of the American Physical Society, between 1907 and 1909 she had begun to participate in both when encouraged by the society's president: "The introduction of the German Smokers was long a bar to the women being present at the banquets of the Physical Society, but finally a number of the younger women joined and under the presidency of Professor Nichols of Cornell we were present at a banquet. I was not quite sure we were welcome for men had not then as now got over the idea that blue smoke and the presence of ladies at banquets were incompatible."[41]

Since most of the societies' presidents were not so attentive to the women members' delicate sense of propriety, any woman receiving an invitation to a dinner or smoker who really wished to attend had to use some veiled initiative. She might, for example, make inquiries in hopes of receiving stronger assurances of her welcome. Mary Calkins of Wellesley was a master of this tactful maneuver. Since she felt awkward about attending a dinner of the past presidents of the American Psychological Association, which was to be held at the Columbia University Faculty Club in 1911, she inquired of host James McKeen Cattell beforehand, not wishing, as she put it discreetly, to attend "if the smokers would be deterred or uncomfortable" by her presence.[42] Cattell understood her plight and apparently reassured her sufficiently, for she went and had an enjoyable evening; in thanking him for his hospitality she revealed how unusual it was: "It was a pleasure and privilege to attend your dinner for the Psychological Association presidents and I enjoyed it—chiefly for its good fellowship and the interest of the topics discussed, but also in its contrast with the usual procedure of admitting a woman to scientific session or committee duty but regarding her as *ex officio* excluded from the professional dinner."[43]

That such reassurances were necessary is clear from the case of the Association of American Geographers, which deliberately held "smokers" or other important discussions in smoke-filled rooms in order to discourage female attendance and avoid having to consider issues of importance to the women. This tactic was used successfully at their Washington meeting in 1915, when a large number of local schoolteachers had wished to come to discuss the status of geographic education in the schools. Since the men planning the meeting desired neither the "prim schoolmarms" nor their topic, they decided to hold a smoker at the appointed time and thus eliminate both vexing problems altogether. One well-traveled committee member even suggested, perhaps in jest, that the women be allowed to come but be assembled behind a lattice, as in some Islamic mosques, so that they could attend without being seen![44] Yet even this solution, outlandish as it may seem, would have been more liberal than the geographers' "smoker," since it would at least have allowed the women to come and listen. Not all geographers were this staunchly antifeminist, however, and at least one, William Morris Davis of Harvard, three times president of the AAG, was almost a feminist. He was sensitive enough to the women's predicament to suggest as early as 1909 an ingenious way around the annual problem: "Can we not have the thing without the name[?;] the name might offend lady members, even if they enjoyed the thing; because the name is so generally associated with men's gatherings. Can we not announce an informal evening session . . . and then have all the fun and smoke we wish?"[45] It was symptomatic of Davis's increasingly isolated position in his field that his fellow geographers ignored his suggestion and persisted in their antifeminist ways. His view was perhaps predictable to those who knew him: he had been brought up a Quaker, was known for frequently taking stands on ethical issues, and was even a grandson of pioneer abolitionist and woman's rights advocate Lucretia Mott.[46]

Enforcement of this unstated territorial ban on women's attendance at "smokers" depended on their understanding the message and their having so accepted and internalized proper standards of ladylike behavior that, like Elizabeth Britton with the male botanists, they voluntarily conformed to the prohibition and stayed away. In this regard the custom was like the expectation (mentioned in chapter 1) that a female college instructor who married would resign immediately: the pattern was so strong that only a rare woman refused. Nor was to defy such established patterns likely to change any attitudes; it was more likely to cause the rebel to be condemned as having a defective character, since it was common knowledge that anyone properly trained would know that she must conform, despite her own preferences to the contrary. Thus in most instances a woman might well lose more of her reputation by going to a "smoker" than she could possibly gain from the conversation there, however scintillating. Yet in the end the decision whether to attend a "smoker" or behave "properly" and stay at home was largely up to the individual, and, as in other areas, there were various reactions. If most women scientists, especially those from the women's colleges (who were, after all, the most likely to be members of these "professional" organizations), were cautious and uncomplaining, a few others could be quite daring. Christine Ladd-Franklin, for example, was not one who could be induced by something as nebulous as social propriety to forgo a psychological meeting voluntarily. Somewhat avant-garde anyway, she considered social conventions that limited one sex more than the other to be mere pretexts (or here, more appropriately, smokescreens) for excluding women. Accordingly in 1914 she insisted, seemingly irrefutably (logic was, after all, one of her specialties), to Edward Titchener of Cornell University that she could not be offended in the slightest by cigar smoke at a meeting of his Society of Experimental Psychologists, since she always smoked in "fashionable society" anyway![47]

Ladd-Franklin's counterpart a generation earlier in the world of amateurs had been Annie Trumbull Slosson, who later regaled the largely male membership of the New York Entomological Society with her reminiscences of her first meeting with them as early as 1893. She had been properly escorted to the meeting by one of the older gentlemen, but when they arrived most of the men already present were not only smoking cigars but also drinking an unspecified foamy beverage. The men were mortified, but Slosson held her ground and selected a comfortable seat. Soon someone with a clear head provided her with a cup of coffee, a beverage deemed suitable for a feminine throat, and the animated discussion of insects resumed. Slosson, who became a stalwart member of the group and one of the major financial backers of its journal, later explained that she was not at all offended at their smoking and drinking and certainly never considered leaving, since she had had five brothers and was well accustomed to holding her own in such a masculine setting.[48] Yet one suspects that even twenty years later such a daring intrusion into a group of etiquette-conscious professionals would have been accorded a far cooler, probably quite icy, reception and caused not only her character but also that of anyone associated with her to be scrutinized seriously. Someone, after all, must have broken ranks and invited her.

Despite an occasional iconoclast, most women rejected by these professional groups or side-stepped in the intense upgrading of the 1880s and 1890s did not protest, and either forwent such scientific associations or worked to expand and strengthen their own subculture of separate clubs and activities. These clubs had been an essential infrastructure for women scientists such as Erminnie Smith as late as the 1870s and possibly the 1880s, but by the 1890s, when the range of scientific attainments between the top women and the more numerous others had broadened greatly, these self-help clubs were more like the contacts available at a woman's college—others who shared a general interest but few fellow specialists—than an adequate substitute for membership in the appropriate professional organization. Although such clubs would seem to have offered little to the most accomplished women scientists of the time, a few of the "professionals" who were not faculty members at the women's colleges, such as Elizabeth Britton, Ellen Richards, and Alice Fletcher, were nevertheless fairly active in them. Perhaps they were glad to help the others and to have in a sense "students" or protégées who might send along an unusual specimen or ask an interesting question and whom they could in turn instruct and exhort to greater efforts. Nevertheless, these clubs seem to have done little more for women in science than enhance their already highly segregated corner of the scientific world.

Washington, already the home of the Women's Anthropological Society of America, and Chicago, finally rebuilt after its fire in 1871, were the two centers of this women's scientific subculture in the 1890s. Washington was the site of the largest and, for a short time, the strongest of the separate science clubs for women, the (Woman's) National Science Club, established in 1891. This group grew out of an evening discussion group organized by ex-Post Master General ("Colonel") and Mrs. Horatio King in the 1870s, which had featured mixed company, something novel at the time. After the death of the "Colonel," his widow continued the discussions for women only. Apparently this group was also quite successful, for in 1891, when the AAAS held its annual meeting in Washington, D.C., Mrs. King and her elder sister, Laura Osborne Talbott, decided to embark on the ambitious undertaking of forming a national scientific society for women, which was to be centered in Washington but was to have local affiliates across the nation. For about eight years the club was active, attracting over 200 members from all parts of the nation, electing a few "fellows" and "honorary members," and counting as its "patron" Phoebe Apperson Hearst, who as the widow of a former senator still had a residence in Washington. The club held an annual meeting in Washington and published many of the papers presented there in its *Proceedings*, three volumes of which appeared in 1895, 1896, and 1897, and its bimonthly *Journal* of 1898 and 1899.[49]

Some of the articles in WNSC publications represented original contributions to knowledge. Others were summaries of scientific work by women in a particular field with exhortations to others to do more, as in Mary Murtfeldt's 1897 "Report on [the] Present Status of American Women in Entomology," and ichthyologist Rosa Smith Eigenmann's 1895 "Women in Science," which made two points: an old one, harking back at least to Almira Hart Lincoln Phelps in the 1830s, that under prevailing social conditions women's lives were so filled with daily tasks that they were more likely to be popularizers than researchers, and the new and related idea that the few women who had contributed to science had been treated like curiosities and overpraised for the little that they had actually done. She was particularly offended, she said, offering herself as an example, at being praised for "doing well, for a woman." Her dissatisfaction with this patronizing attitude toward women in science was a good sign that some women were in the 1890s demanding to be taken seriously and have their work evaluated realistically.[50]

Meanwhile Elizabeth Britton, chairman of the club's division of bryophyta, was less petulant and seized the opportunity to communicate with so many women interested in science to urge them to take up the study of mosses. The need for competent collectors was so great, she told the members condescendingly in 1897, that she had decided to take time "from my [own] more advanced work to teach beginners to help themselves." In elaborating on the nation's need for bryologists, she even sex typed the subject a bit to assure the members that studying mosses could indeed be "womanly":

> The flora is so varied and extensive that we cannot have too many at work; in fact, we are suffering from a lack of American students, and it is the duty of all those who have the means, taste, and facilities to join in helping along the cause of this most interesting science. It is one particularly fitted for women, for it requires patience, keen observation, and much indoor study; the material is easily obtained, and, with exception of a few books and a compound microscope, the expense entailed is not great and the rewards and pleasures are endless.[51]

Before long Britton had a large network of observers and correspondents who sent her local specimens with their own descriptions for her correction. She also encouraged several women to send along drawings, since she was always on the lookout for a good illustrator. Within a year there was sufficient interest in the mosses for Britton and her friend Annie Smith Morrill to start their own highly feminized club, the William Sullivant Moss Society (named for an early collector), later the American Bryological Society, still the main organization in this specialty. Thus Britton's use of the National Science Club gives a glimpse of its potential and what it might have become on a broader scale but did not: the effective organization of women amateur specialists by the only woman professional in the field for the systematic study of a particular part of the plant kingdom. Strangely and unfortunately, however, just as this prospect seemed close to reality, the National Science Club vanished.

In 1899, shortly after launching its bimonthly journal in 1898, the National Science Club abruptly ceased operations for reasons that are unclear. Although it seemed to be flourishing, it may have suffered financial reverses, though there was no sign of this in its *Journal*. More likely, one or more of its leaders may have died. But after giving up national ambitions and the grandiose name, a remnant of the original group reformulated itself as the purely local Eistophos Science Club of Washington, D.C. This group has proven far more longlasting than its predecessor, for its members, local women scientists, often wives of scientists as well, still, over eighty years later, meet monthly to discuss a member's paper.

The other high point of the 1890s for women scientists outside the women's colleges was the World's Columbian Exposition, or "World's Fair," which was held in Chicago in 1893. This had required about two years of preparation and offered a variety of groups and causes a chance to exhibit their recent "progress" to the public. Interestingly, the exposition also revealed how flexible the operation of the "separate spheres" for men and women scientists could be (if desired) in the 1890s: though the segregation was adhered to by some groups, others ignored it without ill effects. In fact the exposition had a dual policy toward women participants: it allowed them to join the men in some meetings and exhibits but also held a series of separate events for women only. Some of this seeming inconsistency was the result of the legislation that set up the exposition in 1891. Offended by the lack of recognition women had received at the 1876 Centennial in Philadelphia, suffragist Susan B. Anthony had worked behind the scenes in 1891 to make sure that the federal legislation and appropriation authorizing and subsidizing the exposition included a provision for a separate Board of Lady Managers, which was to have its own Woman's Building and was to be in charge of all exhibits submitted by women.[52] This was in fact similar to the evolving duplicate administrations under the "deans of women" that were being established at many coeducational colleges and universities in the 1890s.

When this separate board, headed by Mrs. Potter Palmer of Chicago, began to function, however, it felt confident enough of women's potential contributions to soften, but not abolish, this total segregation and to allow most of the women's projects and demonstrations to go into the main buildings with those presented by the men. Thus although there were a few last-minute exhibits by women botanists, geologists, entomologists, inventors, and miners to fill the empty space left in the Woman's Building (the only building at the Fair designed by a woman architect), the more impressive ones, such as the scientific restaurant or "Rumford Kitchen" run by Ellen Richards and the other home economists from Boston and the Mexican archeology exhibit by Zelia Nuttall, were located in the main (liberal arts and ethnology) departments of the exposition. Similarly, several by-now familiar women's organizations, including the Association of Collegiate Alumnae, Bryn Mawr College, the Association for the Advancement of Women, and the National Science Club, sent exhibits to the Education Building. Yet the Board of Lady Managers was not so liberal as to leave such women exhibitors to the mercy of male judges. Instead it retained enough of the concept of "separate spheres" to require that whenever an exhibit by a woman was to be evaluated in any way, another woman had to be present on the panel of judges.[53] (It is not known if this increased the number of women winning prizes, or if so, by how many.)

This dual arrangement also applied to the planning and running of the over 200 special congresses held at the exposition (several each week) on a host of contemporary issues, such as evolution, religion, public health, temperance, and education (including that of "colored persons"). Some were of special interest to women, such as that on woman suffrage, which was considered "notable" by *Science* magazine "for the large number of men present who seemed to enthusiastically support the claims of their sisters," and the Congress of Representative Women, which was held in May 1893. This symposium included an address by Mary Putnam-Jacobi, a prominent New York City pediatrician and neurologist, on "Women in Science" (a history through the ages), another by Helen French of Wellesley College on "Our Debt to [the University of] Zurich," and a third on "Women's Dress," by Professor Ellen Hayes, also of Wellesley, who deplored the students' frivolous tastes.[54]

In addition to these general congresses, several scientific ones were held at the exposition. Some included women speakers, such as Ellen Richards, who spoke on the comparative efficiency of several methods of ventilation at the International Congress on Chemistry; the unknown Hortensia Black, who read a paper on bird protection to the World's Congress on Ornithology; and two astronomers, Williamina P. Fleming of the Harvard Observatory, who spoke on "women's work" in astronomy, and Dorothea Klumpke, an American at the Paris Observatory, who described its Bureau of Measurements to the World's Congress on Astronomy and Astro-Physics. Even more impressive in Chicago, as earlier at the AAAS meetings of the 1880s, was the disproportionately large number of papers presented by women anthropologists. Five women (four Americans and one Englishwoman) presented six papers at the International Congress on Anthropology, where, again according to *Science,* "the paper which attracted the most attention" was that of Zelia Nuttall on the Mexican calendar system, which "presented a highly ingenious theory for the solution of this obscure and famous problem." Although the women's friend Frederic Putnam did not arrange the program of this congress, the women's strong showing there may be attributed to him indirectly—his protégée Alice Fletcher was on the planning committee with six men (despite official rules against such mixed committees), and Sara Yorke Stevenson was there because Putnam, in charge of the ethnological exhibits, had appointed her one of his judges.[55]

Less well publicized (unnoticed by *Science*) was the women's even more impressive showing at the World's Congress of Geology in August 1893. Their performance was overlooked, probably because, being so numerous, they were placed on a program wholly separate from and in conflict with the main (men's) sessions. Speaking at this separate "Woman's Department of Geology" were thirteen women: ten Americans, including Mary Holmes, Ph.D., of Rockford, Illinois, and Mrs. Ada Davidson of Oberlin, Ohio (later a president of the Woman's National Science Club), as well as two Englishwomen and one Irishwoman.[56] One suspects that the male geologists could have included these women on their main program, as the anthropologists had done so well, if they had wanted to. Was the point therefore of all this separation as late as 1893 to camouflage the suddenly large proportion of women participating in geology or to protect the women (or the men?) in some way from the presumed embarrassment or competition of intermingling the sexes? Probably it was both, but this whole subculture of women's clubs and separate groups in science needs more study, especially concerning women's own attitudes toward such groups and their relations, if they had any, with the men's groups in the field.[57]

There were thus quite a few women in science by 1900, if one knew where to look for them, as in the National Science Club, the amateur groups, the whole field of home economics, and among the members (but not the fellows) of the AAAS. For a variety of reasons related to the rise of "professionalism" in science and the discriminatory employment discussed in chapter 3, only a few of even the qualified women were in the most prominent or visible places. Yet it was several years before the women's leaders seem to have realized what was happening and protested that the main thrust of science was passing them by. Before 1910 all but a few women scientists were so impressed with their recent highly publicized advances in "women's work" and their occasional "firsts" in higher education and elsewhere that they were more inclined to congratulate themselves on how far they had come since the 1870s than criticize the remaining obstacles. They could not see (or if they did, they could not believe) that they were caught in the midst of a Sisyphean dilemma: while they had been working hard for several decades to increase the number of opportunities for women in science and had visibly expanded their role in some areas, many new barriers had arisen even faster than the best women of the time could scramble over them. They could thus in all sincerity think that women were making many important "advances" only to discover a few years later that it had all been a cruel illusion—that the men had moved so much higher than they that their "progress" was by contrast virtual stagnation or possibly even regression. The coming of professionalism in the 1880s and 1890s had contained and circumscribed the women and restricted them to the fringes of science, almost as far from the real involvement and leadership as they had been decades before, when they had had to work through their fathers and brothers.

Then in 1910, when quantitative evidence appeared on just how much the women had achieved in science, the expected great rejoicing became disillusionment as it became clear how far women still had to go to reach full equality. It showed that though they had achieved many "firsts" in a variety of fields, they were still only a marginal, even miniscule, proportion of the important scientists. These findings set off a spirited debate about women's abilities and future in science and awakened feminists of both sexes to the prospect that if women were ever to play a sizeable role in science, they needed to move beyond the domain of "women's work," which had served its purpose but did not receive much professional recognition, and into the mainstream, where they would be noticed and counted. The obstacles to this full equality seemed now to lie as much in the popular attitudes and cultural expectations that limited women's role in American society generally as in any hypothetical lack of ability or training among the women. By the early 1910s, therefore, the women's leaders were changing their strategy from one of building and seeming to desire a separate universe for women scientists to the different and more aggressive one of attacking these restrictive social attitudes and practices head on.

5

THE WOMEN'S MOVEMENT, THE WAR, AND MADAME CURIE

The main reason for the women scientists' dissatisfaction with "women's work" and underrecognition around 1910 was their politicization, as they responded and contributed to the burgeoning "women's movement" around them. Still politically weak around 1900, the two components of the "women's movement," feminism and suffragism, developed quickly thereafter and attained major proportions in the 1910s. Women scientists and a few liberal men contributed to both movements by (1) doing research, especially in anthropology and experimental psychology, which questioned and refuted long-standing, even "scientific," opinions about women's abilities; (2) protesting publicly women's inferior status within the scientific community; (3) loyally supporting the war effort, primarily in suitably feminine ways; and (4) engaging in local and national politics to win the vote for women. Only the last two of these bore fruit, however, for though American women did get the vote in 1920, neither feminist research nor protests of discriminatory employment led to any significant improvement in women's status in science. Instead sex-typed employment remained the norm and even expanded after 1920, as the women's few wartime forays into other areas were quickly and quietly obliterated. Nor did the highly publicized visit to the United States by Madame Curie, the Polish-French Nobel Laureate, in May 1921, which in many ways was the ultimate in the new public relations in science, have any lasting favorable impact on how women scientists were regarded and what work they did. Thus although the period from 1900 to 1921 was an active one politically for women in science, it wrought little change in women's scientific employment. Henceforth, though a few liberal and well-placed men might make occasional "exceptions" for particular women, the employment system itself remained highly segregated, both territorially and hierarchically. The strategy of improving women's position in science by changing political attitudes thus failed miserably, and any new strategies after 1920 would have to turn inward and work within the prevailing system.

It is probably a truism of the history of the social sciences that its practitioners can only study, indeed are only aware of, social phenomena after they have become entrenched in some part of society. Thus during the 1890s, when what can now be called "sex typing" of occupations and roles was rampant, few persons were aware of how artificial or contrived it was or were at all critical of it. It just seemed so obvious and *natural* (to use a common word of the time) that it hardly required any justification or aroused any suspicion. To contemporaries, seeking signals on how to behave and function in a rapidly changing world, projecting or extending old categories and labels onto unfamiliar new phenomena must have seemed harmless and even logical. Since the proferred sex typings usually accorded well with everyday observations, such as the one that held that women were weaker and more domestically oriented than men, no one suspected any undue restrictions were being placed upon them. In fact, many if not all of the women scientists of the time, seeking to justify women's employment in science, not only accepted these sexual stereotypes, publicly at least, but even used and capitalized upon their prevalence to ensure that some of the new jobs opening up were reserved for women. Liberal (or "environmental") social concepts such as those of "stereotyping" or "sex typing," which implied that not all women or men fit the roles held out for them, had not yet emerged and might well have been incomprehensible in the 1890s if they had.[1]

But as segregated employment became a regular feature of the American economy and the full ramifications of women's consequent underrecognition became clear to some persons, the largely political notion of the system's "unfairness" began to spread. Now that women were being as well educated as men and were holding jobs, although marginal ones, in science, the remaining limits to their full equality, formerly accepted as inevitable, began to seem intolerable. The root of this new impatience and anger was the political doctrine of "feminism," or the view that women were the equal of men and that any ("artificial") social constraints preventing this should be changed or abolished. Although still a rather radical doctrine to many (it dated back to the eighteenth century), feminism became a strong ideology among quite a few professional women in the 1910s. It should not be confused, however, with the more moderate "suffragism," which made much greater headway in this decade. Because the suffrage movement had the narrower goal of winning the vote for women (rather than changing their lifestyles and much of society), it appealed to the far broader political and social spectrum of middle-class women, including many members of temperance organizations and women's clubs. Suffrage leaders, more conservative and politically realistic than the feminists, generally spurned the idea that women were the equal of men and instead relied heavily on those very sexual stereotypes that the feminists were trying to refute, such as the belief that women, being purer than men, would clean up corrupt politics, or that government (national as well as local) was just an expanded version of housework and was thus well suited to women's "special skills." Yet even these relatively nonthreatening ideas were not sufficient for the final passage of the Nineteenth Amendment; for that to happen, the suffragists had to become increasingly racist in the late 1910s as they cultivated support among anti-Negro and anti-immigrant voters.[2] Since the fight for the relatively modest reform of the vote was so difficult, it is not surprising that the more extreme demands of the feminists were usually not met. Nevertheless, many women scientists were active in both the feminist and the suffrage movements, supporting both the vote and more drastic reforms (such as equal pay), fully aware of the inconsistent rhetoric and politics of the two groups.

Although everyone "knew" around 1900 that women were biologically inferior to men and although the "literature" on the subject was vast and seemingly authoritative, a graduate student in psychology at the University of Chicago fired an opening salvo at the concept in 1898 when she pointed out that the issue had apparently never been put to a valid experimental test. Thereupon Helen Thompson (later Woolley), reportedly one of the most brilliant students that John Dewey had ever seen, proceeded to map out the new psychological specialty of sex differences in her dissertation, "Psychological Norms in Men and Women." She gave a battery of tests (motor abilities, sensitivity to taste and smell, hearing, vision, intellectual abilities, and personality traits, among others) to twenty-five male and twenty-five female undergraduates at the university. She found that the men were slightly better at some things (physical strength and "manual dexterity"—despite what the advocates of "women's work" were saying) and that the women were better at others (memory work and sensory perception). She considered the differences minor, and she broke important new ground when she attributed them to the different training and social environment that society provided for the two sexes. Men were expected to play outdoors and develop greater physical strength, but women were expected to stay indoors and become sensitive to other persons. She considered the fact that the differences in such socially induced traits were minor among the students at a coeducational university added evidence of how easily shaped any inherent so-called sex differences actually were. After being awarded her degree summa cum laude, Thompson spent a postdoctoral year in Paris and Berlin as the ACA's European fellow. In 1901 she joined the faculty of Mount Holyoke College (which, under a new president, was now recruiting women doctorates), set up a laboratory, and prepared her dissertation for publication. But in 1905 she resigned to marry Paul Woolley, a medical student she had known in Chicago, and accompany him to the Philippines. Although they returned to the United States in 1908 and she eventually found stimulating work in the burgeoning vocational guidance movement, "Mrs. Woolley" was never able to regain the prominent place "Miss Thompson" had held in research on sex differences.[3]

Meanwhile, however, a second new current, the revived suffrage movement, was beginning to affect the political and intellectual climate of some portions of the scientific community. Although the American women's suffrage movement had begun long before at a "women's rights" convention in 1848, as late as 1905 it had failed to attract much support. Despite the fact that women had won the vote in certain western states and had suffrage in, for example, school board elections in some states, enfranchising women still seemed to many a radical and threatening step, the need for which was by no means clear. Opinion had long been divided on the issue even among those women scientists who had fought so hard for other kinds of advancement for women, such as higher education and employment. Thus Emma Hart Willard, her sister Almira Hart Lincoln Phelps, and Ellen Richards had all opposed the vote, though Maria Mitchell had favored it, and Graceanna Lewis had worked hard for it. Even the Association of Collegiate Alumnae, which had long promoted the advancement of women in other realms, prohibited the discussion of suffrage at its meetings until the rather late date of 1917 on the grounds that it was too controversial and disruptive. But the divisions and "doldrums" finally ended, as most historians of the movement agree, in 1906, at the annual meeting of the consolidated National American Woman's Suffrage Association (NAWSA) in Baltimore, when for the first time the college women of America, including therefore most women scientists, were called upon to play a strong role in supporting the suffrage movement.[4]

Although neither Bryn Mawr President M. Carey Thomas nor her associate Mary Garrett, the Baltimore heiress, had done much for the women's suffrage movement before 1906, they responded enthusiastically to a personal plea from the now aged Susan B. Anthony to make the Baltimore convention a success. Anthony had probably heard the story of how they had bought woman's way into The Johns Hopkins Medical School in 1893 and knew how many new and influential friends they could bring to the movement. In this she was correct, for Thomas and Garrett proved extraordinarily successful in this, their first foray into suffrage politics. One innovation they made was broadening the convention's appeal to those who had not previously supported votes for women by staging several sessions as a public tribute to Anthony herself, who was eighty-six years old and known to be dying. In addition to capitalizing on Anthony's personal celebrity, Thomas and Garrett cast one key session as a "College Women's Evening" or, as the program put it, a discussion of "what has been accomplished for the higher education of women by Susan B. Anthony and other women suffragists, a tribute of gratitude from representatives of women's colleges," as if to indicate that the leaders of the women's colleges had favored female suffrage all along.[5] Thomas and Garrett also added to the pageantry of this memorable evening by having the "usherettes" (undergraduates from nearby Goucher College) wear their caps and gowns. Several representatives from the major women's colleges and the Association of Collegiate Alumnae gave appropriate addresses, the most inspiring and memorable of which were reportedly those of Thomas herself and Wellesley psychologist Mary Calkins, who urged women to "cease being ignorant or indifferent on the question" of suffrage and to support it wholeheartedly.[6] But the evening was best remembered for its presiding officer, who was none other than chemist Ira Remsen, now the president of The Johns Hopkins University. Although long known by Thomas (he had delivered the commencement address at Bryn Mawr as early as 1890 and had spoken at the ACA annual meeting in Washington in 1902), he had not been expected to accept the proferred invitation to preside, and his presence greatly enthused the women as a sign of official approval and encouragement. Perhaps Remsen just wished to please and cultivate the wealthy Garrett, whose money might help him construct the university's new Homewood campus. But it is more likely that Remsen's concern for the problems of college women was genuine, since it was he who, as mentioned in chapter 2, would remove the university's longstanding ban on admitting women to the graduate school one year later. If he had ulterior motives for cultivating Garrett, she and Thomas were also working on him, and by 1906 many years of influence were paying off.[7]

In a second masterly political stroke Thomas and Garrett induced Professor William H. Welch, the former dean of The Johns Hopkins Medical School and the 1906 president of the American Association for the Advancement of Science, to chair a session on women's role in municipal government at which Jane Addams of Hull House in Chicago was the principal speaker. Welch, who had come to approve of coeducation at the medical school after Garrett's timely gift in 1893, was sufficiently interested and intrigued by the whole notion of woman's suffrage that, as his diary for the week shows, he attended several of the convention's other sessions, dinners, and receptions as well. Apparently bemused by his position as a bachelor among all the spinsters, Welch wrote his sister, "I am to preside at one of the evening sessions—think of it! Miss Garrett and Miss Thomas made a special appeal to me to do so, and I consented." A few days later he added,

> I have been hobnobbing with woman suffragists, Susan B. Anthony, Julia Ward Howe, and others. Last night I presided at their meeting devoted to municipal affairs, which was largely attended. They seemed pleased with what I said, although I did not commit myself on the suffrage question. I told them the administration of a city was largely housekeeping on a large scale, and that the more women's influence was felt in such matters the better for the people. . . . The suffragists are not such a queer lot of women as many suppose, and they could hardly have selected me to preside on account of their supposed preference for long-haired men [Welch was bald]. Still I am told by some Baltimore ladies, not in sympathy with the movement, that I was called in to discover the germs of the disease.[8]

Despite Welch's private humor, Thomas and Garrett must have hoped that the presence of and seeming endorsement of votes for women by such influential figures as Remsen and Welch would soon affect opinion elsewhere both within and outside the scientific community. Thus in one week these political neophytes had managed not only to infuse college women with an ardor for women's suffrage, but also to align two of the most influential men in American science and medicine behind it. Conservative Baltimore was moving rapidly toward supporting votes for women, and the medical and academic "establishment" there, or at least some key figures within it, seemed willing to endorse it as well. In fact, The Hopkins, and especially its medical school, quickly became very liberal on this issue, and in 1909 Florence Sabin, by then an assistant professor of anatomy there, reported triumphantly that all of the full professors at the medical school except J. Whitridge Williams, the professor of obstetrics, favored women's suffrage.[9]

After Thomas's coup in 1906, she and other women academics and scientists worked hard for the women's suffrage movement. Her "College Women's Evenings" became a regular feature of the NAWSA's annual conventions, and in 1908 Thomas was one of the founders of the National College Equal Suffrage

League of undergraduates, faculty, and alumnae whose goal was to politicize the campuses and spread the influence of "college women" still further. This group enjoyed wide success; by 1914, for example, it could count two-thirds of the Mount Holyoke College faculty as members. Thus increasingly after 1906, college women, and so in one way or another almost all women scientists, were exposed to considerable discussion and fervor about women's suffrage. Concern for "the cause" would permeate many of their public and scientific activities until 1920.[10]

In this changed (and charged) political and social climate, interest in and discussion of sex differences, dormant since Thompson's pioneering work a few years before, suddenly erupted as a field of great popular and professional interest. Sigmund Freud's visit to the United States in 1909 added to the clamor, as to a lesser extent did the work on brain sizes and weights by Franklin P. Mall, anatomist at The Johns Hopkins Medical School. (As a good feminist, he reported that there were no measurable differences.)[11] By 1910, however, the age-old arguments of female (and Negro) inferiority had reached even more subtle levels of reasoning. The burden of the theory had come to rest on the statistical concept of greater "variability" in the male of the species, a view that dated back at least to Charles Darwin, who had used it to argue in the 1870s for the male's superior fitness. (Being more variable, males could adapt to more varied environments than females.) Later versions of this theory acknowledged that the brain sizes and abilities of both sexes varied a great deal and overlapped to a large extent, but stressed the fact that statistical studies of elites found more men than women in the top percentiles for ability and achievement. (Similarly, other studies found more men than women at the bottom levels of ability in homes for "idiots" and "imbeciles.") Women, who, in contrast, were less numerous at the extremes, suffered the opprobrium of being considered merely mediocre. Although Karl Pearson, a prominent English biometrician and feminist, had challenged the validity of this theory as early as 1897, leading American social scientists such as G. Stanley Hall of Clark University, Edward Thorndike of Teachers College, and James McKeen Cattell of Columbia University all held tenaciously to the "variability" theory of male superiority. As successful male scientists, they had such strong faith in the fairness of modern social institutions that they considered rankings of eminence based on achievement or positions held as valid indicators of sexual differences in ability. They did not see, let alone give credence to, the possibility that some form of systematic discrimination might be taking place around them or that they might even be contributing to it.

The variability theory was detrimental enough to women's aspirations when it was communicated in professional writings and textbooks, but these men sought to apply it to the burgeoning world of educational and scientific administration, in which they were prominent participants, as well. If statistical studies of the past claimed to have proven women to be of only middling ability (when they had actually found only a few visible achievements), then, these men concluded, there was no reason to think that present or future women would be any more successful in reaching the top ranks of the professions. Administrators interested in efficiency were thus fully justified in channeling ambitious women students into less demanding forms of training—away from medicine and into nursing or away from graduate school and into schoolteaching—since "science" had shown that their chances for successfully completing these lesser programs were greater. Such restrictions were obviously for their own good and would save those who aspired too high from much unhappiness; needless to say, they would also, if enacted, put an end to any further female encroachments on traditional areas of "men's work," including the professoriate.[12]

In this increasingly politicized atmosphere, quantitative (and so "objective") evidence data on sex differences were eagerly seized upon as evidence for or against women's equality. The most important producer of much of this evidence in the early 1900s was James McKeen Cattell, who not only was a professor at Columbia and the editor of *Science* but also, since 1900, had been a specialist in large statistical studies of various elites. One such study in 1903, "A Statistical Study of Eminent Men," noted that there were only 32 women in his top 1,000 persons (as measured by length of entry in certain biographical directories). Eleven of these were royalty, whose claims to fame were thus strongly determined by heredity and marriage, but only 10 women (versus 72 men) were "eminent" in literature, and only 1 (Sappho of ancient Greece) had excelled in poetry or art. Although these findings surprised Cattell, who had expected to find more women in the arts, they did not cause him to suspect his methodology of any innate biases. Instead he asserted that they reconfirmed the variability theory, despite Karl Pearson's recent challenges of it.[13]

Nor was Cattell particularly supportive or encouraging about women's future in those fields in which they were successful, as in schoolteaching, if a 1909 article on "The School and the Family" is any indication of his true feelings. There, on the basis of no evidence, scientific or otherwise, he unleashed a virulent diatribe against the high proportion of women teachers in the public schools:

> The vast horde of female teachers in the United States tends to subvert both the school and the family. The lack of initiative and vitality in our entire school system is appalling. The influence of our half million teachers on the problems of democracy and civilization is entirely insignificant. The attractive and normal girls, and the few able men tend to drop out, leaving the school principal, narrow and arbitrary, and the spinster, devitalized and unsexed, as the dominant elements. Boys get but little good from their schooling and leave it when they can. Girls, who need men teachers even more than boys, predominate in the upper classes. Women are good teachers, especially young girls with their intuitive sympathy for children and mothers who have bred children of their own, and women are cheaper than men of equal education and ability. But the ultimate result of letting the celibate female be the usual teacher has been such as to make it a question whether it would not be an advantage to the country if the whole school plant could be scrapped.[14]

Since Cattell showed so much anxiety about the feminization of schoolteaching and so little understanding of the problems actually facing the aspiring woman teacher in 1909—she had few other job prospects, was paid much less than men for the same work, and would in most cities (including New York City) be forced to resign if she married—he could not be expected to be particularly sympathetic to women's advancement in science. But within two years he had changed his views and tone considerably.

In 1910 Cattell published the second edition of his biographical directory, the *American Men of Science*. Its count of the number of women "contributing to science" had jumped one-third in four years, from 149 in 1906 to 204 in 1910, when it accounted for about 3.5 percent of the total. But Cattell was not interested in this overall growth and did not mention it in his preface or in a statistical article about the new directory in *Science*. Instead, being interested in elites, he focused exclusively on the top 1,000 scientists, as ranked by a selected group of their colleagues and to whom he gave a highly coveted "star." Although 19 women had been included among the top 1,000 scientists in his first edition in 1906 (1.9 percent), this achievement had passed largely unnoticed. But in 1910 interest in the women's performance had increased enough for Cattell to take special notice of it. When only 18 women, or 1.8 percent, ranked in the top 1,000, Cattell was puzzled and claimed that he could not understand the women's unexpectedly poor showing, since they now had good opportunities for both scientific education and employment. But rather than seeing this statistic as evidence of how discriminatory or "unfair" science was to women, Cattell made the suggestion, typical of the biologically oriented sociology of the time, that the reason for these low rankings must be the genetic inferiority of the female sex.

> There are now nearly as many women as men who receive a college degree; they have on the average more leisure; there are four times as many women as men engaged in teaching. There does not appear to be any social prejudice against women engaging in scientific work, and it is difficult to avoid the conclusion that there is an innate sexual disqualification. Women seem not to have done appreciably better in this country than in other countries and periods in which their failure might be attributed to lack of opportunity. But it is possible that the lack of encouragement and sympathy is greater than appears on the surface, and that in the future women may be able to do their share for the advancement of science.[15]

The naiveté in this logic is startling. For example, Cattell did not mention that by refusing to award stars in the field of home economics he might have been diminishing the women's contribution unduly. Nor did he report what now seems apparent to one examining the ranking sheets submitted by the scientists that are to be found in the Cattell Papers in the Library of Congress: those male scientists who included any women at all in their rankings tended to cluster them together at or near the bottom of their lists, a practice that resulted in lower scores for the women than might otherwise have been the case. (Nor did he point out the clear sex differences among rankers—the few women consulted put more women on their lists and ranked them higher than did the men in the same fields.)[16] Thus visibility and a certain interaction with those doing the ranking (generally men), as well as good work, were necessary for inclusion in the increasingly competitive top 1,000, but Cattell seemed unaware of this in 1910. Nor did he seem at all aware that such visibility and interaction were being systematically denied to those scientists restricted to "women's work."

Although such "explanations" of the women's poor showing might have been considered acceptable earlier, by 1910 the feminist doctrine that women were the equal of men and that all social and artificial barriers to this equality should be removed was becoming a full-scale social movement in the United States. It was typical of the new interest in this topic among professional women that at least two prominent women scientists, sensitive to Cattell's charges, wrote immediately to him at *Science* to challenge them. (Others may also have written, but their letters were not published.) Both Marion Talbot, who as professor of household administration and dean of women at the University of Chicago, may also have been miffed at not being eligible for a star, and Ellen Hayes, the radical professor of mathematics at Wellesley College, were surprised and distressed that there had not been a sizeable increase in the number of women in the top 1,000 and that all the recent efforts to help women in science had brought them less than 2 percent of the honors. But these two professors had a different explanation for the women's poor showing and offered numerous social and "environmental" arguments that have since become the standard feminist defense of embarrassing statistics. Professor Hayes made the point that few women entered science because in childhood they had been taught to value their appearance and apparel more than science or learning. Even worse than this early socialization, as both women stressed, and the chief reason why women in science did not become eminent, was the discrimination in hiring practiced by the male and coeducational institutions that controlled the "major positions" in science; they added that "by major position is meant one that a man of the select first thousand would be willing to occupy." Women were "quite welcome," Hayes claimed, "to become experts in washing bottles and adding logarithms and dusting specimens" at these prestigious institutions, but only rarely, if ever, did they go any further. In fact, she concluded sardonically: "There is indeed room for doubt whether we should have any thousand men of science if all gifted and ambitious young men were confronted by such barriers as a young woman is obliged to face today. . . . It appears difficult therefore to avoid the conclusion that other factors besides innate sexual disqualification must be reckoned with in attempting to account for the insignificance of women's share in the advancement of science."[17]

Cattell reprinted their letters without any comment. Evidently the women had convinced him of their view, however, since a month later he reprinted an article from the *New York Post* about Marie Curie's prospects for election to the French *Académie des Sciences*. The article was clearly on her side ("we cannot help feeling that it is at most only a question of time when Mme. Curie's admission will be effected") and concluded by relating the situation of this first woman Nobel Laureate rather cynically to that of women scientists everywhere: "If she is admitted, there will be one woman, out of the handful that devote their lives to scientific research, distinguished by one of the highest scientific honors; if she is kept out [as she was], it will be one more proof of the immeasureable difference between the degree of encouragement and incentive held out to women and that held out to men for sustained devotion to strenuous intellectual labors."[18] Cattell had thus, to his credit, shifted his views rather quickly in 1910 and 1911, and after this he became a strong supporter of feminism and women's advancement in science, printing and reprinting many articles on these issues in his two journals, *Science* and the *Popular Science Monthly* (later renamed the *Scientific Monthly*), and personally helping many women psychologists, such as Mary Calkins and Christine Ladd-Franklin, to enter more fully into informal professional activities. He thus joined Remsen and Welch in the rank of influential men scientists who were sympathetic to at least one aspect of the "women's movement" by 1911.

Meanwhile the less receptive editors of the *Scientific American* commented on Cattell's controversy with his feminist critics in an article entitled "Women and Scientific Research." They agreed that the women's 1.8 percent of the total top 1,000 scientists was not a "fair" proportion, whatever the reasons for it. They did not agree with Ellen Hayes that attitudes implanted in girls during childhood were particularly important in discouraging them from entering science, but they did think that sexual discrimination in university hiring was a major factor later in their careers. Yet in trying to explain the reason for this circumstance, the editors Charles Munn (Princeton, 1881) and Frederick Beach (Yale [Sheffield], 1868) hastened to defend their sex by blaming the women for being "willing enough victims" and to suggest that their poor showing was inevitable, since men and women were after all destined for different "spheres": "But one cannot quite shake off the feeling that the fault, if fault there be, does not rest entirely with the male sex. If women are in this matter the victims of circumstances, one rather suspects that most of them are willing enough victims. . . . There are some walks of life, some missions, which are eminently woman's sphere. The pursuit of scientific research is perhaps not one of these."[19]

This whole controversy over the correct interpretation of Cattell's statistics was both the beginning of a prolonged examination of women's status in science and academia that would last into the 1920s and the culmination of a series of such attacks on women's growing share of college enrollments, especially at coeducational schools. When Marion Talbot sent her letter to *Science* in 1910, she had already been waging a decade-long struggle at the University of Chicago to preserve full coeducation and prevent quotas at the student level while fighting for more fellowships, appointments, and promotions for women at the graduate and faculty levels. Undergraduate women at Chicago had become so numerous that by 1902 they constituted one-half the student body, an alarming jump since 1892, when they were one-quarter of the total. Administrators, afraid of feminizing learning entirely, put sharp limits on the women by organizing a separate college for them within the university. Because of her experience as an embattled dean of women at an initially liberal but increasingly conservative coeduational university, Talbot not only was familiar with the limits being put upon women's achievements by frightened male administrators but also was growing angry. She had, as Cattell and his friends had not, externalized the phenomenon and could see the women's few honors as the direct result of systematic discrimination rather than of hypothetical female inferiority. By the time Cattell's article appeared, she had already written a book, *The Education of Women* (1910), which she concluded with a call to action: "Even allowing for alleged differences in capacity and availability, the discrepancy in academic treatment is unjustifiable. It is time that the full facts should be made known and some of the needless barriers removed from the path of women scholars."[20] Like other Progressive reformers of the time, Talbot thought that the first step should be the collection and dissemination of data on the extent of the problem. Although she did not specify in 1910 how this would lead to the removal of the "needless barriers," within two years she had devised the first organized protest of such discriminatory practices.

A year later Susan Kingsbury, professor of economics at Simmons College, presented quantitative data on the number and rank of women on college faculties in the *Journal of the Association of Collegiate Alumnae*. She discovered that although women were now being hired more often than earlier by coeducational schools, they were almost always restricted to the lowest ranks or to the department of home economics. In short, rather than proving to be a merely temporary phenomenon, hierarchical and territorial forms of "women's work" were spreading to more and more institutions. Kingsbury's methodology was straightforward: she wrote to women faculty members at the twenty-three institutions that were members of the ACA (those whose alumnae met its membership standards). Of her 147 respondents, 56 (38.1 percent) were in science; 23 of them (41.1 percent) held positions in coeducational schools. Of these 23, however, only 5 (21.7 percent) were full professors, and they were chiefly (60 percent) in home economics. Almost one-half (43.5 percent) were at the bottom of the academic ladder as instructors.[21]

In 1911 Charles H. Handschin presented an even more dismal picture in a much larger study published in *Science*. Handschin drew his data on 7,960 faculty members at eighty-one coeducational institutions from the college catalogs at the U.S. Bureau of Education and found that of these almost 8,000 persons just 717 (9 percent) were women. He did not have data on the faculty rank of all these women, but he did present a breakdown of a sample of 149 of them. An overwhelming 73 percent were instructors, and just 19 percent were full professors (the 3 in science were in home economics).[22] Thus, even though these early "status of women" reports lacked a great many controls (such as age, degrees, years of experience, publications, and comparisons with control groups of men), the evidence presented seemed to indicate that women were not being given an equal chance for employment and advancement, particularly at coeducational institutions, and that the only field in which they could realistically expect to rise to the position of full professor was that of home economics.

At the same time that Kingsbury and Handschin were counting women faculty members, Gertrude Martin, the "women's adviser" at Cornell University, was presenting her data on the deans of women at fifty-five universities to the ACA.[23] Her finding that, although they were required to hold a doctorate, these women were no longer considered faculty members confirmed Marion Talbot's suspicions of the widespread decline in their status. A year later, in 1912, Talbot presented the executive committee of the ACA, an organization she had helped to found thirty years before, with a suggestion of how that group might pressure colleges and universities to improve the status of their women. The committee members approved her call to stiffen the ACA's own membership and accreditation policies to require that applicant schools provide at least minimal status to the women on their faculties and governing boards before their alumnae could be admitted to the association. These new stipulations included a "reasonable recognition of women in the faculties," the presence of women on the boards of trustees, faculty status above the rank of instructor for the dean or adviser of women, and, perhaps the most radical of all, "approximately the same salaries [for women] as men of the same rank."[24]

Yet her committee on admissions had to report a few months later that it had been unable to implement these requirements for membership or, as she said, "We found that these tests needed interpretation." They had had to decide that "reasonable recognition" of women on faculties meant only that "there should be at least one woman above the rank of instructor, and at least one woman giving instruction in regular college subjects," and, "in regards to salaries, we took the position that the salaries paid to women shall not be conspicuously less than those paid to men, i.e., the highest paid to a woman shall not be less than the lowest paid to a man." With these weakened provisions the committee could recommend the admission of seven new institutions, including Indiana University and Swarthmore, Grinnell, and Mount Holyoke colleges. To have enforced these new rulings any more rigidly would have greatly limited the number of new schools accepted and would have led to the disqualification of several others already admitted. Thus the sentiment was strong at the ACA in 1912 that it should do something to prod or coerce coeducational and other institutions into hiring, promoting, and paying women faculty equally. But as a private alumnae organization without full accreditation powers, it lacked the leverage to enforce its rules effectively. Despite this weakness, however, the ACA did not give up its criteria for admission. Annually into the 1920s, every time its Committee on Recognition presented its list of institutions approved for membership, the governing boards of the ACA discussed the problem of gaining greater compliance. In some years members questioned the employment practices of specific universities, such as Pennsylvania State University in 1919 and West Virginia University in 1921. Acceptance was by no means automatic, since in 1922 the ACA (newly renamed the American Association of University Women [AAUW]) accepted only five of the sixteen schools that had sought admission.[25]

Meanwhile in 1917, the leaders of the ACA, apparently dissatisfied with their tactic's inability to change academic hiring practices, formed a new committee to reexamine the continuing problem. This group held a discussion in 1918 at the ACA's biennial "Joint Conference of [Women] Trustees, Deans and Professors." Some of the participants at that series of meetings then prodded the recently formed American Association of University Professors to form a new committee to investigate the status of women at academic institutions systematically.[26] (The two reports of this "Committee W" in 1921 and 1924 are discussed more fully in chapter 7.) Thus in just a few years Talbot and others had taken Cattell's statistics and naive speculations about female inferiority, reformulated the issue in terms of an unjustified discrimination in employment that had to be eliminated, and mobilized the ACA and AAUP to study and battle it. Women, according to this new view, did not have the "unique" or "special" skills that had earlier justified jobs as good subordinates or adept home economists. Instead, they were fully equal to the men in science, and it was widespread and systematic discrimination that was keeping them from rising any higher or entering other fields. All discriminatory beliefs and practices should be exposed and abolished, though just how remained unclear.

Meanwhile other feminists besides Talbot were working to change public opinion about women's abilities in various other scientific and quasi-scientific ways in the 1910s. Interest in feminism was particularly intense in the liberal and intellectual circles of New York City, especially at Columbia University (including Teachers College), Greenwich Village, and the *New York Times*. The New York feminists were not involved in Talbot's protest movement through the ACA, but took another tactic of working to liberalize the social thinking, including several social sciences, that justified such discrimination. Among the cluster of active women at Columbia after 1910 was the vigilant Christine Ladd-Franklin, now in her sixties but as outspoken as ever. (In 1920 she suggested to Cattell that he change the sexist title of the *American Men of Science*, "but do say *Men and Women of Science*, or *Scientists*, or, like the English, *Scientific Worthies!*" His successors did not make the change until 1971.) She became a lecturer in the Columbia psychology department in 1909 (when her husband joined the *New York Post*), and continued to write fiery letters to the editor and signed and unsigned book reviews on women and women's history, especially for *Science* and the *Nation*. Meanwhile one of Cattell's students, Cora Sutton Castle, wrote her 1913 dissertation on "A Statistical Study of Eminent Women," using his methods to rank 868 women of the past on the basis of the length of their entries in a variety of encyclopedias and directories. She was apparently pleased to find so many well-known women, but her sample, heavily skewed toward royalty and writers, failed to reach any conclusions about the key issue at hand—whether it was biological or social conditions that had limited women's achievements.[27]

Of wider impact were the several books and numerous magazine articles by the wealthy dowager, ethnologist, and feminist Elsie Clews Parsons, formerly a lecturer in sociology at Barnard, who supported many of Franz Boas's anthropological enterprises at Columbia. Her many articles on such topics as "Feminism and Conventionality," "Feminism and Sex Ethics," and "Anti-Suffragists and the War," and especially her book, *The Old-Fashioned Woman: Primitive Fancies About the Sex* (1913), all acclaimed the new feminism as relieving modern middle- and upper-class women from the social pressures to lead idle and empty lives, a tendency she and other professionally inclined women had successfully resisted. Her technique was the largely satirical one of collecting descriptions of the role of women in various tribal societies and then using this comparative data on sexual customs, roles, and taboos to ridicule many current attitudes and pressures in American life, thus showing that they were simply customs and prejudices, too, and not the "innate," "genetic," or unchangeable characteristics that ethnocentric persons (including most scientists) claimed they were. (Margaret Mead would develop this "cross-cultural" viewpoint further in the 1930s.)[28]

Meanwhile the new feminist view of women's nature was rapidly gaining support from a few other prominent educators and scientists, including the rather unlikely Father John Zahm of Notre Dame University and Simon Flexner, director of the prestigious Rockefeller Institute for Medical Research in New York City. Flexner was less of a surprise since he was, not coincidentally, a protégé of William Welch at The Johns Hopkins Medical School and husband of Helen Thomas, the youngest sister of M. Carey Thomas.

Published in 1913 and immediately linked to the feminist cause, although undertaken many years earlier with no such intention in mind, was *Woman in Science*, by H. J. Mozans (pseudonym), a history of women scientists to about 1900. Although the book had a foreign flavor and contained only a few Americans, it was written by a prominent liberal American priest and educator, Father John Zahm of Notre Dame University. Inspired while on vacation in Athens in 1896 to write a history of women of achievement, Zahm had finally decided to limit his book to science, a field that, unlike art, music, or literature, had seldom been studied in light of women's contributions. In the course of his researches Zahm became convinced of the validity of the feminist viewpoint, persuaded that it had been limiting social conditions and the lack of education and opportunity that had hindered women's achievements rather than any overall lack of ability. The book received favorable reviews from all quarters, including one from the unlikely source of former President Theodore Roosevelt, a man not known for his feminist sympathies but a personal friend of Father Zahm from their travels and hunts together in South America. Even Roosevelt apparently enjoyed the book, reporting himself quite pleased and relieved to find that so many women of accomplishment had not forsaken their primary duty to raise their families and take care of their husbands.[29]

An indicator of the amount of interest in feminism among prominent male scientists at this time and, incidentally, of the pitiful state of social science, which lacked and sorely needed the concepts of "sex typing" and "stereotypes," was a series of three full-page articles in the *New York Times Magazine* in 1913 and 1914. Setting off the debate was a provocative interview with W. L. George, an outspoken British male feminist, that was reprinted from the *Atlantic Monthly* under the title "What the Feminists Are Really Fighting For." George's point was that the feminists wished to erase the concepts of "men" and "women" from society and eliminate what is now known as the "sexual stereotyping" of roles, occupations, emotions, and all human activity. To this

Professor William T. Sedgwick of MIT responded a month later in a rather hysterically antifeminist vein that was reminiscent of the 1870s. He had evidently missed the distinction they made between women's biological and social roles, for he asserted that if the feminists succeeded in their project of eliminating "men" and "women" from society, then human reproduction and therefore all of life would stop. A month after this pronouncement six feminists, Carrie Chapman Catt of NAWSA, and five male scientists and professors (Simon Flexner of the Rockefeller Institute, Professors William H. Howell and Franklin P. Mall of The Johns Hopkins University Medical School, Frederick Peterson of Columbia's College of Physicians and Surgeons, and James Harvey Robinson of the History Department at Columbia) responded that women seemed to them as capable as men, and that many women had in fact already learned professional skills without the serious repercussions that Sedgwick feared.[30]

The major new experimental research on sex differences in the years 1913 to 1916 was the work of Leta Stetter Hollingworth, a graduate student in psychology at Columbia University and the first Civil Service psychologist in New York City. Barred from teaching in the public schools because she was married, Hollingworth became a committed feminist, determined to disprove the pernicious "variability" theory and other sexual stereotypes that were still, a decade after Helen Thompson Woolley's pioneering work, emanating from such influential sources as Teachers College and Columbia University. She was greatly encouraged in her work by her husband (Harry Hollingworth, a graduate student and later professor of psychology at Barnard College and Columbia University) and her feminist friends in Greenwich Village, to whom she gave frequent progress reports. (One member of this informal club, a journalist, was so impressed with Hollingworth's work that she described it in a lengthy article in the *New York Times* in 1915; another did so in the *Masses,* the Village's radical newspaper in 1916.) The result of Hollingworth's researches was a series of articles between 1913 and 1916 describing her strictly controlled quantitative studies of a variety of physical, motor, and intellectual traits among newborn babies of both sexes, college-age men and women, and women during menstruation. In these articles and in a more general overview of the problem ("Science and Feminism") which she and anthropologist Robert Lowie published in Cattell's *Scientific Monthly* in 1916, Hollingworth presented her own data and questioned that of others who had claimed to prove male superiority. She asserted that, contrary to what "armchair psychologists" believed, there was no clear-cut evidence either way, and when there were minor differences, they were the result of social forces, conclusions that most responsible studies since then have tended to confirm. Hollingworth also, like many feminists since, felt strongly the responsiblity to use her psychological findings to rebut, or at least to argue against, the perpetration of sexual stereotypes in employment by vocational guidance counselors. In 1916, the high point of her reformist writings, Hollingworth not only contributed a chapter to her husband's book on vocational psychology (in which she reminded counselors that women need not be restricted to "women's work"), she also published an article in the *American Journal of*

Sociology in which she ridiculed the concept of the "maternal instinct" and justified, but did not quite advocate, birth control. (She never had any children.) Yet not even these activities upset or offended Edward Thorndike and his colleagues at Teachers College. Thorndike respected her work on variability and invited her to lecture on sex differences to his classes, and in 1916, when psychologist Naomi Norsworthy, the first female (and possibly only Jewish) faculty member at Teachers College, died unexpectedly, he appointed Hollingworth to her place. She then went on to make major contributions to the study of gifted children, an important but far less controversial subject.[31]

There was thus by 1916 a growing and increasingly vocal scientific feminism that used anthropological and especially psychological data to undermine and discredit traditionally antifeminist social science, usually on the grounds of inadequate logic or methodology. Henceforth when old-fashioned psychologists persisted in publishing books and articles on such fictions as "The Feminine Mind" (as did Joseph Jastrow of the University of Wisconsin in 1918) Hollingworth, as had Woolley earlier, chastised their sloppiness, often sarcastically, in biennial review articles in the *Psychological Bulletin*. In this way the relatively few feminist psychologists were able to help chip away at the basic tenets of the old "armchair psychology" of William James, whose basic concept of human "instincts" was overthrown by experimentalists around 1920. The feminists thus played an important, though generally unacknowledged, part in the transition from the biological or hereditarian psychology of the early 1900s to the more "environmental" social science of the 1920s; this was several years before Franz Boas and Ruth Benedict popularized the liberal notion of "cultures" and the excesses of the eugenics movement turned many other social scientists away from rigid racial or hereditarian doctrines.[32]

It is somewhat ironic that, just when feminism was having modest success in helping to liberalize the social sciences, it was having almost no visible impact on the leaders of the women's suffrage movement. These leaders may have been aware of what the intellectuals and scientists were proving experimentally, but they were moving more and more to the right in order to secure passage of, first, state laws, and then, when they proved insufficient, a federal constitutional amendment. Despite this inconsistency, which only Margaret Washburn saw, a few women scientists held major positions in the suffrage organizations, and many others defied taunts and slanders (like H. L. Mencken's that only "ninth rate men," if any at all, would "fall in love with lady physicists, embryologists, and embalmers" who were suffragists) to work in the numerous state and federal campaigns in the 1910s. It is not clear what they thought the vote could do for them as women scientists, but as professional women with higher degrees and often responsible jobs, they must have seen the absurdity in the fact that uneducated and even "feeble-minded" men could vote but they could not. Thus entomologist Adele Fielde wrote pamphlets and gave speeches in Seattle in 1910; Lillien Martin, professor of psychology at Stanford University, became the third vice-president of the northern California branch of the National College Equal Suffrage League in 1911, the year of an important victory for women's suffrage in California; and Helen Brewster Owens, a mathematics teacher in Ithaca, New York, where her husband was a professor, left her job for a year to organize the successful "votes for women" campaign in Kansas in 1912 and again, though unsuccessfully, in New York in 1915. In 1913 flamboyant civil engineer Nora Blatch de Forest (a third-generation suffragist, descended from Elizabeth Cady Stanton herself) rode a horse across New York State to campaign for votes—and gained thereby as much attention for her "culottes" as for her cause (see plate 17).[33]

In the later campaigns, psychologist Leta Stetter Hollingworth was a poll-watcher for the Woman's Suffrage Party in New York City and marched in suffrage parades with her husband in 1917; psychologist Helen Thompson Woolley, director of the vocational bureau of the Cincinnati public schools, organized suffrage groups there in 1919; and anatomist Florence Sabin wrote hundreds of letters to Maryland legislators and marched in parades in Baltimore and Washington, D.C., with her friend Goucher physiologist Lillian Welsh. (Sabin later named her first automobile the "Susan B. Anthony.") The most radical was the widow Agnes Chase, a botanist at the U.S. Department of Agriculture and the world's expert on grasses, who not only picketed the White House regularly during the Wilson administration but also was twice arrested and jailed, first in 1918 for making a speech on women's rights in Lafayette Square, and again in 1919 for maintaining a fire in which she intended to keep burning "all Presidential speeches mentioning liberty or freedom until women were permitted to vote." Even the retired women scientists took an interest; Mary Whitney, immobilized from a series of strokes, followed the movement closely, and Susan Cunningham, professor emerita of mathematics at Swarthmore College, was reportedly an "ardent suffragist" until her death in 1921.[34]

A significant factor behind the passage of the Nineteenth Amendment by Congress in 1919 and its rapid acceptance by two-thirds of the states by 1920 was the contribution American women made to winning World War I in 1917 and 1918. Some women took on temporary war jobs, filled in elsewhere for men serving in the military, or did volunteer work at home and abroad. Since the suffragists supported the war loyally while also keeping up their political pressure on Congress and the president, it proved impossible to deny women the vote for long after the war. For women scientists, the war not only opened up practically their first jobs in industry, it also expanded their employment in government and created considerable activity and turmoil on the college campuses. Yet for many reasons the way in which women scientists were utilized in World War I reflected and, it could be argued, even increased the prevailing sexual segregation in scientific employment. Despite incursions into a few areas of "men's work" that were facing desperate shortages,[35] such as the chemical industry, most of the women scientists who served were called into traditional and new

PLATE 17. Engineer Nora Blatch de Forest, an ardent suffragist, is shown here campaigning for the vote in New York City in 1913. (United Press International.)

areas of "women's work" related to the war effort. Not even the Greenwich Village feminists, many of whom became pacifists, criticized this pattern, and their silence is remarkable, considering how sensitive they had been to such limitations as recently as 1916. Most women took this channeling for granted and were thrilled to serve the nation in any capacity.

Since the U.S. government had made little preparation for the war and had not given either industry or its own scientific agencies any advance notice of potentially large manpower needs (in fact, Woodrow Wilson had deliberately lulled the nation into thinking that intervention would not be necessary), the sudden recruitment and emergency planning in the spring of 1917 was chaotic. Because no priorities had been established for the optimal use of technical manpower, certain industries and agencies competed feverishly for the limited men available. Particularly intense was the sudden demand for industrial chemists, not only to increase the American production of war materiel (such as gases and explosives), but also to make up for the cessation of chemical imports from Germany on which the United States had been very dependent. A new American industry was in the making, but there were not enough chemists—male chemists, that is—available in 1917 to get it established. Under these conditions women chemists, who had formerly been excluded from industrial fields and shunted systematically into home economics, were suddenly urged to enter industry. Although few actual numbers are available, later reports indicated that many companies did hire women chemists in 1917 and 1918.[36]

Opinion at the time was largely favorable as to how well the women chemists did the men's jobs. Excerpts from a positive report made by William H. Brady, chief chemist at the Illinois Steel Company, on his seven women chemists were reprinted in the June 1919 issue of the *Journal of Industrial and Engineering Chemistry* under the optimistic title, "The Woman Chemist Has Come to Stay." Brady felt that their performance had been at least as good as that of the men in their groups—they learned as quickly, did their full share (and more) of the work assigned, took the night shift willingly, were *less* often sick, and, in general, "They have added tone to our laboratory by their pleasing personalities. They have proved beyond a doubt that they can do and will do at any hour of the day or night, careful, conscientious, reliable, chemical work. They have passed the crucial test of service. They have been weighed in the chemical balance and not found wanting."[37] Yet even this model behavior by one group of women chemists during the war did not lead to any lasting improvement in women's position in the chemical industry, for just two months after Brady's article appeared, his own supervisors at the Illinois Steel Company overruled his report and announced, with evident relief, in a small news item in the same journal, "The women chemists of the Illinois Steel Company not only made good as chemists but showed their fine spirit by resigning in order to make places for the men returning from war work."[38] These layoffs were just one sign of the backlash under way by 1919 against women chemists in industry. (See chapter 9.)

Although the chemical industry was sufficiently desperate in 1917 to seek young women for temporary jobs, the federal government's recruitment for its wartime projects seems to have been limited to having scientists contact their friends, colleagues, and former students through classic kinds of "old-boy" networks. Most of these recruits were necessarily men, not only because the workplace was already segregated, but also because most of the government science projects in World War I were within the various branches of the armed forces and functioned by inducting (male) scientists into the military and giving them an officer's rank and uniform. Under these conditions it is perhaps surprising that there were any women scientists on the now famous government programs that were started and developed during World War I. Thus although there do not seem to have been any women at either the submarine detection task force at the New London, Connecticut, navy base or the new Chemical Warfare Service at Edgewood Arsenal, Maryland, the National Bureau of Standards hired its first woman physicist, Louise McDowell of Wellesley College, to work on radar in 1918. A few other women scientists worked on army projects, including physicist Frances Wick of Vassar College, who was probably the first woman scientist at the United States Army's Signal Corps when she worked there on airplane radios and gun sights for a few months in 1918, and Louise Stevens Bryant, holder of a University of Pennsylvania Ph.D. in medical science, who wrote statistical reports for the Office of the Chief of Staff from 1917 to 1919 on the food supplies available for both the army and the Allies. Thus the army, which would not hire women as officers in World War I, found a way to hire them as civilian assistants instead. This pattern is a good example of the type of phenomena that Elsie Clews Parsons had described earlier in her 1915 article "Anti-Suffragists and the War," where she had said that in war, as in every sphere of activity, the men did not want the "interference" or competition of women coworkers, but would accept their assistance.[39]

Parsons's analysis also held true for the military's underutilization of women psychologists in World War I. All of the top jobs in the famed Army Psychological Testing Program went to male professor-officers, including many, like its director Robert Yerkes of Yale University, whose previous specialty and experience were far removed from mental testing, which had in peacetime been allowed to become a highly feminized subdiscipline. Nevertheless, only two of his large staff were women: Mabel Fernald and Margaret Cobb, both staff psychologists at the highly regarded but also very conservative Social Hygiene Laboratory of the Bedford Hills Reformatory in New York State. Both were listed as mere "assistants." None of the other fifty-nine women psychologists listed in the 1921 edition of the *American Men of Science,* including such luminaries as Margaret Washburn, Mary Calkins, Helen Thompson Woolley, Leta Stetter Hollingworth, and Grace Fernald (Mabel's sister), who had headed the psychological laboratory at the California State Normal College (later UCLA) for almost a decade, was associated in any way with this project that was reportedly transforming the field. This exclusion had serious consequences, not only for later women in psychology, but also for government social policy in the 1920s. Women psychologists suffered subsequently from their lack of participation in this project, since most of the postwar leadership in science, at the National Research Council and elsewhere, grew out of such contacts developed in Washington during the war. Serious as this was, government policy suffered even more from the absence of such careful testers and politically liberal individuals as Woolley and Hollingworth, for it is hard to believe that they would have devised the same faulty tests or reached the same racist conclusions that Yerkes, Thorndike, Lewis Terman, and Edwin Boring did—such as that one-half of the army's recruits were "feeble-minded" or that this deficiency was concentrated among immigrants and blacks. These findings, which Yerkes publicized widely, had a deleterious impact on immigration and other social policies during the next decade.[40]

The situation may have been somewhat better for women in the tiny field of geography during and after World War I. Like psychology, this was a new social science, one whose opportunity to serve the nation in wartime opened a broader and more visible field of usefulness for it in the postwar period. Though information on its contributions is scanty, a team of geographers named "The Inquiry" worked in 1919 at the American Geographical Society in New York City under the direction of Colonel Edward M. House to prepare reports for the American delegates to the Paris Peace Commission. This important group included two women: graduate student Helen M. Strong, later the first woman doctorate in geography from the University of Chicago, and the fifty-six-year-old Ellen Semple, an expert on the Mediterranean region, whose reports on the Austrian-Italian border and the Turkish empire were read by President Wilson.[41]

The area in which the women scientists made their largest overall contribution to the war and the one field in which they were ranked higher than assistants was that of "home economics." This occupation was, as described in chapter 3, already highly feminized by 1917 when the United States entered the war and when Herbert Hoover, chairman of the (U.S.) Commission for Relief in Belgium, was predicting serious food shortages in Europe in the year ahead. In August of that year President Woodrow Wilson, after a long debate in the Senate, established the U.S. Food Administration as a temporary agency to supplement the Department of Agriculture and to use every means possible (short of rationing) to increase domestic food production and decrease consumption. Increasing production required a staff of (male) lawyers who wrote economic agreements with all branches of the food industry, but decreasing domestic consumption required the best efforts of a staff of (female) home economists who were accustomed to communicating with housewives. But rather than appointing a woman to head the Food Conservation Division, Hoover, the new Food Administrator, appointed his friend, Stanford University President Ray Lyman Wilbur, who in turn appointed two women chiefs of the Home Conservation Division: Sarah Field Splint, editor of *Today's Housewife*, and Martha Van Rensselaer of Cornell University's School of Home Economics. Still the highest ranking women scientists in the U.S. government during the war,* they directed a staff of eight other home economists which included Flora Rose, also of Cornell, Katherine Blunt of the University of Chicago, Isabel Bevier of the University of Illinois, and Abby Marlatt of the University of Wisconsin. Mary Pennington, a refrigeration expert who had been at the USDA, also served as a consultant to the USFA, and Mary Swartz Rose was deputy director of its Bureau of Food Conservation for New York City. As such, she published *Everyday Foods in Wartime* (1918) and put on exhibits and demonstrations in the city, including one on the steps of the New York Public Library in January 1918. All these women nutritionists apparently did their jobs well, since American food consumption dropped about 10 percent in 1918 and thus freed much food for export to the war-ravaged Allies; this was a particularly crucial achievement that year, since a nationwide drought caused failure in the American wheat and corn crops.[42]

Nevertheless, there seems to have been considerable friction between Hoover and his suffragist critics in Washington. They were resentful that he, a mining engineer by training, had been put in charge of the Food Administration, which they considered a woman's sphere. They were also incensed at his assumption that it was the women of the country who were wasting the food in the first place, and they disliked many of his methods for reducing their consumption, especially that of forcing housewives to sign pledge cards. They would have preferred to

*Julia Lathrop, chief of the Children's Bureau since 1912, was the highest-ranking woman, but she was by profession a social worker rather than a scientist.

reach the women in another way, but as Anna Howard Shaw, the suffragist leader in Washington during the war, cynically put it, they went ahead, attempting to "do the impossible, and as usual, doing it."[43]

Yet the home economists did not complain. This may be an indication that the women in some scientific fields were politically more conservative, or at least less likely to be feminists or even suffragists, than the women in other specialties. In home economics the conservative heritage went back to Ellen Richards, who had opposed suffrage in the 1870s and advocated instead more cautious, nonthreatening ways of moving women ahead. By accepting the idea of separate "women's work," the home economists had risen higher in the government than women in other fields, but to feminists this was at the high price of accepting segregation and implied inferiority. Tantalizing as such patterns of political self-selection may be, the information available (at present) is too scanty for any reliable conclusion.[44]

Meanwhile, World War I had also created a great deal of excitement and activity for women scientists on the college campuses, mostly in the area of war-related courses and new training programs in "women's work." Several women's colleges offered courses on nutrition and food preservation, industrial chemistry, bacteriology, mapmaking, and wireless telegraphy. In addition, Vassar organized a training program for nurses with Professor Florence Sabin of The Hopkins on the faculty, and Smith College sent its own "Reconstruction Unit," famed for its special psychiatric services, to France in 1918. (This was the forerunner of the Smith College School of Social Work and the field of psychiatric social work.)[45] Mount Holyoke trained women inspectors of health conditions in industrial plants, and Barnard College took the lead in recruiting leaders and students for the Woman's Land Army, an affiliate of the U.S. Employment Service and the U.S. Food Administration, which sought to replace drafted farmhands with college girls on their summer vacation. Ida Ogilvie, professor of geology at Barnard, was particularly active in this program, speaking on campuses across the country to urge women students to till "war gardens" at school and to spend the summers of 1917 and 1918 as "farmerettes" providing agricultural labor. She even ran a summer camp for them herself in Bedford, New York, and hoped to turn this highly patriotic and romanticized war work into a permanent new occupation for women.[46]

Besides such projects, wartime brought political tensions and even battles on a few campuses when some outspoken women professors were charged with pro-German and other unpatriotic feelings. One such scientist was Mary Fossler, assistant professor of chemistry at the University of Nebraska, who was accused and suspended from the faculty in April 1917 but reinstated two months later when feelings had cooled and the charges had been dropped. Elsie Clews Parsons persisted in her pacifism throughout the war and thereby lost most of her friends. But such causes célèbres seem to have been surprisingly rare among women scientists, when one considers that many of them were pacifists, socialists, prohibitionists, and a great many other "-ists" as well. Perhaps this was a sign of how little their opinions mattered.[47]

Other women scientists, not called to government projects or pressed into training programs but anxious to contribute to the war effort, volunteered their services to the YWCA or the American Red Cross and went abroad as physicians or nurses. Thus Clelia Mosher, M.D., the women's physician and director of hygiene at Stanford University, served abroad as a medical adviser for French refugees for two years; paleontologist Julia Gardner, whose mother had just died, drove an ambulance for the American hospitals unit in France; and geographer Millicent Bingham was a nurse at a YMCA base hospital in Angers.[48] The next world war would have more need of their scientific and technical skills.

By 1920 the women scientists had gone through a rapid series of social and political movements. Feminism had led some of them to challenge old beliefs about women's inferiority, the suffrage movement had called forth active campaigns in many states and the nation's capital, and the war had utilized some of their skills and talents. One result of all the excitement was that for a few years around 1920 many of the women scientists' traditional leaders seem to have been lulled into a complacency or temporary euphoria about their success. They had made so much progress in both education and employment since 1880 that they seemed to think their task had been completed. For example, many members of the Association of Collegiate Alumnae thought in 1916 and 1917 that the time had come to disband their association, since its early task of accrediting women's collegiate education had been taken over by others.[49]

Similarly, a committee of the Naples Table Association met in October 1919 to discuss the group's future (itself a sign of uncertainty), and recommended dissolution because its task of encouraging women in science had been accomplished. This group did not think that a visit by Madame Curie to the United States in the near future could serve any useful purpose, and in 1919 it passed a resolution stating that they did not wish to invite her to come.[50] In a sense this generation's vision of the reforms needed by women for them to enter science had been achieved, and the group was facing retirement and exhaustion. Through their efforts women had achieved a modest presence in science—they now held doctorates, marginal forms of employment, and associate memberships in organizations. Yet the leaders had no effective plan for getting beyond this level. They apparently did not even see, let alone have methods for improving, the worsening postwar employment situation: layoffs were taking place, a backlash was starting, and the statistics on women in academia and government were still not good. The new issues were advancement and recognition, but preliminary attempts to get more of these for women had proven unsuccessful. There was thus around 1920 a certain groping for a new vision, new leaders, and new strategies to take on those very problems that had eluded earlier generations of women in science.

While the women scientists were seeking new leaders and new strategies, another, different, woman planned and executed the most lavish attempt yet to celebrate and capitalize upon the theme of women in science. This was Marie

PLATE 18. Madame Curie's triumphal tour of the United States in 1921 featured this reception by President Warren Harding and a committee of women on the steps of the White House. (Courtesy of the Library of Congress.)

Meloney of New York City, journalist and editor of the *Delineator,* a women's magazine, who visited Paris and Madame Marie Curie in 1920. Aghast at Curie's poverty and poor working conditions, Meloney conceived the idea of an American tour for Curie as a way not only to publicize her achievements but also to raise discreetly a large sum of money to help her continue her work on the medical uses of radium. Other foreign women scientists had visited the United States before (for example, Winifred Cullis, an English physiologist, had come in 1919 to interest Americans in starting the International Federation of University Women), but Madame Curie's visit was to be very different—it was to be a triumphal tour for the "First Lady of Science" and was to capitalize upon the public's interest in cancer and radium and the media's hunger for celebrities.[51]

After gaining Curie's consent to the plan, Meloney began to make arrangements. She appointed committees to raise the money, arranged for more than twenty universities (including Yale) to award Curie honorary degrees, and planned for two large receptions to be held in her honor. These were the opening one at Carnegie Hall in New York City, where Curie would receive the Ellen Richards Prize of $2,000 from the old Naples Table Association (which, having lacked Meloney's vision earlier, now joined the bandwagon), and the second on the White House porch, where, as pictured in plate 18, Curie would accept a precious gram of radium (worth over $100,000) from President Warren Harding on behalf of its donors, the women of America.[52] Grandiose as the plans might have seemed, Meloney was such a master of public relations that she managed to pull it all off successfully, and the trip became one of the great publicity and fund-raising campaigns of the 1920s.[53] Curie's only reaction seems to have been one of exhaustion through most of her three weeks in the United States, as she and her daughters were apparently overwhelmed by the enthusiastic reception they received at every turn. Everyone, it seemed, wished to see the great lady of science and help her continue her work.

Meloney made surprisingly little use of the women scientists of America in her productions. Either she knew of their malaise or, more likely, she did not know many of them.[54] In any case the only women scientists to play even a minor public role in Curie's visit were the medical researchers Florence Sabin of The Johns Hopkins Medical School and Alice Hamilton, recently appointed assistant professor of industrial medicine, a field she had largely founded, at the Harvard Medical School. Both of them spoke briefly at the Carnegie Hall reception. Yet not even these luminaries were included in the fund-raising side of the visit, which was the work of the "Marie Curie Radium Fund" committee, chaired by Francis Carter Wood, M.D., director of the Institute for Cancer Research at Columbia University. He was assisted by two separate groups: the "Executive Committee of Women," which included M. Carey Thomas and such prominent wives (and fund-raisers?) as Mrs. Calvin Coolidge, Mrs. Andrew Carnegie, and Mrs. Herbert Hoover but no scientists; and a "Committee of Scientists," which had no women but did have such friends of the cause as William H. Welch, M.D., and William Duane, a Harvard physics professor who had worked for several years with Madame Curie in Paris.[55] As usual, therefore, but particularly surprising in this case, the women scientists of America were caught invisibly between two almost mutually exclusive stereotypes—that of "women" and that of "scientists," including medical researchers.

Meanwhile Meloney engineered and sanitized all the publicity connected with the trip, even insisting successfully that there be no mention of Curie's 1911 affair with physicist Paul Langevin which had cost her her election to the French *Académie des Sciences*. Meloney also arranged for Curie to write a short autobiography for release shortly after the visit. In fact the whole campaign—the first 1921 visit, a second, quieter one, in 1929, and all the publications, including daughter Eve Curie's best-selling, but distorted, biography in 1938 and a wartime movie based upon it (starring Greer Garson)—went off as smoothly as if Madame Curie had coldly surveyed the latest methods of raising over $100,000 for her researches and personal future and then hired Meloney as her own public relations agent. She could hardly have dreamed up such a successful extravaganza or done better than agree with Meloney's venture.[56]

Such publicity campaigns entered science in the 1920s and have flourished there with varying degrees of sophistication ever since. The basic motivation was that the expense of scientific research was mounting so quickly that it was surpassing even the increasing amounts of money that universities and foundations were providing for it. Raising the extra money required various kinds of public relations and outright salesmanship, such as the glorification of research, the canonization if not actual deification of individual scientists, large fund drives, and a general "hype" of science through the new fields of science journalism and even the history of science itself. That science was noble and beneficial, had always been so, and deserved wide support were messages that influential persons worked hard to communicate to an expectant but unsuspecting populace. Figures such as the frail Curie, who always dressed in black, or the unkempt and otherworldly Albert Einstein, who also toured the United States in 1921, became essential symbols in this emerging iconography of the selflessness of science. Such signs of personal modesty and gentle eccentricity were particularly useful for scientists in the United States chiefly for fund-raising purposes.[57]

But what message did impresario Meloney want Madame Curie's visit to deliver? Were the women scientists of America to benefit from it in any way? As might have been expected, the publicity surrounding this celebrity's tour stressed the sentimental side of Curie's life and pandered to prevailing stereotypes rather than trying to change them. The historical facts of Madame Curie's career were straightforward enough and were spelled out (with embellishments) in innumerable magazine articles: she had overcome childhood adversities in Poland to seek an education in Paris; she had starved in a garret while earning one of the first French doctorates in physics to go to a woman; she had met, married, and worked very hard with her beloved Pierre to discover radium. Now, as his widow, she held his professorship at the Sorbonne, had won two Nobel Prizes (in chemistry and in physics), and had received $50,000 from philanthropist Andrew Carnegie for her laboratory. Sadly, however, postwar inflation had put this indomitable woman in severe straits once again. She might, however, with a gram of radium, continue her researches and discover more of that element's wondrous medical effects. Since she had already overcome so many obstacles and done so much for humanity, for science, for medicine, and for women, how could a compassionate America *not* help?

Inevitably this sanitized and heroic version of the Curie saga gave her almost too much credit and minimized and obscured the remaining problems that even she, the great lady of science, had not been able to overcome. Curie had done great things in science, and American women had made certain of their own advances, but was it fair to say the two were linked in any way? What, for example, might a speaker claim that Madame Curie's career had "shown"? Had she "proven" anything about women's abilities, careers, or prospects in science? Had any of it had any noticeable effects on American women scientists? Should it have? Would it have? Had she really, as so many speakers would say, "opened" any "doors" for women in science? Had she, for instance, reserved a place in her laboratory for women, American or otherwise? A skeptic could point out, for example, that Curie had not opened any doors for women at the American colleges and universities. The women's college had been founded in the 1870s and 1880s, long before Curie had even left Poland. She had certainly "opened the door" at the Nobel Foundation with her two prizes (1903 and 1911), but for whom, since no other woman had yet followed in science? Nor had she succeeded with either the French *Académie* or the U.S. National Academy, both of which had refused to elect her a member. Similarly, Harvard and Princeton (unlike Yale) had both haughtily refused to give her an honorary degree, even on this triumphal tour.[58] Didn't all this mean that even the greatest woman scientist faced a recognition barrier that would probably not have existed for a man? Wasn't that perhaps a problem that would affect lesser women of science even more seriously? If one thought about what Madame Curie had not been able to do, how could he or she be at all euphoric about women's future in science? Curie had achieved more in science than any of her female contemporaries, and yet even she had been sadly limited by her sex.

But an optimist might persist (as indeed they did) in believing that Curie had "opened doors" psychologically—she had proven that women could make great discoveries in science. No one seemed disturbed that apparently this still had to be proven in 1921. Everyone seemed to have forgotten all those American women who, like Maria Mitchell in the nineteenth century or those more recent ones starred in Cattell's series of directories, had, despite all the obstacles to winning even that honor, "proven" the sex's ability to work in science long before. Why then did the burden of proof still exist? And wasn't it symptomatic of a certain kind of persistent and pernicious double standard that it took a Madame Curie with two Nobel Prizes to "prove" that women could work in science, when everyone had assumed that men could do so, even before there had been such prizes? One wonders if any observers of all the hoopla had a hunch that it might all backfire someday in some yet unpredictable way.

Or was the key to Curie's visit perhaps less in the realm of the rational or tangible and more in the inspiration that she gave to other women that they too might enter science and achieve her glory? But if inspiring the young was a purpose of the visit, suspiciously few American women scientists were involved in her schedule, and hardly any visits with students were included in her trips to the women's colleges.[59] We may never know what thoughts went through the minds of her overflow audiences. Did they really think that they could be Madame Curies too? Or that their daughters might be? If so, didn't they then realize that her career was the product of rather unusual circumstances that few Americans would desire (or be able) to duplicate? And didn't it seem odd at least to some of them that Madame Curie and her professorship at the Sorbonne were being idealized at a time when all that her American followers could reasonably expect to obtain was a job in some kind of "women's work"? Or did her audiences not identify with her career specifically and just feel uplifted and rejoice in knowing that other women might reach such heights and that the name of even one woman scientist was fast becoming a household word?

Thus, though one could have taken a more somber view of the status of American women in science in 1921 and shown how little Madame Curie's particular life and career meant for American women (as James McKeen Cattell had done in his 1911 editorial on her possible election to the French *Académie*), this apparently did not happen in 1921. The mood Meloney created was instead glib and optimistic, and speakers and overflow audiences alike preferred inspiration to realism. They apparently wanted to be told that women had arrived in science, that all their problems were now solved, and that they too could aspire to the heights of her glory. The shrewd Marie Meloney had sanitized and manipulated the career of Curie to make it stress her personal sacrifice and heroism and minimize women's problems in science, the feminists' usual theme. From the vantage point of 1921, the trip was good theater and a great financial success; from the vantage point of even a few years later, however, it was, at best, a great opportunity lost and, at worst, a detriment to women's future in science, for before long Curie's glorified image would backfire disastrously on her would-be American followers. By allowing their claims to a position in science to be linked (however illogically) to Curie's exceptional career, they probably thought that they were implying that *any* American woman *could* follow in her footsteps. But before long most professors and department chairmen were interpreting Curie's example far more restrictively and expecting that *every* female aspirant for a faculty position *must* be a budding Marie Curie. They routinely compared American women scientists of all ages to Curie and, finding them wanting, justified not hiring them on the unreasonable grounds that they were not as good as she, twice a Nobel Laureate! One is reminded of the college president who had told Maria Mitchell in the 1870s that he would hire women scientists if they were as good as Mary Somerville (see chapter 1). Thus in the prejudiced politics of science in the 1920s and 1930s, the seemingly innocent and beneficial goal of raising women's ambitions in science was quickly absorbed and transformed into a stringent heightening of its already double standard. The impact of Curie's visit thus came to mean less an "opening of doors" for women in science, as the rhetoric of the time proclaimed, and more (to mix the metaphor slightly) a "raising of the thresholds" to almost unattainable heights. (For more on this phenomenon see chapters 7, 9, and 10.)

More realistic than most other Curie-inspired addresses and an indication of how the visit might have been used to emphasize other themes was Simon Flexner's commencement address, "The Scientific Career For Women," delivered at Bryn Mawr College two weeks after Curie's Carnegie Hall reception. He too talked of "open doors," expressing the hope that "now that the doors of opportunity have been thrown widely open to women, one may expect that many more will pass their portals and enter upon the career of science," but he also concluded by discussing the sex stereotyping, childhood discouragements, and unequal opportunities in later life that afflicted women in science. He hoped these things could be overcome—"As one who would write himself down a lover of opportunity for women, I wish to express the hope that the difficulty may not prove insurmountable"—and he cited Florence Sabin of The Johns Hopkins University and Louise Pearce of his own Rockfeller Institute as examples of successful women in science. A day later the *New York Times* editorial staff, already recovering its balance after the Curie mania, chided Flexner for being too optimistic about women's future in science.[60] There would always be more men than women in science, the anonymous editor wrote, because more men "have the power—a necessary qualification for any real achievement in science—of viewing facts abstractly rather than relationally, without overestimating them because they harmonize with previously accepted theories or justify established tastes and proprieties, and without hating and rejecting them because they have the opposite tendencies."[61] Had the feminist scientists made no impact at all, not even in New York City?

Standard themes emerged again a year later, when William H. Welch of The Hopkins delivered another commencement address at Bryn Mawr, this time to honor its retiring president, M. Carey Thomas. After consulting with her about a suitable topic he too talked of triumph, but in seeking to describe "how complete has been the victory, how great the advance" since the 1880s, he obliged her by discussing the traditional feminist theme that discriminatory hiring by male and coeducational institutions created the difficulties women faced in academic careers. The early faculty at Bryn Mawr had included both men and women, he recounted, but when the men had achieved their early successes in research, they had been called to larger universities to pursue their careers further. Almost all the women, however, had remained at Bryn Mawr, despite their many accomplishments.[62]

Thus by 1921 the women scientists and their leaders had come far from their beginnings with Emma Willard one hundred years before. They had had a particularly active period since 1906, with much discussion of women's position in science, and much activity in the suffrage movement, and the war. In many ways the pioneers' agenda, the gaining of higher degrees and entry positions in science, had been completed, but their successors' ambitions for full equality had so increased and science itself had changed so much in this century that even these first attainments now seemed insufficient for full careers in science. Yet so far the expanded goals of advancement and recognition had eluded the feminists' tactic of trying to change public opinion through research and politics. Women scientists needed new strategies, and some in fact were already being implemented. In the terms of the prevailing rhetoric of the time, many doors to the house of science had been unlocked and opened to them, but women were still on the first floor and still faced many obstacles to their further acceptance.

6

GROWTH, CONTAINMENT, AND OVERQUALIFICATION

Although many persons felt around 1920 that the full acceptance of scientific careers for women was now to be a reality, they were to be sadly disappointed, for women scientists made little progress in the next two decades. The years from 1920 to 1940 were years of boom and depression in much of the nation's economy, of great expansion and then retrenchment in the scientific labor force, and of optimism followed by despair for women in the professions. Thus although women scientists registered, as described in the next several chapters, certain important advances in these years—such as a tremendous increase in the number of women in science and the percentage of doctorates among them; a certain broadening of employment opportunities open to them outside the women's colleges; notable achievements and wide popular reputations of a few, such as Ruth Benedict and Margaret Mead; and the occasional election of a woman president of a professional group or a woman member of an honorary society, such as the National Academy of Sciences—their position within science was as marginal and their chance for recognition as remote in 1940 as both had been in 1920. This situation arose from the fact that although science was growing in these decades, its employment policies continued to follow those social and economic patterns which had been set earlier and which systematically channeled women into secondary roles. Those structures and stereotypes remained solidly entrenched despite several organized efforts by women scientists to improve their subordinate status.

Although the quantitative evidence presented in this chapter indicates that women's place in science improved numerically in these years, it also suggests that this improvement came at considerable cost to them. There were many more women in science but they had had to work extra hard to be there, consistently earning more degrees than the men and suffering more than their share of unemployment, especially if married. Yet rather than continuing the old feminist strategy of trying to change public opinion about women's role by documenting and protesting such "unfair" treatment, the women scientists of the 1920s and 1930s adopted a new, more conservative, and less confrontational strategy of deliberate overqualification and personal stoicism. Under this new tactic they accepted inequity in scientific employment as something that they could not change (and that was perhaps even justified in the aggregate) but that some might overcome by hard work (such as by earning more degrees than the men) and by stoicism (such as suffering any unemployment in silence). The political climate became so intolerant and critics so quick to blame the victims that to mention these inequities could only be counterproductive and confirm the prevailing low expectations of women as poorly trained bad risks who would soon quit to get married and so could never be promoted to management positions. Thus rather than continuing to identify and fight discriminatory conditions, the women scientists of the 1920s and 1930s began to internalize the double standards, and to accept as legitimate the prevailing views that to deserve a place in science they had to be not only better than the men (or "exceptions") but, preferably, "Madame Curies." Sadly, a few even drove themselves into exhaustion or nervous breakdowns in attempts to meet the heavy demands they placed upon themselves. Thus the many figures and tables presented in this chapter give a few preliminary glimpses of a new behavioral pattern, the "Madame Curie strategy," that was emerging in these years and that is discussed more fully in succeeding chapters.

By any measure the scientific labor force in the United States grew tremendously between 1920 and 1940. The major graduate schools were producing more and more male and female doctorates, and the number of positions available to scientists was increasing greatly at both leadership and subordinate levels, as scientific teaching, research, and application attracted new funding from universities, industry, foundations, and federal, state, and even local governments. Yet despite this overall growth, there were sufficient economic differences among the various sciences for some fields to be more attractive to newcomers than others and for specialists in different fields to have dissimilar job experiences. Thus although both sexes benefited to some extent from this overall expansion of training and opportunity in the 1920s and 1930s, they did so in different and unequal ways, largely because the women, including the rising proportion of married ones, were distributed irregularly among the sciences and utilized differently in them. In order, therefore, to understand the variegated position of women in science in the 1920s and 1930s, one must examine both the economic situation in the fifteen major fields and those additional factors that particularly affected the women in them.

Although counting the number of scientists in the United States at a given time can be difficult and is open to many problems of defining who is and who is not a "scientist," the usual starting place for data on this period has become the number of science doctorates awarded annually by American universities. Useful chiefly as indicators of how rapidly the supply of one kind of scientist was increasing, doctoral data have been widely used for manpower studies since the early 1960s, when the National Research Council (NRC) of the National Academy of Sciences in Washington, D.C., made them readily available by publishing a volume of historical statistics. When these NRC totals are broken down by sex they reveal, as shown in figure 6.1, that the number of women

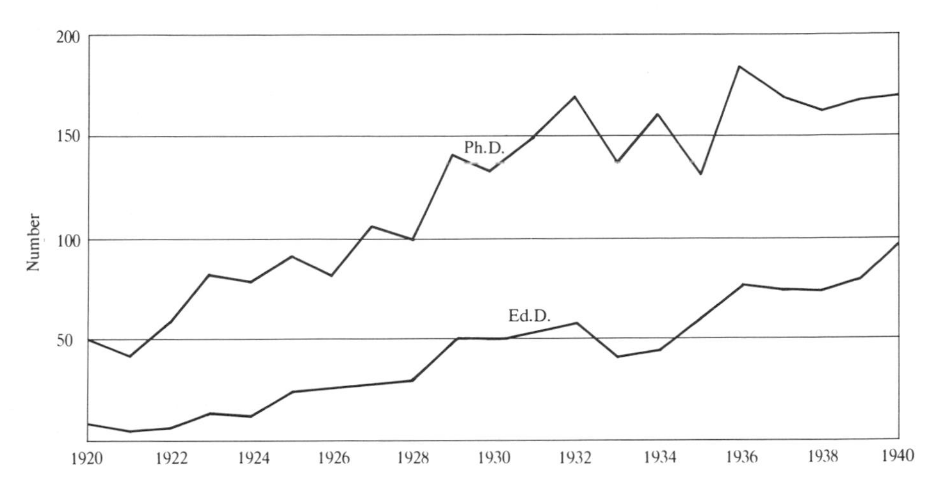

FIGURE 6.1. Number of Doctorates in Science and Education Awarded to Women, 1920–1940.
NOTE: Figures for doctorates in science include degrees awarded in psychology and anthropology, but not engineering.

earning doctorates in the sciences rose sporadically in the 1920s and 1930s from about 50 annually in the early 1920s to about 165 per year in the late 1930s. Yet because the number of men earning doctorates in science increased even more than did that of women in the 1930s, the women's percentage of the total declined after 1932, as shown in figure 6.2, from a high of 15.5 percent in 1920 to about 11.5 percent in the late 1930s.[1] Although the NRC reports give no reason for this decline, they also demonstrate a rising number and percentage of women earning doctorates in education after 1933, which suggests, perhaps, that those women who had earlier earned doctorates in science and then faced the uncertain job prospects there (which are described more fully in succeeding chapters) might have turned, after the mid-1920s, to the schools of education, the Ed.D. degree, and more promising careers in the highly feminized world of professional "education." Their departure for teachers colleges, state colleges, and public school administration would also help explain why the women who did go into science in these years were the highly qualified ones listed in the *American Men of Science*. At present, however, there is insufficient information on the recipients of the Ed.D. degree for this to be more than speculation.[2]

Although doctoral statistics may be the best indicator we have of women's desire to participate in science (an important kind of self-selection), they provide only a limited and even quite distorted picture of what was happening to those women in science in the 1920s and 1930s. They stress the earning of higher degrees, the women's chief accomplishment in academia since the 1890s, when, as shown earlier, the major universities had established the pattern of awarding them graduate degrees but not employing them on their faculties (except in "home economics"). Moreover, the NRC's data say nothing about the women doctorates' subsequent careers and totally omit those women scientists who did

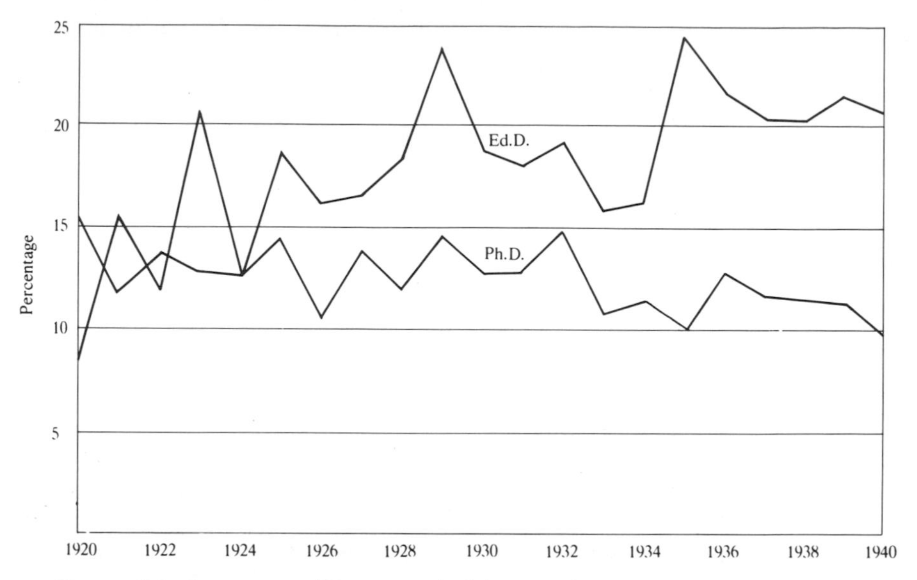

FIGURE 6.2. Percentage of Doctorates in Science and Education Awarded to Women, 1920–1940.

NOTE: Figures for doctorates in science include degrees awarded in psychology and anthropology, but not engineering.

not earn doctorates at American universities in these years, a more striking deficiency in some fields than in others. Thus although doctoral data on women have often been used to reach certain conclusions about the employment and careers, not only of all women scientists, but of all professional and "academic women," they actually give only a limited and unduly favorable impression of how well women were faring in science in these years. The percentages should, therefore, be taken as more of a maximum level of women's potential participation in the top levels of science than as an accurate assessment of their real involvement in the 1920s and 1930s, which other sources indicate was only 5 to 7 percent of the total, or about one-half their proportion of science doctorates.

More useful sources of data on men and women scientists of the 1920s and 1930s (although with defects of their own) are the several editions of the *American Men of Science* that appeared between 1921 and 1938. Because James McKeen Cattell's stated criterion for inclusion in these directories had been the broad and functional one of being a person who had "carried on research in science," rather than an educational requirement, such as holding a doctorate, or an employment one, consisting of a job title, his directories of "scientists" contained a slightly different, though largely overlapping, population from that of the NRC's "doctorates," many of whom made no further contributions to science after earning their degrees. Since Cattell's directories also offer fairly complete entries on the careers of over 27,000 persons by 1938, including almost 2,000 women, and omissions are hard to find, they were, until eclipsed by the

NRC's more readily accessible tables, long considered the standard source for quantitative data on the "scientific community" in the 1920s and 1930s.[3] It should be noted for future reference, however, that despite its extensive coverage and contemporary acceptance, the *AMS* also systematically omitted other persons who, with or without doctorates, would not be accepted as valid members of the scientific labor force—those who (1) held jobs in applied science, especially in industrial chemistry; (2) worked at the middle or lower levels of government bureaus; or (3) had unknown employment but belonged to a major professional organization, such as the American Physical Society, which in 1927 had twice as many members (1,920) as there were "physicists" in the *AMS* (991).[4]

Despite all the definitional problems, which plague any source of data, statistics drawn from the *AMS* directories allow one to document in greater detail than is possible otherwise the different positions of men and women in American science in the 1920s and 1930s. In a sense the *AMS* presents both a more pessimistic and a more optimistic picture of the position of women than does the NRC doctoral data: it depicts their percentage as being consistently lower than in the NRC but as rising over time (rather than as initially high and then falling); women constituted only 4.7 percent of the persons in the 1921 *AMS*, but accounted for 5.8 percent of those in the 1927 edition. In 1938, after the women had earned 11.9 percent of the science doctorates awarded by American universities from 1920 to 1937, they constituted 7 percent of the persons in the *AMS*. Evidently a great many women were earning doctorates in science but were then (for whatever reasons) not listed in the *AMS*, while many men, even those without American doctorates, were included after all, probably because of their more visible positions, immigration, or better publicized contributions. Despite the NRC's evidence that women were a declining percentage of total science doctorates in the 1930s, the highly qualified women who were in the 1938 *AMS* still held a far higher percentage of doctorates than did the men there. This double standard that required a doctorate of more women than men in the *AMS* is a statistical clue to the discriminatory employment conditions at the time: women were still largely restricted to college teaching, for which doctorates were required, and could only get other jobs if they were extra well qualified. Forewarned about the limitations of the *AMS* data, and remembering its clue that the women's relatively high percentage of science doctorates was as much a danger signal of widespread difficulties in employment as a sign of triumph at the graduate schools, we can now examine in some detail the almost overwhelming amount of data on fifteen sciences presented by just two editions of the *AMS*, those for 1921 and 1938.

Data on the distribution of male scientists in the United States in 1921 and 1938 are presented in table 6.1. They are based on two samples drawn from the *AMS*, one of 502 males chosen from the third edition of 1921, and another of 1,015 from the sixth edition of 1938. The totals show that the overall size of the male scientific community almost tripled in these years, from a calculated 9,036 men in 1921 to 25,375 in 1938, a rate of growth that was much faster than that of

TABLE 6.1. Distribution of Male Scientists, 1921 and 1938 (By Field)

	1921		1938		
Field	Number	Percentage of Total	Number	Percentage of Total	Change since 1921 (Percentage)
Chemistry	1,350	14.9	4,675	18.4	+246.3
Medical sciences	1,332	14.7	3,325	13.1	+149.6
Zoology	936	10.4	2,600	10.2	+177.8
Engineering	882	9.8	3,500	13.8	+296.8
Botany	864	9.6	2,550	10.0	+195.1
Physics*	864	9.6	1,825	7.2	+111.2
Mathematics	684	7.6	1.275	5.0	+86.4
Geology*	630	7.0	1,425	5.6	+126.2
Agricultural sciences	396	4.4	950	3.7	+139.9
Astronomy	306	3.4	200	0.8	−34.6
Biochemistry	252	2.8	900	3.5	+257.1
Psychology	234	2.6	1,000	3.5	+327.4
Microbiology	198	2.2	700	2.8	+253.5
Anthropology	108	1.2	225	0.9	+108.3
Nutrition	0	0	225	0.9	∞
Total	9,036	100.2	25,375	99.8	+180.8

SOURCES: *American Men of Science,* 3d ed. (1921), based on a sample of every eighteenth male, and *American Men of Science,* 6th ed. (1938), based on a sample of every twenty-fifth male.

*Between 1920 and 1938, meteorology, a small specialty, moved from geology to physics.

the adult male population (ages 25 to 65) at the time, which grew from 25 million in 1920 to just 33 million by 1940.[5] Table 6.1 also presents data on the sizes of the various sciences in 1921 and 1938 and the growth rates of each. In size the two largest fields for men in 1921 (chemistry and medical sciences) were still among the largest in 1938, but had been joined by engineering, which grew tremendously in the 1920s and 1930s. Overall these "big three" accounted for 45.3 percent of the male scientists in 1938. All the remaining scientists were spread over the twelve smaller sciences; anthropology and astronomy were particularly tiny and precarious fields.

Although the overall trend between 1920 and 1940 was for the largest sciences to get even larger, there were some other interesting changes under way. A few new fields, such as microbiology and biochemistry, emerged from the shadows of the medical sciences to become separate fields, and a few men began to enter the field of nutrition (or food chemistry), from which they had been noticeably absent in 1921, when it was still known to many as "home economics." The field of psychology also grew very rapidly, as it began to permeate all areas of American life. New specialties also emerged within older fields during the 1920s and 1930s, such as aeronautics within engineering and poultry science within agriculture, although these details are not shown in table 6.1. From this perspective the growth rates for the sciences are perhaps as interesting as the actual sizes, for they show that the fastest growing field for men in the 1920s and 1930s was nutrition, followed by psychology, engineering, biochemistry, microbiology, and chemistry, all of which had growth rates of over 200 percent.

At the other end of the spectrum and equally important for the study of scientific careers were seven sciences that had below-average growth rates. Mathematics grew at the slowest rate, and one field, astronomy, surprisingly, did not grow at all but instead had an overall decline from a calculated 306 astronomers in 1920 to 200 in 1938, of whom ⅜ were retired. One wonders why should this happen to just one field. Was there a lack of jobs, interesting scientific problems, or big telescopes? Or is it all an illusion or artifact of the data? And if so, why for astronomy only?[6] But despite the lack of growth in certain areas, only 13 of the 1,015 men in the 1938 sample were unemployed, or roughly 1.3 percent, spread over several fields. Apparently the men were able to avoid the worst effects of slow growth and the depression, or chose not to report them, for it is not always possible to tell the exact professional status of the men in this sample. For example, although the title "consulting" chemist or geologist might have disguised unemployment, most persons so listed had been consultants since the 1920s; their salaries might have dropped somewhat, but their job titles had not changed perceptibly. The official study by the American Association of University Professors, *Depression, Recovery and Higher Education* (1937), also tended to minimize the effects of the depression on faculty members, but recent autobiographies by men who were starting academic careers then present more of the personal anguish—they faced great insecurities for several years, pay cuts of up to 20 percent were common, anti-Semitism was rife, and the contracts of instructors, especially those with unpopular political or religious beliefs, were often not renewed.[7]

Although the sixth edition of the *AMS* did not present data on the marital status of its men, another statistical study of 10,000 male scientists in 1927 by C.-J. Ho of Columbia University reported that 62 percent of them were married. This largely agrees with the percentage given by Jessica Peixotto's study of the faculty of the University of California at Berkeley in 1922 and 1923, which reported that though only 57 percent were married, the rate varied greatly with age and rank, ranging from 19.8 percent of the associates to 47.2 percent of the instructors and 74.6 percent of the full professors. Even these proportions, she pointed out, were higher than had been the case earlier, when she thought most academic men had been celibate. A 1927 study of the Yale faculty found that 74 percent were married, ranging from 52 percent of the instructors (whose median age was 28.5 years) to 97 percent of the full professors (52 years). The economic pressures of the depression may have lowered the marital rate somewhat during the 1930s, but data on this do not seem to be available.[8]

AMS data on women scientists in the United States in 1921 and 1938 presented in table 6.2 stress their substantial overall increase in these years, with the total number of women in the *AMS* jumping from 450 in 1921 to 1,912 in 1938. The percentage of scientists in the *AMS* who were women therefore increased moderately during the 1920s and 1930s, from 4.7 percent in 1921 to 7.0 percent

TABLE 6.2. Distribution of Female Scientists, 1921 and 1938 (By Field)

	1921			1938			Change since 1921
Field	Number (1)	Percentage of Total (2)	Percentage of Field (3)	Number (4)	Percentage of Total (5)	Percentage of Field (6)	Rate of Increase (Rank) (7)
Botany	84	18.7	8.9	256	13.4	9.1 (9)	+204.8 (11)
Zoology	78	17.3	7.7	281	14.7	9.8 (8)	+260.3 (9)
Psychology	60	13.3	20.4	277	14.5	21.7 (2)	+361.6 (6)
Mathematics	42	9.3	5.8	151	7.9	10.6 (7)	+259.5 (10)
Medical sciences	41	9.1	3.0	186	9.7	5.3 (10)	+353.7 (7)
Chemistry	28	6.2	2.0	163	8.5	3.4 (12.5)	+482.1 (5)
Geology	23	5.1	3.5	60	3.1	4.0 (11)	+160.9 (13)
Physics	21	4.7	2.4	63	3.3	3.3 (12.5)	+200.0 (12)
Astronomy	20	4.4	6.1	36	1.9	15.3 (3)	+80.0 (14)
Nutrition	20	4.4	100.0	164	8.6	42.2 (1)	+720.0 (3)
Microbiology	18	4.0	8.3	109	5.7	13.5 (4)	+505.6 (4)
Anthropology	8	1.8	6.9	29	1.5	11.4 (6)	+262.5 (8)
Biochemistry	7	1.6	0.4	129	6.7	12.5 (5)	+1,742.9 (2)
Engineering	0	0	0	8	0.4	0.2 (14)	∞ (1)
Agricultural sciences	0	0	0	0	0	0 (15)	0 (15)
Total or average	450	99.9	4.7	1,912	99.9	7.0 (8)	+324.9 (8)

SOURCES: *American Men of Science*, 3d ed. (1921), and *American Men of Science*, 6th ed. (1938).

in 1938. These percentages rose in all fields except nutrition, and ranged in 1938, as shown in column 6 of table 6.2, from highs (> 20 percent) in nutrition and psychology to lows (< 5 percent) in geology, chemistry, physics, engineering, and the agricultural sciences. It is worth noting that few women dropped out between editions, although many got married or changed their names and thus might at first seem to have left science. Others, however, baffled the editors and were listed twice (such as Ellen Pawling Corsonwhite, under Corson and under White); Marian Irwin Osterhout and A. Louise Palmer Wilson were listed under both their married and maiden names. In addition to these easily corrected artificial increases, there was also, as with the men, a modest increase in the immigration of foreign women scientists, especially eminent ones, in the 1920s and 1930s—either as wives of foreign or American men, such as Gerty Cori in biochemistry, Maria Goeppart-Mayer in physics, or Else Frenkel-Brunswik in psychology, or on their own, such as chemist Maria Telkes, physicist Hertha Sponer (later Sponer-Franck), botanist Katherine Esau, mathematician Emmy Noether, biochemist Gertrude Perlmann, paleontologist Tilly Edinger, and mathematical biologist Dorothy Wrinch.[9]

Here too it is interesting to observe the different sizes and relative rise and decline of the various sciences. As with the men, but to a lesser extent, the women congregated in a few large fields; the rest were scattered over several smaller ones. For the women, however, the "big three" in both 1921 and 1938 were not chemistry, medical sciences, and engineering but the biological and social fields of botany, zoology, and psychology. In fact altogether the many fields that might be considered branches of "biology" (agricultural sciences, biochemistry, botany, geology, medical sciences, microbiology, nutrition, and zoology) accounted for 60.2 percent of the women in 1921 and 61.9 percent of them in 1938. The smallest for women in 1921 was biochemistry, but there were (according to these directories) no women at all in either engineering or agricultural sciences, a situation that had barely changed by 1938. Yet small fields did not always remain so, as shown by biochemistry and microbiology, which, though barely detectable in 1921, grew so dramatically in the 1920s and 1930s, as opportunities opened to women in these fields, that they rapidly surpassed other more traditional and slowly growing fields, such as physics and geology, which only tripled in size between 1921 and 1938. These two thereby not only remained among the smallest of fields for women, but actually suffered a loss in the percentage of all women scientists who were in them. Yet, because an even higher proportion of men were leaving (or were failing to enter) geology and physics (see table 6.1), the proportion of the field that was female actually rose between 1921 and 1938. The decline was even worse in the tiny field of astronomy, though the number of women (unlike that of men) did manage an increase during the 1920s and 1930s (suggesting that women were more inclined to enter an unpopular or unpromising field than men). Thus one can see that though the period was overall one of great growth, the specialties had many important differences that would affect the kinds of persons they could attract and retain. If one assumes that scientists can be somewhat mobile, especially early in their careers, these differences in fields' prospects could result in rather different labor forces among them.

These differences in growth rates of the sciences are stressed here because, as Henry Menard has shown in his provocative and intriguing book *Science: Growth and Change* (1971), the growth rate of a science can have a strong effect, possibly more than the actual size of the field, on the career of an individual scientist. Menard describes the intriguing way in which early advancement comes to a man in a fast-growing field, where there are more jobs available than people to fill them. There, for example, promotions and awards come early in a man's career, and after just a few years in such a field, even a young scientist can be considered an "elder statesman," can be a full professor, or can even be president of the young professional society. But in a slowly growing field, the same persons would be much older before becoming a professor, and perhaps even be emeritus before winning a prize or election to professional office. Such an interpretation of the importance of growth rates would explain certain other findings here (such as the high percentage of retired astronomers in 1938) and in other studies, such as that of C.-J. Ho mentioned above. Finding that astronomers lingered thirteen years as associate professors but that bacteriologists averaged only one year at that level, Ho attributed the difference to the greater intelligence and industry of bacteriologists rather than to external forces of supply and demand.[10]

One can extend Menard's reasoning to certain other characteristics of growing scientific populations. In particular, these growth forces might, besides affecting the timing of individual careers, also influence the recruitment policies and the attitudes of scientists toward women or minority groups trying to enter their field. Menard's model can then be used to investigate why, for instance, some fields were apparently more "open" to women and welcomed them even during the depression when others did not. Yet, suggestive as this model may seem, the data presented in table 6.2 indicate that there is rarely a clear-cut relationship between a field's growth rate and its subsequent percentage of women. Only in small fields that have either a very large or a very small rate of growth is there much effect. Biochemistry, which grew explosively during the 1920s and 1930s, and to a lesser extent microbiology, would be the prime examples of rapidly growing fields whose percentage of women rose significantly, from 0.4 percent (column 3) to 12.5 percent (column 6) in biochemistry, and from 8.3 percent to 13.5 percent in microbiology. One can imagine that in these fields women not only were tolerated but were even sought out, especially since their typically lower salaries as research associates would have made them attractive additions to the staff of a laboratory or institute.

A certain feminization was also taking place at the other extreme, in those fields with very slow growth rates. Women were not totally shunned there, as Menard's model of severe competition in stagnant fields would seem to predict, but instead were actively recruited into astronomy, and to a lesser extent mathematics and anthropology, whose proportion of women rose between 4.5 and 9.2 percentage points in these years despite low overall growth rates. Though their actual numbers were generally small (except in mathematics), the women's proportion of the total entrants was quite high, since relatively more of them were willing to enter these unpromising fields and accept the difficult employment prospects there than were men.[11] In anthropology, for example, the lack of jobs was a constant theme, almost a religion, in the 1920s and 1930s, as recruits were warned repeatedly to be ready for sacrifices. In this way Franz Boas and his colleagues at Columbia University were able to transform the field's lack of jobs into a noble, self-sacrificing cause of great appeal to certain, especially affluent, men and women graduate students, as revealed in this interview with Dorothy Bramson Hammond, a Barnard graduate (class of 1939):

> When she entered the Columbia graduate school, she met with discouragement from all sides; it was, at that time, almost impossible to make a living in the field of anthropology. The graduate students were few in number, and, quoting Mrs. Hammond, "to come in at all they had to feel a real sense of dedication and a willingness to starve."
>
> "I had the sense of dedication all right," she adds, "to say nothing of Papa if it came to starvation. It is quite different now. . . . But then, in that small and very chummy Department, we all felt that we were doing something very special and exciting. And it was fun."[12]

Yet as many anthropologists discovered later, the highly romanticized sacrifices required in a field like anthropology were not shared equally, and most of the actual jobs in the field in the 1930s went to the men, while the women made do on short-term grants and fellowships. Thus when one talks about actual jobs rather than just numbers of persons "in the field," Menard's suggestions prove correct after all. The women might "enter" the depressed areas of science relatively more often than the men and be listed in the *AMS*, but they rarely got jobs there. That was partly because the men were preferred and partly because the women would, if they married, be expected to resign their jobs. The fact that by 1938 less than one-quarter of the married women in the *AMS* had resigned (see text that follows) does not seem to have changed the expectation that individual women would. In a tight market, even this shadow over their future greatly reduced the women's employment prospects and resulted in the self-fulfilling prophecy that, of all scientists, it was the married women who would be the most often temporarily or even chronically unemployed.

Although some authorities have deplored the lack of adequate unemployment information for scientists before the 1970s and have even said that none is available,[13] such data can be obtained from the older editions of the *American Men of Science,* since they retained, as mentioned earlier, entries of even those scientists who did not hold jobs. A wealth of data is presented in table 6.3, where one can see that there was a muted side to the great increase in the numbers of women scientists in the period from 1921 to 1938, since 171 of the 1,912 women (8.9 percent) did not hold jobs in 1938, a year when only 1.3 percent of the men were visibly unemployed. Some of these women may have stayed out of the work force voluntarily, but others, as one can observe in the entries in the 1938 directory, had held a variety of precarious positions in the early 1930s before

TABLE 6.3. Unemployment and Marriage Rates for Female Scientists, 1921 and 1938
(By Field)

Field	1921				1938			
	Unemployed		Married		Unemployed		Married	
	Total Percentage* (Number) (1a)	Married Percentage (Number) (1b)	Total Percentage (Number) (2a)	Unemployed Percentage* (Number) (2b)	Total Percentage* (Number) (3a)	Married Percentage (Number) (3b)	Total Percentage (Number) (4a)	Unemployed Percentage* (Number) (4b)
Anthropology	37.5 (3)	33.3 (1)	50.0 (4)	25.0 (1)	17.2 (5)	0 (0)	34.5 (10)	0 (0)
Zoology	29.2 (21)	61.9 (13)	28.2 (22)	59.1 (13)	10.7 (30)	83.3 (25)	28.8 (81)	30.9 (25)
Chemistry	20.0 (7)	57.1 (4)	14.3 (5)	80.0 (4)	9.8 (16)	62.5 (10)	23.3 (38)	26.3 (10)
Biochemistry	(included under chemistry)				6.2 (8)	37.5 (3)	24.0 (31)	9.7 (3)
Geology	17.4 (4)	0 (0)	8.7 (2)	0 (0)	15.0 (9)	66.7 (6)	31.5 (19)	31.6 (6)
Psychology	15.0 (9)	55.6 (5)	25.0 (15)	33.5 (5)	10.5 (29)	75.9 (22)	36.1 (100)	22.0 (22)
Botany	14.9 (13)	46.2 (6)	19.5 (17)	35.3 (6)	7.4 (19)	57.9 (11)	21.1 (54)	20.4 (11)
Physics	14.3 (3)	0 (0)	0 (0)	0 (0)	11.9 (8)	62.5 (5)	19.0 (12)	41.7 (5)
Medical sciences	12.5 (7)	28.6 (2)	14.3 (8)	25.0 (2)	9.7 (18)	38.9 (7)	26.9 (51)	13.7 (7)
Microbiology	(included under botany and medical sciences)				7.3 (8)	50.0 (4)	27.5 (30)	13.3 (4)
Mathematics	9.8 (4)	25.0 (1)	11.9 (5)	20.0 (1)	5.3 (8)	75.0 (6)	15.9 (24)	25.0 (6)
Astronomy	6.3 (1)	0 (0)	5.0 (1)	0 (0)	13.9 (5)	100.0 (5)	44.4 (16)	31.3 (5)
Nutrition	5.0 (1)	33.3 (1)	15.0 (3)	33.3 (1)	4.9 (8)	87.5 (7)	20.7 (34)	20.6 (7)
Engineering	0 (0)	0 (0)	0 (0)	0 (0)	0 (0)	0 (0)	62.5 (5)	0 (0)
Total or average	16.2 (73)	45.2 (33)	18.2 (82)	40.2 (33)	8.9 (171)	64.9 (111)	26.4 (505)	22.0 (111)
Remainder: employed and/or single	83.8 (377)	54.8 (40)	81.8 (368)	10.9 (40)	91.1 (1741)	35.1 (60)	73.6 (1407)	4.3 (60)

SOURCES: *American Men of Science*, 3d ed. (1921), and *American Men of Science*, 6th ed. (1938).

*Percentages may vary, since retirees (or unemployed persons over age sixty-five) were deducted from totals before percentages were calculated.

finally losing even that vestige of employment by 1938. Many listed themselves as a "guest investigator" or "fellow by courtesy" at a university, and others claimed to be in "private research" or to be doing "ecological investigations," but they specified no institution. They apparently wished others to consider them as somewhat actively involved in science, although they did not hold actual jobs. Yet even this high rate of unemployment was only about one-half what it had been for women in the 1921 *AMS*, when 16.2 percent of the (single and married) women had been unaffiliated or amateur contributors to science.

An important related factor was in the percentage of women scientists of the 1930s who were married. Although the *AMS* is not a totally reliable source of data on the marital status of its women (the thrice-married Margaret Mead listed herself as "Miss" in 1938), it does provide the best guide available to this murky subject. The numbers and percentages of married women scientists presented in table 6.3 must be considered a minimum: only those women who could be shown to have been married were counted as such; otherwise they were considered as single, as apparently they wished to be. As for widows and divorcées (whose numbers also increased, from six in 1921 to thirteen in 1938), they were counted as still married rather than as single, as they must have wanted, since all except Mead retained their married names. In fact, although divorce or widowhood would seem to have increased sharply and suddenly these women's financial needs and incentives for employment, it had little visible impact on their entries in the *AMS*. Almost all of the widows and divorcees listed there who were of working age had been employed before the rupture (five of six in 1921 and ten of eleven* in 1938) and continued to work afterward.

The *AMS* data presented in table 6.3 indicate not only that there was a wide variation in the marital rates among fields but that it was rising, often dramatically, in most fields in these years. In 1921 the percentage of women scientists who were married was roughly 18.2 percent, ranging from a high of 50 percent in anthropology to a low of zero in physics. (Interestingly but inexplicably the percentage was lowest in the three fields that grew least in subsequent years, as if the women in those fields were already prepared for a period of self-denial.) By 1938 the average for all fields had risen to 26.4 percent (an increase of over 45 percent), and striking changes had taken place in certain cases: the few women engineers were most often married (62.5 percent), usually to fellow engineers in whose companies they worked; the women astronomers were now very frequently married (44.4 percent); and the physicists, geologists, medical scientists, and psychologists were also much more likely to be married than they had been in 1921. Only in anthropology was there a decrease after 1921 in the percentage of women who were married (or divorced), as apparently the young single

*Two widows in 1938, Anita Newcomb McGee and Mary Vaux Walcott, were aged seventy-four and seventy-eight, respectively, and so considered as past employment age, although another widow, Agnes Chase, was still working at age sixty-nine. Helen Thompson Woolley, age sixty-four in 1938, had worked through most of her marriage but had been so shaken up by her divorce that she had a nervous breakdown and had to retire early. In her case, divorce did not bring her back to employment but upset her enough to take her from it.

professionals supplanted the older married women amateurs. In fact, scientific couples had become numerous by the 1920s and 1930s; table 6.4 identifies the more notable pairs.[14]

When the *AMS* data on female unemployment are broken down by marital status, and vice versa (those on marriage by employment status), as they are in table 6.3, the large difference in the 1938 rate of unemployment for married women (22.0 percent) and single women (4.3 percent) becomes evident. In fact, one can say that by 1938 unemployment had become largely a married woman's problem. A comparison of several columns in table 6.3 gives some clues as to what had been happening: columns 1a and 3a show that the proportion of women who were unemployed dropped by one-half in most fields between 1921 and 1938 (16.2 percent to 8.9 percent), but columns 1b and 3b indicate that the percentage of these unemployed women who were married rose dramatically (45.2 percent to 64.9 percent). In other words, in 1921 married women, who constituted 18.2 percent of all the women scientists in the *AMS,* were 45.2 percent (33 of 73) of the unemployed women there; by 1938 both overall percentages had increased by a remarkably consistent 45 percent—26.4 percent of the women were now married, and they constituted 64.9 percent (or 111 or 171) of the unemployed women. Thus although over three-quarters of the married women scientists were employed in some fashion in 1938, the one-quarter who were not working constituted almost two-thirds of all the unemployed women, as shown graphically in figure 6.3. This large discrepancy confirms the contemporary attitude and practice, widely discussed in the popular magazines of the time, that job discrimination against married women, especially schoolteachers and other professionals, was justifiable, since they were presumably blatantly taking good jobs away from destitute family men. Even the U.S. government encouraged the practice by passing its notorious section 213 of the Economy Act of 1932, which specified that spouses (almost always wives) of government employees were to be the first dismissed in any necessary reductions in force. Many university antinepotism rules also date from these difficult times (see chapter 7). Under such hostile conditions it is perhaps less surprising that 22 percent of the married women scientists were unemployed in 1938 than that 78 percent did hold onto jobs of some sort.[15]

Of particular interest among the still overwhelming numbers of single women of science (still almost three-quarters of the total) was one rapidly growing subgroup, the Catholic nuns. This group increased in size from just one in 1921 (Sister Angela Dorety, professor of botany in the College of St. Elizabeth in

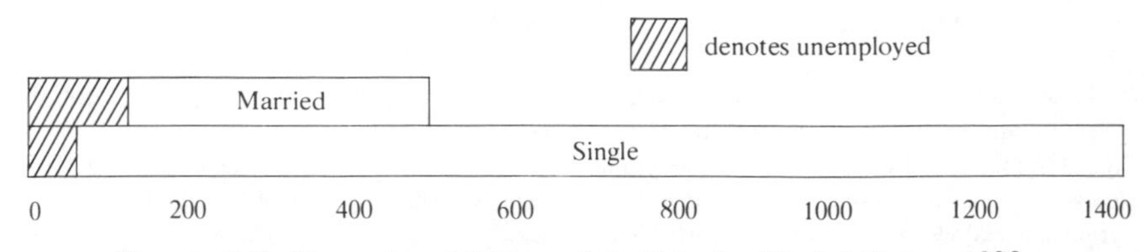

FIGURE 6.3. Unemployed Women Scientists, by Marital Status, 1938

TABLE 6.4. Notable Couples in Science before 1940
(By Wife's Field)

Anthropology
Matilda Coxe and James Stevenson (geology)
Ruth Fulton and Stanley Benedict (biochemistry)*
Margaret Mead and Reo Fortune*
Margaret Mead and Gregory Bateson
Frances Shapiro and Melville Herskovits

Astronomy
Charlotte Moore and Bancroft Sitterly
Cecilia Payne and Sergei Gaposchkin
Helen Sawyer and Frank Hogg
Hannah Steele and Edison Pettit

Biochemistry
Gerty Radnitz and Carl Cori
Helen Tredway and Evarts Graham (pathology)

Botany
Miriam Carpenter and Forrest Strong
Elizabeth Knight and Nathaniel Britton
Edith Schwartz and Frederic Clements

Chemistry
Ellen Swallow and Robert Richards (engineering)
Mary Rising and Julius Stieglitz

Engineering
Nora S. Blatch and Lee de Forest*
Lillian Moller and Frank Gilbreth

Geography
Delia Denning and Carl Akeley*
Mary Jobe and Carl Akeley

Geology
Eleanora Bliss and Adolph Knopf

Mathematics
Anna Johnson and Alexander Pell

Medical Sciences
Gladys Henry and George Dick
Marian Irwin and W. J. V. Osterhout
Margaret Reed and Warren H. Lewis
Jane Sands and R. Cumming Robb

Microbiology
Rebecca Craighill and Donald Lancefield
Katherine Golden and A. W. Bitting

Nutrition
Mary Swartz and Anton Rose (chemistry)

Physics
Maria Goeppart and Joseph Mayer (chemistry)

Psychology
Augusta Bronner and William Healy (psychiatry)
Luella Cole and Sidney Pressey
Catharine Cox and Walter Miles
Christine Ladd and Fabian Franklin (mathematics, journalism)
Gertrude Rand and C. E. Ferree
Leta Stetter and Harry Hollingworth
Helen Thompson† and Paul Woolley (pathology)*

Zoology
Anna Botsford and J. H. Comstock
Ethel Browne and E. N. Harvey
Agnes Claypole and Robert Moody (anatomy)
Gertrude Crotty and Charles Davenport
Sally Hughes and Franz Schrader
Florence Merriam and Vernon Bailey
Margaret Morse and L. Blaine Nice
Marcella O'Grady and Theodor Boveri
Susanna Phelps and Simon Gage
Lillian Sampson and T. H. Morgan
Adelia Smith and P. P. Calvert
Helen Thompson† and Frederick Gaige

NOTE: A husband's field, if it differed from a wife's, appears in parentheses.
*Indicates subsequently divorced.
†Different women with the same maiden name.

Convent Station, New Jersey) to thirty-one in 1938. Ten of these women were in chemistry, seven in botany, three each in physics and zoology, and the rest spread over biochemistry, mathematics, medical sciences, nutrition, and psychology. There were no nuns in anthropology, astronomy, engineering, geology, or microbiology, perhaps because these subjects were rarely taught in the Catholic colleges, where all these women were employed. Twenty-six were themselves graduates of Catholic colleges, and thirty of the thirty-one held doctorates, most often (nine) from the Catholic University of America, whose graduate school first admitted women in 1928. It is hard to evaluate the careers and work of these women, their motivation to undertake higher studies, and the treatment they received from other nuns and other members of the scientific community, but their very existence and the rapid increase in their number in the 1920s and 1930s would seem to modify certain assumptions in the literature on the Catholic women's colleges, which stresses their "finishing school" atmosphere, their few graduates who became scientists, and their anti-intellectual atmosphere. They and their work deserve further study.[16]

There are two more aspects of women's careers in science in the 1920s and 1930s to which a statistical examination of the *AMS* can contribute. These are the patterns of their education, the seemingly well-worked areas of "baccalaureate origins" and "doctorate production." Although the subject of undergraduate backgrounds seems at first to have an extensive literature, one soon discovers that little of it pertains to women scientists or women at all. The standard source on the subject, the Robert Knapp and H. B. Goodrich study, *The Origins of American Scientists* (1952), which was based on an extensive sample of scientists in the college classes of 1924 through 1934, deliberately omitted women from consideration, despite signs in their data (published in an appendix) that the women's colleges had produced many scientists. The subject remained untouched until recently, when M. Elizabeth Tidball and Vera Kistiakowsky collected and published data on the undergraduate backgrounds of women who subsequently earned doctorates from 1920 to 1973 (in any field, including education and other nonsciences). Although their sources were the records of the U.S. Office of Education and the Earned Doctorate File of the Human Resources Commission of the National Research Council–National Academy of Sciences, their conclusions as to which undergraduate institutions were the major "producers" of achieving women before the 1970s are similar to those reached here for women scientists in the 1938 edition of the *AMS,* whether or not they held doctorates.[17]

Table 6.5 presents data on the undergraduate "origins" of the 1,821 women (95.2 percent of the total) in the sixth edition of the *AMS* who reported baccalaureate degrees. These women graduated from 291 different institutions, but, as usually happens, a high proportion came from just a few institutions. For instance, the 5 most common or popular or "productive" schools were, as for the years before 1920 in table 1.1, eastern women's colleges—Mount Holyoke, Barnard, Smith, Vassar, and Wellesley colleges—which together accounted for 408 women, or 22.4 percent of the total. The top 10 schools—these 5 plus 2 private universities (Chicago and Cornell), 2 state universities (Wisconsin and California), and 1 more woman's college (Bryn Mawr)—account for 661 women, or 36.3 percent of the total. The total "production" of the top 24 schools (those from which 20 or more women had graduated), was 1,072, or 58.9 percent, with the long string or "tail" of 267 other institutions making up the remaining 40 percent. Taking all the women's colleges together, including the Catholic colleges and the numerous "state colleges for women," one finds that 60 such schools (20.6 percent of the institutions) contributed 714 women (39.2 percent of the graduates), or roughly twice as many women scientists as the institutional average. It is also worth noting that the small coeducational colleges that some studies have found extraordinarily productive of male scientists, such as Reed, Swarthmore, Oberlin, and Carleton colleges, did rather poorly here with women scientists (as they did in the Tidball and Kistiakowsky study). Only Oberlin College "produced" more than 20 women scientists in the 1938 *AMS*, and the others account for fewer than 10 each.

Table 6.5 also documents the strong influence exerted by some colleges in particular fields: Barnard College was quite strong, almost dominant, in anthropology, psychology, and to a lesser extent nutrition; Wellesley in mathematics, physics, and botany; and Mount Holyoke in chemistry, biochemistry, nutrition, and zoology. Such patterns of influence also occurred among schools and specialties not shown in table 6.5: Kansas State College was quite active in nutrition (eight graduates) but in no other field; Bryn Mawr College and the University of Missouri trained almost all the women in cytology but were negligible influences in other areas of zoology; and within geology, the University of Chicago accounted for six of the sixteen geographers (six of the thirteen single ones), while no other school "produced" more than two.

It is relatively easy to account for some of these patterns. Thus when these data on the baccalaureate origins of the women scientists in the 1938 *AMS* are compared with the list of notable women professors at many of these colleges (as presented in table 1.2 in chapter 1), certain lines of influence become clear. Thus Barnard's strength in anthropology was apparently due to the strong presence there of first Franz Boas and later Gladys Reichard, who, as mentioned in chapter 1, not only taught anthropology to undergraduates but also chaired one of the few anthropology departments at a woman's college. Wellesley's contributions were perhaps due to the strong influence of botanist Margaret Ferguson and physicist Louise McDowell; Mount Holyoke's to chemist Emma Perry Carr (plate 19) and zoologists A. Elizabeth Adams and Ann Haven Morgan; Vassar's to psychologist Margaret Washburn (plate 20); and Bryn Mawr's to geologist Florence Bascom (plate 21). The strong showing of the University of Kansas in zoology may have been due to the presence of Ida Hyde on its faculty (until 1920), and that of Kansas State College to Margaret Justin, longtime director of its nutrition program.

Yet a comparison of tables 1.2 and 6.5 can also be interpreted to show that often even these model professors were not enough to produce a string of scientist alumnae, and that in fact many women went into science without such

TABLE 6.5. Baccalaureate Origins of Female Scientists, 1938
(By Field)

Institution	Total	Psychology	Zoology	Botany	Medical sciences	Nutrition	Chemistry	Mathematics	Biochemistry	Microbiology	Physics	Geology	Astronomy	Anthropology	Engineering
Mount Holyoke	99	9	18	4	13	8	22	9	8	1	1	2	3	1	0
Barnard	87	25	4	9	2	13	7	2	6	5	2	3	1	7	1
Smith	79	9	9	16	14	2	3	6	8	3	1	2	4	2	0
Vassar	73	18	5	5	4	5	11	6	2	7	3	2	2	2	1
Wellesley	70	7	6	12	7	3	1	13	4	6	7	2	2	0	0
Chicago	68	7	8	10	4	11	5	1	7	7	1	7	0	0	0
Cornell	48	8	12	6	2	2	4	2	2	4	4	1	1	0	0
Wisconsin	48	2	7	10	5	5	0	5	5	6	0	3	0	0	0
California at Berkeley	45	6	8	4	7	3	3	4	1	3	1	2	1	2	0
Bryn Mawr	44	7	6	0	11	0	2	2	2	2	4	7	0	1	0
Goucher	42	3	12	5	5	2	2	5	2	1	3	0	1	1	0

Michigan	41	10	6	9	2	2	3	1	3	3	2	0	0	0	0
Nebraska	32	8	1	11	0	3	3	0	2	2	0	2	0	0	0
Minnesota	32	7	1	6	5	5	2	3	2	0	0	1	0	0	0
Ohio State	31	11	4	2	1	5	2	0	2	1	0	3	0	0	0
Kansas	29	3	13	0	4	2	3	2	1	1	0	0	0	0	0
Illinois	29	6	1	4	3	7	3	0	2	3	0	0	0	0	0
Radcliffe	27	3	5	3	2	1	3	2	3	1	0	0	2	2	0
Stanford	27	6	4	6	4	2	2	2	0	1	0	0	0	0	0
Brown	26	1	6	4	2	0	0	5	0	4	1	2	1	0	0
Washington (Seattle)	26	8	2	6	1	2	1	2	3	0	0	1	0	0	0
Oberlin	24	3	5	5	2	0	1	3	2	0	3	0	0	0	0
Missouri	23	3	6	3	1	2	3	2	2	0	1	0	0	0	0
Hunter	22	1	0	1	6	0	4	3	2	2	2	0	0	1	0
267 other institutions	749	96 from 67	117 from 82	106 from 73	60 from 48	76 from 59	69 from 54	65 from 52	53 from 40	39 from 36	23 from 18	18 from 16	18 from 10	4 from 4	5 from 5
Totals: Number	1,821	267	266	247	167	161	159	145	124	102	59	58	36	23	7
Percentage	100.1	14.7	14.6	13.6	9.2	8.8	8.7	8.0	6.8	5.6	3.2	3.2	2.0	1.3	0.4

SOURCE: *American Men of Science*, 6th ed. (1938).

PLATE 19. Chemist Emma Perry Carr of Mount Holyoke College was known for the many students of hers who became chemists and for her uncompromising advice on how to overcome employers' hostility. (n.d.) (Courtesy of Mount Holyoke College Library/Archives.)

PLATE 20. Psychologist Margaret Washburn of Vassar College, known for her conservative views, was the second woman president of the American Psychological Association (1921) and the second woman member of the National Academy of Sciences (1931). (Courtesy of Vassar College Library.)

PLATE 21. Geologist Florence Bascom, shown here in 1910, was the first woman to receive a doctorate from The Johns Hopkins University. In her thirty-three years on the faculty at Bryn Mawr College, she trained many of the nation's women geologists. (Permission of the Sophia Smith Collection [Women's History Archive], Smith College Library.)

strong female support. Thus the long line of Vassar astronomers produced only two of the important women in that field in the 1930s, and the mathematics department of Bryn Mawr College accounted for many more doctoral students (see table 6.6) than its just two undergraduates in the *AMS*. As for the existence of scientists without obvious teachers, it should be remembered that table 1.2 omitted the many male professors at the women's colleges who, like E. B. Wilson and T. H. Morgan in cytogenetics at Bryn Mawr, surely played a role in stimulating the young achievers. Another complication of some importance arises from the fact, not controlled for here, that some women scientists switched fields after graduation from college. Since table 6.5 classifies the women by their alma mater and their ultimate field (that given in the *AMS* rather than their actual undergraduate major), the department that "produced" them would not get the credit here if they changed fields. Margaret Morse Nice, for example, graduated from Mount Holyoke in psychology but was listed in the 1938 *AMS* as an ornithologist. She thus appears in table 6.5 as one of the many Mount Holyoke

TABLE 6.6. Institutions Awarding Doctorates to Twenty or More Female Scientists, 1938 (By Field)

Rank	Institution	Total	Psychology	Zoology	Botany	Medical sciences	Chemistry	Nutrition	Mathematics	Biochemistry	Microbiology	Geology	Physics	Astronomy	Anthropology	Engineering
1	Chicago	234	31	19	43	22	17	27	28	15	14	7	4	4	4	0
2	Columbia	212	68	23	12	8	19	31	6	19	8	4	4	1	9	0
3	Cornell	91	10	18	18	11	6	5	11	2	3	3	4	0	0	0
4	Johns Hopkins	88	17	12	0	21	10	0	7	4	11	3	3	0	0	0
5	Yale	87	7	14	1	6	7	25	7	12	3	1	3	1	0	0
6	California at Berkeley	71	8	21	8	6	2	6	5	1	1	2	2	6	3	0
7	Wisconsin	60	4	7	25	3	0	3	4	4	9	1	0	0	0	0
8	University of Pennsylvania	53	10	12	6	7	0	1	3	4	6	0	3	0	1	0
9	Minnesota	52	12	1	11	5	7	6	3	4	2	1	0	0	0	0
	Illinois	52	2	11	5	2	10	2	7	9	4	0	0	0	0	0
11	Michigan	46	7	6	7	5	3	0	3	5	2	1	2	5	0	0
	Bryn Mawr	46	8	11	0	0	10	0	7	2	0	7	0	0	1	0
13	Radcliffe	42	7	7	5	6	2	0	6	1	1	1	2	4	0	0
14	Ohio State	37	13	7	4	1	2	1	3	3	2	1	0	0	0	0
15	Iowa	32	14	2	0	3	3	5	1	2	1	1	0	0	0	0
16	Stanford	22	7	5	4	1	2	1	2	0	0	0	0	0	0	0
17	Washington (St. Louis)	21	0	4	8	6	0	1	0	1	1	0	0	0	0	0
	86 other institutions	345	36 from 19	45 from 26	37 from 22	56 from 30	39 from 25	20 from 13	20 from 15	32 from 21	20 from 17	13 from 8	19 from 12	2 from 2	2 from 2	4 from 3
	Totals: Number	1,591	261	225	194	169	139	134	123	120	88	46	46	22	20	4
	Percentage	99.9	16.4	14.1	12.2	10.6	8.7	8.4	7.7	7.5	5.5	2.9	2.9	1.4	1.3	0.3

alumnae zoologists, though her debt to psychologist Helen Thompson (Woolley) was also great. This switching factor may be significant in those fields where such movement was common, such as from chemistry to nutrition or physics to astronomy.

But one should not use these patterns to become too deterministic about the strength of a few special teachers and some particular baccalaureate institutions. In fact, the more one studies such patterns and notices the strong channels that existed in certain fields and all the advantages that must have gone with this, the more curious one becomes about the motivation and strength of purpose of those achieving graduates of the 267 "other" institutions that had no such traditions or patterns of influence that might have eased their passage. For example, Oakland City College in Indiana had only one woman graduate in the 1938 *AMS*, yet she was Melba Phillips, later professor of physics at the University of Chicago and recent president of the American Association of Physics Teachers. Certainly an institution's rank in the list of "baccalaureate origins" of women scientists is no accurate measure of the quality of the women it "produced."[18]

Table 6.5 can also serve the purpose of providing evidence on how widely or commonly taught a particular science was in the 1920s and 1930s. This is fairly important, since there were apparently wide differences among the sciences. Such data can give some measure of the relative openness or ease of entry into the various fields and, to a lesser extent, can help estimate the size of the potential job market for women faculty members in those years. Thus, for example, the women zoologists graduated from 105 different baccalaureate institutions, indicating that that field was the most widely taught and therefore most accessible of all the sciences to women. The botanists came from 95 schools and the psychologists 92. By contrast, the women physicists came from only 33 schools, the geologists 31, and astronomers just 20, though these included a wide variety of types of institutions: state and private universities as well as women's colleges. The women anthropologists, however, came from just 13 schools, 9 of which were women's colleges, making it a very rare and relatively elite subject. This evidence also offers striking confirmation of Ruth Benedict's frequent lament (in *Time* and elsewhere) that there were hardly any teaching jobs in anthropology in the 1930s, and practically none for women.[19]

These patterns and data also suggest some clues as to a possible mechanism by which young women were recruited into careers in science in these years, and give some indication of the importance of maintaining a certain critical mass or number of women professors in a field to keep the process alive. This recruitment pattern also helps to explain how the larger fields tended to get larger and the smaller ones to falter and have trouble even surviving. One is struck by certain recurring patterns in the data. Can it be a coincidence that the largest fields of 1921 were still the largest in 1938 and that the women in these three fields (zoology, botany, and psychology) came from the greatest number of schools? One suspects that some of the women in these fields in 1921 had recruited many of the others over the intervening years. Thus a cycle of recruitment was set up whereby the largest fields, with the largest number of teachers spread among the nation's colleges and universities, were able to grow even larger relatively easily. However, in those fields with (for whatever reason) smaller numbers of women in a given generation, there would be far fewer of them to teach their subject to the next generation, thus perpetrating the idea that women did not take the subject or have careers in it. Below a certain "critical mass" of women in a field, therefore, their position can become quite precarious, and they run the risk of being totally driven from a field. It may perhaps be relevant that it was in the 1930s that the first articles examining the women's low enrollments in physics, even at women's colleges, began to appear.[20]

The *AMS* also provides new and useful data for the well-worked subject of "doctorate production." The extensive literature on this subject shows that a large proportion of doctorates have always been granted by a small number (less than 20) of historic graduate institutions.[21] Since the NRC's data on women doctorates are not broken down by the universities that granted their degrees (but only by year and by field), it is worthwhile to present here the institutional data on the 1,591 women in the 1938 *AMS* (83.2 percent of the total) who reported holding doctorates. Their entries, as tabulated in table 6.6, show that, as earlier (table 2.1 in chapter 2), relatively few universities have granted most of the doctorates to women over the years.

Although these women earned doctorates at 103 different institutions in the United States and abroad, by far the most popular graduate schools for them (as for the men) were the University of Chicago and Columbia University, which rapidly outstripped the earlier leader, Yale University. Together Chicago and Columbia awarded 446 doctorates (28 percent of the total) to women in the 1938 *AMS*. Both universities had strong programs in almost every science, but Chicago was particularly effective in botany, mathematics (L. E. Dickson), medical sciences, and the biochemical cluster of chemistry (Julius Stieglitz), biochemistry, and nutrition. Columbia (including Teachers College) was not only very strong in the nutritional sciences (Henry Sherman and Mary Swartz Rose) and zoology,* but especially in anthropology (where its 9 degrees constituted 45 percent of the total granted to women in this field in the entire country) and psychology (where its 68 doctorates were the largest number granted to women by any department in any science in the nation). Far behind these two giants were the trio of Cornell (91), The Johns Hopkins (88), particularly important in the medical sciences, and Yale University (87), primarily strong in nutrition, biochemistry, and zoology because of Lafayette B. Mendel, Treat Johnson, and Ross G. Harrison.

In addition, the *AMS* data provide a clue as to the importance of foreign doctorates in some fields in these years, a subject that the NRC data, which are limited to American universities, do not consider. Of the 103 institutions that

*Of the nine women whose specialty was genetics or cytology and whose degree was awarded before 1928, when T. H. Morgan moved to Cal Tech, only one (Mary B. Stark) seems to have been included in the activities of his famed "Fly Room."[22] His years at Bryn Mawr and many women students there had not made him a feminist.

awarded doctorates to women listed in the 1938 *AMS*, 26 were in foreign countries. Altogether these schools granted seventy-eight degrees (4.7 percent of the total), sixty-nine of them from just four countries: Canada (thirty-five), Germany (eighteen), and Great Britain and Switzerland (eight each). Table 6.7 provides data on the 11 universities which granted more than two doctorates to these women. This reveals that only 2 institutions, the University of Toronto and McGill University, granted more than ten degrees each, and that foreign doctorates were most commonly held by women in biochemistry (fourteen) and zoology (twelve). The single most productive foreign departments were Toronto in zoology (four) and Göttingen in physics (four), the latter known internationally for its outstanding faculty in the years before Hitler rose to power. When Göttingen's total is compared with those of the American physics departments shown in table 6.6, its quantitative importance is further apparent, since its four doctorates tied the Chicago, Columbia, and Cornell departments as the most "productive" of women physicists. Göttingen might also be thought to have surpassed the other departments, since one of its four female doctorates was Maria Goeppart-Mayer (plate 22), the only Nobel Laureate in the group.

It is important to remember when reading either table 6.6 or table 6.7 that although large numbers of women earning doctorates in a particular program or at a certain university probably indicate that the program was a large and longstanding one where the women were relatively welcome, a small number of women doctorates may mean several things. It may indicate that rampant discrimination was taking place (as for instance seems to have occurred in the chemistry department at the University of California, from which only two women listed degrees); it may also indicate that the program was weak and attracted few students of either sex (as in the anthropology departments competing with Columbia) or that the program was late in starting (such as the psychology department at Yale or astronomy at Harvard, which were not formed until 1927) and thus could not have granted many doctorates by 1938.

As with the women's baccalaureate degrees, one can attempt to trace some of these patterns to women on the faculty of some of these doctoral universities, although there were far fewer of them so employed. Thus one might suspect that the presence of Ruth Benedict in the anthropology department at Columbia, Mary Swartz Rose in the nutrition department at Teachers College, or Florence Goodenough in the psychology department at the University of Minnesota would have helped to increase the number of women doctorates coming from those departments.[23] This may have been the case, but one would have to imagine that any department that would elect or even tolerate a woman associate or full professor in the 1920s and 1930s was already more liberal and tolerant than many others. The by far more common pattern was that even those departments that were known for their large numbers of female doctorates, for example, botany at Chicago or biochemistry/nutrition at Yale, had no women faculty.[24] These departments were liberal and tolerant up to a point and no further. (This pattern and particularly the exceptions to it are discussed more fully in chapter 7, and are illustrated in tables 7.4 and 7.5.)

TABLE 6.7. Foreign Universities Awarding Doctorates to American Female Scientists, 1938 (By Field)

Institution	Total	Biochemistry	Zoology	Medical sciences	Physics	Botany	Chemistry	Psychology	Mathematics	Microbiology	Nutrition	Engineering	Anthropology	Astronomy	Geology
1 Toronto	19	2	4	3	2	3	0	0	2	1	1	0	0	0	1
2 McGill	13	3	2	2	0	2	2	0	0	0	1	0	0	1	0
3 Zurich	6	0	3	1	0	0	1	0	1	0	0	0	0	0	0
4 Berlin	5	2	0	0	1	1	0	1	0	0	0	0	0	0	0
Göttingen	5	0	0	0	4	0	0	0	1	0	0	0	0	0	0
6 Cambridge	4	2	0	0	1	0	1	0	0	0	0	0	0	0	0
7 Munich	3	1	0	2	0	0	0	0	0	0	0	0	0	0	0
8 Vienna	2	1	0	0	0	0	0	1	0	0	0	0	0	0	0
London	2	0	1	0	0	0	0	0	0	0	0	0	1	0	0
Moscow	2	0	1	0	0	0	1	0	0	0	0	0	0	0	0
Würzburg	2	0	0	1	1	0	0	0	0	0	0	0	0	0	0
15 others	15	3	1	2	0	2	1	4	0	1	0	1	0	0	0
Total	78	14	12	11	9	8	6	6	4	2	2	1	1	1	1
Percentage of total women doctorates (U.S. and foreign) in *AMS*	4.7	10.4	5.1	6.1	16.4	4.0	4.1	2.2	3.1	2.2	1.5	20.0	4.8	4.3	2.1

SOURCE: *American Men of Science*, 6th ed. (1938).

PLATE 22. Physicist Maria Goeppart-Mayer, who won part of the Nobel Prize in 1963, is shown here with Enrico Fermi and her husband, Joseph (obscured), in 1930, shortly after the couple moved to the United States. (Courtesy of Mandeville Department of Special Collections, University of California, San Diego.)

Although one might like to compare the rankings of the women's favorite American graduate schools as given in table 6.6 with the men's most frequent choices, this can only be done imperfectly, since the cumulative institutional totals published by the NRC for all doctorates awarded from 1920 to 1937 are unfortunately not broken down by sex. Nevertheless, a comparison of these two lists may give a rough indication of which universities were more receptive to women doctoral candidates and which were less than would have been predicted by the NRC data alone. This comparison is made in table 6.8, but yields a rank correlation of only r = .37 (insignificant at < = .10 level), indicating (if anything) that the women were only somewhat likely to go to the institutions most popular with the men, despite the ranking of Chicago and Columbia at the top of both lists. More receptive to women than the NRC data would have predicted were the universities of Pennsylvania and Minnesota, but they were more than countered by the relative hostility shown by Harvard, New York University, and of course Princeton University, which refused to admit women to its graduate school until the 1960s. Harvard had created a separate graduate school for women in 1902, as explained in chapter 2, but Radcliffe served only as a partial substitute and, by ranking only thirteenth among graduate schools for women, it

TABLE 6.8. Most Productive Doctoral Universities, 1938 (By Sex)

Institution	Men and Women	Women Only
Columbia	1	2
Chicago	2	1
Harvard	3	13
Wisconsin	4	7
Cornell	5	3
Yale	6	5
Johns Hopkins	7	4
California at Berkeley	8	6
Illinois	9	9.5
Michigan	10	11
New York University	11	23
Iowa	12	15
Ohio State	13	14
Minnesota	14	9.5
University of Pennsylvania	15	8

SOURCES: Men and women: Lindsey Harmon and Herbert Soldz, comps., *Doctorate Production in United States Universities, 1920–1962* (Washington, D.C.: National Academy of Sciences-National Research Council Publication no. 1142, 1963), p. 19. Women only: *American Men of Science*, 6th ed. (1938).
NOTE: r = .37.

failed to play the prominent role in women's graduate education that Harvard did for men.[25] Thus though the women attended many of the same doctoral institutions as the men, there were enough variations in the patterns of the sexes to make the women's mental map of their favorite graduate schools somewhat different from the men's.

Perhaps the most suggestive data on doctorates derived from the *AMS* are those that demonstrate the important role this degree played in the women's career strategies in the 1920s and 1930s. As shown in table 6.9, the percentage of men and women in a field who held doctorates in 1921 and in 1938 varied widely by field and by sex over time. Those women in the directory were, on the average, 13 percentage points more likely to hold a Ph.D., M.D., or Sc.D. than the men, not only in 1921 (when the two percentages were 71.8 percent and 58.2 percent), but again in 1938 (83.2 percent and 70.3 percent). For there to be this much of an increase in the cumulative figures in *AMS* entrants, practically all of the younger women and most of the new men added to the *AMS* between 1921 and 1938 would have been earning doctorates, which does seem to have been the case.

To a certain extent the men's overall lower likelihood of holding a doctorate reflects the dissimilar distribution of the sexes among fields. The men were far more likely than the women, as shown in tables 6.1 and 6.2, to end up in what might be termed the "low doctorate" or fieldwork sciences, where less than 35 percent of the scientists held doctorates. The women, by contrast, were more likely to be in the "high doctorate" fields, such as the medical sciences, psy-

Table 6.9. Preponderance of Doctorates, 1921 and 1938
(By Field and Sex)

Field	Men*				Women			
	1921		1938		1921		1938	
	Number	Percentage	Number	Percentage	Number	Percentage	Number	Percentage
Medical sciences	1,368	98.7	3,325	100.0	51	91.1	169	90.9
Psychology	216	92.3	975	97.5	56†	93.3	261	94.2
Biochemistry	(under chemistry)		875	97.2	(under chemistry)		120	94.6
Physics	612	79.8	1,625	74.8	14	66.7	46	73.0
Mathematics	504	73.7	950	74.5	35	83.3	123	81.4
Chemistry	1,062	65.6	3,775	80.7	29	82.8	139	85.3
Zoology (except entomology)	378	56.8	1,375	71.6	48	69.6	203	80.9
Entomology	36	13.3	400	53.3	7	77.8	22	73.3
Anthropology	54	50.0	100	44.4	2	25.0	20	69.0
Astronomy	126	41.2	75	37.5	4	20.0	22	61.1
Botany	360	40.8	1,775	69.6	48	55.2	194	75.8
Geology	216	34.3	575	53.5	14	60.9	46	76.7
Agricultural sciences	126	30.4	450	47.3	0	0	0	0
Engineering	198	22.4	925	26.4	0	0	4	50.0
Microbiology	(under medical sciences)		550	78.6	(under medical sciences and botany)		88	80.7
Nutrition	0	0	150	66.7	15	75.0	134	81.8
Total	5,256	58.2	17,850	70.3	323	71.8	1,591	83.2

Sources: *American Men of Science*, 3d ed. (1921), and *American Men of Science*, 6th ed. (1938).

*Based on samples of every eighteenth male in the third edition and every twenty-fifth in the sixth.

†One could add Mary Calkins and Christine Ladd-Franklin here and make the total even higher (see chapter 2).

chology, and biochemistry, where over 90 percent of the persons held the degree. These different doctoral patterns, already a sign of different career strategies of the two sexes, can be detected even within the same field in 1921. Thus the women in the "low doctorate" field of entomology did not leave graduate school after the master's degree and join industry or the government, as the men usually did, but instead 77.8 percent of them (six times the men's percentage) earned doctorates and headed for academic careers. Their tendency to stay on for doctorates was, in effect, a result of the restrictions placed upon them in the job market, which forced them to congregate in academia, where doctorates were already necessary for women in 1921.

Lest one think that this prodoctorate "academiotrophic" strategy was common to all the women in the 1921 *AMS*, however, table 6.9 also shows that the women astronomers and anthropologists did just the opposite and earned doctorates only 50 percent as often as the men in their field who were listed in the *AMS*. This pattern may reflect the continued segregation of women in these fields in various kinds of "women's work," such as astronomical assistants or as free-lance, anthropological "collaborators," where degrees were not yet necessary in 1921. In fact, one might expect that those women astronomers or anthropologists who did seek doctorates before 1920 would have been met with much discouraging "advice" to the effect that such degrees were unnecessary as well as refusals to admit them to the few doctoral programs then available. But without examples, this remains only a speculation.

By 1938 this wide variation in doctoral patterns among the various sciences had narrowed greatly, as also shown in table 6.9. Almost all fields became highly doctorate-conscious, and some, such as botany and geology, showed a net increase of 20 percentage points or more in the number of men holding doctorates (40 points in entomology) by 1938. The women likewise made striking net gains in botany (of 20.3 percentage points) and more than caught up with the men in anthropology (up 44.0) and astronomy (41.1). Although the percentages of doctorates actually dropped slightly between 1920 and 1938 in three other cases—for men in astronomy (by 3.7 percentage points) and anthropology (5.6) and for women in entomology (4.5)—they represent anomalies in the otherwise very rapid acceptance of the Ph.D. as a prerequisite for almost all professional careers in science by the 1930s. Yet despite this large overall increase by the 1930s, the women, as noted earlier, still held a higher percentage of doctorates than did the men in almost every field, with particularly large differences in anthropology (24.6 percentage points), astronomy and engineering (23.6), geology (23.2), and entomology (20.0). In only one field, medical sciences, did the men's percentage exceed that of the women by more than 3 percent (100 percent vs. 90.9 percent), perhaps because some of the women were staff assistants and associates without doctorates. These differences, so large and persistent, are particularly remarkable for the late 1930s, since they occurred at a time when, as shown in the NRC doctoral data, the women's proportion of new doctorates had been declining. This pattern suggests circumstantially that what was earlier termed the "Madame Curie strategy" was continuing and had started before 1921. Many women scientists evidently perceived that to be considered "equal" to men, and even to be included in the *AMS,* they had to be demonstrably "better." The most determined of them (the "survivors") therefore internalized this double standard and settled down to becoming "Madame Curies"; earning an extra degree was just the first step in a lifelong struggle to overcompensate for being women in a man's occupation. In time such women became the extraordinarily well-qualified, uncomplaining, and utterly self-sacrificing paragons that abound in the obituaries of women scientists.

This quantitative overview of the strong forces affecting women's careers in science in these important decades suggests that, though there were great differences among the sciences, the period in general was one of rapid growth, with record numbers of women entering science. Even if one uses only *AMS* data, by 1938 one can find almost 2,000 women scientists from almost 300 undergraduate colleges (though from some more than others), most of whom were earning doctorates at major universities in their fields. Far more were married than had been earlier, but a disproportionate number of these were unemployed in 1938. Nevertheless, by the 1930s the women in the *AMS* had become so successful in surmounting earlier educational restrictions and barriers that they were, on the whole, even better educated than the men in their field. Yet there was a danger signal hidden in this "success," an indication that this strategy of deliberate overqualification had been thrust upon them: restrictive and discriminatory employment meant that they had to have far better credentials than did men in order to be included in the *AMS* or to get even subordinate positions. Thus behind all these figures of growth and expansion in the 1920s and 1930s there were signs of the familiar tale of restriction and discrimination.

Yet reactions to it varied. Some women scientists, especially those in academia, were angry at the restrictions that still held even the best women of the time from much advancement or recognition, and they continued into the 1920s their old strategy of collecting statistics and publicizing the unequal pattern of women's careers. Others, however, especially those in industry, were less optimistic and advised a lessening of goals and a greater acceptance of and adaptation to the inevitable. Neither strategy produced much notable change in the next several decades, but each offered the women scientists not only a way to survive in science a while longer but also a chance to feel that their personal struggle might help later women go a bit further.

7

ACADEMIC EMPLOYMENT: PROTEST AND PRESTIGE

Although women scientists, to use their own phrase, had "opened the doors" to careers in science by 1920, their jobs there were still largely limited to faculty positions at the women's colleges, instructorships and jobs in home economics at the coeducational institutions, and other classic kinds of "women's work" at research institutes and government bureaus. This segregation persisted with few modifications over the next two decades, as most women took this by then established pattern for granted. Some idealists, however, were sufficiently discontent with the remaining inequalities facing women in science to work in a variety of ways to increase their advancement, pay, and recognition. Largely unsuccessful, their efforts nevertheless help illuminate the attitudes toward women in science and the limits put upon them in the 1920s and 1930s. Since the contexts in which these women (and some men) struggled first to overcome but then just to adapt to the entrenched resistance were very different in academia, government, and industry, it is convenient to treat these three areas separately—that of the more numerous academic women first, in this chapter, and those of the others in the next two. The women's imaginative response to the almost total lack of recognition for any women in science is considered in the final chapter.

Women scientists in this period were still a predominantly academic group. Over three-quarters (76.2 percent) of the 1,652 employed women listed in the 1938 *American Men of Science* remained, despite new opportunities for them elsewhere, in higher education. Academic women scientists, in fact, were so numerous and, though dispersed across many fields and types of institutions, were facing such distressingly similar circumstances around 1920 that many of them began to criticize what seemed to them pernicious employment practices. Unlike the women scientists in industry (who are discussed in chapter 9), academic women still held in the early 1920s the optimistic liberal faith of the Progressive Era that an evil once documented would be largely self-correcting, or that once notified that their behavior was offensive to others, moral, well-behaved persons would take corrective steps of their own accord. Thus if deans and department chairmen were alerted with properly documented studies to the fact that their employment practices were unfair to women, they would, it was hoped, accept the findings and voluntarily take steps to change their ways. In retrospect this may seem to have been quite naive, especially for women whose predecessors had been so politically astute, yet these women did not explore any alternatives. Since a version of this faith had worked in the 1890s, when academic feminists had successfully wielded the idea of "fairness" to get doctorates for women at the major graduate schools, it is no wonder that these weaker feminists of the 1920s expected their academic officials to be equally obliging about employing women on their faculties once they had been informed how prejudiced and "unfair" their practices had become.

The inability of the women's strategy to reform academia was quickly evident, however. The women had neither the power nor the will to do anything stronger than publicize evils; in the 1920s, when institutional autonomy was sacrosanct and accreditation procedures weak and largely voluntary, they certainly had no way to coerce reluctant administrators into hiring women. Similarly, state or federal government intervention on behalf of women faculty, despite women's recent acquisition of the vote, was well-nigh unthinkable at the time. Influencing employers in other ways, such as through statistical studies that demonstrated inequalities and thus appealed to their good will or sense of fair play, was the only strategy that the women of the 1920s were able to employ. Yet though many academic officials responded to the feminists' numerous surveys and questionnaires of the period (many others did not), there is no sign that any changed their employment practices because of the women's findings. Nor, apparently, could the women find even one obliging employer whom they could then play off against the others.

Although, as mentioned in chapter 5, the newly established American Association of University Professors had formed in 1918 its own committee to examine systematically the status of academic women, this group did not publish its first report until 1921 because its chairman, A. Caswell Ellis, professor of the philosophy of education at the University of Texas, prominent Democrat, and social reformer, was already busy with numerous other projects, including fighting the influenza epidemic at his university, rebuffing an attack on his department, and pressing for the ratification of the suffrage amendment in his state.[1] In the interim, the Association of Collegiate Alumnae, which had long been concerned with the problems of women faculty, maintained the momentum with a steady barrage of strident articles on the topic in its *Journal*. These show the women's rising anger at their mistreatment, their growing sophistication in handling statistical data, and their increasing willingness to single out for criticism institutions whose policies were particularly discriminatory.

The women's anger and improved ability to manipulate statistics in their favor are shown quite dramatically in two articles in the *Journal of the ACA* in 1918. The first merely reported data from the U.S. Bureau of Education on the distribution of 10,034 faculty members which showed that sixty coeducational schools employed more women than did nine major women's colleges.[2] But the second, published a few months later and written by an angry assistant professor at the University of Montana, recast these same data to document the almost total lack of advancement for women at such coeducational schools, where promotional opportunities lagged far behind those at the women's colleges. Helen Sard Hughes calculated that women held 59.4 percent of the full professorships at the women's colleges but a paltry 3.5 percent of those at the coeducational schools. A high proportion of all faculty women were instructors—47.3 percent of those at the women's colleges and 62.5 percent of those at the coeducational schools. Moreover, both these percentages were twice as high as those for men—just 26.5 percent of the men at the women's colleges and only 32.5 percent of those at the coeducational schools were instructors. Apparently what was fairly common practice for women at both kinds of institutions was quite rare for the men, who were not expected to start at the bottom, or if they did, to stay there very long. The author was so incensed at this deliberate discrimination that she concluded her article by reproducing a particularly insulting letter a friend had recently received rejecting her for a job at a coeducational school solely because of her sex.[3] Surely the ACA should be doing something to help eliminate such an affront to the dignity of women scholars.

Two years later the angry members of the Lincoln, Nebraska, branch of the ACA took this data-gathering tactic a step further and used it to publicize both nationally and locally some particularly offensive discriminatory practices in their state. This approach at least had a specific target, and it had the benefit of embarrassing university administrators, whose sly evasions and hot denials could then be refuted in another round of unfavorable publicity. Since most members of the Lincoln ACA were graduates of the University of Nebraska and many were or had been faculty members there, the group had much firsthand experience with its hiring and firing policies. Although the group's report, blandly and euphemistically entitled "Opportunities and Salaries of Women in the Teaching Profession in Nebraska," covered several other colleges, three normal colleges, and several public school systems as well as the university, it was the discussion of UN's "tendency in the past five years to replace women on the faculty by men" that aroused the greatest public interest and response. The ACA women claimed that in the school year 1915/16 there had been 68 women on the university's staff of 295 (23.1 percent), but that by 1919–1920 the number had dropped to just 53 of 328 (16.2 percent). Likewise, a new salary scale had recently been enacted which did not reward either higher degrees or advanced work by women and actually offered female associate professors a *lower* maximum ($2,500) than it did female assistant professors ($3,000).[4]

This report was published in the April 1920 *Journal of the ACA* but also received so much local publicity in Nebraska and at the university's Alumni Day in June 1920 that P. M. Buck, Jr., dean of the College of Arts and Sciences, issued an angry response, which was reprinted in the July–August *Journal*. After asserting that the absence of new women on the UN faculty was due to a lack of female applicants for such positions, Dean Buck went on to compare the proportion of women at his institution very favorably with those of six other midwestern universities. He did not explain, however, the rapid drop in the number of women faculty at UN, the policy behind the decline, or the pernicious salary scale that seemed designed to weed out the stronger and more accomplished women. The ACA women responded by pointing out that open applications for academic jobs were still a rarity for both sexes in 1920, by rejecting the institutional comparisons as irrelevant to the issues at hand, and by citing specific examples of ACA members who had been laid off rather than advanced as men automatically were at the university. This spirited exchange in the public press got the women a hearing and some satisfaction and brought the university some negative publicity, but otherwise led to little immediate change. The situation was at an impasse there as elsewhere, for the women had no other way to fight even a public institution in 1920. And as the dean had pointed out, the University of Nebraska was by the standards of the time one of the region's most favorable institutions toward women (see table 1.3). In time the protest may have helped somewhat, since by 1938 the number of women scientists the university had listed in the *AMS* was up from three in 1921 to nine.[5]

By 1920 such seemingly objective, but in fact politically motivated, fact-gathering had become the main tactic among women's groups trying to publicize and deplore their inferior position in academia. Though the sample sizes in these studies tended to grow over time and the methodological rigor to increase (for example, age or length of service was never included in the early studies, although the women's low rank might have been the result of shorter careers in academia), the conclusions hardly ever varied—the women were always found to be lagging far behind men in academic status and salaries. Although most of these studies were aimed at female audiences, like those for commencement addresses at women's colleges or the readerships of ACA (later AAUW) publications or the *Journal of Home Economics,* a few appeared in *Science*. Most authors, however, were reluctant or were unable to identify, even with these friendly audiences, what was causing the situation or to suggest possible solutions, and none specified or called for corrective action. The clearest analysis of the situation from the feminist perspective (which held that the system was at fault and was treating women unfairly) was William H. Welch's commencement address at Bryn Mawr College in 1922, which, as mentioned in chapter 5, was written with much guidance from M. Carey Thomas. It appeared in *Science* and clearly indicted the discriminatory hiring practices of academic institutions; though it was, with the reports of the AAUP's Committee W, perhaps the high point of the feminist critique before 1940, it suggested no remedies or way to enforce improvements.[6] The problem, once exposed and documented, was apparently supposed to solve itself.

In 1921 the long-dormant Committee W of the AAUP finally sprang to life and prepared the most comprehensive survey on the status of academic women to date. Although none of the committee members was a specialist in research on sex differences, women scientists, especially those from the women's colleges, were especially well represented on this committee (as they were generally in the AAUP) by geologist Florence Bascom of Bryn Mawr; astronomer Harriet Bigelow of Smith; zoologist Cora Beckwith of Vassar; Anna McKeag, professor of education at Wellesley; Marion Talbot, professor of household administration at the University of Chicago; and Louise Stanley, professor of home economics at the University of Missouri. Among the several men were the feminist and philosopher John Dewey of Columbia University, who was away in China much of the time, historians D. C. Munro of Princeton University and Herbert E. Bolton of the University of California, and economists Walter F. Willcox of Cornell University (brother of Mary Willcox, a Wellesley College zoologist) and A. B. Wolfe of Ohio State University (formerly of the University of Texas). Yet the committee's records in the University of Texas Archives indicate that most of the work fell upon chairman Ellis, who complained to the AAUP's secretary in 1924, "I deeply regret being so slow with this [report], but I have done the best I could under the press of much other work and with a committee scattered from Maine to California, three fourths of whom cared nothing about the work. I rather suspect that the men except Dewey and Wolfe were afraid of it!"[7]

Using data collected on almost 13,000 faculty members at 145 institutions that belonged to the AAUP, Committee W found that there were no women at all on the faculties of 27 of the 100 coeducational schools and that the women who were employed at the others held only 4 percent of the full professorships (and less than 3 percent if the highly feminized fields of home economics and physical education were omitted) but 23.5 percent of the instructorships (1,019 of the 1,646 women in the study [61.9 percent] held this position). Yet in the year 1921 a majority of the committee could find grounds for optimism in even these dismal statistics, since many of the women had been hired in the previous two years, a circumstance respondents attributed to the recent war and the suffrage movement. Part of the committee's optimism also grew from the recent appointments of the first women to both Harvard and Yale faculties.[8]

Yet a similar report eight years later by Marion Hawthorne of Northwestern University found that little had changed. When in 1929 she examined the advancement of 844 faculty women at 122 graduate institutions in the American Association of Universities, she found that, as earlier, they held lower status and had smaller salaries than men. More interestingly, she tried to correlate this employment status with their advanced degrees and their length of experience, but not even these factors could account for the women's low rank, or, as Hawthorne concluded, "It would appear that the field of college teaching holds comparatively little promise for women. The majority of the women reporting seem to have been retarded or advanced by reason of factors in their institutions, over which they had relatively little control. Training represented in degrees and years of teaching experience contributes little to the advancement of women in the college teaching field, as shown by the low correlations between rank and degrees (.38) and between rank and years of teaching experience (.47)." Although primarily statistical, her study also revealed some of the anxiety and despondency of the women themselves: "Women with exceptional ability and proper influence testify to the fact that they were able to rise to a position equal to that of male colleagues, but the rank and file of the respondents seem to have developed a defensive attitude bordering on martyrdom, and complained, waxed bitter, and voiced resentment toward the conditions of which they were victims."[9] Thus what is often considered a "golden age" for women in academia, a time when the overall percentage of women on college faculties, as calculated by the U.S. Bureau of Education, was relatively high by later standards (26–28 percent), was filled with inequalities and discrimination that relegated most women at the coeducational schools to positions of instructor and offered them little hope of advancement.[10]

Yet these numerous reports and studies apparently convinced no one (except perhaps the already converted) of the need to treat women equally or just better in academia, perhaps because of what was revealed in a second, more slippery, kind of status of women report, the opinion survey. Not content with having collected so many data and circumstantial evidence of discrimination, the AAUP's Committee W moved on in 1924 to "discover what opinions" were held of women faculty members by their male and female colleagues. In trying to determine whether their performance was really so weak as to justify their low ranks and salaries on university faculties, the committee sent a lengthy questionnaire on sex differences in teaching and research to one administrator and one male and one female faculty member (where there was one) at each of the 176 institutions in the AAUP. This produced 152 responses from 86 coeducational and 13 women's colleges (109 men and 43 women). From these responses the committee found that ". . . it appears that in the matters of conscientiousness, of efficiency in committee service, and of developing a social point of view in the students, the women have equalled or excelled the men in the opinion of the overwhelming majority (96%, 80% and 80% respectively) of their colleagues. In the matters of interest in the larger problems of college life, 66% regard them as equal or superior, while in their knowledge of, and interest in, civic and social problems 58% considered them equal or superior" (p. 72). Regarding the specifics of teaching and research, the women again did well by most criteria: ". . . it appears that while 81% of their male colleagues regard women as equally successful with men in the ordinary college teaching, only 54% think that they are as successful with advanced classes, 59% regard them as continuing to advance in their own scholarly life," except in the critical area, where "only 31% regard them as keeping pace with the men in productive scholarship" (p. 70). But having uncovered this alleged deficiency, the committee was quick to point out: "To what extent the above opinions are in accord with the facts, no one can say, for no accurate and adequate data have been collected. The Committee proposes next to undertake to secure some reliable data on the scholarly productivity of faculty women, as this is open to scientific study . . . [unlike teaching]." But Committee W, which had once asserted confidently that "mere prejudices must die if exposed to the light," never published this proposed study, and by 1927 the committee ceased to exist and was not revived until 1970.[11]

Meanwhile Ella Lonn, associate professor of history at Goucher College, who had undertaken her own survey of the attitudes of academic employers toward women, reported her results to the AAUW in 1924. Because she wished to fathom "the masculine mind," her report reveals more of the prevailing mentality than does the more cautious and cryptic AAUP report. She mailed questionnaires to 202 department chairmen (including two women) at seventy universities in the Association of American Universities. (Relatively few of the chairmen—the number was not given—were in the sciences, and she omitted home economics entirely as not sufficiently scholarly.) Responses came back from 129 chairmen at fifty-three schools, but twenty-five were blank (usually because there were no women in that department). As usual these chairmen blamed the few women in their departments for not being Madame Curies—for not being more scholarly, for not holding more doctorates, for not publishing more, for not being better informed, and even for not speaking up enough at faculty meetings. Collectively they said that, though they had no personal prejudice, they did not wish higher education to become effeminate. Therefore, they preferred women in lower positions (if in any), where their role coincided neatly with tradition, the men's preference, and the women's own demonstrated ability at routine tasks. Besides this, the chairmen thought that students of both sexes preferred male faculty. Although Lonn hinted to the AAUW members that this projection of all the desireable traits onto the men and all the undesireable ones onto the women was unreasonable, her only recommendations were for the women and their organizations to work even harder to overcome their admittedly "vulnerable" traits. If the chairmen would promote only women who were so outstanding that their advancement could not be avoided, then some and perhaps all women must try to reach that level. The burden was clearly upon them to become "exceptions."[12]

Although the strategy behind the opinion surveys had been the dispassionate one of trying to separate valid "facts" from "mere opinions," in practice neither Committee W nor Professor Lonn was able to maintain such distance on what turned out to be a highly emotional subject. They did not realize that when their surveys had simply asked chairmen what their male and female faculty members were like, they had put no controls upon one important variable—faculty rank. Thus it apparently never occurred to the members of Committee W or to Professor Lonn that many of the negative traits that the male employers had attributed to women and blithely considered immutable sex differences were probably related more to the women's predominantly lower ranks as instructors (and so might change if the women were given more opportunity). Nor did any of these analysts challenge the officials' extraordinarily high expectations that females (usually instructors) achieve at the same level as males (often full professors) as still more evidence of systematic discrimination. Nor had Chairman Ellis taken the precaution of including on his committee any of the feminist psychologists, like Helen Thompson Woolley or Leta Stetter Hollingworth, who were, as shown in chapter 5, experts in sex-difference research. This was a grievous omission, since they would have reveled in saving the committee from just this kind of logical error, the fallacy of the suppressed variable.[13]

Thus instead of turning the chairmen's responses back upon them as still more evidence of their prejudice and bias, these analysts, apparently sensing the employers' intransigence and even anger and desiring to maintain their stance as objective, moderate reformers rather than politically strident feminists, hastened to lessen the conflict; they became apologetic and advised the women to work harder to overcome their evident weaknesses, especially in publications (the assumption being that if they did publish more, their employers would promote them then). This turnabout marked the end with minimal results of several years of protest by some relatively outspoken academic women. Not only had they not changed employment practices there noticeably, but they ended up accepting and internalizing their superiors' view that the system was treating women fairly and that only "exceptional" women deserved a rank higher than instructor. The extra burden of proof, of having to overcome initially low expectations, had been put upon them, and they had accepted it. In this way the academic feminist movement of the early 1920s collapsed by 1924.

Thus women became what Marion Hawthorne termed in 1929 "martyrs" to the prevailing academic mentality of the 1920s and 1930s; they left unchallenged that ideology that was applied to women, that can be termed the "logic of containment." Under this way of thinking the basic social desire to restrict women to the lower levels of the academic hierarchy subtly distorted the presumably meritocratic, "objective" logic and terminology applied to academic careers. Since no upward mobility was desired for women, almost regardless of their actual performance or accomplishments, the "logic" that prevailed in the thinking about women's promotion did not provide for any. Alleged sex differences were therefore often used to override a woman's achievements and were used to justify blocking her further advancement. Thus when a woman did good work as an instructor, she was rarely rewarded in these years with promotion to a higher rank. Instead she was told that being good at such menial work meant that she could do nothing else and would therefore have to remain an instructor for the rest of her career. Instructorships were not for women "stepping stones" or "entering wedges" to higher positions, but were the upper limits of their careers, as all the statistics had shown. (Industry had, as is shown in chapter 9, its own variation on this theme of "low ceilings"; there women were told that they could not manage.) It is important to note that the "proof" of this restrictive doctrine was not that women had been full professors (or managers) and had failed, but rather that, since no one in charge could imagine them being one at all, let alone a successful one, they were not given the chance to try. This practice thus institutionalized employers' lower expectations for future women scholars, who, seeing that they would not be promoted either, must have lessened their aspirations accordingly.[14] Thus although most of the barriers to women's advancement that one finds documented are administrative or procedural, at root they were cognitive and perceptual. The belief in women's inferiority was so strongly embedded in everyday experience and stereotypes that although some feminist psychologists had shown that under laboratory conditions women and men were equal, most of society preferred not to accept its full implications.

Moreover, the powerful, self-righteous, and highly conservative academic mentality of the 1920s and 1930s was so preoccupied with status and prestige that it incorporated them into its employment and promotion practices. It prejudged who would be good prospects for recruitment and, presuming women (and blacks, Jews, Catholics, and other nonelites) to be inferior, systematically relegated them to the marginal positions and poorer institutions. It defended such restrictions as meritocratic on the grounds that if such a person was so outstanding as in time to show herself the equal of the best men in her field, despite all the obstacles placed in her way, she might deserve some greater role and recognition. Since, however, most women were so placed as to be unable to jump this high hurdle, the system's rationale of low expectations was generally supported by its own self-fulfilling prophecies, and few women ever reentered the mainstream. Moreover, those who did were quickly labeled "exceptions," or women whose performance indicated nothing about other women, especially any conclusion that they too might merit better jobs and opportunities. Thus neither the many "status of women" reports of the time nor the hardworking "exceptions" made a dent in the academic mentality that continued to exonerate the system, to talk in terms of "prestige," "exceptions," and "extra burdens," and to blame women instructors for that very appropriate behavior and cluster of subservient personality traits that so well suited their positions. By the end of the 1920s this tendency to blame the victim had become so common that most academic women were treated contemptuously as deserving their marginal positions and low status, and officials (of both sexes), including those of the women's colleges, were deliberately hiring men to increase their schools' prestige.[15]

This prestige-linked antifeminism at the women's colleges, whose faculties had once been so enthusiastic about Committee W, occurred, interestingly (and necessarily?), at a time of relative prosperity at the women's colleges, if one can make broad generalizations about these decades on the basis of a few imperfect college histories. Bryn Mawr, Mount Holyoke, and Wellesley colleges, for example, were able to hire several new faculty members, raise faculty salaries (except from 1933 to 1935), and erect several new science buildings in the 1930s. Numerous other women's colleges, especially the many "state colleges for women" and the Catholic institutions, also managed to expand in these years and to upgrade their faculties to include better-trained and more accomplished scientists. (As mentioned in chapter 6, the appearance of nuns with doctorates in the sciences on the faculties of the Catholic colleges for women was a post-1920 phenomenon.) In addition, Hunter College in New York City, formerly a teachers' college, was taken over by the city government and given such increased support that its faculty increased tenfold between 1916 and 1928. Similarly, Connecticut College for Women, which was established in 1915, prospered during the depression, and built almost its entire campus under President Katherine Blunt from 1929 to 1942.[16] Thus the women's colleges shared the general growth and upgrading of the period, and both enrollments and faculty size reached new heights despite the depression.

In a broadly quantitative sense women scientists shared a large part of this general growth and upgrading at the women's colleges. The number of women scientists in the *AMS* employed at these colleges more than tripled between 1921 and 1938—from 99 women at twenty-seven such colleges in 1921 to 378 at eighty-five in 1938. Moreover, as happens so often in scientific careers, most of these faculty members were congregated at a relatively few institutions, as shown in table 7.1, which lists the major employers of women faculty in the 1938 *AMS*. Although it includes instructors to give coeducational institutions as large a representation as possible, the four largest employers of women scientists were still the women's colleges (Wellesley, Vassar, Mount Holyoke, and now Hunter), as were twelve of the twenty-four schools whose faculties had more than 8 women from the 1938 *AMS*. Furthermore, these top twelve women's colleges employed almost one-half (187) of the women scientists at all of the women's colleges in 1938; the other half (191) were distributed over seventy-three such colleges, most of which had not employed any women listed in the *AMS* seventeen years earlier. A few departments could even count 5 or more women in the *AMS*, such as the zoology and botany departments at Wellesley College, which had 9 and 5 women listed, respectively, zoology and chemistry at Mount Holyoke College (6 and 5), and mathematics at Hunter (6). Numerous other departments had 3 or 4 women scientists in the 1938 *AMS*, which was often more than the entire college had listed there in 1921, and which might compare favorably with the situation in many of these same departments today.

Some women's colleges, it should also be noted, began in these years to show a modest tolerance toward retaining on their faculties those women who married. President Virginia Gildersleeve of Barnard has recounted how readily she allowed a woman physicist to keep her job as an assistant despite her upcoming marriage, apparently unaware of the Harriet Brooks episode in 1906 (see chapter 1). By 1938, 40 of the 378 women scientists listed in the *AMS* as employed at women's colleges were or had been married (10.6 percent). Interestingly, they were not uniformly distributed among all eighty-five women's colleges, but taught at only twenty-one. The largest number was at Hunter College (6 of 25 women, or 24 percent), followed closely by Barnard (5 of 12, or 41.7 percent) and Wellesley (4 of 33, or 12.1 percent). At Mills College in California 2 of its 6 women were married, and at Radcliffe College, which had no real faculty since it rented the services of the Harvard professors, 3 of its 4 women tutors were married.[17]

Despite all this evidence of growth and diversity at the women's colleges in the 1920s and 1930s, there were also many problems and inequities for women scientists at these schools. In fact, there are certain signs that this expansion and prosperity were not enough to keep pace with the even greater growth and emphasis on research elsewhere; nor were they even necessarily in the women's best interests. Tables 7.2 and 7.3 show that, despite a threefold increase in the number of women scientists in the *AMS* working at the women's colleges, their percentage of the total employed by academic institutions dropped sharply, from 40.4 percent to 30.0 percent; furthermore, the decline was more pronounced in some fields than in others. Although these colleges introduced some instruction in anthropology, biochemistry, and microbiology in these years and increased it somewhat in nutrition and parts of the medical sciences, they were still not major employers in these areas.[18] Otherwise they retained their prominent role only in physics (69 percent, down from 100 percent in 1921), chemistry (55.1 percent,

TABLE 7.1. Largest Employers of Female Science Faculty, 1938
(By Field)

Institution	Total	Zoology	Botany	Psychology	Mathematics	Nutrition	Medical sciences	Chemistry	Biochemistry	Microbiology	Physics	Geology	Astronomy	Anthropology	Engineering
1 Wellesley	33	9	5	2	4	0	0	3	1	0	4	4	1	0	0
2 Vassar	26	3	2	3	4	2	3	2	1	0	4	0	1	1	0
3 Mount Holyoke	25	6	3	0	2	0	1	5	1	0	2	3	1	1	0
3 Hunter	25	3	5	0	6	0	4	2	2	0	2	0	0	1	0
5 Columbia	22	1	0	5	1	3	4	0	2	4	1	0	0	1	0
6 Smith	21	4	4	1	2	0	1	2	0	2	2	1	2	0	0
7 California	16	0	1	1	3	5	4	0	1	1	0	0	0	0	0
8 UCLA	15	2	1	5	2	4	0	0	0	0	0	1	0	0	0
Wisconsin	15	1	3	0	2	2	1	0	1	4	0	0	0	1	0
Minnesota	15	0	1	3	3	4	2	1	1	0	0	0	0	0	0
Chicago	15	0	0	3	1	3	5	1	2	0	0	0	0	0	0

12 Iowa State	14	0	1	0	4	7	1	0	0	1	0	0	0	0	0
13 Barnard	12	3	2	1	0	0	0	2	0	0	1	2	0	1	0
Cornell	12	1	0	1	0	6	3	0	1	0	0	0	0	0	0
15 Connecticut College	10	2	1	1	1	2	0	1	1	0	1	0	0	0	0
Goucher	10	0	1	2	2	0	1	1	0	1	2	0	0	0	0
17 Pennsylvania Woman's Medical	9	0	0	0	0	0	5	0	2	2	0	0	0	0	0
Ohio State	9	0	1	3	1	2	0	0	1	0	0	1	0	0	0
Nebraska	9	0	3	1	1	1	0	0	0	0	0	3	0	0	0
20 Kansas State	8	1	0	1	1	3	1	0	0	0	1	0	0	0	0
Johns Hopkins	8	0	0	1	0	0	6	0	0	1	0	0	0	0	0
Illinois	8	0	1	0	2	2	1	0	0	2	0	0	0	0	0
Florida State College	8	1	0	3	1	2	0	0	1	0	0	0	0	0	0
Sophie Newcomb	8	0	0	1	3	0	0	2	0	0	2	0	0	0	0
359 other institutions	523	111	91	81	69	60	57	64	31	20	14	9	9	4	3
Total	976	148	126	119	115	108	100	86	49	38	36	24	14	10	3

SOURCE: *American Men of Science*, 6th ed. (1938).

NOTE: "Faculty" here signifies instructor and above.

TABLE 7.2. Female Scientists Employed in Academia, 1921 (By Field and Type of Institution)

Field	Total Number Employed	In Academia		Coed Colleges or Universities		Women's Colleges		Research Institutes	
		Number	Percentage of Total Employed	Number	Percentage of Total in Academia	Number	Percentage of Total in Academia	Number	Percentage of Total in Academia
Mathematics	37	35	94.6	11	31.4	24	68.6	0	0
Astronomy	14	13	92.9	1	7.7	6	46.2	6	46.2
Home Economics	19	17	89.5	16	94.1	1	5.9	0	0
Physics	18	15	83.3	0	0	15	100.0	0	0
Anthropology	5	4	80.0	0	0	0	0	4	100.0
Zoology	51	40	78.4	13	32.5	18	45.0	9	22.5
Botany	72	55	76.4	26	47.3	19	34.5	10	18.2
Chemistry	28	21	75.0	7	33.3	14	66.7	0	0
Psychology	51	38	74.5	24	63.2	10	26.3	4	10.5
Medical sciences	49	31	63.3	19	61.3	2	6.5	10	32.3
Geology	18	11	61.1	5	45.5	4	36.4	2	18.2
Total or average	362	280	77.3	122	43.6	113	40.4	45	16.1

SOURCE: *American Men of Science*, 3d ed. (1921).

TABLE 7.3. Female Scientists Employed in Academia, 1938 (By Field and Type of Institution)

Field	Total Number Employed	In Academia		Coed Colleges or Universities		Women's Colleges		Research Institutes	
		Number	Percentage of Total Employed	Number	Percentage of Total in Academia	Number	Percentage of Total in Academia	Number	Percentage of Total in Academia
Mathematics	132	121	91.7	66	54.5	49	40.5	6	5.0
Astronomy	29	26	89.7	8	30.8	6	23.1	12	46.2
Medical sciences	163	137	84.0	78	56.9	22	16.1	37	27.0
Biochemistry	118	99	83.9	32	32.3	17	17.2	50	50.5
Physics	51	42	82.4	7	16.7	29	69.0	6	14.3
Anthropology	21	17	81.0	6	35.3	4	23.5	7	41.2
Nutrition	153	117	76.5	94	80.3	14	12.0	9	7.7
Botany	215	163	75.8	73	44.8	53	32.5	37	22.7
Chemistry	147	107	72.8	27	25.2	59	55.1	21	19.6
Microbiology	99	62	62.6	29	46.8	9	14.5	24	38.7
Psychology	241	142	58.9	82	57.7	37	26.1	23	16.2
Geology	45	26	57.8	11	42.3	13	50.0	2	7.7
Zoology	230	196	41.7	82	41.8	66	33.7	48	24.5
Engineering	8	3	37.5	3	100.0	0	0	0	0
Total or average	1,652	1,258	76.2	598	47.5	378	30.0	282	22.4

SOURCE: *American Men of Science*, 6th ed. (1938).

down from 66.7 percent in 1921), and geology (50 percent, up from 36.4 percent). By 1938 they had also lost their former importance in astronomy to the observatories and in mathematics and zoology to the coeducational colleges, especially the teachers' colleges, which were expanding rapidly in the 1920s and 1930s.

One reason for this decline may have been that the climate for research remained difficult at the women's colleges. Even at the wealthiest ones, research funds, facilities, and faculty time remained scarce or nonexistent, making a career there increasingly difficult for those women who were strongly research oriented. Equipment and assistants had to be paid for out of one's own often meager salary, and even proven researchers were rarely given reduced teaching assignments. Even such an outstanding scientist as Margaret Washburn at the relatively well-endowed Vassar College could not get any assistance or a reduced load when she was the only woman editor of the *American Journal of Psychology*. She complained to the editor-in-chief in 1931, "I doubt if anyone else on the board is teaching eighteen hours a week, as I am. I simply must cut down my work somewhere," and she submitted her resignation.[19] Younger, more mobile women seriously interested in research were beginning to forsake the women's colleges for the seemingly better opportunities that were becoming available elsewhere. Marjorie O'Connell, for example, a hardworking and accomplished young paleontologist, preferred in 1921 to remain a research assistant at the American Museum of Natural History in New York City rather than teach geology at Vassar College, and Olive Hazlett, whom her professors termed "one of the two most noted women in America in the field of mathematics," eagerly left Mount Holyoke in 1925 for an associate professorship at the University of Illinois. Yet in neither case did the move lead to lasting productivity.[20]

Nor were many of the women scientists at the women's colleges successful in getting much of the increasingly available outside support for their research. Although the Mount Holyoke trio of chemist Emma Perry Carr and zoologists A. Elizabeth Adams and Ann Haven Morgan were master grantswomen and were able to obtain a series of grants in the 1930s from the Bache Fund of the National Academy of Sciences, Sigma Xi, the American Association for the Advancement of Science, several committees of the National Research Council, and the Rockefeller Foundation, all were for sums of less than $5,000. Similarly, Gladys Reichard at Barnard was able to get occasional support for her summer fieldwork from the Southwest [Archeological] Society and Columbia University, but it was perhaps symptomatic of the women's situation that at least the former (and perhaps both) was simply a camouflage for the continued personal philanthropy of Elsie Clews Parsons.[21]

This lack of institutional support for research at women's colleges was so prevalent that Waldemar Kaempffert, the science editor of the *New York Times*, wondered in 1937 why the administrators, especially those at the more prestigious women's colleges, who knew that their faculties were being watched in this area, allowed this situation to persist and were so reluctant to support their women's research. He pointed to a "certain illogicality" in their position: they often hired their faculty on the basis of their dissertations and professional reputations and took considerable pride in their achievements, once completed, but offered little help and did not reduce teaching loads to help them. They wanted the "prestige" of research but did not have the financial strength to support it. Kaempffert thought these benighted institutional policies hurt not only the women at the women's colleges but women scientists everywhere, because, as he concluded his article, ". . . it is as plain as Venus in the evening sky that women will not be recognized by the universities as the academic equals of men until they have demonstrated by great discoveries what they can do in the laboratories of their own colleges."[22] Yet for all his glib concern and sympathy for women in science, there was a "certain illogicality" in Kaempffert's own argument. It did not seem at all unfair or unreasonable to him in 1937 that the women's equality to men in science still was not assumed but had to be demonstrated again and again, years after Maria Mitchell in the nineteenth century and Madame Curie even more recently both had "proven" it. Moreover, he expected the women to prove this equality at institutions that had far fewer resources, as he himself described, than the men generally had available. Thus his "logic" reflected the prevailing academic mentality that seemed open and meritocratic and yet repeatedly put the women into no-win situations where they were expected to struggle against great odds to "prove" once again what should have been assumed from the start. And even if they "succeeded" in proving women's ability, their victory would be so undercut by being labeled an "exception" that later women would have to prove it all over again.

An even more likely reason for the relative decline of women scientists at the women's colleges than the lack of research support was the increasing tendency for even formerly militant women's colleges to hire men, especially for advanced and senior positions. To a certain extent this was the result of economic forces that made the women's colleges more acceptable to male faculty in the 1930s—their rising salaries were making them increasingly competitive with other institutions, and in depression years, any job, even one at a woman's college, could suddenly be desirable. Even more importantly, interest in feminism and a commitment to hiring women, which were especially needed during a depression, were both waning at most of the women's colleges in these decades, as several of the major ones began to stress the antifeminist concept of "prestige" more than they had earlier. Thus despite the colleges' expansion and the vigilance of the remaining old-time feminists, there was substantial backsliding at some of the older institutions, and many lost opportunities for women at the newer ones. Much such possibly deliberate antifeminism occurred at Bryn Mawr College, which had formerly been one of the leading employers of women scientists—seven were listed in the *AMS* in 1921. But by 1938, after sixteen years under President Marion Park and a certain growth in faculty size, the number was down to five (of which three were mathematicians: two in teaching and a third in administration). This practice of replacing women with men, which seemed to persons on the scene to be intentional, can be seen in operation at close hand in Bryn Mawr's physics department in the academic year 1935–36. A. L.

Patterson, an X-ray crystallographer, has reported in his autobiography how pleased he was to be invited to Bryn Mawr in 1935, since he had already spent two years looking for a job in his specialty. Meanwhile, correspondence elsewhere reveals how upset Jane Dewey, a highly regarded spectroscopist, former National Research Council Fellow, and associate professor in the same department since 1933, was when she discovered a few months later that she was to be let go. Writing to her former professor for help and advice, she mentioned that it was commonly understood on campus that the authorities were "appointing men to all the vacancies left by women."[23] (She was subsequently unemployed for four years before obtaining a temporary night job at Hunter College in 1940. It is not known what happened to her after that.)

Other examples of opportunities lost to women include numerous positions at Douglass College, which, despite tremendous growth between 1918 and 1930, had only five women scientists by 1938. It also had had an employment policy so lacking in any commitment to careers for women that it had encountered public criticism in the mid-1920s from the New Jersey State Federation of Women's Clubs and other women's groups that had been among its original supporters. Even those women who were employed at Douglass were often not happy with conditions there; Hettie Morse Chute, a botanist there from 1930 to 1959, was remembered years later as being resentful of the school's lack of research facilities.[24] Other women's colleges from which one might have expected more were Skidmore, established in 1921, but with only four women scientists in the *AMS* in 1938, and Rockford College, established in 1882, but with only three women scientists fifty-six years later.

The reasons these colleges began to hire male faculty were rarely spelled out, since it was assumed to be a common preference that required no explanation. Administrators of both sexes apparently liked to have academic men around and saw their presence as a good way to "upgrade" the college's "image," if not its actual instruction. The preference for men was often quite explicit, as for example in the exchange of letters in 1932 between Professor William H. Longley of Goucher College, who wished to hire a male instructor of biology (and mentioned the desired "man" or "gentleman" four times in one letter), and Professor Ross G. Harrison of Yale University, who was enthusiastic about having women graduate students in his department and had trained many of them. Harrison presented Longley with the names and qualifications of three men who were available but added that three women were also finishing that year, including Dorothy Hewitt, a Mount Holyoke graduate whom he considered "to be one of the best students [we have] ever had." Then, apparently irritated at Longley's insistence on a male, Harrison concluded with a challenging rhetorical question, "If you who are running colleges for women don't employ them on your faculty, what are we going to do with them?" The question remained unanswered, and Goucher hired its man.[25]

The men's faculty rank became a sticking point for some women, since some administrators at the women's colleges were particularly eager to hire them for the top positions, especially chairmanships, from which positions they might, like Longley, then hire other men. Vigilant feminists saw this happening around them, considered it unjust, tried hard to prevent it, but were usually powerless to do more than complain to their friends. As early as 1905 the perceptive Christine Ladd-Franklin had written Mary Whitney at Vassar College, her alma mater, to suggest Charlotte Scott of Bryn Mawr as a candidate for the vacant headship of the mathematics department. Whitney was enthusiastic about Scott but dubious that President J. M. Taylor would appoint her, since he "questions her age [forty-seven] and her peculiarities and hesitates in regard to her adaptability." She added sadly and humorously, "I note that these personal characteristics count a great deal with a woman candidate, but are barely asked about with the male candidate. Certainly if they were, we should hardly have had among us certain officers of our group." But Whitney hoped that another strong woman could be found, since she feared that the appointment of a man would set a precedent difficult to break. Many women, she wrote, who had originally been instructors at Vassar, had by 1905 been promoted to associate professorships. Appointing a man to head the department, she added, "disturbs me somewhat as I see therein a likelihood that in [the] course of time all our heads of departments will be men, and no position beyond that of associate professor will be regarded as available for a woman. I do not wish, of course, that an inferior woman should be appointed to head of a department, but that headships should be closed on the basis of sex and ratio, would be deplorable in my opinion. For that reason I wish with all my heart that a very strong woman might be found."[26] Thus she wanted a very good woman but was unclear how to compromise if one could not be found. Apparently a mediocre woman would be preferable to a mediocre man, but not necessarily, since even Whitney might think such a woman not up to standard and find it difficult to fight for her. Thus to become a chairman at Vassar, a woman not only had to be so outstanding in her field as to be clearly superior to all the male candidates, she had to have an unimpeachable personality as well. One who was merely "equal" or even superior but offensive in some way was not good enough for Vassar—such a woman would lose out, since no one would be particularly enthusiastic about her. The double standard about who should hold the higher faculty ranks was already in place at Vassar College by 1905, and even its strongest women were ambivalent about its fairness.

In succeeding decades this tendency to appoint men to the higher positions on the faculties of the women's colleges was acknowledged and alternately deplored and defended by even the most feminist of presidents. In 1913 M. Carey Thomas of Bryn Mawr described the pressures:

> In all coeducational colleges and universities the number of women holding even subordinate teaching positions is jealously limited. Presidents of coeducational universities have sometimes told me that they would gladly advance women scholars were it not for the opposition of men teaching in the same departments. Even in a woman's college like Bryn Mawr there is a steady, although I believe almost unconscious, pressure exerted by some of the men on our faculty to prevent the appointment of women to vacant professorships. Men who have taught women at Bryn Mawr (and sometimes only women and never men) for as many as fifteen years will say to me on accepting a call elsewhere: "Of course, my

position can only be filled by a man. I do not know why it is but my subject can be best taught by men."[27]

Years later Barnard's President Gildersleeve admitted that she hired men chiefly for the higher positions—since the women would accept jobs at lower levels, to keep the sexes roughly balanced, she brought men in at the top. Undoubtedly the Columbia University faculty also had some influence in this decision.[28]

Two well-documented searches at Vassar College in the late 1930s reveal even more than does presidential rhetoric of the careful sexual politics involved in senior faculty appointments at the time. Although the institution's relative poverty could work to the advantage of a woman scientist on occasion, that was not always the case. Vassar officials were far more sensitive to the opinions of prominent outsiders, especially prestigious ones, than to actual economics. Thus when in 1936 Vassar College needed to find a successor to Caroline Furness, the recently deceased chairman of the astronomy department and director of the Vassar College Observatory, President Henry Noble MacCracken consulted Henry Norris Russell, professor of astronomy at Princeton University and friend of the college. After speaking at the college and talking with its faculty, Russell wrote MacCracken a long letter in which he described the anomalies of such a research position at a woman's college. Since the position required so much elementary teaching (of the sort his colleagues dubbed pejoratively "girls' college astronomy"), anyone who got an outside offer would tend to leave. Since only men got such opportunities, Vassar should appoint Maud Makemson, already on its staff, as the department's new chairman and director. Although "mature" at age forty-four, she had done good work, was likely to do more, and would continue to reach the students successfully. Given Vassar's economic constraints this appointment was the best one possible, and, as Russell assured MacCracken, who was evidently worried, "Vassar would *not* lose prestige" (his italics) by her appointment.[29]

A second faculty search a year later shows that without such strong support from a prestigious outsider, there could still be many psychological obstacles to the women's colleges' hiring seemingly well-qualified women scientists. In 1937, after a series of strokes had felled the great Margaret Washburn, Vassar College had to seek a new chairman of its psychology department. Despite her many students, Washburn had apparently not chosen her protégée and successor well, since Josephine Gleason had not published enough to be suitable as chairman. Yet she was in charge of the search committee. She wrote Edwin G. Boring, who was professor of psychology at Harvard University and was known for his wide contacts in the field, and he responded with a list of twenty-one names and a series of artless and revealing comments. Although there was only one woman on his list (Florence Goodenough, professor of psychology at the University of Minnesota, who was well known for her "Draw-A-Man" test and studies of gifted children), Boring immediately denigrated her by adding that though he did not known Goodenough personally, she "is perhaps what her name implies, although she was described to me recently as having more drive than brains." He added in subsequent letters that she was too old for the job (age fifty-one) and had a "somewhat grotesque sense of humor." Then, having disposed of perhaps the top woman psychologist after Washburn (who might not have come anyway), Boring proceeded to report that he had asked several young men around Harvard about their thoughts on the position "on the ground that it is the bright young men who know best about the bright young men, and that it is the bright young men who think they know how to run the world and ought therefore to have a chance to try"—apparently even at Vassar College. However, he had to admit that he was in a quandary about how to proceed from there. How could he "advise you to consider a man when I would advise him not to accept?" Such a bright young man, if found, would be better off at a university with more laboratory equipment than Vassar could offer.

By now, with the top woman and the most promising young men eliminated, the search entered its critical stage. The only woman anyone seemed able to think of in the entire field of psychology (where women comprised about one-third of the total by 1937), was Helen Peak, age thirty-eight, a Yale doctorate and chairman of the department at Randolph-Macon College for Women in Virginia. Yet everyone found her impossible as well, for various personality reasons (and prided themselves on their honesty for telling her so). The most specific complaint against her was that while lecturing to the Vassar students she had spent twenty minutes cajoling them to study the field, a trait that made it hard for Gleason's colleagues to "see her as one of themselves." Finally Lyle Lanier, age thirty-five, a doctorate from George Peabody College and an assistant professor at Vanderbilt University, was appointed chairman. As Boring, a gossip as well as skeptic of women's abilities, confided later to Helen Peak, the situation had indeed been delicate. No one would admit to preferring a male, yet the situation, Boring felt, demanded it: Gleason (age forty-five) would hire neither a good woman nor a young one to the top job since either one would eclipse her, but she would accept a "bright young man" for it, since "women get accustomed to men being rated higher than women for equal ability and so do not resent it so much." Peak, who had earlier agreed that with men in the running she had never really had a chance, replied meekly, "I do not blame her [Gleason] in the least," and said that she was glad to have been considered.[30] Yet had Boring written the kind of letter for her that Russell had for Makemson a year earlier, the result might well have been different.

In analyzing these episodes one can see the strong internal and external resistance among both men and women to appointing women to top positions at the women's colleges. There seem to have been strikingly few female candidates, even in psychology, in which the 1938 *AMS* would soon be listing over twenty who were teaching at other major institutions. Although some might have come, little attempt was made to seek them out, apparently because most were in child psychology or child study, which Washburn had scorned. Nor does Gleason seem to have sought advice from persons at Columbia or Chicago, where, as shown in chapter 6, most women were earning their doctorates, or to have prepared a list of Washburn's other students. Then the few women who were listed seem to have been eliminated rather quickly, with hardly any mention of the positive things that they might contribute to the college. Finally, as for Mary Whitney over thirty years before, when the ideal woman was not available and it came down to the mediocre man versus the as yet undistinguished woman, the woman was rarely seen as presenting any strengths, her personality was criticized harshly, and she rapidly lost out. Nor was there much evidence of feminist activity by the few faculty women involved in these searches.[31] Yet if this was life at the women's colleges, the traditional bastion of the woman scientist, in the 1930s, one can perhaps imagine how few senior appointments were made elsewhere, in the coeducational colleges and universities, where there was not even a pretense of feminist feelings or traditions, and where the presumed superiority and favoritism Boring had expressed for the "bright young men" was even more pronounced.

So many other women scientists of the 1920s and 1930s had joined the faculties of the nation's coeducational colleges and universities that, as shown in tables 7.2 and 7.3, their representation there jumped almost fivefold. One hundred and twenty-two women scientists at 57 such institutions in 1921 became 598 at 298 institutions in 1938, by which time they constituted almost one-half (47.5 percent) of those women scientists listed in the *AMS* who were employed by academic institutions. Yet despite this numerical preponderance at the coeducational schools, the women there were still in 1938 far less visible (almost invisible, in fact) than those at the women's colleges because they held, for the most part, the lowest and least desirable positions and were distributed thinly over the numerous public and private universities that comprise the vast and miscellaneous American educational system. This pattern was not accidental, but grew up as the result of a severe negative selection of being forced for decades to take the lowest-ranking, least-contested, and poorest-paying jobs in academia. The most successful of these women found it advisable (to use a biological analogy) to develop certain "adaptive behavior" in order to survive; they would seek out a strong protector and, on occasion, take on the coloration or attitudes of the dominant group in an attempt to "pass" as something they were not. With these techniques and within certain boundaries, the women found themselves tolerated and able to eke out an existence at the coeducational institutions in the 1920s and 1930s. But it was at best a marginal and precarious existence, full of deliberate and daily reminders of their subordinate status, and psychologically very far removed from even the relative warmth and welcome of the women's colleges.

Table 7.4 presents the names and rank (by field) of the 112 women scientists listed in the 1938 *AMS* who held faculty appointments at the twenty largest doctoral universities in the nation. If one estimates that each of these twenty universities had a department of from five to twenty persons in each of these fourteen sciences, or that table 7.4 represents a potential total of from 1,400 to 5,600 jobs, the low proportion of women in these departments (2 to 8 percent) in

1938 is evident. Even this would be too high, however, since many of these women were not in the main science departments, but, as the footnote symbols indicate, in others that were more closely related to traditional "women's work," such as education, social research, home economics, and child welfare. In addition, these women were often distributed among several campuses and institutes: two of the four women at the University of Pennsylvania were at the Henry Phipps Institute (for tuberculosis research), at least three of the twelve at the University of California were in San Francisco at the medical school, and of the fourteen at Columbia University, three were at the College of Physicians and Surgeons and nine at Teachers College.[32]

Table 7.4 also shows that there was a pattern even to these "exceptions." The largest numbers were in the three fields of psychology (especially at the new child welfare institutes on several campuses), nutrition (at the schools of home economics), and perhaps surprisingly, most often in the medical sciences (usually taught at medical schools). (One reason those faculties were more receptive to women than other institutions is suggested in what follows.) The fields offering the bleakest prospects to women were the departments of physics and engineering, where there were actually no women faculty at these twenty institutions, followed closely by geology, astronomy, and chemistry, which had just two "exceptions" apiece. Even zoology, which was the largest field for women and had the most women in academia in 1938 (table 7.3), had only five women on these top faculties that year, one-half as many as in the much smaller field of mathematics and little more than in all of anthropology, which had only one-tenth as many women! Obviously the men in zoology were fiercely resisting even the slightest tendency toward feminization.

It is also possible to compare the departments that employed women faculty in 1938 with those that had been training the most women doctorates in that field to see whether they overlap. Table 7.5 seeks to organize the considerable data on this topic. It lists in one column those twenty-one departments that had, as shown in table 6.6, produced a sizeable proportion of the female doctorates listed in that field in the 1938 *AMS*, and in a second column those ten that employed three or more women faculty, as listed in the same *AMS*. One can see quite readily that there was little duplication in these two lists; only three departments appeared in both columns (nutrition at the University of Chicago, psychology at Columbia University [which includes Teachers College], and medical sciences at The Johns Hopkins University). The bottleneck was in segregated faculty employment, for of the ten departments that did hire three or more women, nine were in nutrition, psychology, and the medical sciences. The tenth department was that of botany at the University of Wisconsin, which trained few women doctorates but employed three female faculty members (one associate professor and two assistant professors) in 1938, a highly unusual pattern. By contrast, the University of Chicago, which produced most of the women doctorates in botany by a wide margin, employed no women faculty. (If one lowers the criterion for inclusion in the right-hand column of table 7.5 to those departments that had two or more women faculty, thirty programs would be listed, of which only six were among

TABLE 7.4. Female Science Faculty at Twenty Major Universities, 1938 (By Field and Rank)

Institution	Total	Medical Sciences	Nutrition	Psychology	Biochemistry	Mathematics
Columbia	13	Neal (B) Scott (D)	M. Rose (B)† MacLeod (B)†	Hollingworth (B)† Meek (B)† McDowell (C)† Whitley (C)†	Caldwell (D)	Walker (C)†
California at San Francisco and Berkeley	12	Simpson (C) Perry (D) Lucia (D) Montgomery (D)	Morgan (B) Okey (C) Fyler (D) Gillum (D)	Bridgman (B)	Anderson (D)	Sperry (C) Levy (D)
Minnesota	11	Ziegler (D) Boyd (D)	Biester (C) Leichsenring (C)	Goodenough (B)‡ Foster (B)‡ Shea (C)	Kennedy (C)	Gibbens (D) Carlson (D)
Chicago	10	Slye (C) Humphreys (D)	Roberts (A) Halliday (C) Brookes (D)	Koch (C)§	Eichelberger (C) Sandiford (D)	Logsdon (C)
Wisconsin	10	Hellebrandt (C)	Marlatt (A) Parsons (B)		Wakeman (D)	
Iowa	8	Cooper (C)	Daniels (B) Daum (C)	Wellman (B)‡ Kemmerer (C)‖ Updegraff (D)‡	Stearns (C)	
Cornell	8	Farrar (D) Evans (D)	F. Rose (A) Monsch (B) Hauck (B) Pfund (B) Fenton (D)	Wylie (B)§		
Ohio State	7		McKay (B) Griffith (D)	Rogers (C) Stogdill (D)	Wikoff (D)	Bareis (D)
Washington (St. Louis)	6	Trotter (C) Smith (D) Wolff (D)			Graham (C) Ronzoni (D)	
Bryn Mawr	4					Wheeler (B) Lehr (C)
Yale	4	Van Wagenen (C) Kennard (D)		Miles (B) Washburn (D)‡		
Johns Hopkins	4	Hines (C) Clark (C) Richards (C)		Rand (C?)		
University of Pennsylvania	4	Zeckwer (D)			Seibert (C)	
Michigan	3	Crosby (B)				
Illinois	3		Woodruff (B) Outhouse (C)			Hazlett (C)
New York University	3	Hoskins (C)		Holden (D)		
Stanford	2	Layman (B)		Merrill (C)		
Catholic	0					
Princeton	0					
Harvard	0					
Total	112	26	24	20	11	10

SOURCE: *American Men of Science*, 6th ed. (1938).

NOTES: Faculty listed were at the level of assistant professor and above. A = department chairman or dean or director of a school or college; B = professor, research professor, or clinical professor; C = associate professor, research associate professor, or clinical associate professor; D = assistant professor, research assistant professor, or clinical assistant professor.

the major training programs for women doctorates in that field. These six included, in addition to the three departments already starred, biochemistry and medical sciences at the University of Chicago and nutrition at Teachers College. They are marked in table 7.5 with a dagger.)

TABLE 7.4.
(*continued*)

Zoology	Microbiology	Botany	Anthropology	Chemistry	Geology	Astronomy	Physics	Engineering
	Broadhurst (B)†		Benedict (D)					
	Seegal (D)							
				Cohen (C)				
				Link (D)				
	McCoy (D)	Fisk (C)	Gower (D)					
	Holford (D)	Walker (D)						
		Shands (D)						
Slifer (D)								
					Stewart (C)			
						Stephens (D)		
Gardiner (C)					Wyckoff (D)			
King (B)			Carter (D)*					
Woodward (D)						Losh (D)		
Howland (C)								
5	4	3	3	2	2	2	0	0

*in department of social research
†in education department or Teachers College
‡at institute of child welfare
§in department of home economics
‖in charge of correspondence courses

There is another way in which to utilize these two sets of data. This is to compare the schools' overall ranks as producers of women doctorates and as employers of women faculty, as is done in table 7.6. When the second rank is subtracted from the first, a final score, ranging from +9 to −6, is obtained. This

TABLE 7.5. Universities Producing Female Science Doctorates and Employing Female Science Faculty, 1938

Highly Productive Departments		Employers of Women Faculty	
Bryn Mawr	Geology	California	Medical sciences
California	Astronomy		Nutrition
	Zoology	Chicago	Nutrition*
Chicago	Biochemistry†	Columbia	Psychology*
	Botany	Cornell	Nutrition
	Chemistry	Iowa	Psychology
	Geology	Johns Hopkins	Medical sciences*
	Mathematics	Minnesota	Psychology
	Medical Sciences†	Washington (St. Louis)	Medical sciences
	Microbiology	Wisconsin	Botany
	Nutrition*		
	Psychology		
Columbia	Anthropology		
	Biochemistry		
	Chemistry		
	Nutrition†		
	Psychology*		
	Zoology		
Johns Hopkins	Medical sciences*		
	Microbiology		
Yale	Nutrition		

SOURCE: *American Men of Science,* 6th ed. (1938), as presented in tables 6.6 and 7.4 herein.
*Appear in both columns.
†Are among the thirty departments with two or more female faculty.

indicates perhaps as well as any numerical data can that five major midwestern, one western, and one eastern university were far more likely to hire women faculty than to have trained them; four other schools, with scores of +1 to −2, were about even in this regard; and seven others were likely to train them but quite unlikely to hire them. Since neither Princeton University nor the Catholic University of America employed any women scientists listed in the 1938 *AMS* (though the latter had awarded nine doctorates), their scores at the bottom of table 7.6 are only approximate. All this is in conformity with conclusions presented earlier—the women were trained at elite institutions but worked at land-grant ones or medical schools, largely in nutrition, psychology, and the medical sciences.

One can hypothesize from the distribution of the "exceptions" shown in table 7.4 that behind each such woman there stood a mentor. In the highly feminized world of nutrition this might on occasion be a female, such as Mary Swartz Rose of Teachers College. Elsewhere it was a man, since few of the women at coeducational colleges and universities were in the position to do much hiring. It is also necessary to distinguish between two types of "mentors": the first would com-

TABLE 7.6. Twenty Major Universities Producing Female Science Doctorates and Employing Female Science Faculty, 1938
(By Rank and Order)

Rank (for Number of Female Doctorates Awarded)	Institution	Rank (for Number of Female Faculty)	Difference
15	Iowa	6	+9
23	New York University	14	+9
17	Washington (St. Louis)	9	+8
14	Ohio State	8	+6
9	Minnesota	3	+6
6	California at Berkeley	2	+4
7	Wisconsin	4	+3
2	Columbia	1	+1
11	Bryn Mawr	10	+1
16	Stanford	17	−1
8	University of Pennsylvania	10	−2
1	Chicago	4	−3
3	Cornell	6	−3
11	Michigan	14	−3
13	Radcliffe/Harvard	18	−5
5	Yale	10	−5
9	Illinois	14	−5
4	Johns Hopkins	10	−6
103	Princeton	599	−496
28	Catholic	599	−571

SOURCE: *American Men of Science*, 6th ed. (1938), as presented in tables 6.6 and 7.4 herein.

prise the graduate school professors in several fields, like Lafayette B. Mendel and Ross G. Harrison at Yale or F. R. Lillie at Chicago, who trained many women and wrote many letters urging others to hire them but did not do so themselves; the second kind were, like Edwin B. Fred at the University of Wisconsin or Liberty Hyde Bailey at Cornell University, willing and able to hire women themselves. Several of this second kind were at medical schools, where employment practices were apparently still so patriarchical and autocratic that one strong professor could engineer a desired appointment without all the faculty committees and votes of the more democratic faculties of arts and sciences. Thus Florence Sabin (plate 23) was repeatedly promoted at The Johns Hopkins Medical School until her mentor, Franklin P. Mall, died in 1917, and Elizabeth Crosby rose steadily through the ranks at the University of Michigan Medical School, starting as an instructor in 1920 and reaching a full professorship in 1936, because of G. Carl Huber. Similarly, Gertrude van Wagenen was an assistant professor at the Yale School of Medicine because John Fulton, Sterling professor of physiology, had invited her to come there. Under such a system of personal patronage, about all the favored women could do was keep up the good

PLATE 23. Among the many honors received by anatomist Florence Sabin of, first, The Johns Hopkins Medical School and, later, the Rockefeller Institute for Medical Research, was election to the National Academy of Sciences in 1925 as its first woman. (Courtesy of the American Philosophical Society.)

work, be loyal to their benefactors, and stay out of "politics." So grateful were such women for this opportunity to do research, they tended to outdo even themselves.[33]

There were pitfalls to this kind of personal patronage as well, however. On occasion it led to the unfortunate assumption (for female but apparently not for male protégés) that the young woman was merely helping the professor with his research rather than doing her own work, which in van Wagenen's case was pioneering endocrinological studies of rhesus monkeys. Although perhaps perturbed by such assumptions, van Wagenen persevered and was soon promoted to an associate professorship. Yet for another female protégée this kind of close personal relationship led to charges of favoritism and her eventual resignation. A. Elizabeth Verder's rank as an assistant professor of bacteriology at the George Washington University Medical School and her salary of $3,300 upset the rest of the staff there, because she worked chiefly on the research of Professor Earl B. McKinley, who was dean of the medical school at the time. McKinley was aware of the problem, since he wrote a colleague in 1933 that "she feels there is a little resentment from some members of other departments regarding her rank and salary and the protection which she naturally enjoys working with me." McKinley's advice to Verder was that she must "try to curb her sensitiveness," but apparently the situation was so intolerable and McKinley's self-proclaimed "protection" so inadequate, that she left for a job at the Maryland State Health Department in 1935 and moved on to the National Institute of Health (which she had formerly thought a step downward) in 1936.[34]

For those women scientists without strong personal patrons, full professorships were still, as at Cornell in 1911, available only in certain separate and designated "womanly" fields. Accordingly, twenty-three of the twenty-six full professors listed in table 7.4 were in nutrition (thirteen), psychology (seven), and the medical sciences (three). Even the three others support the pattern: they were Anna Pell Wheeler, professor of mathematics at Bryn Mawr College; Helen Dean King, professor of embryology at the Wistar Institute at the University of Pennsylvania (not its main department of zoology); and Jean Broadhurst, professor of biology at Teachers College, again not in the zoology department.

As for department chairmanships, little had changed since 1920 when Flora Wambaugh Patterson, a prominent plant pathologist at the U.S. Department of Agriculture, had written,

> The writer sincerely wishes it could be said that for women pathologists there is always room at the top. There is no such pleasing prospect in view, but the exceptional woman may secure such a coveted position[,] and time proving woman's efficiency the old-fashioned prejudice may be overcome.
>
> The head of the department of plant pathology in one of the leading coeducational institutions [Cornell?] writes: "There is nothing in our university regulations or departmental practice to bar women from any position in the staff. However, in our department, the highest position thus far held by a woman has been that of instructor. I should not hesitate to recommend a woman for assistant professor or even professor, if her qualifications warranted it. . . ."
>
> The disadvantages may be reduced to one, competition with men; not that the qualifications of men are necessarily superior; but that much of the old-fashioned prejudice against women in professions still exists and the popular demand is still for the men.[35]

Only three women in table 7.4 were heads of departments or deans of colleges, and all of these were in nutrition. In fact, although some other women in the *AMS* were chairmen at the teachers colleges and junior colleges, where they may even have been the only member of their "department," only two women scientists were chairmen at even minor coeducational universities in these years in fields outside home economics. The circumstances around both cases are so atypical as to demonstrate just how rare such an appointment was. The first was June Downey, chairman of the psychology department at the University of Wyoming from 1915 until her death in 1932. The unique circumstances behind her appointment include a father who was not only a founder of the university but also president of its board of regents for many years (an auspicious start to anyone's academic career), and her own position as one of the few doctorates and researchers on the University of Wyoming faculty in 1915. The circumstances behind the appointment of the second woman chairman (bacteriologist Mary E. Caldwell of the University of Arizona—not to be confused with Mary L. Caldwell, biochemist at Columbia University) were even more remarkable, since they would have been more likely to work against her: although her husband was a professor of zoology at Arizona and became head of his department in 1933, this apparently did not limit her advancement, as it might well have, since she became a chairman of her own department in 1935 and a full professor in 1937. Like Downey, she was a graduate of the institution (and hence well known to many of its senior staff) who had come back to teach after earning a doctorate from the University of Chicago, apparently a winning combination.[36]

Despite these two precedents, other universities still refused to promote even the top women of the time to department chairmanships. Such denial of promotion was especially painful when the woman was the obvious heir to the position. Florence Sabin, professor of anatomy at The Johns Hopkins Medical School, suffered this indignity in 1919, when she was passed over for department chairmanship by Lewis Weed, who at age 33 was fifteen years her junior; Ruth Benedict (plate 24) underwent a similar experience at Columbia in 1937 when Ralph Linton was chosen to succeed her mentor, Franz Boas. In both cases the resulting status inconsistencies proved so uncomfortable that someone eventually left—Sabin moved to the Rockefeller Institute in New York City in 1925 and Linton left for a Sterling professorship at Yale in 1946. At this point Benedict, by now acclaimed for her award-winning book *The Chrysanthemum and the Sword,* finally became chairman but died shortly thereafter.[37]

An associate professorship was the highest rank to which most women faculty could aspire realistically, regardless of how strong their aging mentor had been or how highly acclaimed their own work was. Some did not even get that far: Alice Hamilton, founder of the field of industrial medicine, was appointed (at age fifty) an assistant professor at Harvard Medical School in 1919 (its first). She retired sixteen years later without a single promotion. Similarly, anatomist Florence Sabin wrote a protégée that if she came to The Hopkins, she would probably expect to be an associate professor someday, but never a full professor, since they were reserved for the very few. As for a department chairmanship, there would be a "real struggle," as she had learned, before one ever went to a woman. She added, "you may make it," but generally advised Marion Hines to come to The Hopkins and expect to get her satisfaction from the work, the facilities, and the students there rather than from any further academic advancement.[38]

But if this successful woman scientist had sympathetic and realistic advice for her protégée, other equally successful women claimed (publicly at least) to be unaware of any problems or barriers. Thus another eminent anatomist, Elizabeth Crosby, minimized, when interviewed as an octagenarian in 1971, any obstacles in her long career at the University of Michigan Medical School. Not only, she claimed, had she never been bothered by any such problems, she insisted she had been unaware of their existence:

> "Sure, sometimes there was a pay difference and I was an instructor for a long time, but a title and money don't keep you from being fulfilled or happy with your work—those are just tokens.
>
> I've always been happy with my status and have never sought or requested a promotion or raise. And as a human being, I've always been treated equally and fairly. . . .

PLATE 24. Because anthropologist Ruth Benedict's partial deafness limited her ability to do fieldwork and learn native languages, she made her greatest contributions in the more theoretical or thematic realm of "patterns of culture." (Courtesy of Vassar College Library.)

I think it is all in one's attitude. If I were conscious of the difference, I'm sure that would make others conscious of it too. But I don't ever think about it so the people around me are never aware of how they treat me except as a human being." . . .

She then added that as a young woman she "probably wouldn't have known discrimination or resentment even if it had developed. I was such an unsophisticated little country girl, thrilled just to be there. I don't think or even know to think about those things.

It's been that way all along. Each time I've been promoted or given a raise, I've been so surprised and pleased that I haven't stopped to think or care about whether it's equal to others."[39]

What both women anatomists were saying, though with strikingly different personal styles and attitudes toward other women, was that however deserving a woman might be, promotions were gifts from one's colleagues, particularly the powerful ones. If a woman expected to be promoted at such institutions merely because her work was good, she might well be disappointed, because a promotion required the initiative and intervention of someone strong enough to override the resistance and criticism that would come at all levels of the appointment process, from chairmen to deans to presidents and boards of trustees. Her promotion would be a personal gift. She could take no initiative, such as to threaten to leave to accept an offer elsewhere, since she would only rarely get one. Nor was her status transferable—to leave one job was not to go to an equal or better one elsewhere—it was, rather, to fall to the bottom, as a middle-aged lecturer, instructor, or assistant professor, or, if married, perhaps to unemployment, because of antinepotism rules. Thus most successful women, as those in table 7.4, seem to have spent their entire career at one institution, where a strong benefactor got their promotions for them.

Under such a personal patronage system, organized feminist protest was futile. A woman was dependent on the good will and tolerance of those around her for the opportunity to work; advancement was a luxury that was to be hoped for but that was not subject to demand. To suggest that conditions might be unequal or that women might be facing barriers that men were not could only worsen the situation. Such criticism was far more likely to draw attention to oneself as an ingrate or "troublemaker" and lead to negative consequences than it was to alleviate or rectify the problem. It was better to ignore the problem of discrimination, which an "exception" might not believe existed in any case. If a woman wished to help other women (and their attitudes toward this differed) there were other, less risky, ways of doing it, such as within women's groups or through separate awards (see chapter 11). But it might be prudent to stay away from even these activities, whose intent might be misinterpreted. Thus, as Marion Hawthorne had found in 1929, promotion at a major university in the 1920s and 1930s was an unpredictable and uncontrollable process that generally coopted women scientists and isolated them from the other, less favored, women of their time. The results, as shown in table 7.4, said more about a woman's colleagues and how liberal and powerful *they* were than it did about her own ability, which, one suspects, was in all these cases quite high.

One particularly well-documented and revealing case occurred at Duke University in 1936 and called forth such a full presentation of views that were widely felt but rarely articulated that it is worth examining in some detail. The discussion started when President W. P. Few of Duke appointed, or seemed to appoint, a woman to a full professorship in his university's struggling physics department. (In fact, though her title and laboratory were in the main physics department, she was to be paid partly by the Rockefeller Foundation and partly by Duke's new coordinate college for women.) Hertha Sponer, a German refugee, was reputed to be the third greatest woman physicist of the time (after Madame Curie and Lise Meitner), though her rank among male physicists was never mentioned. Shortly after appointing her, President Few received a letter from "one of the three most distinguished physicists in the country" questioning how the appointment had come about and whether the job was a permanent one, and indicating obliquely that her record in Germany might not have been as good as Few had apparently thought. Somewhat intimidated by this letter, Few asked the Rockefeller officials for advice, saying "I should ordinarily pay no attention to a thing of this sort, but I must pay attention to this coming as it does from so high a source."[40]

When pressed for more details, Few revealed that his unnamed correspondent was none other than Professor Robert Millikan of the California Institute of Technology (who had long refused to have anything to do with women physicists, even as postdoctoral fellows, and had rejected Sponer in 1925). Millikan later apologized that his letter had meant "no reflection on the individual concerned, but is a mere expression of opinion as to the policy of bringing women into a university department of physics." He then went on to spell out and justify what might be called the "bright young man" theory of academic advancement. Millikan's letter is reproduced in toto in plate 25 and is well worth reading, since it illustrates the attitudes and reasoning of the time.[41] Millikan sincerely believed that the future of physics in the United States depended on extending every opportunity to the "bright young *men*" (his italics) in the field, and allowing the talents of the rest to go undeveloped. He seemed to start with the assumption that although some men were better than others, all were superior to the women, and to adjust all subsequent "reasoning" to support that view. Although Madame Curie and Lise Meitner had made worthwhile contributions to physics, he did not think that this indicated that other women would also. In fact, rather than "opening the doors" to other women to follow, these two examples had the opposite effect, and justified continued exclusion, for, to Millikan at least, they showed how *unlikely* it was that any other women physicists would ever attain their high, but now the minimum acceptable, level. Millikan then used this unlikelihood of past women in physics to be Curies or even National Research Fellows (see chapter 10) to argue that no good job should ever be wasted on a woman, which table 7.4 shows was instead the general practice in 1938. Thus past performance and past discrimination were to be used to justify more exclusion and to deny an opportunity to a talented woman who had already jumped several hurdles and shown herself almost the equal of Curie and Meitner. Most importantly here, Millikan equated the presence of women faculty in a physics department with a lowering of its prestige. Any male physicist of whatever ability would add to it, but a woman of even great ability and proven accomplishments, including the future Nobelist Maria Goeppart-Mayer, whom Millikan's letter overlooked, would lower it. Millikan apparently did not share the view of the Duke department chairman that Sponer should be hired because she was more qualified than anyone else there (she was, for example, the only member of the department who was a Fellow of the American Physical Society), and that therefore her appointment would actually *raise* (not lower) their collective prestige.[42]

Yet Millikan could not bring himself to make an "exception," even when circumstances clearly required it, and continued to justify the antifeminism or sexism (as well as agism, anti-Semitism, racism, elitism) and other prejudices and provincialities that abounded in academia in the 1920s and 1930s. With beliefs like these held by the most influential men in science, and with the threat to a department's "prestige" that he admitted the appointment of a woman implied, it is not hard to see why all the "status of women" reports and even solid performance by many women scientists could not budge the misogynist academic mentality. It justified every indignity—low rank, low pay, unemployment, lack of recognition—which table 7.4 shows abounded in physics in 1938. The subject of women's abilities was obviously not open to rational discourse—

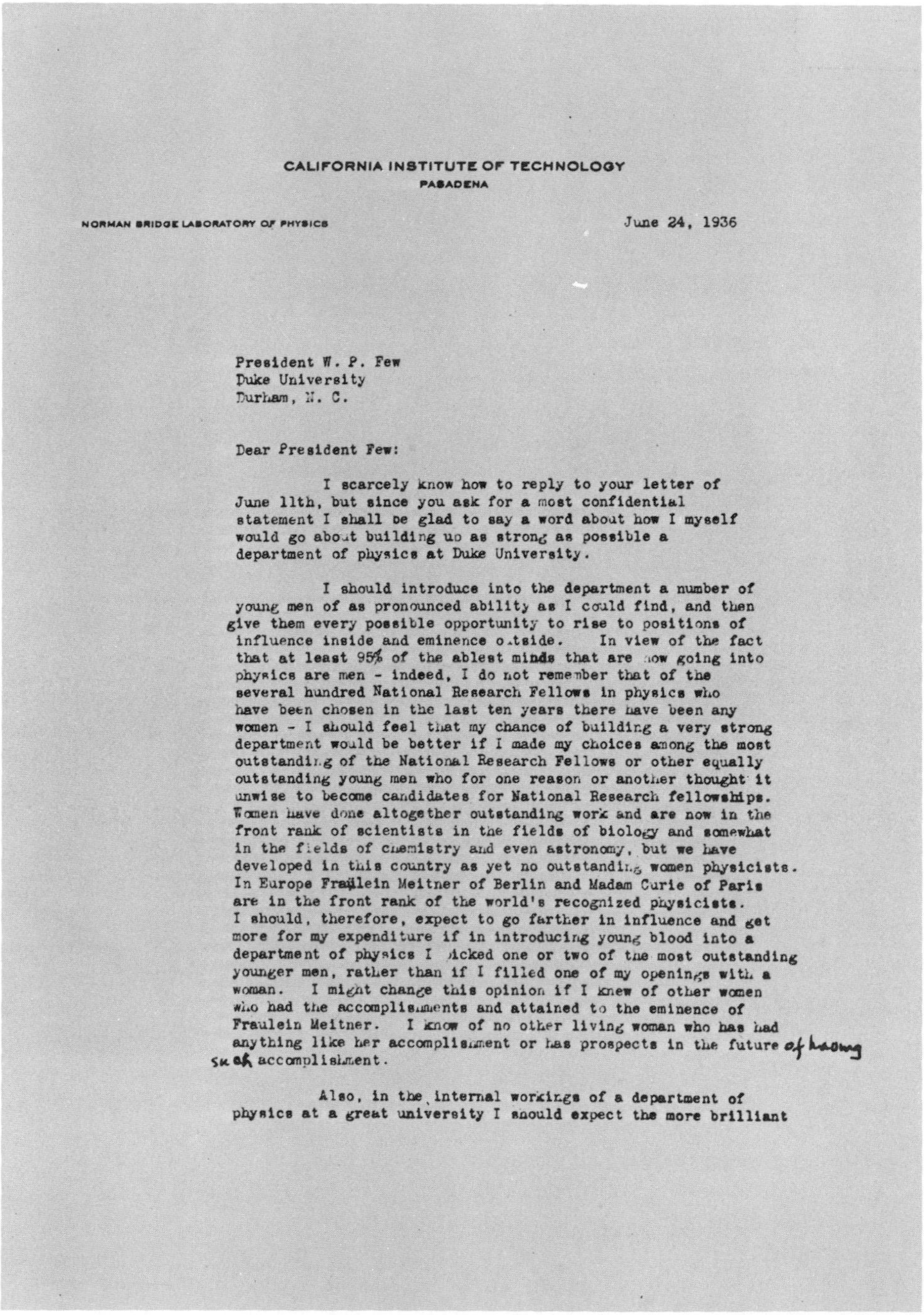

CALIFORNIA INSTITUTE OF TECHNOLOGY
PASADENA

NORMAN BRIDGE LABORATORY OF PHYSICS

June 24, 1936

President W. P. Few
Duke University
Durham, N. C.

Dear President Few:

I scarcely know how to reply to your letter of June 11th, but since you ask for a most confidential statement I shall be glad to say a word about how I myself would go about building up as strong as possible a department of physics at Duke University.

I should introduce into the department a number of young men of as pronounced ability as I could find, and then give them every possible opportunity to rise to positions of influence inside and eminence outside. In view of the fact that at least 95% of the ablest minds that are now going into physics are men - indeed, I do not remember that of the several hundred National Research Fellows in physics who have been chosen in the last ten years there have been any women - I should feel that my chance of building a very strong department would be better if I made my choices among the most outstanding of the National Research Fellows or other equally outstanding young men who for one reason or another thought it unwise to become candidates for National Research fellowships. Women have done altogether outstanding work and are now in the front rank of scientists in the fields of biology and somewhat in the fields of chemistry and even astronomy, but we have developed in this country as yet no outstanding women physicists. In Europe Fraülein Meitner of Berlin and Madam Curie of Paris are in the front rank of the world's recognized physicists. I should, therefore, expect to go farther in influence and get more for my expenditure if in introducing young blood into a department of physics I picked one or two of the most outstanding younger men, rather than if I filled one of my openings with a woman. I might change this opinion if I knew of other women who had the accomplishments and attained to the eminence of Fraulein Meitner. I know of no other living woman who has had anything like her accomplishment or has prospects in the future of having such accomplishment.

Also, in the internal workings of a department of physics at a great university I should expect the more brilliant

PLATE 25. Robert A. Millikan justified in this 1936 letter to W. P. Few, president of Duke University, his strong belief that no woman deserved a good job in physics. (Courtesy of the Duke University Archives.)

Pres. Few, 6/24/36 - 2

and able young men to be drawn into the graduate department by the character of the men on the staff, rather than the character of the women.

These considerations relate more to the graduate work than to the undergraduate. In a coeducational institution where there are many women students it is undoubtedly also desirable to have for pedagogical purposes women instructors, but only in very exceptional cases would I think that the advance of graduate work would be as well promoted by a woman as by a man.

Also, I have heard the report that the general feeling which I have herein expressed had appeared to some extent in Franck's Laboratory at Göttingen, Miss Sponer having got herself into a larger position of influence on the administrative side of the research laboratory than was best for its effectiveness. This report may be only an unjustified rumor, perhaps worth looking up, however, if you are assigning her graduate responsibilities.

If the Rockefeller Foundation as a part of its effort to assist in meeting the very unfortunate exile program has supplied the funds for taking care of one of the exiles, the University may well profit by that act. If, however, I had administrative responsibility at the University I should want to watch developments very carefully to see that antagonisms were not aroused, since women instructors in physics in the long run might react unfavorably upon the prestige of the department, unless they were there solely because of their merit as physicists.

Very sincerely yours,

Robert A Millikan

RAM:IH

facts and credentials, such as Sponer's, were dismissed as irrelevant in favor of stereotypes, fears, and long-cherished views. Sponer was evidently up against not only all the other applicants for a job at Duke in 1936 but also certain physicists' collective views and misconceptions about all of womankind.

Yet unless some women were given a chance to do well and break these stereotypes, there never would be any change. Fortunately President Few held firm, and Sponer kept her job at Duke. But there would have to be a real shortage of the desired "bright young men" and some legislation (both unthinkable in the 1930s) before many more would get the chance to prove both themselves and their kind worthy of much attention and further opportunity. Although Millikan and his colleagues did a great deal to promote research opportunities for scientists in the 1920s and 1930s, their insistence both in words and in deeds that these be restricted to men also had the opposite effect and greatly reduced the nation's potential for scientific excellence.

This elitist and sexist philosophy of restricting the good opportunities in a given generation to a few favored men was reinforced in the 1920s and 1930s by many contemporary academic practices—the "tenure-track," the antinepotism rules, and the salary scales, as well as the systematic channeling of women into separate, less prestigious parts of the university (such as the schools of home economics, the institutes of child welfare, positions as deans of women and research associates, and the separate women's faculty clubs)—all of which put and kept women in places where they would be least noticed and their efforts least rewarded. This system seems to have been quite effective in hiding women's talents since, as is explained in chapter 10, the recognition process managed for years to overlook these many women and their accomplishments.

There does not seem to be any history of how the "tenure-track" came to be a systematic feature of academic employment, but it was introduced in the 1930s, when jobs were scarce, to eliminate relatively early in their careers as painlessly and impersonally as possible (by making it a rule) those persons at the bottom of the faculty ranks who were not developing as scientists or scholars. These were the faithful and agreeable colleagues of no great distinction who had been tolerated earlier but who during the depression were lingering too long and crowding out others of greater "promise." Such embarrassments now had to be moved "up or out." Although highly controversial when introduced at Harvard University in the mid-1930s, where many chairmen at first refused to enforce it, the new policy was adopted by the American Association of University Professors as part of its official statement on academic freedom and tenure by 1941.[43] Because scientific ability and achievement were presumed to be evident early in a career and seven years to be sufficient for it to be both demonstrated and assessed, this new procedure seemed to be fair and to affect everyone equally, but one suspects that a high proportion of its victims must have been the female instructors, lecturers, and assistant professors whose chances to move "up" were already remote. Many women in the 1938 *AMS* had spent decades or whole careers as instructors or assistant professors, until by the time of their retirement they were outranked by everyone in their department, including the young men just starting. Often these women taught the large "service courses" to nonmajors, as Sarah Atsatt did for the biology department at UCLA in the 1920s and 1930s, and as Olive Hazlett complained her task was in the mathematics department at Illinois. Sometimes this was the fate of faculty who were held over when a university was upgraded, as was UCLA from a state normal school in 1919. Before the coming of the tenure-track, such tasks rarely merited either raises or advancement, however well they were performed, as the women frequently learned.[44] After 1941 even these jobs were to be eliminated.

Another way in which the good employment opportunities were restricted to the men, and the women (especially the married ones, a rising percentage of the total) were forced to bear the brunt of academic unemployment, was through the enforcement of what are known as antinepotism rules, a genteel form of discrimination, which afflicted many wives in college towns but rarely husbands or other relatives. The history of these "laws" is also hard to recreate, mostly because they were local, ad hoc, rulings that grew up informally as needed among administrators, and were often not written down. Though few schools (only thirteen out of eighty) admitted to having such rules in the AAUP survey published in 1924, the practice was, Committee W felt, more widespread than that, and it became increasingly so during the 1920s.[45] How these "rules" arose and quickly became sacrosanct, despite what seems in retrospect their dubious legality, has apparently never been studied, but they make a rather unpleasant chapter in the history of American higher education.

The earliest scientist-wives from the 1870s through 1890s were often employed at the same institutions as their husbands—such as Ellen Richards at MIT, Anna Botsford Comstock at Cornell, and Louisa Reed Stowell at Michigan—though usually in low-ranking positions as instructors, lecturers, or assistants.[46] Yet by the 1900s faculty women who married were, as Harriet Brooks had learned at Barnard College in 1906, expected at most institutions to resign their positions whether their husbands were employed there or not, and those who came already married were not given an equal chance at employment.[47] How these informal practices and expectations became rules is not clear, but three theories have been presented as explaining and perhaps justifying them: (1) ethical in origin, they have existed since time immemorial and serve to prevent married couples and other relatives as distant as cousins from evaluating each other to the detriment of other colleagues; (2) political in origin, they grew up at state universities in response to pressures from state legislatures that unqualified persons be appointed to university jobs; and (3) economically inspired, they arose during the depression to protect men's jobs from female competition.[48]

Though these "rules" and practices were widespread and undoubtedly kept many women scientists from using their talents to the fullest, they were chiefly methods of status deprivation, since many wives were allowed to become paid research associates instead of faculty members at their husband's universities (see what follows; apparently there were no rules against this form of employment). Others taught at less prestigious colleges nearby. Many, however, lacked even these alternatives, and one can find whole clusters of these trained and talented but underutilized scientists during the 1930s in such academic neighborhoods as St. Paul Street, Baltimore; Berkeley, California; and Hyde Park, Chicago. A few of these women had such strong drive and motivation that they were able to continue their scientific work without any academic position, as Maria Goeppart-Mayer apparently was able to do while unemployed ("volunteer associate") in Baltimore from 1931 to 1939 and in Chicago from 1946 to 1951, and Gerty Cori at Washington University in St. Louis from 1931 to 1944. Since both these women later won Nobel Prizes for work done under these conditions, one wonders what they might have achieved with proper faculty support. But some universities were inconsistent and made "exceptions" for some women. Thus the department of pharmacology at Washington University Medical School found it possible to hire and promote, if not Gerty Cori, then Helen Tredway Graham, who advanced from instructor in 1926 to associate professor in 1937 and full professor in 1954, although her husband was already a professor of surgery there. Her appointment apparently meant so much to them that when The Johns Hopkins Medical School offered him a professorship in 1939 but did not offer one to her (she was seen as a "complicating factor"), the couple refused to move.[49]

Such antinepotism rules were for many decades a common administrative device used by universities to protect themselves from having to consider employing qualified wives and perhaps to protect the employed husbands from the legitimate professional competition from them as well. What the loss of this professional opportunity meant to these women or to science as a whole, which by this wasteful and systematic disqualification lost the contributions of many highly talented persons, was apparently of little consequence to anyone but the married women, who found themselves blamed first for marrying and leaving science and then for trying to get back into it! Years later, however, some colleagues would admit that it had been an unfortunate mistake not to have put some of these women on the faculty. Eleanora Bliss Knopf, for example, a geologist for the U.S. Geological Survey stationed in New Haven and frequently at Yale where her husband was a professor, would have made a good addition to the faculty, one colleague suggested in her obituary in 1977, since she had had many ideas for dissertations and was very helpful with the students (often with private courses in her specialties).[50] Yet senseless as these rules may seem in retrospect, they were believed in so strongly and held so rigidly that they were not loosened even during World War II, when trained scientists were at a premium, and many other practices were temporarily suspended. These rules were not seriously questioned until the 1960s, when the AAUW began to publicize their unfairness to its members.

A comparison of these antinepotism rules with the earlier prohibition against doctorates for women (see chapter 2) may provide an example of how bureaucracies enforce restrictive policies and help to explain why the women were unable to break through the barrier for so many years. In both cases a prevailing social practice or prejudice had become university "policy" without any alternatives apparently even being considered. But in the case of employment, once there was an accepted policy, all administrators down the line to the department chairmen upheld it, even if, in rare cases, they did not approve of it personally. Occasionally cases might arise where enforcement brought sufficient embarrassment for "policy" to be temporarily overridden, and exceptions had to be made. But since university employment, unlike instruction or even formal admission to a graduate program, was tightly controlled by chairmen and other administrators, the women could not use their earlier tactic of "infiltration," which had been so effective in opening the graduate schools to them in the 1880s and 1890s. They could find no sufficiently independent members of the faculty to disagree with institutional policy and make the informal exceptions for them that might in time convince the governing boards to change or drop their antinepotism policies. Instead, the resistance to employing couples on university faculties remained strong. Society's attitudes toward the employment of married women would have to change dramatically before the antinepotism rules would be loosened.

Since it would be tedious to recount the numerous early studies of the salaries of academic women from 1900 to 1930 (with their ranges, averages, and medians for each rank by sex), suffice it to say that the women's salaries were lower than men's not only because they were instructors rather than professors (as was usually the case) but even when they occupied the same ranks; that these salary discrepancies were greater for those few women at the associate and full professor levels; and that opinions were divided on the issue of giving women equal pay for equal work.* Two interesting accounts from Nebraska, where the salary issue was apparently a particularly sensitive one in the 1920s, illuminate the circumstances and reasoning of the time. Although men's salaries were not high in the 1920s (and in fact many academics were in a form of "genteel poverty"), women could (and would) accept even lower pay, and were thus employed by some schools as cheap replacements or substitutes for men. Although the Nebraska men apparently feared incipient feminization of the profession and angrily accused the women of "underbidding" them, the women were not happy with the situation either. The ACA Nebraska study found that they wanted higher ("equal") pay for their work and thought, rather naively it seems in retrospect, that the best way to bring this about was not through legislation but by having the ACA encourage younger women to enter fields other than teaching and thus restore some balance to the economic forces of supply and demand.[51]

This simplistic view of "overcrowding" and subsequently depressed salaries was only one of the prevailing rationales for women's lower salaries in the

*Marion Hawthorne found some exceptions among the women "assistants" (one step below instructor), who earned more than men in the same rank because of their longer experience. Rather than being promoted to higher rank they had been given salary increases (pp. 152–53).

1920s, according to Committee W of the AAUP, which discussed the phenomenon in its 1921 report. The other two were the necessity for (most) men to support families and the social desirability of keeping men in the teaching profession. Feminists pointed out repeatedly the hypocrisy of the "married man" argument by showing that most colleges paid single men more than they did single women and that many unmarried women had dependents (especially aged parents) but were not given added stipends for their support. The committee also quoted one woman professor, a former chairman of her department with a doctorate and many years of experience, on the insult she felt at receiving a lower salary than a young, inexperienced male who just happened to be married.[52]

On a related financial matter, Professor Amy Hewes of Mount Holyoke College pointed out in a 1919 article that the new Carnegie pension plan for professors (forerunner of the current TIAA-CREF system) discriminated badly against women by giving them only two-thirds of the amount it gave to men. Moreover, most of the men were married and had children who could be expected to support them later on. The women, were, by contrast, single, and despite lower salaries usually had to support aged parents (whom the married men did not aid as often). A retired woman professor might therefore be left destitute and without anyone to support her when she retired.[53]

But despite these criticisms, the principle of equal pay for equal work, which gained a limited acceptance in the federal government in the 1920s (see chapter 8), was not paid even lip service in academia until the 1970s. Instead, as shown in a second Nebraska study done in 1928 (and published in 1930), opinion seemed to favor even *lower* wages for academic women. This study of the spending habits of 29 faculty women at the University of Nebraska found that they spent twice as high a percentage of their income (not actual amounts) on savings, gifts, churches, charity, recreation, and travel as did a mixed sample of 155 faculty members. The two female authors concluded, "From this comparison it is obvious that the expenditures of the unmarried women are larger in every item which might be termed luxury. The enjoyment of such luxuries may in a measure, at least, be considered compensations for their lack of family life. At the same time, such a comparison may raise a question as to the justice of our present method of salary determination and present an argument for some system of adjusting the salary to the number of persons dependent upon it."[54] Thus by 1930, even before the depression hit academia, some women were finding arguments to reduce women's salaries even further.

Likewise, when some pay cuts and layoffs were necessary at many coeducational universities during the depression, the few women on the faculty apparently bore the brunt of them. Not only was their unemployment rate much higher than that of the men (see chapter 6), but their salaries, already lower, were frequently the first to be slashed. Since payroll records are usually not open to scholars, salary reductions are difficult to document, but a few instances have come to light in other ways. Ruth F. Allen, a distinguished plant pathologist ("the most cited woman of the past 30 years, mainly for her cytological studies of rust infections") was a joint employee of the U.S. Department of Agriculture and the genetics department of the University of California at Berkeley. But she had her troubles during the depression, as her obituary in 1964 described it: "During the economy drives of the depression years between 1933–36, her salary was reduced by half. Many considered this an unjustified discrimination against women in science, but Miss Allen accepted it as inevitable and without complaint." She retired in 1936 at age fifty-seven and devoted her energies to playing the stock market, from which she built up a sizeable fortune. In 1965, her heirs gave a fund to the American Phytopathological Society for the Ruth Allen Award for outstanding work in the field by either men or women, thereby perpetrating her memory in her field.[55]

Another kind of depression cutback where the women seemed to take more than their share occurred at The Johns Hopkins School of Hygiene and Public Health, which had been established in 1919 by the Rockefeller Foundation to be a "West Point" for the public health field. This school employed many research associates in the 1920s, as described in what follows, but it also had a few female faculty members, including two in its small department of "physiological hygiene" (or industrial or occupational medicine): Janet Howell Clark, daughter of famed Hopkins pioneer W. H. Howell, and Anna Baetjer, an outstanding physiologist and toxicologist. Although the department was small (it had fewer than five faculty members), it offered courses that were taken by almost all candidates for the M.P.H. degree. For a long time W. H. Howell, the school's director, protected this department's interests, but when he retired in 1931 he was not replaced, and the department's budget was greatly reduced as part of an economy drive. By 1935 the department was suffering so severely that its very existence was threatened. When Janet Howell Clark found that she had, like the other tenured associate professors, been put on a possibly terminal five-year contract, she decided that despite her strong ties to Hopkins, she could perhaps do more to preserve the department and bring the administrators to their senses by resigning than she could by staying and arguing with them. She thus left to become headmistress of the Bryn Mawr School in Baltimore and later dean of women at the University of Rochester. A recent historian of The Hopkins has credited her move with saving the department and has termed this period of its history "extraordinary" for the little financial support it received from the university. It is unclear why this department was treated so much worse than many others, but one can surmise that the presence of so many women on its faculty may have made it particularly vulnerable.[56]

Fortunately for women scientists trying to stay in academia in the 1920s and 1930s, there were several other kinds of employment or "women's work" reserved for them there. They could enter, as everyone wanted them to, one of academia's highly feminized fields, such as home economics or child psychology, or they could leave the faculty proper and become a "dean of women" or a "research associate." Women were far more acceptable at the coeducational universities in these slightly removed fields and roles, where, as shown in table

7.4, there were opportunities for advancement. (Such women could also leave academia altogether and enter government, industry, or some other form of nonacademic employment, as discussed in the next two chapters.)

One particularly strong area for women scientists in this period, as shown in the tables and the other studies mentioned here, and one that stands in sharp contrast to women's overall marginal position in academia, was the still predominantly female field of "home economics" or "nutrition" (as the biochemical portion of it was now being called to distinguish it from the field's new and less chemical areas of textiles, housing, children, and economics). Like the other agricultural sciences with which it was usually grouped, nutrition benefited greatly after 1914 from its strong ties to the land-grant universities and the experiment stations. It thus shared in the increased appropriations and new federal legislation such as the Smith-Lever Extension Act of 1914, the Smith-Hughes Vocational Education Act of 1917, the Purnell Act (for agricultural research) in 1925, and the Bankhead-Jones Act (for more research and extension work) in 1935. With its outside funding assured and increasing, home economics was a rather prosperous field in the 1920s and 1930s, and many programs and departments grew to whole "schools" and "colleges," as their staffs trained the teachers, dietitians, and "home demonstrators" needed by the public schools, by government agencies, and, increasingly in the 1930s, by business, especially the food industry.

Several women nutritionists, as listed in table 7.4, found positions not only as full professors but also as chairmen and deans in these rapidly growing schools and colleges at a time when such opportunities were rare for women in other fields. In fact, they were among the few women scientists of the time who faced the more usual male problem of having high administrative positions take them from their teaching and research. (Several, as mentioned earlier, became presidents of women's colleges.) Vying with Yale as the nation's most prominent graduate program in nutrition was Teachers College, Columbia University, where Henry C. Sherman, Mary Swartz Rose, and Grace McLeod trained many of the faculty members of the land-grant home economics programs. Also strongly research-oriented were the programs at the University of Chicago (Lydia Roberts, Katherine Blunt, and economist Hazel Kyrk), California (Agnes Fay Morgan), Wisconsin (Helen Tracy Parsons), and to a lesser extent Kansas State University (Margaret Justin), Pennsylvania State University (Pauline Beery Mack), and Cornell (Martha Van Rensselaer and Flora Rose).[57]

But all this prosperity and expansion did not necessarily bring home economics high status on the university campus. In fact in many ways it brought the opposite, since most of the large departments at state universities were forced to emphasize teacher training programs for the public schools, a task that caused them to be scorned as "trade schools" by their colleagues in the more traditional academic subjects (many of whose students must also have become schoolteachers).[58] Nor was the home economists' government-supported "research" much appreciated by the many administrators who considered it tainted not only for its service orientation but also for its seeming triviality. For example, Agnes

Fay Morgan's specialty, the vitamin contents of cooked and uncooked vegetables, was seen as making less of a "contribution to science" than the work of her colleague Herbert M. Evans on hormones in a rat's metabolism, which almost won him a Nobel Prize. The field was thus always on the defensive about its academic standards, and as revealed in a 1937 criticism of the home economics curriculum, frequently overcompensated by requiring its students to take additional science courses that would increase its "academic respectability":

> We have loaded our college courses with prerequisite requirements in physical and biological sciences and filled the students' days so full of laboratory hours that they have none of the leisure we talk so much about, in which to develop desireable personal qualities or to think about how to use what they are learning. We have been so concerned about academic respectability—about upholding scholastic standards, about living down the reputation among some of our academic colleagues that home economics offered a haven for the intellectually unfit, that we have failed to do what we should have done for the majority of our students.[59]

Ruth Okey, a biochemist and later chairman of nutrition at the University of California at Berkeley, has described vividly the difficulties Agnes Fay Morgan (plate 26) faced in building up the California program:

> In the early years in Letters and Science, she had to deal on the one hand with university administrators, such as President Benjamin Ide Wheeler, who were strongly oriented toward high academic standards and had little respect for home economics, and, on the other hand, with a state Department of Education which demanded teachers trained in the practical aspects of home cooking and sewing, and dietitians who could deal with problems of quantity cookery and food management as well as therapeutic dietetics. Dr. Morgan literally fought to maintain high standards of scientific training in the undergraduate majors in Nutrition and Dietetics and to develop a research program in nutrition which compared favorably with other graduate programs in the university. That she succeeded is evidenced by the fact that in 1955, when the university decided to concentrate general Home Economics training at the Davis and Santa Barbara campuses, the specialized programs in Food Science and Nutrition, together with the research and graduate training in those areas, were reorganized to become the present Department of Nutritional Sciences at Berkeley.[60]

The results of the countervailing pressures on home economics programs were twofold. There were on the one hand continuing battles about the qualifications of new personnel who, the administrators often claimed, did not meet their standards, and on the other hand a constant rhetoric among the leaders of the field about the value of research that exhorted their followers to do more of it. There are in fact so many references to research in obituaries of academic home economists that one can detect a three-way division of roles within the field. A few women really were researchers (the "nutritionists") and were greatly honored for this, since they enhanced the field's image and prestige; others, who were usually chairmen and deans of home economics, were praised for having valued the importance of research and having urged others to do it, although they apparently did very little themselves. For example, it was said of Florence

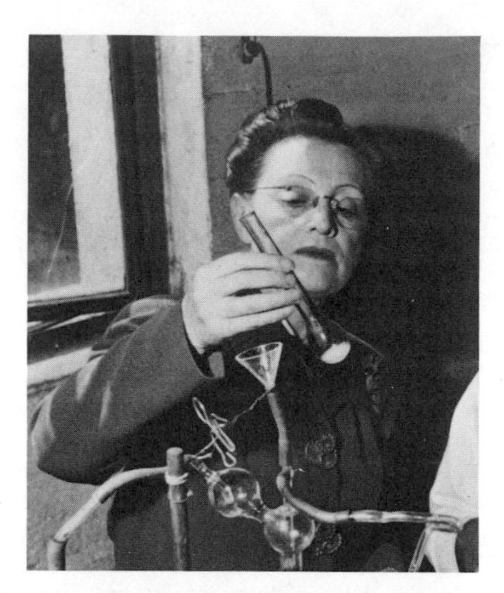

PLATE 26. Nutritionist Agnes Fay Morgan of the University of California at Berkeley, shown here in 1946, was such a strong department chairman she frightened many others in her field. (Courtesy of the Bancroft Library, University of California, Berkeley.)

Fallgatter, for twenty years the head of the home economics education department at Iowa State University, "During her years at the Iowa State College she has built a strong undergraduate program of teacher preparation. But most of all she has encouraged research by staff and graduate students and has built a growing program of graduate education in the department. Her ability to inspire and facilitate the work of others while herself staying in the background is well appreciated by her colleagues."[61] Thirdly, there were those home economists whose contributions were in other areas, as in extension work, teaching, fund-raising, or lobbying with the state legislature. Their obituaries were tactfully silent about their attitudes toward research and stressed instead their other often colorful accomplishments, such as Abby Marlatt's inviting the Wisconsin legislature to breakfast at the university just before they voted on her budget, or Mary Gearing's contributions to the University of Texas where, as the first woman on the faculty building committee, she oversaw the design of fifteen new buildings.[62]

Yet despite what obituaries (known for their generosity) say, other evidence, such as the oral history memoirs by their coworkers, indicates that many nutritionists of the 1920s and 1930s had mixed feelings about these strong and effective women executives among them. Few nutritionists were sympathetic to the problems these female administrators faced, and they were often critical when such women were caught in an administrative about-face and were not able to behave as ethically as they and others might have wished. Thus Agnes Fay Morgan had extended a job offer to one of Lafayette B. Mendel's students in 1923, but had been forced to withdraw it, leaving the woman temporarily without a position. Mendel advised later students to go elsewhere and avoid Morgan, who was nevertheless known for the high-caliber faculty she did attract. Others had less rational or specific reasons for feeling put off by these women. In particular Agnes Fay Morgan, the battler at Berkeley, was feared by many (perhaps most) of the home economists of her time, including Frances Zuill, later the powerhouse at Wisconsin. Zuill not only got her staff the latest in scientific equipment but also had her "department" upgraded into a "School of Home Economics" with five departments, two things that her predecessor, Abby Marlatt, had never been able to do. Yet the Wisconsin staff clearly liked Marlatt better than Zuill; Helen Tracy Parsons, their biochemist and research professor, was afraid to complete a projected history of the program when she realized how her preference for Marlatt would hurt Zuill. More work needs to be done on these women, who were often nicknamed "The Generals" of "women's work."[63]

Thus, though a woman nutritionist could expect to be far more "successful" in both rank and salary than her female colleagues in other fields—the Wisconsin home economists earned more than women in other fields but less than all men—she would still have her battles cut out for her both on campus and in the profession at large. In many ways these pioneering home economists were reminiscent of the first professors at the women's colleges in the 1870s and 1880s—they were institution-builders who urged others to do the research and earn academic respectability while they created the opportunities for them to do so.

Psychology was a second highly feminized area of science in the 1930s, though (as shown in table 6.2) to a lesser extent than was nutrition. Yet even here competition was deliberately minimized by the creation of two separate labor markets within the field—the men took the academic positions (for the most part) and left the "clinical" ones, in hospitals and social welfare agencies, to the women. Even those women who did go into academic jobs tended to head for certain specialties; thus educational psychology, especially "intelligence" and "testing," flourished at the teachers' colleges, where many women were employed in the 1920s and 1930s, and child psychology, or more broadly "child development," which included the nutrition and physical growth as well as psychological development of the child from birth to age six, was a popular field for women. Like home economics, this highly femininized branch of academic psychology was institutionally somewhat separate from the major university departments in the field. It was also supported differently, being largely underwritten in the 1920s and 1930s by the Laura Spelman Rockefeller Memorial Foundation, which established many research professorships and other positions at several "institutes of child welfare" across the nation. Its better-known institutes were those at Teachers College (Columbia University), the University of Iowa, the University of Minnesota, the University of California at Berkeley, and Yale University, although several other, older, institutions also contributed important work, such as the Merrill-Palmer School in Detroit and the Fels Institute at Antioch College. Among the many women prominent (either as "research professors" or "research associates") in this field, whose history is just beginning to be written, were Helen Thompson Woolley of Teachers College, Florence Goodenough of Minnesota, Amy Daniels of Iowa, and Frances Ilg, Louise Bates Ames, and Katherine Wolf of Yale University. Yet women were rarely chosen to fill supervisory positions at these institutes, and only Woolley and her successor, Lois Hayden Meek (later Stolz), ever directed one.[64]

Another continuing form of "women's work" at the large coeducational universities in the 1920s and 1930s was that of the "dean of women," whose origins were discussed in chapter 3. Although young women scientists were less often chosen for such a position by the 1920s and 1930s, a few "mature" women scientists were. Thus Agnes Wells, professor of mathematics at Indiana University for twenty years, was far better known as the school's dean of women (an appointment she held concurrently) and for her success there in building up one of the nation's largest dormitory systems for women. When she resigned in 1938, she was replaced by Kate Heuvner Mueller, a married psychologist whose husband's position on the faculty barred her from teaching there. Although Wells approved her appointment, it was also influenced by the growing politicization of the position in the 1930s, when there was a new stress on psychological counselling and a rising antispinster sentiment in academia. The latter was intense in the appointment of Janet Howell Clark to the deanship at the University of Rochester in 1938. Though the university was seeking a "scholarly" dean for its new coordinate women's college (and assured applicants that they would be able to continue their researches), the search committee evidently cared more about candidates' personalities than their scientific accomplishments, or as Professor Detlev W. Bronk wrote the president: though Clark was no Annie Jump Cannon or Florence "Sabine" [*sic*], "I have seen no evidence of neurotic tendencies which, as you know, are not always lacking in [single?] women scholars." By then being a widow and a mother were considered good qualifications for a deanship, and Clark was appointed. But even these appointments of senior women with a touch of downward mobility grew rarer over time, and by the 1930s, such deans had become part of the burgeoning field of educational administration with their own professional societies and special concerns. By the 1940s few women scientists would find such positions attractive or even be eligible to hold them.[65]

But there was one increasingly common university position to which women scientists were cordially welcomed, even by the all-male universities in the 1920s and 1930s—that of the "research associate."[66] The numbers of women holding the title "research assistant," "research associate," "collaborator," or "fellow" increased from 45 in the 1921 edition of the *American Men of Science* (or 16.1 percent of the total employed) to 282 in the 1938 edition (22.4 percent). By the 1920s and 1930s these female assistants, who had long been employed in subordinate positions in museums and especially observatories (as described in chapter 3), had become even more numerous in the medical schools, universities, and private research institutes (often attached to universities), where they were integral parts of major research teams. In fact, data on such research personnel shown in table 7.3 indicate that in 1938 they were most common, both numerically and by percentage, in the new field of biochemistry (where they constituted over one-half of the academic women), followed by the smaller fields of astronomy (46.2 percent), anthropology (41.2 percent), and microbiology (38.7 percent). Such positions were quite rare, however (less than 10 percent), among the women in the differently structured and funded fields of nutrition, geology, and mathematics.

The rise of the female research associate in the late nineteenth and early twentieth centuries was, as indicated in chapter 3, just one result of a whole cluster of intellectual, technological, financial, and social forces that were transforming science into an increasingly team-oriented enterprise in those very years when women were seeking to enter it. But even more attractive to the employer than the women's availability was the economic reality that they were willing to work harder for lower salaries than were men.[67] This was an important factor in scientific budgeting at a time when small amounts of money were becoming available for research projects. Because ambitions and plans always outstripped the sums available, much such stretching was always necessary and desirable. But the practice continued even when research funds became more plentiful, reaching sums of 30 and 50 thousand dollars. Dollar for dollar the female research associates remained a good investment for a project or institution that required first-rate work, but that offered a low salary, an uncertain future, and no prospects for advancement. It was a situation that offered much to certain women in a depressed job market, but it could and did easily slip into exploitation.

Although the funding of these new positions is not always clear, their appearance seems to reflect a shift as well as an increase in the type of support given research in the 1920s and 1930s. Some fields, especially those in which the federal government was interested (for example nutrition), were, as mentioned earlier, relatively well funded in these decades, and universities were able to create new professorships, schools, and colleges for them. But other fields were dependent on private philanthropy, especially the large foundations and interest groups (such as the Rockefeller Foundation, the Carnegie Institution, and the National Tuberculosis Association), which all held to the elitist philosophy of helping the proven investigator—either by bringing him to their own research institution or else by providing him with additional funds, in the form of an institute and team of assistants, to continue and expand his work at his own university. Similarly, several other lesser philanthropists who wished to perpetuate their names, like the Spragues, the Phippses, the McCormicks, and the Crockers, followed suit and established small institutes for specific persons and researches at the better-known universities across the nation.[68]

As these research opportunities increased in the 1920s and 1930s and tempted many professors to move, universities found it necessary to offer their faculty equivalent research funds in order to retain them. Thus most of the money for research in fields such as medical research, zoology, botany, and astronomy was channeled through the established investigators at the major universities, who then used it to strengthen their positions, hire more assistants, and increase their empires still further. It was a vicious cycle that, it can be argued, narrowed rather than widened the choice of directors of large research projects, and few women, even those in major faculty positions, were able to obtain their own grants. (Nutritionists Helen Tracy Parsons at the University of Wisconsin and Icie Macy Hoobler, director of research laboratories at the Children's Fund of Michigan, are two possible exceptions.) Not even Leta Stetter Hollingworth (plate 27), professor of psychology at Teachers College, Columbia, and an authority in the relatively well-funded field of research on gifted children, received any such research support. One of her colleagues made it a point to mention this omission in an obituary: "She gave most generously of her time and personal funds for the support of research when institutions delayed or refused financial support. . . . Only her closest associates could know that no funds, though often sought, were ever received by her in support of her research. While her personal sacrifices were heavy, she received her reward in the respect and affection of her superior children, in the esteem of her colleagues, and in the international recognition of her achievements."[69]

The research associates listed in the *American Men of Science* in 1938 were thus, not surprisingly, most often found on team projects or in small institutes at the large research-oriented universities, such as the University of Chicago (twenty), University of California (sixteen), Columbia University (fifteen), University of Pennsylvania (thirteen) and The Johns Hopkins University (twelve). These were, it will be remembered, the same schools at which women most often earned their doctorates. Some stayed on at their doctoral institution for the rest of their careers, but more often they moved around among these relatively few institutions. For instance, Charlotte Moore Sitterly, an outstanding astrophysicist, earned a doctorate at Berkeley, and became a fellow at the Lick and Mt. Wilson observatories and then a research associate at Princeton University, where she stayed until World War II.[70]

Yet at even the most popular of these institutions the women were dispersed over so many different fields and departments that they went almost unnoticed. Thus at the University of Chicago they were located at Sprague Memorial Institute, the McCormick Institute, the Nelson Morris Memorial Institute, the Elizabeth McCormick Memorial Fund, and the D. Smith Fund, as well as several medical and biological departments. At the University of California, sixteen women were employed by several departments as well as the Institute for Experimental Biology (two), the Institute for Child Welfare (three), the botanical garden, the Museum for Vertebrate Zoology, and the Hooper Foundation for Medical Research (in San Francisco). The ten female research associates listed at Harvard University in 1938 were employed by the observatory (three), the Arnold Arboretum, the Bussey Institution, the Gray Herbarium, the Fogg Art Museum (a chemist), the medical school, the school of public health, and the Museum of Comparative Zoology. The thirteen women at the University of Pennsylvania (this figure may be low, since it does not include the women who had regular faculty titles at the Phipps and Wistar Institutes, which were affiliated with the university) were employed by the anthropology museum (four), the

PLATE 27. Psychologist Leta Setter Hollingworth, shown here in her later years, published much feminist research between 1913 and 1916. After joining the faculty of Teachers College, Columbia University, she turned to the study of gifted children. (Courtesy of the University of Nebraska Press.)

medical school (four), the zoology department (three), and one each in the botany and chemistry departments.

Quite a colony of talented women scientists, employed and unemployed, formed around The Johns Hopkins University in the 1920s and 1930s. Besides Maria Goeppart-Mayer, the "voluntary associate" in physics mentioned earlier, there were three women staff members at the medical school and eight at the new School of Hygiene and Public Health, giving that school one of the largest numbers of women scientists outside the women's colleges or Teachers College, Columbia. Among these women were Margaret Merrell, research associate in biostatistics who reportedly carried most of her department's teaching load; Bessie Moses, a leader in the birth-control movement and director of a clinic; Eleanor Bliss (not to be confused with Eleanora Bliss Knopf, the geologist, mentioned earlier), who worked on sulfa drugs with Perrin Long; Anna Baetjer, who was active in the early work on radiology and radiation in the environment; Nina Simmonds and Helen Tracy Parsons, research associates in nutrition with E. V. McCollum and Julie Becker McCollum, his wife and coworker; Gertrude Rand Ferree, formerly at Bryn Mawr College, a pioneer in the field of illumination engineering and assistant director (under her husband) of the Wilmer Institute for Physiological Optics; Maud DeWitt Pearl, biologist and managing editor of *Human Biology*, founded by her husband Raymond; Margaret Reed Lewis, research associate at the Carnegie Institute for Embryological Research at The Hopkins and world-renowned authority on tumors; Margaret Gey, her husband's collaborator in virology and oncology; Calista Eliot Causey and Linda B. Lange, in the bacteriology department; and Justina Hill, head of the bacteriological laboratory at the Brady Institute for urology, a male bastion. On the faculty in the 1920s and 1930s were Janet Howell Clark, associate professor of physiology; Esther Richards, associate professor of psychiatry and director of the Phipps Psychiatric Clinic; and Ella Hutzler Oppenheimer, who taught in the pathology department for many years. There were also some interested and supportive faculty wives and widows in the area: Theodora Abel, Mabel T. Mall (who was Florence Sabin's close friend), and numerous members of the nearby Goucher College faculty. Altogether the women scientists in Baltimore in the 1920s and 1930s would seem to have been an unusually rich group whose members must have provided a certain psychological support for each other.[71]

Although biographical information (beyond their entry in the *AMS*) on these research associates is rather hard to find (since many are still alive and others often did not receive even the standard obituary in their professional journal), their lives can be glimpsed indirectly in the biographies and bibliographies of the prominent scientists who employed them. This is most readily seen in the *Biographical Memoirs of the National Academy of Sciences*, where the bibliography, if not the obituary itself, attests to their constant presence and helpful assistance as coworkers and coauthors. Most invisible of all are the wives of these high achievers, who are usually mentioned only once, with very high praise, but then ignored, while their husbands' achievements are discussed in full. Thus the biography of Melville Jean Herskovits, anthropologist at Northwestern University, mentions his wife, Frances, a professional anthropologist in her own right, as "his lifelong collaborator and co-worker whose contributions are not to be measured solely from the list of works which she co-authored with him."[72] But what she accomplished on her own or contributed to his career is left unexplored. It is, after all, "his" obituary.

A prime example of this sort of conjugal collaboration, where the contributions volunteered so readily at the time later went unrecognized, is that of Frederic and Edith (Schwartz) Clements, pioneers in the field of plant ecology. Both earned doctorates in botany at the University of Nebraska around 1900. It was only through her marriage that Edith Clements had any career "in science" at all, since her chief alternative was to quit graduate school and teach school locally in order to support her parents and put several sisters through school. After her marriage, she was, from all evidence, the paragon of a devoted wife and helpmate, contributing enthusiastically her artistic, photographic, secretarial, and even automotive skills to help her husband in "their" lifework. As she put it both humbly and proudly, "Frederic supplied the guiding intellect and I supplemented by my mechanical skills in all that we did." But after his death in the 1940s, and when few persons seemed to know who she was, she began to crave a little more recognition. As her papers at the University of Wyoming show, she culled those statements from his obituaries which stressed her contributions to "his" work. She especially liked (and underlined) one colleague's comment that "together they formed one of the *most illustrious husband and wife teams since the Curies of France*." Realizing how few of her contributions were being remembered, she wrote an autobiography about the joys and frustrations of their fieldwork on midwestern backroads (or no roads) in the 1920s. There one is often reminded that it was usually *her* talents that got them out of the ditch.[73]

It is also apparent from the National Academy's *Biographical Memoirs* that some male scientists had a greater tendency and opportunity to hire female research assistants than did others. Some, like Millikan, refused to have any, but others were more receptive, and some took full advantage of the women's many talents. Among the notable employers and master grantsmen of the time were Lewis Terman of Stanford University, whose comprehensive and well-funded* studies of gifted children employed a large staff of women psychologists between 1920 and 1940—nineteen are listed in the preface to *The Gifted Child Grows Up* (1947). Herbert Evans of the Institute of Experimental Biology at the University of California at Berkeley (whose work was supported by the National Research Council, the University, the USDA, the Merck Company, and the Rockefeller Foundation) had the assistance of at least eight female research associates in his work on hormones and vitamins between 1921 and 1940. Elmer V. McCollum, the discoverer of vitamins A and D, had at least four female associates in his years at The Johns Hopkins University, and he worked with numerous other women in the field of home economics. Vincent du Vigneaud, a biochemist at the Cornell University Medical College in New York City and a Nobel Laureate in 1955 for his synthesis of two hormones, hired many women assistants and started at least four women chemists onto careers of distinction. George Hoyt Whipple of the University of California Medical School and later the University of Rochester also employed several women assistants in the 1920s and 1930s. More flamboyant than any of these academicians was William Beebe, naturalist and explorer at the American Museum of Natural History, who took his staff of up to nine women with him on his voyages to the tropical islands in the 1920s and 1930s. He included photographs of some of these attractive swim-suited assistants (like Gloria Hollister and Jocelyn Crane, who helped with his undersea explorations in the 1930s) in his more popular publications, thus adding a certain raciness to the public image of this branch of science.[74]

Since it is hard to locate material on these women, knowing how they were treated and how they reacted to their role is also difficult. They usually had impressive research backgrounds, almost always held Ph.D.s, and would a few decades earlier have become professors at the women's colleges. But now they

*Terman's studies were financed by the Commonwealth Fund of New York City, the Carnegie Corporation of New York, the Committee for Research on the Problems of Sex (National Research Council), the Columbia Foundation of San Francisco, the Thomas Welton Stanford Fund of Stanford University, and the Marsden Foundation of Palm Springs, California, for a total of almost $150,000.

found these new positions as research associates at the larger institutions more desireable, for they offered them the opportunity to work for at least a few years on important projects and earn a certain, though limited, professional reputation before moving on to a professorship at a teachers college or school of home economics or a staff position at the Bureau of Home Economics of the USDA. This attitude was very much like that of the male research associates (who have not been studied here in any detail), who tended to use the position as an advanced postdoctoral fellowship and moved on to head their own institutes or laboratories within a few years. But many women, lacking such prospects, clung to these temporary positions and made them their careers. This was unfortunate, since the women associates rarely received any advancement, security, or autonomy, even after years of faithful service (even the title *senior research associate* is relatively new), for there was no career plan or line of promotion built into these essentially temporary and dependent positions. Some of the women were restricted geographically, for instance, if they were married or widowed; or ethnically, if they were Jewish,[75] for example; or physically, such as when they were handicapped.[76] When their employer no longer needed them, they would either retire or hope to find another such research position nearby. On occasion a research associate would marry the professor late in life, after his first or second or (as in the case of Evans) third marriage had ended. But otherwise the women associates were left unemployed if the professor moved to another university, lost his grants, retired, or died suddenly. In fact, dispersing his laboratory staff and finding places for his longtime associates was one of the somber final tasks of a retiring professor.[77]

For one other, probably atypical, woman scientist, the retirement of her director freed her to get on with her "own" work. Libbie Hyman (plate 28), now internationally known for her monumental works on the classification of the invertebrates, earned her doctorate at the University of Chicago under Professor Charles M. Child in 1915 and stayed on as his assistant until 1931, when he was approaching retirement. By then she had accumulated enough money from the sale of two popular laboratory manuals to travel abroad for eighteen months, and she used the opportunity to rethink what she wanted to do with her life. Admitting that she had never been very interested in Child's metabolic studies, she moved to New York City where, completely on her own, she began to write her books on the invertebrates. In 1937 the American Museum of Natural History made her a titular research associate and gave her an office, but she apparently never had a secretary, an illustrator, an assistant, or even a salary. Nevertheless, she managed to produce 100 more articles and six hefty volumes. Though she received many honors for her acclaimed works, no university, not even the University of Chicago, would hire her, since she was Jewish, a woman, and reportedly tart and abrasive. She was one research assistant who, having outgrown her position, did something about it, and when no university would hire her, made her own opportunity instead.[78]

Another, less drastic, solution was for the frustrated research associate to complain to a trusted outsider in hopes that he would talk her supervisor into some improvements. The talented astrophysicist Cecilia Payne (later Gaposch-

PLATE 28. Although zoologist Libbie Hyman was widely acclaimed for her monumental works on the invertebrates, no university department would hire her because she was Jewish and was thought to be "abrasive." (Courtesy of James V. McConnell.)

kin) put her feelings into words in 1930 after seven years at the Harvard College Observatory, when Barnard College approached her about replacing the retiring Margaret Maltby as head of its physics program. Because she did not want to go to Barnard, Payne wrote a long forthright letter to Henry Norris Russell of Princeton University, seeking advice and some intervention. Her dissatisfactions at Harvard were considerable:

> First I should mention that I have during the past four years had a very unhappy time at Harvard; the chief reasons have been (a) personal difficulties within the Observatory, particularly with Dr. Shapley, and usually arising out of personal jealousies because he seemed to like others more than myself. (b) disappointment because I received absolutely no recognition, either official or private, from Harvard University or Radcliffe College; I cannot appear in the catalogues; I do give lectures, but they are not announced in the catalogue, and I am paid for (I believe) as "equipment"; certainly, I have no official position such as instructor. Presumably this is impossible, and so I have always thought it; but I have felt the disappointment nevertheless. (c) I do not seem to myself to be paid very much, quite honestly I think I am worth more than 2300 dollars to the Observatory. (d) In the seven years I have spent at Harvard I have not got to know any University person except through my work (which confines my acquaintance to the Observatory staff and Professor Saunders); whereas the wife of any Harvard man of my status is called upon by the wives of dozens of others.

Despite all these problems, Payne was sure that "on the whole I do not want to leave the Harvard Observatory." Chief among her reasons was her affection for the director, or as she put it, "I cannot face the thought of leaving Dr. Shapley . . . for I am devoted to him." Nevertheless, she did want to improve her position at Harvard and hoped that Russell could talk to Shapley.[79] He apparently did, since Payne spent the rest of her career at Harvard. Her diplomatic protest and hint of an outside offer may even have had some lasting effect on Shapley and others at Harvard, for after World War II, Payne-Gaposchkin began to move toward faculty status and eventually became chairman of Harvard's astronomy department, the university's first woman in that role.

Although many employers were, like Shapley, generous to their staff and treated them kindly, such employment was nevertheless liable to exploitation. Thus some professors became overly dependent on their assistants and tried to retain them as subordinates even when better opportunities beckoned elsewhere. For example, Lewis Terman wished Florence Goodenough to return to Stanford during the summers to help him with his data on gifted children, even when she was an assistant professor at Minnesota with projects of her own, and Peter Olitsky at the Rockefeller Institute was quite hurt when his associate, on whom he relied heavily, Isabel Morgan (who was the daughter of Thomas Hunt and Lillian Morgan, both scientists), decided in 1944 to leave him for the poliovirus group at The Johns Hopkins University. Other professors were derelict in their teaching duties and had their associates carry their load for them; Carl Huber reportedly required Lydia DeWitt to teach for him in anatomy at Michigan, and Raymond Pearl required the same of Margaret Merrell in biostatistics at The Hopkins.[80]

A graver kind of exploitation and one that angered the women even more than a heavy work load was the allocation of inadequate credit for their share of the team's researches. Determining who "really" deserves the credit for a discovery is one of the most delicate and difficult of historical tasks, since a great deal depends on the personalities of the workers and the circumstances of the discovery as well as the changing ethics or etiquette of this form of recognition. Absolute fairness is especially difficult to come by in any team research, since the basis for scientific awards and honors has long presumed that discoveries are made by individual scientists, even though this belief was anachronistic in most fields at least as early as the 1920s. Further, it has been observed that with the modern division of labor in science, some styles or roles are more "equal" than others and more likely to receive honors or recognition—for example, in modern physics, the specialists who run and perfect the machines that collect data are less likely to get credit for "their" discoveries than are the generalists or theorists on the same team who put the discovery into a coherent framework and give it meaning.[81] Nevertheless, despite all these sociological reasons for absolute justice not to prevail, scientists of a given period can have a rough notion of the proper norms and be upset when they see them violated. Several such cases involving female associates in the 1920s and 1930s have come to light.

Thus, for example, William Beebe of the American Museum of Natural History hardly ever acknowledged the assistance in his explorations and discoveries of a large number of female associates over several years.[82] Far more complex was the notorious case of Herbert Evans of the University of California, whose difficult personality and temperamental relations with his subordinates led to a very high turnover at his Institute of Experimental Biology. (As one foundation executive put it, "I suppose that I have met individuals with more complex personalities, but when?") Put most positively, he seemed to have a "way" with women, or, as George Corner described him, "He liked especially to talk with intelligent women, who responded warmly to his deferential manner. In more formal conversation and in talk about scientific matters or book-lore, though expressively courteous, he could not conceal an air of superiority of which he may not always have been conscious, an air evinced by allusions to facts or personalities beyond his hearer's acquaintance, introduced in such a way as to suggest that one really ought to know about them." But others who worked with him were less euphemistic. Nellie Halliday had to leave after two and a half years, when she had a nervous breakdown "due entirely," as she told Frank Blair Hanson of the Rockefeller Foundation, "to the difficulties of being associated with Evans," and insisted that "the kindest possible thing she could say about Evans is that he is completely and wholly insane." Similarly, Florence Sabin, the most prominent woman anatomist of the time and one who had worked with Evans in Baltimore earlier, reported to a close friend, "You have sized up Evans to perfection. Nobody can ever tell about him and I judge that he wasn't really fair to Katherine Scott [who had assisted in the discovery of vitamin E in 1922]. She did lots of his good work, made lots of the good observations and lost out in the end. He used her and then let her go." In fact, Evans reportedly made a common practice of firing his associates after a few years—or whenever they sought a pay raise.[83]

Few women were in as good a position to say *no* to Evans as Agnes Fay Morgan, chairman of Berkeley's department of household science, as it was known at the time. When in 1930 Evans proposed to Ruth Okey of her department that they undertake some "cooperative research" on the chemistry of biological tissues, Morgan opposed this on the grounds that Okey's share of the credit might be "overlooked." Astonished and incredulous, Evans asked Morgan what she meant. She, who had known him for years and shared a building with him, claimed tactfully that she had a longstanding opposition to any joint research on the grounds that the "entire independence of each person" was "essential to the satisfaction of the workers and the success of the enterprise," a policy she diplomatically attributed to her department's vulnerability rather than to his personality: "I do cherish the independence of the humble attempts at research carried on by this department as probably our most important asset. To this we may sometimes sacrifice longed-for opportunities."[84] Since she had no objection to joint research within her department, her "policy" may have been designed just for Evans.

Another case, one in which the system of recognition was probably as much to blame as the individuals, involved Frieda Robscheit-Robbins, for thirty-six years the associate of George Hoyt Whipple. First at the University of California in San Francisco and later at the University of Rochester Medical School, where Whipple became the first dean in 1921, they coauthored practically all of "his" numerous articles on pernicious anemia and other liver and iron-deficiency diseases for several decades. Thus when he and two male colleagues at other institutions won the Nobel Prize for this work in 1934, Whipple was acutely embarrassed that Robscheit-Robbins had not shared in it as well. He seized every chance in his comments to the press to praise her contributions in the highest terms, and although he was known as a penny-pincher he even went so far as to share the prize money with her and two other female assistants. Her reaction is unknown, but she was probably grateful and glad for his success and at the same time more than a little upset that she had not shared the prize itself and thereby become the first American woman and the first woman since Madame Curie to win a Nobel Prize in science. When interviewed by the press she insisted modestly that "a scientist never seeks publicity and expects no reward." The work is the important thing, and, she continued, "If you are successful you really deserve no great credit." Had she been a professor at the University of Rochester her situation might have been somewhat different, but the whole point of a research associateship was that, unlike professors, its holders were less than equal and structurally invisible.[85]

Whether "invisible" like the research staffs or segregated into "women's work" on and off the faculty, the women's presence at the coeducational universities in the early twentieth century created one final social problem. Were these women, whose role had been so carefully subordinated to and segregated from the men's, to be invited to join, or more bluntly, to be allowed into, the previously all-male faculty and research clubs on campus? Since the answer was almost always *no,* some of the more spirited and energetic of these early women formed their own clubs; Lydia DeWitt, an instructor in histology at the University of Michigan, formed the Women's Research Club there in 1902 one week after she and several others were rejected from both the Faculty Research Club and the Junior [Faculty] Research Club. Eighty years later her group is still in existence and is still encouraging the work of campus women, a feat that is all the more remarkable when one considers what happened elsewhere.[86]

The more common solution was for the excluded women to form their own separate "faculty clubs," which for a variety of reasons tended to become purely social "women's clubs." The Stanford University Women's Faculty Club (in which archaeologist Hazel Hansen was later active) was apparently the first when it was formed in 1896, since other campuses did not yet have enough women either to need or to be able to support such a club. About twenty years later several were formed in rapid succession: that at UCLA in 1918, at Berkeley (in which anatomist Katharine Scott Bishop was a charter member) in 1919, at the University of Michigan in 1921, at the University of Texas in 1922, at Columbia University in 1925, and at Indiana University (of which Dean Agnes Wells was a longtime president) in 1930. Although most of these clubs have survived into recent times, they had, like the women themselves, rather precarious existences on most campuses. Unlike the Women's Research Club at the University of Michigan, however, the chief function of these other groups was not so much to support research as to enhance friendships and improve morale and communication among faculty women on campus. Because there were barely enough such women at these institutions to keep the club solvent (and not all of them joined), various other women, such as staff members, wives of faculty members, other professional women in the community, and sometimes even graduate students, had to be invited to join. Unfortunately, since there were always far more faculty wives (and widows) than women faculty, this move usually confirmed the style and purpose of the club as purely social ones, similar to those of traditional "women's clubs," such as welcoming new wives to the area, running language tables and card groups, and holding social events like Christmas parties. Many of these clubs felt the need for a clubhouse of their own; once undertaken, however, such a project required either prolonged fund drives or long periods of work and worry to pay off the debts incurred. Even so, rising costs and dwindling membership were perennial concerns.[87]

Meanwhile, despite the existence of these women's faculty clubs, the major academic social functions, such as faculty meetings and dinners and receptions for visiting scholars, continued to take place at the men's faculty clubs. Since almost all of these excluded women (while claiming to represent the whole faculty) until the 1970s or so, stories abound of the indignities that their exclusionary policies inflicted upon the most eminent women of the time; famed nutritionist Icie Macy Hoobler reported in her oral history memoir that she was forbidden to eat dinner at the Michigan Faculty Club, even when she was the after-dinner speaker, and was refused admission to another such club when she was again the honored guest. Although a few women managed to protest their ban and show their disapproval by embarrassing (and so sensitizing) those male colleagues who were members, such gestures were usually futile. Thus Edwin G. Boring never forgot the fact that in 1934 Margaret Washburn of Vassar had insisted on eating in the men-only dining room of the Harvard Faculty Club rather than moving to the separate one for ladies, as he would have preferred her to have done. It is perhaps typical of them and the times that it took several subsequent letters to thrash out the issue. Finally she explained her behavior as not a burst of feminist protest but a simple mistake—she had gotten lost and, being in pain at the time as well as mortified, she had not felt able to get up and leave. Boring, evidently greatly relieved, replied by return mail, "I have absolved you completely from the aggressive feminism which I did momentarily attribute to you."[88] Though he remembered the issue years later and retold it to other female psychologists, he seemed never to think the club's rules should be changed.

From this survey of the prospects and strategies of women scientists in academia in the period before 1940, one can see that hundreds of women were taking up careers, not only at the women's colleges, their traditional bastion, but also on the faculties of the coeducational institutions. In a sense this was "progress," but the women quickly discovered that there were many hierarchies and invidious distinctions within faculties and that despite their greatly increased numbers and percentages, they were not advancing normally. Though a few were in fortunate circumstances and did manage to rise to the top, most were getting lost in labyrinthine passages that worked in a variety of ways to channel them into certain fields, keep them in low ranks, pay them lower salaries, and direct them to adjunct positions as research associates and deans of women. For a few years, roughly from 1910 to 1924, the women tried to protest this unequal treatment with a variety of pressure tactics through the ACA (later AAUW) and the Committee W of the AAUP, but little came of these efforts, since a whole system of prestige, advancement, and logic had arisen which minimized their contributions and blamed them for not doing more. Then the economic depression of the 1930s weakened still further the marginal position women had been granted in the 1920s and put them on the defensive; they were too demoralized to think of reforming the system, of confronting the now ubiquitous stress on the antifeminist concept of "prestige," or of opposing such new restrictions as the antinepotism rules or the tenure-track. Instead of criticizing these new measures, academic spokeswomen could only urge other women to work harder and publish more in order to jump the ever-higher hurdles. In general then, the period from 1920 to 1940 was for academic women, despite all their initial political protests and overall numerical expansion, one of social and psychological containment.

Although it has been possible to document this phenomenon of discrimination, it is more difficult to penetrate into the reasons behind this deliberate policy. We can see the actors and actresses dancing around a certain forbidden territory, but since reasons for this behavior were only rarely articulated, it is difficult to penetrate the mentality that perpetrated it. If the scanty reports of the Cornell faculty meeting in 1911 and Millikan's 1936 letter to the president of Duke are valid indicators, the men in positions of authority seem to have been deathly afraid of losing their status or lowering it in the eyes of other academics elsewhere. They were intent on impressing each other (and any interested outsiders) with their importance, and in this competition, women, even "brilliant" ones, had no place. Thus in the end the treatment afforded women scientists in academia in the 1920s and 1930s says much less about their actual abilities, which were usually considerable, than it does about the fears, insecurities, and punitive personalities of most of the men running the universities. One suspects that, as with the "professional" societies in the 1890s, the economic and social position of the professoriate was so precarious in the 1920s and 1930s that the equitable treatment of an influx of women was more than its leaders could face.

The heart of the system seems to have been the sociopolitical concept of "prestige," which linked one's assessment of the quality of the work done more to the gender of the worker than to his or her actual talents. Gender was such an overriding consideration that once it had entered the calculation, university faculties showed considerable ingenuity in institutionalizing the women's presumed inferiority into separate programs and roles that allowed them to "do science" but that guaranteed that whatever the quality of their work, its value would be diminished and their share in a career's normal recognition and glory withheld.

After the collapse in the mid-1920s of the academic feminists' confrontational tactics for improving women's position in academia, most of the women's energies went less into economic and more into psychological remedies, as in building up their own separate subculture of clubs and prizes, which are discussed in chapter 11. Before discussing them, however, it is necessary to explore the situation of women scientists outside academia who were discovering many of the same sorts of barriers to advancement in government, industry, and wherever else they could find employment in the 1920s and 1930s. They too were developing strategies first to protest, but then just to accept and endure, the prevailing employment practices.

8

GOVERNMENT EMPLOYMENT: PAPER REFORMS BUT EXPANDED SEGREGATION

The experience of women scientists employed in government in the 1920s and 1930s was similar to that of their counterparts in academia: though their numbers increased greatly, they remained clustered at the lowest levels, often in various kinds of "women's work," where they were grossly underpaid and from which they were only rarely promoted. Like most other women working for the government, the scientists remained subordinates rather than directors, were assigned to ongoing projects rather than choosing their own, and were so immobile that most spent their entire careers with one agency. Once again, almost their only advancement came in the field of home economics, in which a few women rose to top managerial positions.

Although this overall pattern of discriminatory employment and advancement was similar to that prevailing in academia at the time, the actual circumstances in government work and the women's reactions and strategies there were somewhat different. At the federal level, at least, women's political groups were strong enough in the immediate postwar years from 1919 to 1923 to get certain Civil Service reforms enacted, but they were unable to prevent still other congressional actions and narrow administrative interpretations from immediately nullifying these hard-won changes. The result of this official turnabout was that, despite the women's short-lived political strength, little changed structurally in the federal government's employment of women scientists between the wars. In fact, the best the women could do at this time was work to extend the classic pattern of sex-segregated jobs that already existed in home economics and parts of botany to the new fields of statistics, and, particularly at the state and local levels, bacteriology and psychology.

Although a few women scientists, as mentioned in chapter 3, had been working for the federal government before 1890, they had been either isolated individuals or, when clustered within certain receptive bureaus, such as the Bureau of Plant Industry at the USDA, the prototypes of workers in sex-typed "women's work." Several other pre–World War I women continued these same patterns, which were largely fostered by the Civil Service rules of the time, by becoming not only "firsts" at their agencies but, on occasion, the forerunners of still other areas of "women's work" that would emerge after World War I. Among these pioneers at the federal level was Florence Bascom, the Johns Hopkins doctorate and Bryn Mawr professor of geology who in 1896 became the first woman scientist at the U.S. Geological Survey, when she spent her first of several summers in the field mapping certain portions of Pennsylvania, New Jersey, and Maryland for the agency. Another notable federal scientist was Mary Engle Pennington, an early doctorate from the University of Pennsylvania, who joined the USDA's Bureau of Chemistry in 1905. After the passage of the Pure Food and Drug Act a year later, bureau chief Harvey Wiley appointed her head of the new Food Research Laboratory, then located in Philadelphia, which was to help enforce the law. Remaining there until 1919, she became famous for her specialty, which was determining the correct safe temperature for refrigerated foods in freight cars.[1]

Meanwhile in 1906 a young bacteriologist named Alice Evans became the first woman scientist at the USDA's Bureau of Animal Industry. Although her coworkers, as she later put it, "almost fell off their chairs" when they discovered that "A. Evans" was a woman, she was later highly regarded for her work on bacteriological diseases carried by milk, especially brucellosis. Another pioneer was Eloise Gerry, who with two degrees in botany from Radcliffe College became the first woman scientist at the U.S. Forest Service in 1910, when, because of a shortage of men in wood science, she was hired by the newly established Forest Products Research Laboratory at Madison, Wisconsin. In time she became an authority on gum production and the turpentine industry. One of the last of the prewar pioneers was Ida Bengtson, initially a librarian at the U.S. Geological Survey, who around 1910 met an unidentified woman scientist from another agency (Evans?). Inspired to become a government scientist too, she left her job, earned a doctorate in bacteriology, and then returned to the federal government in 1916 as the first woman scientist at the Hygenic Laboratory (later National Institute of Health) of the U.S. Public Health Service. There her greatest among several accomplishments was the isolation in 1922 of a new strain of the botulism bacillus. Thus even before World War I the precedents had been set for the later employment of women in some federal health and natural resource bureaus.[2]

With the coming of World War I, women's employment in the federal government entered a new phase. Before the war women had been largely limited to typing and clerical positions, and some departments had refused to hire them, as several suffragists complained in 1916, even for positions as government stenographers. The war created such a sudden need for trained office personnel, however, that old restrictions were swept aside, and the federal government hired about 250,000 clerks and stenographers between 1916 and 1919, most of them women. (The shortage of this type of worker was so great during these years that the Civil Service Commission gave tests for these jobs around the clock in Washington, D.C.) One repercussion of this wartime feminization of the government's clerical force was the opening of a new route to a professional career for self-supporting women. Thus although many of the women clerks, especially those at the War and Navy departments, were laid off after the war, others stayed on, and a few even proceeded to work their way up the Civil Service rankings to become professional scientists. Jewell Glass, for example, who retired in 1960 as a full "geologist" at the U.S. Geological Survey, had started her career as a clerk in the War Department in 1917 and subsequently earned degrees in science at nearby universities. Wartime clerical work could, with a little luck and a lot of extra study, prove the beginning of a career in science, for one woman at least, who would not have qualified initially as a scientist.[3]

This sudden growth and partial feminization of the federal bureaucracy also revealed how intolerably inefficient and unjust the government's rudimentary personnel system had become. Each agency was largely independent and had its own rules, tests, job descriptions, and even salary scales. Since an agency's practices had often been set much earlier, in its original "organic act," and rarely changed thereafter, some agencies had significantly higher wages and more liberal policies than others. Women employees were often the victims of such differences; some agencies, like the Children's Bureau, which employed mostly females, were required to pay them lower, essentially clerical, salaries for doing the same tasks that other departments paid men professional salaries for performing. To regularize women's careers within the government by systematizing this hodgepodge of personnel procedures and eliminating the most blatant discrimination, several women's groups, including the League of Women Voters, the General Federation of Women's Clubs, and the Association of Collegiate Alumnae, worked hard for Civil Service reform in the years from 1916 to 1923. (The ACA, which was especially interested in government jobs for college women, was in 1916 so new to political lobbying that one activist reported to her fellow members that the recent actions of the Washington branch would "prove a revelation to the members of the A.C.A.")[4] The goals of this combined Women's Joint Congressional Caucus included opening all Civil Service tests to women, establishing some kind of retirement plan, standardizing ("reclassifying") job descriptions, and even instituting equal pay for equal work. To an astonishing extent the women's political efforts paid off, and within a few years they had succeeded in having all their reforms enacted. A new day seemed to be dawning for women in the Civil Service.

The women's first reform, the one they achieved most easily and the one of greatest interest to women scientists, was in the area of open testing. Although the ACA had been aware of the fact since 1913 and had been complaining since 1916 that many Civil Service tests were closed to women, it was not until the fall of 1919 that anything came of this. Then the combined women's groups finally induced the newly established Women's Bureau in the Department of Labor (whose formation they had also strongly supported) to issue a report on the women's working conditions in the federal service itself. Investigator Bertha Nienburg quickly exposed the highly uneven practices that surrounded women's employment in the various agencies. For example, in analyzing the 260 Civil Service examinations held in the first six months of 1919, she found that 155 (or

60 percent) were closed to women and that many of these were in the scientific and professional fields. When Nienburg examined the scientific fields further she found striking inconsistencies: women were allowed to take the examinations for jobs in the study of human diseases (U.S. Public Health Service) and plant diseases (USDA's Bureau of Plant Industry), but not animal diseases (USDA's Bureau of Animal Industry). Even within the Bureau of Plant Industry (BPI), they could work on potato but apparently not on tobacco diseases. Equally absurd was the navy's willingness to let women study "materials . . . suitable for naval purposes" while the Forest Service refused to let them study wood (even though Eloise Gerry was already doing so). Neither would the Smithsonian let women classify reptiles; nor would the USDA let them classify samples of cotton or corn. The report was effective, and ten days after Women's Bureau chief Mary Anderson had taken an advance copy to the Civil Service Commission, that group, which claimed to have been unaware that all these differences constituted discrimination, agreed to open all its tests to women. Thus the Women's Bureau could rejoice in what seemed like an important victory. In November 1919, even before its report was officially released, its major recommendation had been enacted.[5]

Although hailed at the time as a major breakthrough, this suspiciously easy administrative reform (which had cost the government no money) was to have little effect on women's actual careers in government, since two other Civil Service rules greatly limited its impact. One such stumbling block was the new "veterans' preference" rule, approved by Congress just a few months earlier, in July 1919, which allowed veterans (almost all of whom were male) five extra points (ten if they were disabled) on most Civil Service examinations. As a result of this rule, even though women might now be taking more tests, veterans would outscore all but the most highly skilled women in their field, and, ranking higher on the final list, be the most likely to be hired. But even if a woman did outscore the veterans taking the same test, a second Civil Service rule, the real "Catch 22," would prevent her from getting a job in almost any area popularly considered "men's work." This rule was the longstanding one that allowed the "appointing officer," almost always a man, to express a preference for a male or female employee. Thus no matter how high a woman scored on the examinations, she would not be hired for any job unless the person in charge had already requested a woman. Mary Anderson professed to liking this rule, since it protected established areas of "women's work" and allowed her as chief of the Women's Bureau (and thus one of the few female "appointing officers" in the government) to pass over any high-scoring veterans and continue to appoint only women to her highly feminized agency. Similarly, most other federal employers sought women only for traditionally "feminine" jobs, but might accept them reluctantly in others when there was a shortage of men in a particular specialty.[6]

Thus the effect of the several 1919 "reforms" was to allow the feminists a certain broadening of opportunity in testing, which they had been seeking for years, but at the same time to perpetrate, and with veterans' preference even to increase, sex segregation within actual federal employment. Inconsistent as it may seem, the Civil Service Commission within just a few months had agreed that it was unfair to prohibit persons from taking tests solely because of their sex and had been forced by Congress to insure that even very high scores were not in themselves enough for a government job. As elsewhere, the desire to maintain a segregated work force, aided here by the political need to reward the veterans for their sex-linked work of defending the nation militarily, took precedence over test scores and "merit." The dichotomy between theory and practice that Leta Stetter Hollingworth and the other feminist psychologists had already noted—that even groups who paid lip service to sexual equality often shrank from carrying it to its full social consequences—was quite evident in the inconsistent congressional and Civil Service reforms of 1919.

Government's ability to accommodate conflicting interests was again evident in 1923 and 1924, when Congress finally passed but then quickly undercut the feminist-inspired Reclassification Act, which initially had promised two sweeping reforms: standardizing Civil Service job descriptions and instituting equal pay for equal work. Although this long-delayed bill finally passed both houses of Congress in 1923, passage was gained only at the high price of allowing the minority considerable say in the appointments to the Personnel Reclassification Board, which was to make the actual changes. Within a year the women discovered how costly this concession had been, since the minority, led by arch-conservative Senator Reed Smoot of Utah, was so eager to contain the rapidly rising cost of government that it had the board members renege on advocates' earlier promises of reform through pay raises. Instead the board instituted equal pay and job reclassification by downgrading some job titles to fit their already low salaries! (For example, if an agency had employed a male "scientific assistant" at $2,500 per annum, a female "scientific assistant" at $1,500, and a female "clerk" at $1,500, the new "reform" did not give the female assistant the $1,000 raise she had been led to expect, but instead gave her no raise and even reduced her title to that of "clerk"!) In this way the Personnel Board deftly achieved its goal of saving the government considerable sums while also bringing the women's goal of "equal pay" to the Civil Service. Thus though women's groups could muster enough congressional votes to pass a reform bill in their interest in the years immediately after getting the vote in 1920, they could not prevent the minority from double-crossing them in the bill's implementation. By 1924 it was clear that legislative reforms were not going to bring about any major changes in government personnel practices and that if women's opportunities in the federal service were to improve they would have to do so through other, less formal, means.[7]

The statistical data available on women workers in the federal government in the 1920s and 1930s confirm the prediction that, despite these various administrative and legislative "reforms," little changed in these decades in patterns of sex-segregated employment. Women, scientists included, were generally hired only for jobs considered "womanly," and men for all the others. Fortunately for the women the demand for "feminine" social services and clerical work greatly increased in these years. A subsequent Woman's Bureau bulletin entitled *Employment of Women in the Federal Government, 1923–1939,* reported that their percentage of the government service jumped more than one-quarter, from 14.9 percent in 1923 to 18.8 percent in 1938, a period when the total number of government jobs almost doubled, to 809,000.[8] Yet relatively few of these women were scientists. Only 4.2 percent of the women in government in 1938 (versus 10.3 percent of the men) were, according to another study by the Civil Service Commission and the Bureau of Labor Statistics, in the "technical, scientific and professional" category, and most of them were librarians, social workers, "social economists," and other nonscientists.[9] Even this percentage may have been high, since other Women's Bureau data indicate that through the 1930s, when there was a freeze on some female employment, less than 2 percent of the newly hired women were in the professional or scientific fields. (The remaining more than 90 percent were in the traditional subprofessional, clerical, and custodial jobs.) Thus, after almost twenty years of open testing and regardless of women's scores on the newly opened Civil Service tests or their personal desires and training, which would have favored "women's work" to a certain extent, few women were being hired for nontraditional jobs in the government. They were in 1938 still 93 percent of all its home economists and 97 percent of its nurses, and even as late as 1939 there was not one female engineer in the entire federal government.[10] (This would change—slightly—during World War II.)

Yet when one looks at other statistics on women scientists in the federal government in these years (imperfect and variable as they are), one can see not only that the number of women doing traditionally "womanly" science (plant pathology and home economics) increased, but that there was also a modest broadening in the number of other fields open to them. Some of these niches appeared because a few broad-minded "appointing officers," impressed with their first woman scientist hired earlier, now added others. Usually, however, these appointments came in those areas of governmental concern, such as statistics and public health, that were already highly feminized at the subprofessional level. As these fields became more scientific in the 1920s and 1930s, women scientists found an increasing acceptance there, not only at the federal level, but even more noticeably at the state and local levels, as discussed later in this chapter.

Wherever one finds data on which federal agencies did or did not employ women scientists in the 1920s and 1930s, the same basic pattern recurs. Though the extent of these studies varied as greatly as did their definition of a *scientist,* they all found that about two-thirds of the women scientists were working for just one agency, the U.S. Department of Agriculture, while the rest were distributed among a varying number of others. This highly skewed distribution was evident in the two sources of data on the 1920s, in the number of women listed in the 1921 edition of the *American Men of Science* and in a 1925 report by the Women's Bureau on the employment of women in the federal government. The former found twenty-nine and one-half women scientists employed, as shown in table 8.1a, at eight agencies. Nineteen of these women (or 64.5 percent) were at

TABLE 8.1. Federal Agencies Employing Female Scientists, 1921 and 1925 (By Field)

	(a) 1921						(b) 1925			
Field	Total	USDA	USGS	USPHS	SI	Other	Total	USDA	NBS	Other
Botany	16	16	0	0	0	0	32	32	—	—
Geology	3	0	3	0	0	0	0	—	—	—
Zoology	3	1	0	0	2	0	8	8	—	—
Medical sciences	2½	0	0	1	0	1 Army Medical Museum ½ Bureau Labor Statistics	0	—	—	—
Home economics	2	2	0	0	0	0	8	8	—	—
Microbiology	1	0	0	1	0	0	1	1	—	—
Physics	1	0	0	0	0	1 NBS	17	—	17	—
Psychology	1	0	0	0	0	1 Surgeon General's Office, War Department	0	—	—	—
Chemistry	0	0	0	0	0	0	20	14	5	1 USGS
Mathematics	0	0	0	0	0	0	1	—	—	1 Coast and Geodetic Survey
Total	29½	19	3	2	2	3½	87	63	22	2

SOURCES: (a) *American Men of Science*, 3d ed. (1921), and (b) *The Status of Women in the Government Service in 1925*, Women's Bureau Bulletin no. 53 (1926), pp. 10–12.

NOTE: Fractions denote one part-time appointment.

just one department, the USDA, and sixteen of these were in its Bureau of Plant Industry (BPI), continuing the tradition of "women's work" started there in the 1880s and discussed in chapter 3. Two were at the Office of Home Economics and the location of one was not specified.

This high concentration was even more evident in the Women's Bureau survey, taken four years later. More complete in some ways than the *AMS* (but quite deficient in others), this survey found eighty-seven women scientists employed at just four agencies, as shown in table 8.1b. Strikingly, it found sixty-three women scientists (72.4 percent of its total and three times as many as had the *AMS*) at the USDA. Thirty-five of these sixty-three women (55.6 percent) were at the BPI, and thirteen were at the new Bureau of Home Economics. Seven others were in 1925 in the Bureau of Chemistry, but women were still only neglible presences in the USDA's other branches, such as the Bureau of Entomology, with three female employees, and four others, with one each. (Again, the location of one female was unspecified.) After the USDA, the second largest employer of women scientists in 1925, according to this survey, was the National Bureau of Standards, with twenty-two women (seventeen physicists and five chemists), all of whom had been hired since April 1918. But there this survey petered out, for although it found two more women at two other agencies, it did not report the clusters of women scientists at still other agencies, especially the U.S. Public Health Service and the U.S. Geological Survey, which are discussed in the following.[11]

Over the next decade and a half, the number of women scientists in the federal government increased greatly, as its responsibilities expanded in areas that required trained scientists, including women: home economics and agriculture, public health, social statistics, and to a lesser extent, natural resources. By 1938 the number of women scientists employed by the federal government had, according to the 1941 report by the Civil Service Commission and the Bureau of Labor Statistics cited earlier, jumped dramatically, to 1,044. Although this study did not break its data down by which agencies were employing the men or the women, it did give their fields, as shown in table 8.2. Again, the largest proportion of the women may have been working for the USDA, since exactly one-half (522) of those counted were in the agricultural and biological fields, which were usually only a part of the USDA's scientific work force. Interestingly, although these high totals must have included everyone in a subprofessional as well as a professional scientific job, even this broad coverage did not uncover any large proportion of women at the lower levels of scientific work. They were in 1938 still just 2.4 percent of the total number, which was by 1938 an astonishing 43,000 scientists and engineers, or almost 50 percent more than the 29,000 entries in the 1938 edition of the *AMS*![12]

By contrast, the 1938 edition of the *AMS*, which counted only those persons "contributing to science," listed just ninety-eight women scientists in the federal government, only one-tenth as many as found by the Civil Service and the Bureau of Labor Statistics, but still more than three times as many as the twenty-nine and one-half listed in the 1921 *AMS*. Most of these women were, as shown in table 8.3, concentrated in the fields of botany, nutrition, and zoology,

TABLE 8.2. Female Scientists in the Federal Government, 1938 (By Field)

Field	Women	Total	Percentage of Total	Percentage of Women
Agricultural and biological scientists	522	12,225	4.3	50.0
Home economists	(250)	(270)	(92.3)	(23.9)
Agronomists, botanists, and bacteriologists	(131)	(3,450)	(3.8)	(12.5)
Zoologists and naturalists	(50)	(650)	(7.7)	(4.8)
Entomologists and husbandmen	(49)	(1,150)	(4.3)	(4.7)
Foresters and range scientists	(31)	(3,900)	(0.8)	(3.0)
Veterinary scientists	(11)	(2,805)	(0.4)	(1.1)
Other social scientists (noneconomists)	200	2,025	9.9	19.2
Medical and dental scientists	162	5,220	3.1	15.5
Statisticians and mathematicians	85	855	9.9	8.1
Chemists and metallurgists	49	1,455	3.4	4.7
Geologists and other physical scientists	26	1,215	2.1	2.5
Engineers	0	19,820	0.0	0.0
Total or average	1,044	42,815	2.4	100.0

SOURCE: Malcolm L. Smith and Kathryn R. Wright, "Occupations and Salaries in Federal Employment," *Monthly Labor Review* 52(1941), table 8, p. 83. (Data collected by the U.S. Civil Service Commission and the Bureau of Labor Statistics of the Department of Labor.)

and sixty-one (or 62.2 percent of the total, again almost two-thirds), two less than the Women's Bureau had found there in 1925, were working for the USDA. The remaining 37.8 percent were distributed among twelve smaller agencies (versus eight in 1921). These were the U.S. Public Health Service (fourteen), the U.S. Geological Survey (six), the National Bureau of Standards (five), the Smithsonian Institution (three), the Children's Bureau (two), and seven others (including the Naval Observatory and the Army's Chemical Warfare Service) that employed one apiece. Yet despite this modest broadening of the employment open to women scientists since 1921, their opportunities in the government service were still in 1938 both narrower and different from the men's. Thus there were no women listed in the 1938 *AMS* as working for the Bureau of Mines, the Weather Bureau, the Coast and Geodetic Survey, the Bureau of American Ethnology, and the Naval Research Laboratory, all highly scientific agencies. Although comparative data for the most frequent employers of male government scientists have not been calculated and are not available in the Women's Bureau bulletins, other evidence (based only on the elite males "starred" in the 1938 *AMS*) indicates that the strongest agencies for male scientists in these years were not the USDA and the USPHS, as for the women, but rather the Geological Survey, the National Bureau of Standards, and the Smithsonian Institution.[13]

Nevertheless, when one looks more closely at those particular agencies and employers that did hire several women scientists in these years, one can see what was necessary for even this partial feminization of the scientific corps to take place and glimpse the very different atmospheres at friendly and unfriendly

TABLE 8.3. Federal Agencies Employing Female Scientists, 1938 (By Field)

Field	Total	USDA	USPHS	USGS	NBS	Smithsonian	Children's	Others
Botany	30	28	0	0	1	1	0	0
Nutrition	14	14	0	0	0	0	0	0
Zoology	14	8	3	0	0	2	0	1 Bureau of Fisheries
Microbiology	9	1	8	0	0	0	0	0
Geology	7	2	0	4	0	0	0	1 Social Security Board
Biochemistry	5	3	2	0	0	0	0	0
Chemistry	5	0	0	2	3	0	0	0
Medical sciences	5	1	0	0	0	0	1	1 Army Chemical Warfare Service 1 Department of Labor 1 Bureau of Indian Affairs
Physics	3	2	0	0	1	0	0	0
Mathematics	3	1	1	0	0	0	1	0
Anthropology	1	1	0	0	0	0	0	0
Astronomy	1	0	0	0	0	0	0	1 Naval Observatory
Psychology	1	0	0	0	0	0	0	1 National Youth Administration
Total	98	61	14	6	5	3	2	7

SOURCE: *American Men of Science*, 6th ed. (1938).

agencies. Generally feminization was a two-step process. The first woman hired (for whatever reason) was usually, by all accounts, outstanding, which perhaps had justified making an exception for her. Once hired she both urged other women in her field to take the appropriate Civil Service tests and performed so well in her own job that she convinced the "appointing officer" to request more women from the Civil Service lists. There could also be two extreme variations in this basic pattern. An enthusiastic employer of a fledgling agency, like G. W. McCoy of the Hygenic Laboratory of the U.S. Public Health Service, might not only hire very good women but also go out of his way to help them (and so his laboratory) to play a major public role in their field. At the other extreme, a particularly hostile agency, as the Canadian Geological Survey in Ottawa seems to have been, could not only not hire other women but perpetually punish its lone one for having intruded there in the first place. Thus a woman's career in government depended not only on her own scientific ability but also on her predecessor's reputation, her supervisor's personality, and her agency's attitudes, for altogether they played a greater role in determining how successful and happy her experience there would be than has generally been recognized.

The U.S. Department of Agriculture was not only the largest federal employer of women scientists in the 1920s and 1930s, as all the data showed, it was also by far their largest employer of any kind in the nation. By contrast, the largest of the academic employers was Wellesley College with, as shown back in table 7.1, just thirty-three women scientists on its faculty (as listed in the *AMS* in 1938). Even if one restricts the comparison with academic employers to the single field of botany, the USDA's twenty-eight is more equivalent to the number of women botanists at all the women's colleges (over twenty-three) than to the number at any single institution. Yet the growth within the massive USDA between 1921 and 1938 was far from uniform, and some definite shifts were under way.

Only a modest part of the department's overall growth from nineteen women scientists in the 1921 *AMS* to sixty-one in that for 1938 (tables 8.1a and 8.3) came at its Bureau of Plant Industry, which had earlier been famed for its large number of women scientists. Although this bureau's count rose from fourteen in 1921 to twenty-four in 1938 (and may have been even higher in the interim, since the 1925 Women's Bureau report had counted thirty-five women scientists there and the BPI had imposed a freeze on "female eligibles" in 1933), its percentage of the women at the USDA dropped sharply in these years, from 73.7 percent in 1921 to just 39.3 percent in 1938. Even before Erwin Frink Smith died in 1927 and his highly feminized laboratory was discontinued, women scientists had been moving both from it and directly from the outside into the BPI's many other divisions. This was part of the bureau's overall decentralization of plant pathology in the 1920s from one separate research laboratory into closer connection with the divisions that focused on particular crops, thereby making it easier to breed new strains for disease resistance. Thus women plant pathologists who had once worked for Smith could by the late 1930s be found in the tobacco, citrus, sugar plant, and especially the cereal crops and diseases divisions of the USDA. Yet even before their dispersal, these twenty-four women scientists had received strikingly little publicity in the popular or women's magazines. At a time when other women scientists in the government were the subject of occasional feature stories romanticizing their work and at a time when the public was intrigued with the far-flung travels of the USDA's plant explorers and other "hunger fighters," it is remarkable that these twenty-four women (or the sixty-one at the USDA in general) received so little attention. The public was left to believe that, though agricultural science was largely botanical, only men were doing it.[14]

Meanwhile another highly feminized branch of the USDA was undergoing rapid growth in the 1920s and 1930s. This was the new Bureau of Home Economics (BHE), which grew from just a small "office" with two women scientists in the 1921 *AMS* to a full "bureau" with fourteen in 1938, including its chief, Louise Stanley, the highest-ranking woman scientist in the federal government. A Yale doctorate and former chairman of the home economics division at the University of Missouri, she had been, as chairman of the AHEA's legislative committee, such a strong supporter of increased federal aid to home economics that Agriculture Secretary Henry Wallace appointed her the BHE's first chief in 1924. Her bureau grew dramatically in the 1920s and 1930s not only because the social legislation of the time greatly enlarged the federal government's nutritional and economic programs but also because technological innovations in consumer products were expanding the scientific content of the home. For example, two of Stanley's staff in 1938 were physicists, Ruth O'Brien and Margaret Hays, who were pioneering in the new field of "textile physics," or the study and testing of the new synthetic fabrics.[15]

Besides the increasing number of women at the BPI and the BHE the other half of the USDA's growing number of women scientists came from their new acceptability in its many other bureaus, where the number of women jumped eightfold, from just three in the 1921 *AMS* to twenty-four in that for 1938. These included increases from zero to five each in the Bureau of Entomology and that of Animal Industry and from zero to three at the U.S. Forest Service (then a part of the USDA); two others were included in the new Soil Conservation Service. Six other bureaus, including the Food and Drug Administration and the Bureau of Dairying, also all hired their first woman scientist in these years.

Meanwhile, in response to many epidemics of infectious diseases among humans, the tiny Hygenic Laboratory (later the National Institute of Health) of the U.S. Public Health Service had begun to expand its research on bacteriological and viral diseases after 1918. Its chief, George W. McCoy, a surgeon and researcher on leprosy, chose well when he hired as his first two women scientists Ida Bengtson and particularly Alice Evans, who came to the laboratory in 1918 from the USDA. Evans was immediately controversial, since she had recently discovered that the organism that caused Bang's disease in cattle was also the source of certain previously unidentified undulant fevers in humans (now known as the disease brucellosis). Her discovery, however, was resisted by many, including the eminent Theobald Smith of the Rockefeller Institute, on what Evans considered the specious (and sexist) grounds that if the two organisms were the same, someone else (such as Robert Koch in Berlin) would have discovered that fact already! Undoubtedly Evans's lack of a doctorate and her government employment as well as her female gender undercut her credibility. Nevertheless, by the mid-1920s her discovery had been confirmed, as more and more cases around the world were diagnosed as brucellosis. Her vindication brought not only her but the Hygenic Laboratory great fame, and her professional colleagues repented by electing her the first woman president of the Society of American Bacteriologists in 1928, as shown in plate 29. Behind this professional accolade as well as most of her later career, one can see the hand of her supportive director, McCoy, who urged her to go to meetings and take on professional responsibilities (even representing the agency at international gatherings), allowed articles to be written about her in popular periodicals, and was loyal during her long illnesses (she caught the lingering brucellosis herself and had several relapses), keeping her spirits up with accounts of work in the lab and using some ingenious tactics to keep her on the payroll after her sick leave expired. He thus not only allowed but even helped her to become one of the foremost women scientists in the government, a great credit to her agency, and an example of a career in government service at its best.[16]

McCoy was evidently so impressed with his outstanding women scientists (and his budget was so tight) that as the laboratory grew in the next two decades he hired more of them. The number of women scientists in his laboratory that were listed in the *AMS* jumped from two in 1921 to fourteen in 1938. Yet, as in his treatment of Evans, he not only chose well initially but also treated his scientific personnel with care, or as one obituary put it, "perhaps his greatest contribution to science came from his unique method of handling students and young scientists." He was so self-effacing that even though he was the director of the laboratory "he considered himself to be the servant rather than the master of the bench workers."[17] Under these conditions the persons he hired tended to thrive. It was also perhaps typical of his gentle, unassuming personality that when in 1928 an epidemic of meningitis required that he add staff in that area immediately, he traveled to the University of Chicago (where most of the NIH women bacteriologists had earned degrees under Edwin O. Jordan) to hire Sara Branham, who was just finishing her degree. Excited by the challenge, she came quickly and soon became one of the world's experts on that dreaded disease. Also joining the NIH staff in a major expansion in 1936 were Sarah Stewart, later famous for her work on cancer viruses, Eloise Cram, a parasitologist from the USDA's Bureau of Animal Industry who came to head the helminthological section of NIH's Laboratory of Tropical Diseases, and Margaret Pittman, famed for her work on the standardization of the pertussis vaccine for whooping cough, who came to the NIH from the New York State Health Department.[18] All this hiring of important women scientists was possible not only because of McCoy's personality and enlightened (but tight-fisted) employment practices but also because public health work had perennially been so poorly supported that it was already highly feminized at the state and local levels, as discussed later in this chapter.

Another area in which the federal government reportedly hired many women in the 1920s and 1930s, although the *AMS* did not yet consider it a science, was

PLATE 29. Microbiologist Alice C. Evans of, first, the U.S. Department of Agriculture and, then, the U.S. Public Health Service was the preeminent woman scientist in the federal government before World War II. She is shown here in 1928, the year she served as the first woman president of the Society of American Bacteriologists. (Courtesy of the National Library of Medicine.)

that of statistics. Even before 1920 there had been many women practicing as "statisticians" or "statistical clerks" in the government because of the field's roots in clerical and social service work. Prospects were so good for women in this work that in 1921 the Bureau of Vocational Information (BVI) of New York City, an independent group with close links to the women's colleges, published an optimistic report on women's future in this field, *Statistical Work: A Study of Opportunities for Women*. This reported that the largest employer of women statisticians was the federal government. Although it had hired relatively few of the women who had passed the Civil Service tests recently, even that percentage had so improved in 1919 as to justify optimism. Of the nineteen women who took the test for "statisticians" in 1918, all passed, but only two were appointed to positions (10.5 percent); in 1919 all forty-nine of those women tested passed, and twelve (24.5 percent) were appointed. The BVI then supplemented this government data with its own questionnaires to women in the field. Of the seventy-seven who responded, almost all had started as schoolteachers, clerks, computers, editors, assistants, and other kinds of office helpers. After beginning to specialize in statistical work on the job, they had then moved into the federal government, where their chief duties were collecting data and writing reports.[19]

The demand for such services remained brisk in the next two decades, as several old agencies (like the Treasury Department and the Bureau of Labor Statistics) expanded and several new ones (like the Social Security Board, the Central Statistical Board, and the USDA's bureaus of Agricultural and Home Economics) were created. All these hired women statisticians, though often under the rubric of "social economist," a designation so clearly sex typed to contemporaries that most of the persons taking Civil Service tests for it in the 1920s were women. The 1930s proved to be particularly exciting times for government statisticians, since several bureaus were introducing what has since been termed the "revolutionary" technique of reliable sample surveys. Particularly important in this movement was Aryness Joy Wickens, who worked first for the Federal Reserve Board and then the Central Statistical Board before joining the Bureau of Labor Statistics in 1938 as assistant to the commissioner. There she pioneered in the development of statistically reliable cost-of-living indices, despite considerable political pressure from the labor unions to inflate them. Women were so acceptable in statistical work that a 1939 article in *Independent Woman,* which was published by the National Federation of Business and Professional Women's Clubs, reported that the employment of female statisticians in the federal government had increased tenfold since World War I. The data presented in table 8.2 (which were collected by the Civil Service Commission and the Bureau of Labor Statistics) confirm this optimistic assessment, since, although they found only eighty-five women statisticians in the federal government in 1938, they may well have been undercounting the many others in "clerical" positions. Nevertheless, women constituted 9.9 percent of the field, making statistics the most highly feminized (and therefore attractive to still more women) of all the sciences in the federal government after home economics.[20]

Geology, by contrast, did not present such an attractive field to women in the 1920s and 1930s, even though the U.S. Geological Survey doubled its number of women listed in the *AMS* between 1921 and 1938 from three to six. Even the more comprehensive data presented in table 8.2 found only twenty-six women (of 1,215 geologists, or 2.1 percent) working in geology. Nevertheless, the few women geologists who were at USGS in the 1920s and 1930s exemplify certain aspects of recruitment into government jobs rather well, since most were close friends who had been classmates at Bryn Mawr (class of 1904) and students of its pioneer geology professor, Florence Bascom. Although one or more of these women might have followed her into a teaching position at a woman's college, had those institutions taken more interest in geology and paleontology, this did not happen, and Bascom urged her best students to take the federal (and apparently some state) Civil Service examinations for geologists. They passed, and perhaps because of Bascom's influence or example, were eventually hired by the U.S. Geological Survey. Eleanora Bliss (later Knopf), who started as an aide in 1912 and became a full geologist in 1928, was known for her work on the structural petrology of the northeast; Julia Gardner, a paleontologist who came in

1920, did detailed studies of the Tertiary mollusks of the Atlantic and Gulf coasts; and Anna I. Jonas (later Stose), a petrologist known for her controversial interpretation of Appalachian structures, came in 1930 after many years on the Maryland, Pennsylvania, and Virginia state surveys. All three spent twenty-five or more years with the survey, which even obliged Knopf by allowing her to do most of her work in New Haven, Connecticut, where her husband was a professor at Yale University. In return these women were a great credit to their agency, since Knopf and Gardner were among the very few women geologists ever "starred" in the *AMS* (see chapter 10). Other early women at the survey included Margaret Foster, its first woman chemist, who arrived in 1918 and worked for many years in the Water Resources Laboratory, where she perfected methods for detecting minerals in natural waters, and the dashing Russian emigrée Taisia Stadnichenko, an expert on coal, who joined as an associate geologist in 1931.[21]

Thus in those agencies where the appointing officer would request or accept women scientists, their numbers increased in the 1920s and 1930s, as they were recruited by their college professors, desperate officials, friends already in government, and perhaps even by hortatory articles on the opportunities in particular fields. But several other agencies at this time had either no women scientists or were at the stage of hiring their first one. Ruth Underhill, for example, may have been the first woman anthropologist hired by the U.S. Bureau of Indian Affairs when she joined in 1934, and Rachel Carson, with a master's degree in zoology, was probably the first woman scientist at the Fish and Wildlife Service when she became a "junior aquatic biologist" there in 1936.[22]

Despite these inroads into a few federal agencies by the 1930s, the women scientists there got little advancement either in job titles or in salaries at even the friendliest of bureaus. An appendix to the 1941 Women's Bureau report indicated that although women then composed 19.6 percent of the government service, they still constituted only 3.5 percent of the supervisory personnel, and hardly any of these were scientists.[23] As a bureau chief, Louise Stanley of the USDA's Bureau of Home Economics was, as mentioned earlier, the highest-ranking and best-paid (at $5,000) woman scientist in the entire federal government in the 1920s and 1930s. Behind her came the medical scientist Martha Eliot, M.D., of the Children's Bureau. A pioneer in the prevention of rickets in children, she became director of the Division of Child and Maternal Health in 1924 and was promoted ten years later to assistant chief of the bureau.[24]

But outside of home economics and child health such promotions rarely happened, even in those agencies that had hired the most women scientists and in which women had begun to amass large amounts of seniority, such as at the Bureau of Plant Industry and the Hygenic Laboratory. Thus despite the long tradition of women plant pathologists in the BPI, none was promoted until 1929, when C. Audrey Richards was made "pathologist-in-charge" (but not chief?) of the Madison, Wisconsin, branch of the BPI, a job that a colleague later described as "the highest position of authority ever achieved by a woman in the bureau."[25] Of the several women at the National Institute of Health in the 1930s, only Eloise Cram reached even the lowest level of management, that of chief of a section of a laboratory. (Interestingly, her entry in the *AMS* listed her simply as "with" the NIH.) Moreover, Helen Dyer, a biochemist at the Hygenic Laboratory in the 1920s, later complained that McCoy, for all his virtues, would not give raises to single women (most of the NIH women were single). The only other woman scientist to hold even a minor managerial position was Joanna Bussey, a physicist who had joined the National Bureau of Standards in 1918 and became a section chief of thermometry in 1929.[26]

Since so few women were ever promoted, even after decades of service and then usually only to positions in charge of other women, some experienced women began in the 1930s to complain about the discrepancy, which had become a major sore point. In 1939 Ruby Worner, an assistant chemist at the National Bureau of Standards, described the usual rationalizations given for failure to advance government women in her "Opportunities for Women Chemists in Washington, D.C." The reasons were strikingly similar to those given in academia for not making women full professors. First there was the economic argument that men were heads of households and so needed the better jobs more than did women, who presumably lacked dependents. (Worner then, like so many of her academic predecessors on this topic, pointed out the frequent fallacy of this argument in practice: single women received no extra payment when they had dependent parents or contributed to the support of other relatives, as they often did. Nor did single men without dependents receive lower salaries than married men. Gender was more important for determining an employee's salary than either marital status or actual number of dependents, official rhetoric to the contrary notwithstanding.) Secondly, Worner thought that government women were never given a chance "to develop or exhibit their executive abilities." Managers assumed they could be only assistants and kept them at that level. Thirdly, she said that there were still "men who object to working on an equal basis [with] or under women," though she suggested that they could probably get used to this if they had to. Her conclusion from all this was that the obstacles to women's advancement in the federal service were so great that it was "much easier for women to enter the service in a higher classification than to advance from a lower grade to a higher one," though few women ever reached the higher levels either way.[27]

This pervasive lack of advancement in the government service in the 1930s helps to explain why, despite some reforms, the women's salaries remained far below those of government men of the time. Although the Civil Service Reclassification Act of 1923 had, as explained earlier, introduced the idea of "equal pay for equal work" for men and women in federal employment, its implementation had not alleviated the wide differences in the overall averages for the two sexes. One of the reasons for this was the continuing occupational segregation, which meant that only rarely did the men and women actually hold the "same" job title. Since those jobs held by women were paid less than those assigned to men (home economists were the lowest paid of all scientists), the women's average salary remained well below the men's. But even when men and women did work in the same fields, such as bacteriology or chemistry, a second factor, the women's almost total lack of advancement, meant that they remained in entry-level positions ("junior" or "assistant chemist") long after the men were promoted to "senior chemist" or even "assistant chief." Thus despite the lip service to "equal pay," the women's overall salary averages remained far lower than the men's (and apparently even worsened) in the 1930s.

By the late 1930s, however, the Women's Bureau was not as willing to document and protest the salary inequity as it had been earlier. In its 1925 report on women in the federal government, when the salary issue was still a fresh topic, the bureau had devoted considerable space to recent advances. By contrast, its 1941 report made no direct comparisons between the men's and women's salaries. Instead, it avoided controversy and compared women's salaries only with those of other women. This approach allowed it to stress the optimistic news that although home economists were the lowest-paid scientists, as a whole women in the "scientific, technical and professional" category were still the best-paid women in the federal government, with an average salary of $2,299 in 1938. However, one can calculate from the data presented in a separate report by the Bureau of Labor Statistics and the Civil Service Commission that the average salary for men in this category was $3,214 in 1938, or almost $1,000 and 40 percent more than that for women. It is not clear why the Women's Bureau refused to speak out about this when earlier it had protested even smaller discrepancies, but by the late 1930s such feelings were rarely expressed. Like the academic women who had protested their low salaries vociferously in the 1920s (see chapter 7) but had grown silent on the issue after 1929, even Ruby Worner, who had criticized the lack of advancement for women in government work in 1939, now concluded grudgingly that government work did provide women scientists with certain advantages and benefits, such as sick pay, vacation leave, reliable pensions, and above all, job security, a big advantage during the depression, when many felt glad to have any job.[28] Economic insecurity and a rising antifeminism were an intimidating combination.

Moreover, during the depression not even the Civil Service Commission could guarantee the much-vaunted security of women's government jobs. Married women were particularly vulnerable and were singled out as legitimate scapegoats in Congress's notorious section 213 of the Economy Act of 1932. It specified that spouses of government employees who were also employed would be the first discharged in any necessary reductions in force. This provision was repeatedly protested by women's groups and such notables as Eleanor Roosevelt and Frances Perkins, who became the Secretary of Labor in 1933, but it remained in effect until 1937. Of the approximately 1,600 women laid off under this ruling, most were clerks and postmistresses (as were most women government employees) rather than scientists. But everyone felt vulnerable when the budgets were being cut, and many scientific agencies, such as the National Bureau of Standards, the Geological Survey, and the Smithsonian Institution, suffered major reductions in the 1930s.[29]

Some women (or their friends) felt that they were being forced to bear the brunt of their agency's declining budgets. Since women tended to be in the most junior and most precarious positions, this was often the case, although it also happened to senior women who should have been more secure. Ruth F. Allen, for example, a plant pathologist at the USDA since 1918 who was working at Berkeley on rust-resistant plants, found, as mentioned in chapter 7, that only her salary was cut in half in 1933. Although she put up with this and then in 1936 retired early, her friends still thought it such an outrage that they mentioned it in her obituary thirty years later. Ethnomusicologist Frances Densmore of the Bureau of American Ethnology (BAE) at the Smithsonian Institution, where she was famous for recording and transcribing the fast-disappearing tribal music of the native American Indians, found her small salary and expense money cut off in 1933. Although this put an end to her annual fieldwork, she continued to transcribe at her own expense the songs she had recorded earlier. In 1936 the Works Progress Administration hired her as an instructor in Indian crafts, but she was still so badly off in 1937 that she had to sell her back copies of the *American Anthropologist* to buy a new typewriter to keep on with her real work, the transcriptions. Eventually the BAE's fortunes revived and its officials could afford to buy Densmore's completed manuscript for the cost of her expenses. Under such circumstances, a woman needed not only scientific ability but also a great deal of persistence to carry on with her science despite the government's shortsighted economies.[30]

Probably the most difficult experience endured by a woman scientist in government in the 1920s and 1930s, because the most deliberate and long-lasting, was the persecution Alice Wilson suffered at the Canadian Geological Survey in Ottawa. Even though it is not within the scope of this work to follow Canadian developments in any detail, her situation there exemplifies well the all-too-common battles that some women in "men's work" had to fight during the 1920s and 1930s. Although Alice Wilson is mentioned only in passing in the Canadian Survey's own history, she was one of its more distinguished employees, and her career there has received a full chapter in the Canadian Federation of University Women's chronicle of Canada's greatest women. This account offers a particularly well-documented case of what was probably a widespread phenomenon at the time (and is reminiscent of the administration of the antinepotism rules in academia mentioned in chapter 7)—the endless refusals to promote a qualified woman (while promoting the men) and the steady stream of phoney reasons and rationalizations, which, when overcome (often at considerable sacrifice), were then dismissed as irrelevant or insufficient. This process continued until the woman left or exhaustion set in. Rarely, and more luckily, the officials were finally so pressured and embarrassed by overwhelming outside recognition that they relented and made an exception, as happened after several decades in the prolonged case of Alice Wilson.

Alice Wilson worked briefly for the Canadian Geological Survey in 1909 while an undergraduate and returned to it in 1911 as a museum assistant, the first woman to hold any scientific position there. Although two older brothers, who were both geologists and had worked for the survey, managed to clear the path for her initially, one biographer has written, "Many of Miss Wilson's colleagues and superiors resented her ambition to be a geologist, and recognized as one, and did all they could to prevent her from joining their masculine circle. Miss Wilson was a lady, and although her opponents proved that they were no gentlemen, it was not in her nature to fight meanly. But neither was it in her nature to give up. Through years of frustration and humiliation she quietly persisted, doing the work assigned to her and carrying on her Ordovician studies as best she could."[31] When told a few years later that she could not be promoted without a doctorate (despite her fieldwork and numerous publications on the rocks and fossils of the Ottawa Valley), she applied for a leave of absence to earn one, but her appeal was rejected for ten successive years. (Such leaves with full pay were apparently granted routinely to men who sought them.) Then in 1926, at age forty-five, she won a graduate fellowship from the Canadian Federation of University Women, but still the director of the survey refused to grant her a year's leave. Finally he agreed to let her go for six months on the condition that she earn the leave by doing a year's work in six months! After accomplishing this heroic task, Wilson went to the University of Chicago, completed her course work, and by 1929 had her Ph.D. But the director still would neither promote her nor increase her salary, although others received promotions and raises regularly. Her friends at the federation tried to help by petitioning the director, but were unsuccessful. However, her luck began to change in 1935, when a new political party came to power in Ottawa and was looking for a woman in government to honor. On the advice of the federation, it chose Alice Wilson to be one of its first Members of the Order of the British Empire, a rare and highly coveted honor for anyone. Shortly after this Wilson became the first woman elected a Fellow of the Royal Society of Canada (1938) and the first Canadian woman elected a fellow of the Geological Society of America. Finally, in 1940, a new director of the survey decided that the time had come to promote Wilson to associate geologist, and in 1944, the year before her retirement, to full geologist. Though it would be very difficult to determine how widespread such practices were in the 1920s and 1930s, one has the feeling that either they or variations upon them must have been fairly common, especially in the slower-growing fields and agencies that had limited budgets. Promoting women or helping them use their talents professionally was not accorded high priority. Few persons of either sex would have had the perseverence and stamina as well as the influential connections and political savvy of Alice Wilson. (Another geologist, Grace Stewart, left the survey in 1923 for a teaching job at Ohio State because of the hostile conditions.) But few victims of these practices and attitudes could have gotten the satisfaction and eventual vindication that came to Alice Wilson, Ph.D., M.B.E.[32]

In the United States, there was a large increase in the number of women scientists working for state and local governments in the 1920s and 1930s, as these agencies also increased their services, especially in the "womanly" areas of public health, home economics, and social welfare. Although the little that is known about the state civil service systems indicates that, like the federal one, they long limited women formally and informally to jobs already deemed "womanly,"[33] official statistics are lacking for confirmation. Nevertheless, the experience of those women scientists who did work for state government suggests that this was generally the case. Although one of the earliest was Mary Murtfeldt of Saint Louis, Missouri, who served as the (unpaid?) assistant to the state entomologist, C. V. Riley, from 1876 to 1878 and then as acting state entomologist from 1888 to 1896, her example and seeming high rank did not set any precedents for other women at the many state agricultural experiment stations that were founded in the 1870s and 1880s. Although probably the largest employers of scientists at the state level, these institutions generally hired women only for clerical or library positions. The first few women scientists who did work there tended to be botanists, who stayed only a few years, like Katharine G. Bitting and Freda Detmers, who served as "assistant botanists" at the Indiana and Ohio stations, respectively, from 1890 to 1893. After 1900 women in other fields were hired, and a few even started what turned into long careers at certain stations: Elizabeth H. Smith was an "assistant plant pathologist" at the California agricultural experiment station at Berkeley from 1903 until her death in 1933, and Cornelia Kennedy was an "assistant agricultural biochemist" at the Minnesota experiment station from 1908 until her retirement in 1948. Even they may have had trouble finding these jobs, for other women scientists encountered station directors who refused to hire them. When, for example, entomologist Edith Patch first sought a job at an experiment station in 1901, she was rebuffed by several station directors who thought entomology "unwomanly," before finally being hired two years later by the one in Orono, Maine, where she spent the rest of her distinguished career. When in 1909 the New Haven, Connecticut, station hired chemist Edna Ferry, a graduate of Mount Holyoke College and later holder of Yale's first master of science degree (1913), she was its first woman scientist.[34]

Even more restrictive than the agricultural experiment stations were the state geological surveys, which seem to have banned systematically female scientists before (and even to a large extent after) World War I. The only woman in the first three editions of the *American Men of Science* to work for these agencies was paleontologist Carlotta Maury, who was an assistant geologist for the Louisiana State Geological Survey from 1907 to 1909 and prepared reports on the region's rock salt and petroleum deposits. The second woman hired by a state survey was Eleanora Bliss, who, although she was already established at the U.S. Geological Survey, served on the Maryland State Survey as a temporary "assistant" between 1917 and 1920. Then in 1919 the Maryland and Pennsylvania surveys hired Anna I. Jonas, the only woman geologist to make what might be considered a career on the state surveys. She also worked for the survey in Virginia both before and after joining the federal U.S. Geological Survey in 1930.[35]

By contrast with these reluctant state science agencies, the state boards of public health had long been more receptive to women scientists. Ellen Richards was in fact the first in this as in so many other areas when she served as chemist to the Massachusetts State Board of Public Health as early as 1872 (until 1875)

PLATE 30. The New York State Department of Health, still the largest science agency at the state level, has employed many women scientists over the years. Among its staff members in 1915 were the first director of its then-new division of laboratories and research, Augustus B. Wadsworth (seated, center), and his longtime assistant director, Mary B. Kirkbride (seated, second from right). (Courtesy of the New York State Department of Health, Division of Laboratories and Research.)

and again as that state's water analyst from 1887 to 1897. The field grew rapidly after 1900, and especially after 1908, because a series of well-publicized epidemics of infectious diseases (typhoid, tuberculosis, pneumonia, diphtheria, and poliomyelitis, among others) so terrified the citizens that state legislatures voted massive increases in their funding of public health agencies. Because these agencies needed many bacteriologists but could rarely afford to hire men, they were highly feminized as early as 1915, if the accompanying photograph of the staff of the New York State Department of Health (plate 30) is any clue.

The women's reception into bacteriological public health work before World War I was laden with the same qualifiers and restrictions typical of such "women's work." This compromise reflected the conservative social views of William T. Sedgwick, professor of biology at MIT and former president of the Society of American Bacteriologists, who was the leading figure in inviting women into public health work—but in carefully circumscribed ways. In 1902, with the help of Ellen Richards, his MIT colleague who shared his conservative philosophy about sex roles, Sedgwick established at Simmons College in Boston (which he called "The Women's Tech" and of which he was a trustee) a program in "general science," which emphasized training women in chemistry and bacteriology for positions as laboratory technicians. He was probably also behind the Women's Educational and Industrial Union of Boston's 1911 leaflet entitled *Bacteriological Work as a Vocation for Women,* which predicted that there would soon be many opportunities for women employees on city boards of health, in private laboratories, and in most hospitals. The highest level to which such women could ever aspire, however, was that of "assistant director" under the supervision of the eminent (male) physician who would always head such institutions.[36]

Within a few years the general shortage of public health personnel that Sedgwick had been expecting erupted into a full-scale crisis, as both World War I and the influenza pandemic of 1918 greatly magnified the need for bacteriologists. As a result, public health leaders appealed frantically to trained and patriotic women to enter the field. In February 1918 Sedgwick, who had four years earlier been battling the feminists in the *New York Times* (see chapter 5), published an editorial entitled "A Welcome to Women in Public Health Work" in the *American Journal of Public Health.* He was so concerned about the serious shortages that he (or an associate) had even toured the eastern women's colleges to recruit science majors into summer internships that would lead to "professional positions in laboratories, inspectorships, and the like" by fall. His prediction that there would still be a need for such persons after the war was borne out over a year later by his second plea in the May 1919 *Journal,* "Women's New Fields of Work." By then the situation was even worse than during the war, since the demand for public health agencies continued to expand very rapidly. Not only was legislature after legislature voting increased appropriations for the field, but also whole new specialties like child welfare were opening up, for which, Sedgwick thought, women were particularly well suited. Besides, he admitted, the men who had held public health positions before the war were in many cases not returning to them, since they had found more challenging or lucrative positions elsewhere. Sedgwick felt that if the nation was to have the health care it needed and deserved, it was up to the patriotic women trained in these fields to enter the government laboratories and accept the jobs that the men would not.[37]

The women's only verbal response to this left-handed invitation came in 1921 after the *New York Times* repeated these same arguments in an editorial entitled "Women in Public Health Work." It reported that a recent survey by the U.S. Public Health Service had found that thirty-nine of the forty-seven state public health agencies employed "nearly two hundred" women in child hygiene, public health nursing, and vital statistics work. More were needed, because salaries were so low that men generally found better-paying jobs elsewhere. Important work would go undone unless women stepped in and took the positions that the men would not. To this Gertrude Martin of the *AAUW Journal* responded in an editorial of her own. If, she protested, the salaries were so low, why did anyone think that women could take these jobs either? The cost of their medical training had been as high as the men's, their financial sacrifice and that of their families equally great, and if women's previous experience in public health work was any guide, it was not "such as to encourage their sisters to follow in their footsteps."[38] If the citizens wanted greater health protection they should pay for it.

Despite the validity of Martin's points, many women bacteriologists responded to this far from roseate invitation and joined the state health departments in the next two decades. Twenty-five women in the 1938 *AMS* were employed by nine state public health agencies, and, as shown in table 8.4, the number of women employed at the state level in bacteriology had far outstripped the numbers employed there in all the other sciences. Coming next were nutrition and botany, both specialties of the state agricultural experiment stations, thirteen of which employed eighteen women scientists in the 1938 *AMS* (seven in botany, six in nutrition, and five in several other fields but still only one in entomology). Since by 1938 Anna Jonas Stose had joined the federal USGS, the *AMS* listed only one woman as working for a state geological survey: Helen Plummer of Austin, Texas, was listed as a "consulting paleontologist" for the Texas Bureau of Economic Geology.

What is particularly surprising about these women in the *AMS* is, as at the federal level, their highly skewed distribution. Here twenty-three of the sixty-one worked for just one state, New York, making it by far the largest employer of women scientists at the state level. (Michigan was a distant second with only four.) Why New York should have taken such a strong lead is difficult to

TABLE 8.4. Government Employment of Female Scientists, 1921 and 1938 (By Field and Level of Goverment)

	1921					1938				
Field	Total	Federal	State	Local	Foreign	Total	Federal	State	Local	Foreign
Botany	18½	16	2½	0	0	42	30	9	0	3 Canada
Medical sciences	10	3½	2	4½	0	10	5	3	2	0
Psychology	8	1	4	3	0	44	1	8	35	0
Geology	6	3	2	0	1 Brazil	11	7	2	0	2 Canada
Zoology	4	3	1	0	0	21	14	5	0	2 Canada
Nutrition	3½	2	1½	0	0	23	14	9	0	0
Chemistry	2	0	2	0	0	8	5	2	0	1 Canada
Physics	1	1	0	0	0	3	3	0	0	0
Microbiology	0	0	0	0	0	27	9	16	2	0
Biochemistry	0	0	0	0	0	9	5	3	0	1 England
Mathematics	0	0	0	0	0	6	3	3	0	0
Anthropology	0	0	0	0	0	2	1	1	0	0
Astronomy	0	0	0	0	0	2	1	0	0	1 Canada
Engineering	0	0	0	0	0	1	0	0	1	0
Total	53	29½	15	7½	1	209	98	61	40	10

SOURCES: *American Men of Science,* 3d ed. (1921), and *American Men of Science,* 6th ed. (1938).

NOTE: Fractions indicate that some persons held two appointments at once, only one of which was a governmental one, such as a staff position at a state experiment station with a junior faculty appointment at a state university, or a hospital appointment or city health department appointment with another position.

determine and, with the almost total lack of literature on scientists in state government, is open to speculation. It could be the result of some selection factor (or "artifact") in the preparation of the directory—New Yorkers might have been more visible or eager to be included than, say, Californians or Texans, but there is no way to check this. The answer would seem to lie in the political, economic, and scientific climate in New York. It was apparently one of the wealthiest of states in the 1920s and 1930s and one of the most willing to spend its revenues on social services for its many immigrants and other citizens and on scientific assistance for its industries. The explanation for this may lie in the state's liberal politics, which supported scientific reforms and legislation to a far greater extent than did even other progressive states, such as Massachusetts, Pennsylvania, or Illinois, whose governor had earlier appointed Alice Hamilton and the other women at Hull House to state commissions. But even if New York were hiring more scientists than other states, why was it so willing to hire women? Were its Civil Service rules more open than those elsewhere, its male bosses more liberal, or its pay scales more attractive to women? Or were the jobs in the more feminized fields or the skills those that women were more likely to have? All were true to a certain extent, as shown in the history of the New York State Department of Health, which was the women scientists' best experience in state government.

New York State had had its share of epidemics of contagious diseases when in 1913 a commission of experts recommended that a separate laboratory or research division be set up to make safe vaccines and take advantage of the latest medical advances. Among the first steps taken by the division's new director, Augustus B. Wadsworth, was hiring two bacteriologists, Mary Kirkbride and Ruth Gilbert (see plate 30). Both served as his assistants for over thirty years, Kirkbride in, and eventually "acting director" of, the Division of Laboratories and Research, and Gilbert as bacteriologist and assistant director of the Diagnostic Laboratories from 1921 until her retirement in 1949. By 1938 the New York State Department of Health (NYSDH) had eleven women scientists listed in the *AMS* as well as three others working for related state agencies, making the NYSDH almost as frequent an employer of women scientists as the entire U.S. Public Health Service; but it was an even better one, since it gave some women more advancement than did the federal agency. Also important at the NYSDH, besides Kirkbride and Gilbert, was Hilda Freeman Silverman, a biostatistician in the Division of Communicable Diseases, who did important epidemiological studies in the 1930s on the effectiveness of the new pertussis vaccines and in 1944 on the spread of poliomyelitis during an epidemic in New York. A car accident in 1953 cut short a career of great distinction.[39]

Proud of its position as the premier state health department in the nation, the NYSDH gladly lent its vaccines and highly talented personnel to other states in time of emergency. Sometimes its scientists stayed at the new location and helped that state start or build up its own agency. One of these recruits from New York was Pearl Kendrick, who left the NYSDH in 1920 for the Michigan State

Health Department and stayed on to become its associate director in 1932. Her first project in Michigan was to find new methods for the detection of syphilis, but she soon switched to the work on whooping cough and the development of a safe and effective vaccine for it, one of the major public health projects of the 1930s. Although her appointment as associate director came just as the state was threatening to cut the department's research budget, Kendrick saved the precious vaccine project by inviting Eleanor Roosevelt to visit. The First Lady came to the laboratory and, impressed with what she saw, had several WPA workers added to the laboratory staff in time to complete the vaccine studies.[40]

Even more remarkable than New York's hiring of women scientists was its willingness to promote a few of them to the top administrative positions that were so often closed to them in the federal service. (It was also in New York State that both Frances Perkins and Frieda Miller had long distinguished careers in social service before moving on to prominent positions in the federal government.) Interestingly, however, these top positions in New York did not come in those bureaus that had the most women listed in the *AMS*, such as the state health department, but in some of the smaller, less-feminized, agencies. Among these outstanding women employees of the state of New York was Emmeline Moore, a leader in a brand new field. She left her job as an assistant professor of biology at Vassar College in 1919 to join the New York State Conservation Department, where she pioneered in the field of fish culture, especially fish diseases and pollution studies. She rose rapidly to the position of chief aquatic biologist and became director of the State Biological Survey before her retirement in 1944. She was also in 1928 the first woman president of the American Fisheries Society, and had, among her other honors, a ship in the New York State Merchant Marine named for her in 1958.[41]

Another remarkable New York woman was Winifred Goldring, an outstanding paleobotanist at the State Museum in Albany who rose through the ranks from 1914 until her retirement forty years later as state paleontologist, by far the highest position ever attained by a woman in paleontology at that time. But her extensive correspondence reveals that her career was far from easy. After majoring in botany and ranking first in her class at Wellesley College (possessing one of the finest minds that Margaret Ferguson and her colleagues there had ever met), Goldring stayed on to earn a master's degree in geology and hold an instructorship. In 1914 she took a temporary position as scientific expert at the New York State Museum in Albany to prepare educational exhibits. Encouraged by some of her superiors, she stayed on and began to work seriously on the rare Devonian plants that were relatively abundant in New York State (see plate 31). Her numerous publications were highly acclaimed, but in 1926, overworked and suffering from the nervous strain of being a woman in such a hostile setting, she had a mental and physical breakdown. (Although she was given tasks that the men would not perform and was paid less than one-half the salary of the museum's clerical staff, Goldring was too proud to complain or ask for what she called "concessions." Instead, like the women chemists to be discussed

PLATE 31. Paleobotanist Winifred Goldring, shown here in 1929, was known for her work on the Devonian plants. She rose through the ranks at the New York State Museum in Albany to become its first woman "state paleontologist" in 1939. (Courtesy of Joyce Goldring.)

momentarily, she determined to prove that she could do paleontology as well, and even better, than the men on the staff, despite the unequal treatment she was receiving.) Fortunately she recovered within the year and returned to her scientific work, but she never regained her earlier idealism. Although the obvious candidate to succeed her boss, whom she revered, Goldring long doubted that she would ever be promoted. Nevertheless, the museum's regents surprised and delighted her in 1939 with the coveted promotion. (Professor Charles Schuchert of Yale University, her longtime supporter and confidant, apparently played a key role in the appointment.) Like Alice Evans in the federal government, she too was distressed at the second-class treatment accorded women in her field, and as one obituary put it, "it irked her immensely that there were so many outstanding male geologists and paleontologists who were vocally prejudicial to women in science." She spoke out against this practice whenever she could, although this meant that she was known to many for "her mildly acidulous comments on the pomposity and self-satisfaction of certain geologists (mostly male)." Even so, in 1949 she was the first woman elected president of the Paleontology Society, and she served as vice-president of the Geological Society of America in 1950. Thus, though women scientists in state government still faced many problems, a few, like Moore, Goldring, and the public health women, rose to positions of authority in the 1920s and 1930s (in New York and Michigan at least) and found their accomplishments recognized by some of their professional societies.[42]

Similarly, city and local governments increased the number of women scientists they employed between 1920 and 1940. In fact, the increase at this level was proportionally even greater than at the federal or the state levels in these years, quintupling, from seven and one-half in 1921 to forty in 1938, as shown in table 8.4. Again one city led the way: New York City far outstripped all other municipalities in the number of women listed in the 1938 *AMS,* having ten females employed by its Board of Education, Department of Health, and numerous other courts and hospitals.[43] Cleveland, Ohio, was second, with four psychologists employed by its Board of Education.

Most of this growth at the city level was, as elsewhere, in established areas of "women's work," such as public health, social service, and schoolteaching, which were growing more scientific in these years. One typical area was city health departments, which, like their equivalents at the state and federal levels, also began to hire women bacteriologists for routine testing, diagnosis, and on occasion, research. One of the first and most outstanding of these early bacteriologists was Anna Williams, who joined the New York City Health Department as early as 1895 and became an assistant director in 1905. In her almost forty years with the department she isolated a strain of diphtheria bacillus, prepared a vaccine and a rapid test for rabies, and worked on streptococcal, pneumococcal, and meningococcal infections. A colleague, physician S. Josephine Baker, established in 1908 and directed until 1922 the department's pioneering Bureau of Child Hygiene, an agency subsequently copied by many cities and states. Although relatively few city bacteriologists were listed in the *AMS,* these jobs seemed to have existed in abundance, since political scientist Avis Marion Saint found the female bacteriologist a standard figure in city government in the 1920s.[44]

By 1938, however, psychologists were by far the most frequently employed women scientists at the local level. This was the result of the tremendous increase between 1920 and 1940 in all areas of "applied psychology," not only in the public school systems, where testing and "guidance" had come to stay, but also in the numerous governmental and quasi-governmental hospitals, clinics, courts, and "homes" for the blind, retarded, orphaned, delinquent, abandoned, and others in unfortunate circumstances across the nation. Although these may well have been the least desirable jobs in the field (little is known about their salaries), a few women, especially married women who were channeled into these positions, struggled to make from them important contributions to applied psychology. Helen T. Woolley, for example, the founder of sex-difference psychology discussed in chapter 5, started a psychological clinic in the Cincinnati public schools, and her pioneering studies of the problems of working children led to revisions in the state's child labor laws; Augusta Fox Bronner pioneered in the study of juvenile delinquents while she was associate director of the Judge Baker Guidance Clinic in Boston.[45]

Because of a dearth of manuscript materials on women employed in all these governmental positions, it is hard to know how they reacted to them. In a sense they were fortunate that the expanding governmental role in the peacetime and even depression economy of the 1930s offered so many openings for microbiologists, nutritionists, botanists, and psychologists—positions in which women were relatively numerous and relatively welcome. But on the other hand, many of these jobs were open to women only because the similarly qualified men had other "better" opportunities and did not find these jobs as attractive or desirable. (This sex typing of jobs backfired a bit on men scientists during the depression when many male psychologists could not find suitably "manly" [academic] employment. When some well-meaning persons advised such men to accept clinical positions instead, a professor at Northwestern University spoke for many when he protested publicly that such jobs "were not the work for a man."[46])

This raises the broader issue of what makes some jobs attractive to men and others not. To a certain extent low pay, as was pervasive in many city, state, and perhaps even federal jobs, drove men away, and yet academic salaries, which they generally accepted, had never been very lucrative either. Thus something besides money was involved. This was recognition or status, since, even if the pay was very low, some men would still take a job that offered the elusive "prestige," that artificial currency or commodity designed to compensate the ego if not the purse. Since men would take any jobs that had either high salary or high prestige (but, preferably, both), the resulting sex-typing of scientific jobs followed a pattern that can be presented graphically, as in figure 8.1. Men would consider suitable three of the four kinds of jobs described here but leave the fourth, those with neither high pay nor high status, to the women. They, having no alternatives, accepted them and thus completed the linkage among low-paying, nonprestigious, and highly feminized jobs that one finds so often in the history of work, scientific or otherwise.

FIGURE 8.1. Typology of "Men's" and "Women's" Work, by Salary and Status

STATUS	SALARY: *High*	SALARY: *Low*
High	Men's work	Men's work
Low	Men's work	Women's work

Even successful woman clinicians got little prestige. The inadequacy of the usual forms of professional recognition to appreciate important contributions to clinical work permeates David Shakow's obituary of Grace Kent, an outstanding clinical psychologist at the Worcester State Hospital and later that at Danvers, Massachusetts, in the 1920s. Initially she had worked on the development of certain psychological tests (especially her Kent-Rosanoff Association Test of 1907–1910), but upon switching to clinical work in the 1920s, she stood firmly for nonstatistical, personal, and "truly clinical" methods. Yet despite her excellence and inspiration to the many persons who came to work with her, neither her work nor her self-effacing personality were given professional awards, a circumstance that she had professed not to mind but that her male biographer missed for her.[47]

It was during the 1920s and 1930s that the employment of women scientists became firmly established in the federal, state, and local governments. It followed, however, most of the patterns of sex typing that were prevalent in academia at this time: the women were generally at the lowest levels, where they were underpaid and, except in home economics, rarely advanced. Despite certain weak administrative and legislative reforms in the years from 1919 to 1923, this pattern persisted and even spread beyond home economics and botany and into statistics, bacteriology, and psychology. Such sex typing fell far short of the reforms that the women had sought initially, but when compared with their employment experience elsewhere in the 1920s and 1930s, government work was a relatively serene haven for many women scientists.

9

INDUSTRIAL AND OTHER EMPLOYMENT: STOICISM, VERSATILITY, AND VOCATIONAL GUIDANCE

If the women scientists' experiences in government and academia were roughly comparable in the 1920s and 1930s, their situation was far worse in industrial and other employment. Although the women did carve out a few areas of "women's work" in industry in these years, even this modest presence generally created such intense hostility that, if one can judge from articles in the professional journals of the time, bare survival—getting a job and holding onto it—was a greater concern for these women than was the relative luxury of fighting for a promotion. Accordingly, they made no attempts to change the system, such as by agitating for legislative reforms like the government women or by preparing confrontational "status of women" reports like the academic women. Instead, the industrial women scientists' chief tactic was to advise each other how to cope with their difficult situation and avoid making it even worse for others. This advice or vocational guidance, which appeared in a series of books and articles over twenty years, was of two types. The first was the optimistic, perhaps idealistic, "Madame Curie strategy," as it was termed earlier, of quiet but deliberate overqualification, personal modesty, strong self-discipline, and infinite stoicism—the classic tactics of assimilation required of those seeking acceptance in a hostile and competitive atmosphere, the kind of atmosphere women heading for bastions of "men's work" encountered at every turn. When this strategy failed to produce measurable improvement and the depression brought on a major employment crisis, other women scientists began to preach a second, more "realistic," and almost cynical strategy of versatility and conformity to sex-typed employment. They advised young women not to try to overcome sexual stereotypes in employment but to accept them and adapt to them. A woman should no longer, for example, drive herself to become a research chemist, but should choose between the many "womanly" jobs offered her, such as chemical secretary or librarian. Thus in industry as elsewhere, those few women scientists of the 1920s and 1930s who started off with an optimistic and aggressive strategy for attaining some kind of equality had before long to fall back upon the very sex-typed employment they had been trying to avoid. Although sexual stereotypes have many limitations and disadvantages in scientific employment, in a depression they can offer the desperate one kind of survival tactic.

During World War I many chemical companies, as described in chapter 5, had hired women "chemists" to fill in for the departed and sorely needed men. At the time the women seemed to do well, despite their lack of experience and training for such positions, and their efforts were applauded in the chemical journals. The Armistice had barely been signed, however, when the managers laid these women off and rehired the returning veterans and other men fresh from the universities. Accompanying this action and seeking to justify it was a strong verbal "backlash," or, as one source put it, "a flood of criticism" in which even those who had apparently been pleased at the time with the women's ability to do "men's work" now turned on them, ridiculed their unreliability and lack of experience, circulated stories about their ineptitude, and then used this allegedly bad record as a reason for not hiring any more women. (The contrast with the women's postwar reception into public health work, where even neophytes' wartime contributions were used to justify recruiting more women after the war, is striking. The difference was that, rather than foreseeing a continuing and even worsening shortage of qualified personnel, chemical executives feared postwar unemployment, perhaps accurately,[1] and were also far less willing to consider women as scientific employees than were the more flexible and hard-pressed public health officials.)

By 1921, when the chemical industry was in a postwar depression, the women's employment situation had become so bad that the AAUW formed a committee of several women chemists, led by Professor Emma Perry Carr of Mount Holyoke College, to work with the staff of the Bureau of Vocational Information (BVI) in New York City to investigate women's employment in chemistry and recommend a solution. This group sent out about six hundred questionnaires (about two hundred and twenty-five of which were returned) to women chemists and wrote ninety-eight department chairmen (of whom forty-one responded) and contacted over one hundred other chief chemists and experts. A year later the BVI published its book *Women in Chemistry,* a comprehensive survey of the status of women in the field. It sampled managers' opinions and prejudices, expressed a few sexual stereotypes of its own (women made good technicians), surveyed employment opportunities and salaries, discussed Civil Service examinations, described graduate programs, fellowships, and prizes, listed journals of interest to chemists, and provided, as Charles Herty, a former president of the American Chemical Society, wrote in an introduction, other "sage advice" on the difficulties and opportunities encountered by female aspirants to a career in chemistry. The book seems in retrospect to have overestimated the future opportunities in industrial chemistry and to have understressed those in biochemistry, but when compared with other works of vocational guidance that have appeared since, it was exceptionally sober and realistic.[2]

In fact this volume can be viewed as presenting the most complete statement available of the prevailing attitude and strategy among prominent women scientists in the 1920s of how to survive and perhaps be equal in "men's work." None of these women had shared the naive optimism so many others had exhibited during and after Madame Curie's visit in 1921, and they now outlined a much more realistic, almost grim, picture of women's second-class position in chemistry. The strategy outlined here was a very defensive, conservative one, which reflects the intention of the authors—seven professors of chemistry at women's colleges, two biochemists in medical fields, and five women with industrial companies—to meet and overcome employers' criticisms of past women employees. Although its tone is thus somewhat similar to that of the surveys of chairmen's opinions by academic women in the 1920s, its political thrust is quite different since these women chemists believed that there were already too many women seeking employment in the chemical industry! They agreed that many of them were poorly trained, and not wishing to overcrowd the field any further with weak or immature women, they firmly dissuaded any more of them from entering it. Their approach was thus not the Pollyanna-ish, "the-barriers-have-been-broken" one that is common in career literature for women or even the feminist (perhaps idealistic, under these circumstances) goal that women, even average ones, should have an equal chance to hold the same jobs as men. Instead these women sought to forewarn young women of the unrealistically high expectations and latent hostilities of industrial employers, who would rather not employ them. Readers were also informed of the heavy pressures riding on those women who did take up such positions. They should be aware, the women said, that many observers would judge all women chemists by the performance of a few (probably the weakest ones). If some women quit to get married or acted immaturely or irresponsibly, the behavior would not be minimized and forgotten as the weakness or foible of just one person but would be remembered and used to condemn many other women as unstable and unworthy of either employment or promotion. In short, the woman chemist was caught in a classic double bind: if she did poorly, she was "typical" of all womankind, confirming once again the stereotype that most of them were incompetent or unstable (or both); if, however, she did well, she would be called an "exception" whose performance implied nothing about the abilities of other women. (In this way the work of even the best women chemists was undercut and the stereotype maintained.) If this seemed an unreasonable situation, one in which employers expected a worker to fail (and then gloated when she did), that, the advisers warned, was the way it was. The women were there only on sufferance. They had to accept this hostility and work from there.

The chief means of overcoming the prejudices against women's employment in science was, these women felt, to acknowledge the presence and power of employers' stereotypes of women and work hard to become an "exception" to them. To do this a woman chemist had to show that she was not only as good as most of the men in the laboratory but better than all of them. Once she had proven this, she would, they were confident, be accepted and applauded as being "one of the boys." Meanwhile she had to be more conscientious, hard-working and reliable than her male coworkers and take the precaution, which experience had shown desirable, of having more advanced degrees than anyone else or than was really necessary. In general it was her responsibility to prove that she was up to their standards and did not fit their stereotypes. Unless she did this, her male coworkers would be justified in assuming that she was not their equal.

Unfortunately this was the easiest part of the strategy, for in addition to this deliberate overqualification, women had to be stoic and not become bitter at the unequal treatment they would continue to receive even after they had worked so hard to prove themselves "equal." They should not be upset to learn that they still would be paid less than the men (whose work their own surpassed) and passed over for promotions by inferiors (whom they might even be asked to train), but rarely would be given credit for their accomplishments. To give in to bitterness and anger at this point was to undo much of what they had accomplished—or so the strategy said. Anger or, worse, outright protest, was futile and could only hurt oneself and others. It was, these experienced women thought, best for the welfare of all present and future women chemists (a particular concern of the professors) that all women chemists know that acceptance of these unequal conditions, rigorous self-discipline, and stoic preseverence were the only legitimate means to women's final acceptance, or, as the authors put it, "The limitations based purely on tradition can only be overcome through the quiet demonstration by capable and well-trained women of their ability to do the work. . . . In short, the present is a good time for those women who will consider chemistry *seriously* [their italics] as a career [,] although tradition and prejudice against them are still to be reckoned with, especially in industry . . . it is ability which will count in the future [,] as better efficiency in industry is demanded."[3]

Probably this strategy was the only one available to the women chemists of the 1920s that had any chance of success. Yet it seems to have been built upon the naive and optimistic belief prevalent among some employers in the early 1920s that the women's acceptance was only a matter of their demonstrating their competence—that once they had shown they could consistently equal or better the men in those laboratories to which they were admitted, their value would be evident and the barriers against them would fall everywhere. This optimism was reminiscent of the reports of Committee W of the AAUP (discussed in chapter 7), which stated that once prejudices had been documented and exposed to the light they would fade away. The women chemists did not suggest, publicly at least, what they must have feared privately: (1) that the prevailing prejudices might be irrational and that no amount of rational proof would ever change them, and (2) that women were not wanted in industry and might never be wanted, whatever their qualifications and ability, except perhaps in wartime. Nor did these women acknowledge that the well-trained and long-suffering Amazons they were going to produce might in fact be the type of woman least welcome in the average laboratory. They did not see that their strategy might more easily lead to exploitation than to equality. Instead, these women preferred to believe that after they had waged a rigorous kind of guerrilla warfare (perhaps like that their predecessors had waged successfully at the graduate schools in the 1880s and 1890s), prejudices against women chemists would melt away and they would be accepted for their abilities and potential contributions to the field. With this optimistic goal in sight, they could accept the fact that at present they were isolated and unwelcome in the industrial world and had little prospect of improvement. Perhaps their plan would be effective. The women's colleges would do their best to upgrade the chemical training of their students, broaden their experiences, and instill in them professional attitudes, in all of which Emma Perry Carr was, as mentioned in chapter 6 and shown in table 6.5, strikingly successful, although few of her students ever held positions in industry.[4] But the rest of the strategy placed a heavy psychic burden on the young women scientists of the time, more than a few of whom must have, like Winifred Goldring at the New York State Museum, had nervous breakdowns or abandoned science after trying to measure up under such hopeless conditions.

Unfortunately there is little evidence that this starkly realistic but also naively idealistic strategy of assimilation or deliberate overqualification was very effective in broadening women's opportunities in industrial chemistry in these years. Although there are almost no statistics available on women chemists' employment in industry in the 1920s or 1930s, what numbers are available indicate that women chemists maintained only a modest foothold in the field, ranging from 2 percent to 5 percent of populations such as chemists in the *AMS,* members of the American Chemical Society, and "chemists, assayers and metallurgists" in the 1920 federal census. Nor did this strategy provide much room for the great influx of women college graduates who majored in chemistry in the 1920s, when the field was widely portrayed in the media as offering a secure and expanding future in the "chemical century."[5]

In the resulting job crisis of the 1930s, which seems to have hit chemistry particularly hard, even very well trained young women found themselves unable to find positions. Their plight roused several older women, who had worked on the fringes of the chemical industry for several years, to fire between 1933 and 1939 a fresh round of updated career advice for aspiring women chemists. Although their desired audience was the same as it had been in 1922—women at the baccalaureate level and their teachers—the communication this time took place not through a single volume but through several symposia and vocational articles in major professional journals which described for all how the whole discriminatory system worked and prescribed the proper behavior and attitudes for women entering it. These offer interesting reading, for in no other field does there seem to have been a tradition of quite this kind of public and well-meant, yet pointed, advice in the 1930s.[6] (Women psychologists apparently learned in graduate school or in other relatively private settings that their "place" was in clinical work.) In addition, the women chemists' discussion in the 1930s demonstrates the kind of protection occupational segregation could offer women scientists who were, whether overqualified or not, on the verge of being driven from their field entirely.

By the 1930s, when even a top-flight chemical education could not get a woman a chemical job in industry, the advisers found it harder to blame her own inadequate preparation or lack of serious purpose. They now stressed, therefore, in addition to the traditional virtues of patience and stoicism, versatility, breadth of preparation, and a willingness to take any kind of job, especially those in several newly emerging areas of "women's work." A woman was still a chemist, they said, even if she was not actually doing chemistry, and they recommended new hybrid positions such as "chemical librarians," "chemical secretaries," bibliographers, and abstractors. Thus in the 1930s the women chemists' strategy for gaining employment in industry changed sharply. They gave up the hope of ever being equal to or accepted as research chemists and no longer recommended struggling hard to measure up, to become "exceptions," and so be unlike other women. Instead the new view was to accept the discrimination and take advantage of prevailing sexual stereotypes that labeled some positions in chemical companies and on chemical journals as typically feminine.

These new roles for women in chemistry were the result of the growth and bureaucratization of the American chemical industry since World War I. Before then, the industry had been quite small, but during and after it many companies had (with the help of many confiscated German patents) undergone a great transformation. One such change was the merger and consolidation of many small companies into large conglomerates that needed more office workers—secretaries to chemical executives and librarians in company-run libraries and patent collections. Since these roles were by the 1920s and 1930s already highly feminized in the United States, women, preferably those with some chemical background, ranging from a college major to a doctoral degree, were favored for them. Meanwhile chemical journals and *Chemical Abstracts* were also hiring more paraprofessional personnel in order to process the growing number of publications on chemistry. Again women chemists, known for their attention to detail and willingness to take low salaries, had a comparative advantage in these positions and reportedly found greater acceptance in them than they did in laboratory jobs.[7]

Although such realism (or modest ambitions) may have been what women needed at the time, advisers differed as to whether these hybrid positions were the only kind of "women's work" available to chemists. Florence Wall, a consulting chemist in New York City and informal placement officer for the American Institute of Chemists, warned in 1934 that two or three years in a library position would end anyone's prospects for a career in research. She recommended another kind of "women's work" instead: positions in the new and reportedly booming cosmetic industry, which was offering opportunities especially suited to women chemists in New York City and Hollywood, California. Any such jobs were quickly filled by men, however, for by 1939 women's prospects even there were considered quite limited, and Wall too ended up recommending secretarial jobs to women chemists. The field of cosmetic chemistry never became as feminized as one might have expected.[8]

Wall's reasoning in 1939 was the classic justification of "women's work": that in a depression, when employers would hire women only for those jobs that men would not take, a woman should not waste time in preparing or applying for jobs in male bastions like metallurgy or the pharmaceutical industry. Not even women's traditional (and well-documented) acceptance of lower wages would be much help. Instead, Wall advised, such a woman should take advantage of employers' willingness to hire women for proven areas of "women's work," either within chemistry or outside it, such as secretarial work, which she now hailed as a superb "opening wedge" for a woman chemist; library work, which now seemed a noble calling requiring rare and precious talents; and editorial work, since most male scientists needed "ghost writers" for their papers and speeches. If all these failed, a young woman chemist should take some practical courses and plan to move into related areas such as home economics or bacteriology, where there were not only more jobs in general but many specifically for women. The chemical industry thus offered minimal prospects for a serious woman professional in the 1930s, and any women aspiring to do research must have left industry for research associateships in biochemistry or professorships in home economics.[9]

Occasionally symposium organizers invited male spokesmen to give management's view on how women scientists might make themselves more employable. One wonders why anyone bothered, since the views were always the same and delivered with such condescension and ignorance that they just made everyone angry (as, for example, when male leaders denigrated the training given by the women's colleges, which, studies showed, were producing not only the most but also the best women chemists). Industry's attitude, however, was that it was fair and open minded, but that women employees had so many faults that they could not be promoted above a certain minimum level. In its view, the women's chief defect was not their work (which most thought acceptable and even praised) but their personalities. At the very least they (all) lacked "mechanical ingenuity," "imagination," and "aggressiveness," which one executive defined as a willingness to work unlimited overtime, design new processes and products, and generally outperform one's peers. Likewise, "experience" showed that women did not work well with others, would just get married and leave, and could not be promoted in any case (whatever their personality), because neither women nor men liked to work for a woman boss. The "evidence" for all this was based on management's personal experiences with women office workers and laboratory technicians whose conversation seemed "petty" to the males in charge (who were willing to admit that there were also many "troublemakers" among their male employees). Though Emma Perry Carr and her colleagues had raised the marriage issue in 1922 and pointed out that men left their jobs, too, statistical data on this issue were apparently never collected for the chemical industry.[10] But it does not seem likely that the results of such a survey would have changed the employers' minds. They continued to give the same arguments about women's poor showing in industry right into the 1960s, oblivious to contradictory studies and evidence. Thus Ruby Worner of the National Bureau of Standards wondered in 1939 how anyone could say women did not work well together when the Bureau of Home Economics, staffed and run almost completely by women, was such a conspicuous success. Nor would management pay any noticeable attention to studies made during World War II on the Women's Army Corps (WAC) which showed that workers were far more interested in the boss's personality (especially loyalty and support for them) than in his or her sex. Despite all this, industry's only solution to the marriage "problem" remained the rather draconian one of keeping the women (regardless of their personalities and ability) at the bottom levels until the "marriage age" (whenever that was) was safely passed—only then would promotion be considered. By then of course, management's negative descriptions and pessimistic predictions would have come true—the years of routine work without a promotion or even a hope of one would have taken their toll, and most of the women would have left. If the managers had really wanted the women and their abilities, one wonders whether they would have let such wasteful practices continue. One also wonders what "evidence" would have convinced them to change their ways.[11]

One particularly callous article in 1941 unwittingly illustrated the illogic and hypocrisy that prevailed in management circles in these years on the topic of women's advancement. It also shows how the two groups, the women workers and the male managers, were talking past each other. Within two paragraphs this article said that (1) women made very good technicians, since they would stay at the job far longer than men, who, considering such positions as mere stepping-stones on their way to more glorious careers, left such positions quickly; but that (2) "women in chemistry have always been a problem" because of their high turnover and inability either to fit in or to advance at the industrial laboratories![12] Thus women were blamed if they stayed in menial or subprofessional positions (perhaps trying to overachieve, as Emma Perry Carr had recommended) and blamed again if they left them. Their turnover seemed unduly high, because they would not be leaving for better positions elsewhere, as the men might, since few women were ever promoted in the chemical industry, no matter how good their work. Promotions in industry meant managerial positions, for which women were considered unfit sui generis. Thus the basic problem for women in industry as elsewhere was the almost total lack of any kind of advancement open to them. Like the instructor or research associate in academia or the "junior chemist" in government, the woman who did well in her low-level industrial job was never promoted to a more challenging position, but was kept at the same low level (and blamed for a lack of initiative or aggressiveness). But if she did poorly or left, she and all other women were blamed once again as being incompetent or unstable.

It was a highly circular, no-win situation: since the young woman was expected to get married and leave (or leave anyway), she was assigned to an essentially temporary position that offered little incentive to stay. If, however, she did stay, she still could not be promoted since there was no provision for advancement or career line for her to follow. If she grew angry at not being promoted (when less-qualified males were automatically), she was labeled as "difficult" or "hard to work with" and was probably laid off or assigned to a remote corner of the plant. By contrast, one of the advantages of the "chemical secretary's" position was that she was expected to stay with the company and rise up through the ranks with her boss. By the time he reached the executive suite, she had grown from a young typist into the mature and cultivated executive secretary that he now needed, who, rather than being at all "difficult," was gracious and considered a distinct asset not only to him but also to the company that had trained her. Thus there was a realistic role and career line laid out for her—she was expected not only to stay with the company as she grew older (whether married or not, apparently) but also to add managerial skills and grow in her job, two key expectations that were withheld from women chemists. One wonders why industry was so resistant on this point and so unable to admit women scientists to normal career patterns, since they were cutting themselves off from some very highly talented employees of exceptional motivation. One is also staggered at the thought of how much suffering and distress was caused generations of women in these dead-end jobs who were blamed for phoney personal inadequacies when it was the system that refused to let them grow with their jobs and reward them for it.[13]

These low and negative expectations appear even more discriminatory when they are compared with management's strikingly different attitudes toward the employment and advancement of young male scientists in the 1920s and 1930s. Rather than presenting "problems" for management, as women at all levels seemed to, young male scientists were seen in these years as nothing less than the salvation of the scientific industry. When discussing the future prospects of these scientists, the officials' tone was, by contrast, positive and optimistic, and even reached that lofty and elitist tone reminiscent of Professor Robert A. Millikan's references to the "bright, young men" of the academic world. His equivalents in industry, especially Frank B. Jewett, president of Bell Telephone Laboratories, and Willis R. Whitney, director of research at General Electric Company, spoke so often and so loftily of industry's need to recruit and motivate these desired employees that grandiose articles on this theme became a standard genre of the period. According to this view, the talents of these men were so rare and precious that industry was justified in giving them every incentive and inducement to keep on with their research in an industrial setting—to hire them from the major universities, give them whatever staff and facilities they needed, and reward the successful ones with larger laboratories and staffs, more time for their beloved "pure" research, and permission to publish in professional journals. Any women, black, or Jewish scientists reading these articles must have grown even more cynical, since such high-minded sentiments about the value of scientific ability hardly ever reached the personnel offices of these corporations, which were widely known at the time for their highly discriminatory employment practices. Even when the phrase "[white] male Christians only" was not printed in the advertisement, everyone knew that they were the only ones wanted.[14]

One can also see, after reading a certain amount of this rhetoric, how self-fulfilling it could become: corporate attitudes could affect management styles, which could, in turn, determine employee behavior (and productivity) sufficiently to seem to justify the initial prejudice. For example, when the workers were women, who were thought to be poorly motivated and lacking in talent, they were assigned to the lowest-level jobs and managed by highly authoritarian bosses who were given the full support of the higher echelons of management if any problems should arise. Under such circumstances, the resulting lack of originality and high turnover could be blamed upon the (women) workers themselves, since it was, in fact, expected of them and planned for. Yet the management of talented (male) employees was presumed to be a more sympathetic and complex process. Now the manager was transformed from a stern and infallible taskmaster to a gentle and imperfect guide who was responsible for motivating his workers and nurturing their creativity. At the same time this new role so greatly diminished his authority that if these favored workers failed to live up to the high expectations placed upon them, it was *the manager* who was blamed for *their* nonperformance. Since they could do no wrong, the fault must lie with him for not motivating them! Thus prejudice and management style could reinforce each other and seem to confirm management's initially high or low expectations.

There is also evidence in much of this rhetoric that management (in some industries at least) thought more highly of physicists and physical chemists and treated them better than it did other chemists, a practice that helped the few women physicists in industry at the time. Thus Irving Langmuir, assistant director of the General Electric Research Laboratories and a prize-winning physical chemist (including the Nobel Prize in chemistry in 1932), had as his assistants two of the best-trained women physicists of the time, Katharine Burr Blodgett and Lucy Hayner. Blodgett, a Bryn Mawr graduate with a master's degree in physics from the University of Chicago, joined GE (where her father headed the patent department) in 1918. She went on leave from 1924 to 1926 to earn a doctorate in physics at the famed Cavendish Laboratory in Cambridge, England, and was the first woman to earn that degree there. She returned to assist Langmuir in his researches on atomic structure and thin films and was later widely publicized for her own discovery of "nonreflecting" glass. Meanwhile Hayner, a graduate student in physics at Columbia University, had also spent the academic year 1924/25 at the Cavendish completing her dissertation. She must have met Blodgett there, for upon her return to the United States she joined GE, where she worked on electron emissions from vacuum tubes. After just three years there, however, she returned to Columbia University, where she spent most of her career as an instructor. Apparently the only woman chemist to do even this well at GE in these years was Dorothy Hall (later Brophy), a Michigan doctorate who worked on methods of detecting impurities in tungsten from 1920 until 1932. Nevertheless, it was typical of industry's attitude toward women chemists that her marriage and departure (to become a consulting chemist in New Jersey) received more mention in GE's official history than whatever work she did and patents she developed in twelve years on the job. Similarly, little is known about either Florence Fenwick's eight years in U.S. Steel's research laboratory at Kearny, New Jersey (1928 to 1936) or Marion Armbruster's eight years there, from 1935 to 1943.[15]

Nevertheless, a few remarkable women did manage not only to hold jobs in the chemical industry but even to be promoted in the 1920s. One such woman was Betty Sullivan, who joined the Russell-Miller Milling Company in Minneapolis in 1922, became chief chemist at the company in 1927, earned a doctorate at the nearby University of Minnesota in 1935, and after winning several awards for her work in cereal chemistry (she appeared twice on the cover of *Chemical and Engineering News*), was promoted to vice-president and director of research in 1947. Several women seem to have held technical and management positions in the paper industry in the 1920s and 1930s. Jessie E. Minor, a Bryn Mawr doctorate who had taught chemistry at Goucher College, worked as chief chemist, department head, and director of research for several such companies in the 1920s and 1930s. Helen U. Kiely and Lena Kelly were technical director and chemist, respectively, at the American Writing Paper Corporation in the 1930s, and R(osalie) M. Karapetoff Cobb became chemist-in-charge of the research laboratory at the Lowe Paper Company in Ridgefield, New Jersey, in 1926, holding that and related positions until her retirement in 1965.[16]

Although little is known about the personal lives of any of these women, brief descriptions of the job experiences of two others suggest that women paid a high psychological price in such "men's work" in chemistry. Helen Wassell found a way to do important research in applied chemistry, although she was not quite in industry itself—she did research on emulsions at the Mellon Institute for Industrial Research in Pittsburgh from 1919 to 1952. Although her work on floor waxes and hair shampoos reportedly made a fortune for the Carbide and Carbon Chemical Corporation, which sponsored her work, she later admitted that she had once been so discouraged that she had almost quit to become a secretary, which was the "realistic" advice of the 1930s.[17] Louise McGrath, on the other hand, rose to become a vice-president of the Booth Chemical Company in the 1930s. Whether she felt any doubts or discouragements is unknown, but the cryptic way in which a female fellow chemist later described her successful adaptation gives one pause: "Louise has mastered the difficult art of subduing her own personality while working under the most adverse conditions, so much so that she is accepted as 'one of the boys.' On the outside, she is a charming hostess and a good cook."[18]

By contrast, the women scientists' greatest success in the 1920s and 1930s was in the food and home products industry, which not only hired many of them, but, like home economics departments at universities and bureaus in government, even promoted some of them as well. Several companies, such as the Public Service Electric and Gas Company of Newark, New Jersey, hired young women with bachelor's degrees in home economics to promote the sales of stoves and other appliances to housewives. Other companies hired professors of home economics to develop new products or new markets: Mary Barber, formerly of Teachers College, Columbia, achieved fame in the 1930s for her work at the Kellogg Company of Battle Creek, Michigan, in developing the foods for and serving the first meal on an airplane in flight. A few other women home economists even held middle-management positions in the food industry in the 1920s and 1930s. For example, at General Mills in Minneapolis, Janette Kelly and Marjorie Child Husted directed the company's Home Service Bureau where, under the collective name of "Betty Crocker," a team of fifty home economists devised and tested recipes for housewives. Other women found influential posts at radio stations, which were beginning to hire trained home economists for their women's programs, and on the larger women's magazines, which also featured them as columnists and editors. Especially prominent in this field of service cum advertising were *Good Housekeeping Magazine,* which employed a staff of chemists and home economists to test new products for its coveted "Good Housekeeping Seal of Approval," and the *Ladies' Home Journal,* one of whose consulting editors was Christine Frederick, who was credited with standardizing the height of kitchen counters in homes across America. By the 1920s and 1930s advising women about consumer products had become a specialized kind of advertising that offered many women nutritionists or home economists a prominent role.[19]

Women scientists had a more temporary and limited success in the 1920s and 1930s in the suddenly booming field of commercial geology. Their opportunity in that field came in a most unexpected quarter—in micropaleontology and the oil fields of Texas. There had been a large oil strike in West Texas in 1918, and several oil companies were anxious for others. In 1919 Esther Richards Applin, a geology graduate of the University of California, had a summer job with one of the companies in Texas and discovered that the microfossils obtained in exploratory drillings could be used to determine the age of the subterranean formations and so predict the likelihood of petroleum in that locality. When the commercial applications of her discovery became known, other oil companies began to hire young women paleontologists to sit at a microscope in a field office and read off the contents of the samples given to them. Several, such as Fanny Carter Edson, Alva Ellisor, Dollie Radler Hall, Louise Jordan, Virginia Kline, Dorothy Palmer, and Helen Plummer, made careers and reputations in this new kind of "women's work," and some mapped the stratigraphy of the region to facilitate further oil exploration. Many of them, as their obituaries recount, longed to get out and do their own fieldwork, but only rarely did they succeed. In fact, rather than having their roles within the companies expanded or being promoted to higher (managerial) positions because of their expertise, many of these women scientists were laid off during the depression, as the oil companies either merged to consolidate their activities or hired men to use new seismographic detection instruments. By then, however, some of the women were well enough known to be hired as "consulting geologists" by other companies throughout Texas, Oklahoma, and the Gulf Coast region. Thus though the women micropaleontologists were not able to parlay a commercial technique or specialty in which they were readily accepted into management positions in the oil industry, several did use it to build careers as professional consultants, a common practice and role in geology. It is not clear, however, how financially successful any of them were.[20]

Women geographers found a niche in the 1920s and 1930s as editors and librarians. Since geography was only rarely taught at the women's colleges, used by government agencies, or needed by industry, such paraprofessional positions were the only jobs (aside from schoolteaching) open to women in the field. During World War I, when the field got its main impetus in America, several women scientists were added to the staff of the American Geographical Society in New York City, where they soon managed some of its most important (and most "womanly"?) functions—editing its publications and running its extensive library. The highest ranking of these several women was Gladys Wrigley, a Yale doctorate and former research assistant to Professor Isaiah Bowman there. When he became director of the AGS, she followed and edited the *Geographical Review* for thirty years, a position, to judge from her own account, that was one of the most fascinating desk jobs in the subject. She reveled in seeking out the best work and pressing her authors, many of whom were not professional geographers, to "do the impossible." Wrigley's unremitting stress on high standards reminds one of the stronger professors at the women's colleges and the department chairmen in home economics; lacking even such opportunities in geography she was fortunate that Bowman appreciated her talents and allowed her the opportunity to use them so effectively.[21]

The AGS librarians were also a distinguished and energetic group. Elizabeth Towar Platt was known for her "zest" for geographical literature. In addition to performing "the unremitting labors of the librarian, so largely anonymous," Platt also published geographical articles and even started *Current Geographical Publications* singlehandedly in 1938. (Unfortunately, she died in her forties.) Also at the AGS was map librarian Ena Yonge, who had come as a typist in 1917 but who was within a year put in charge of the society's large map collection. There she developed such a "nose for maps" and passion for collecting that by the time she retired forty-five years later she had increased the society's holdings almost sevenfold, from 37,000 maps and 800 atlases in 1918 to 280,000 maps and 4,000 atlases in 1962. Even though these women's jobs were not professorships or bureau chiefs, they were, in some cases at least, superior to whatever other jobs they could have gotten in geography.[22]

Besides the women in chemistry, home economics, geology, and geography who had managed to establish small beachheads of "women's work" in their fields in the 1920s and 1930s, there were a few other women who also held positions in industry or other private employment. These women were so few, however, as to remain "exceptions" or examples of the outer limits of women's employment in science in these years rather than evidence of a broad movement. For example, Elsie O. Bregman was one of the first women in the field of industrial psychology when, as a graduate student at Columbia University from 1919 to 1921, she developed employment tests for the personnel department of the R. H. Macy Company in New York City. Ten years later Helen Blair Barlett, a mineralogist, became one of the first women in a technical position at the AC Spark Plug Division of General Motors. Working on aluminum insulators from 1931 until 1966, she took on several patents and was one of the few women in the American Ceramic Society.[23] Clippings and obituaries report how successful all these "exceptions" were and how important their work was, but even such paragons were unable to convince employers to hire other women to follow them, a testament to the continuing inability of "exceptions" to change stereotypes.

Similarly, women engineers were still so rare in these years, in industry as well as elsewhere, as to be back at the stage of "firsts," where practically every professional activity they undertook was publicized as such in professional journals and women's magazines. Though their impact would be minimal before World War II, and only eight were listed in the 1938 *AMS*, a few did have careers of interest in the 1920s and 1930s. For example, Margaret Ingels was a mechanical engineer who specialized in the new field of air conditioning for the Carrier-Lyle Corporation of Newark, New Jersey; Olive Dennis was a safety engineer for the Baltimore and Ohio Railroad whose job was to suggest improvements in passenger comforts in railroad cars; Edith Clarke was an electrical engineer who first did calculations for the American Telephone and Telegraph Company and later worked on power transmissions for General Electric from 1922 to 1945; Elsie Eaves, a graduate of the University of Colorado in coal engineering, joined the *Engineering News-Record* in 1926 and spent thirty-seven years on its staff, pioneering in many kinds of construction industry statistics; and Mary Sink, one of the first women chemical engineers, was hired by the Chrysler Corporation in 1936 to work on automotive engineering and fuel emission exhausts. If the women engineers in England seem to have made greater gains in these decades, it was only because their employment was oriented more toward home economics and the electrification of the home rather than big business, which was unreceptive to women engineers in both countries.[24]

Meanwhile two other women scientists were making careers in the new scientific fields of agribusiness and health physics. In the 1920s Elizabeth L. Clarke (daughter of the longtime secretary-treasurer of the Naples Table Association and of Samuel Clarke, professor of zoology at Williams College and founder of the American Society of Naturalists in the 1880s) became one of the very few women scientists in the expanding world of agribusiness. After graduating from Smith College in 1916, young Clarke spent a year at the Massachusetts Agricultural College (now University of Massachusetts at Amherst) and then during World War I joined the Woman's Land Army. Finding that she liked agriculture very much, she then got a job at the scientifically run Dimock Farms in Vermont. Within a few years its owner established the Dimock Potato Corporation, which specialized in breeding disease-resistant, high-yield plants for the potato industry, and Clarke became chief inspector of the seeds. In 1929 she was the first woman field inspector for the Vermont Seed Potato Certificate Service, making sure that only properly bred seeds were certified.[25]

Also in the 1920s Edith Quimby became one of the pioneers in hospital physics in the United States. Because she was married and held just a master's degree in physics when she sought work in New York City in 1919, the only job she could get was as an assistant to the radiologist at the New York Memorial Hospital (for Cancer and Allied Diseases, as it was then known). Yet she thrived on this opportunity in a relatively new field, and between 1920 and 1940 she wrote about fifty articles on the correct dosages for radium and X-ray treatments of cancer patients. In 1941 she became the first woman since Madame Curie to win the Gold Medal of the Radiological Society of North America. When in the

1950s radioisotopes made chemotherapy a common treatment and nuclear medicine a respectable specialty, Quimby became a full professor of radiology at Columbia University's College of Physicians and Surgeons.[26]

Other women scientists who could not find even these unusual jobs in the 1920s and 1930s took up the independent career of "consulting." A few, such as the chemists Calm Hoke in metallurgy and jewelry and Mary Pennington in refrigeration (who was also one of the few women scientists ever to quit a government job), were reportedly very successful financially. Yet as Hoke herself pointed out rather sadly, her success as a platinum expert came largely because she hid her gender and used only her initials in her professional correspondence.[27] Less is known about the success of many others who became "consulting psychologists," or, in one case, a "consulting entomologist," occasionally in teams with their husbands. However, one enterprising botanist, the Wellesley graduate Cynthia Westcott, has recounted how she came to be one of the nation's first "consulting plant pathologists." After earning her doctorate at Cornell University in 1933 and ranking number one on the federal Civil Service examination for plant pathologists, Westcott suddenly discovered that she had no chance for government employment after all, since, as her notice put it, there was "little or no demand for female eligibles." H. H. Whetzel, her graduate professor, advised that she try to pioneer as a "plant doctor" and private consultant to wealthy women with large gardens, even though, he admitted, the income from such work was so seasonal and precarious that he would not recommend it to men in the field, who were, in any case, still getting government jobs. Westcott, an exuberant and energetic person, succeeded in making a successful career out of this meager beginning by supplementing her modest and irregular earnings by speaking to garden clubs and writing several books on plant diseases for the home gardener, a relatively new consumer then. One wonders what she might have accomplished at the USDA had it not been so discriminatory in 1933. Probably the best-known woman consultant in these decades was the psychologist-turned-engineer Lillian M. Gilbreth, who, rebuffed for industrial employment, wrote books on human management and designed efficient kitchens, specializing after World War I on adapting homes to handicapped persons, again a new and previously untapped market.[28] Although these women were relatively successful and resourceful entrepreneurs, for others who were either less imaginative or just less lucky, the consulting role may not have paid off and may have been just a thin protection from the hard reality of unemployment, which lurked everywhere in the 1930s and, as mentioned in chapter 6, hit women, especially married women, particularly hard.

Countless other women scientists (the over 100 listed in the 1938 *American Men of Science* should be considered a bare minimum) were unable to find employment in any of the three sectors considered in these three chapters on higher education, government, and industry, and worked instead in the more traditional areas of women's employment—schoolteaching, editorial work (for scientific journals, abstract journals, the Science Service, and the *American Men of Science* itself), translating, writing, librarianship, and office work as administrative assistants and executive secretaries. The size of this cluster of underemployed women scientists (in addition to the 181 others who were unemployed in 1938) is an indicator of both the job crisis of the 1930s and women's continuing precarious position in the sciences.

It should be noted that even this modest overall penetration of women scientists into the various nooks and crannies of the industrial world in the 1920s and 1930s was just a part of some larger alterations under way within women's employment at the time. Previously college women had counted on becoming schoolteachers or librarians upon graduation, but around 1910, when the number of women graduates began to increase sharply and these traditionally female jobs became harder to find, a few economists and college vocational advisers saw the need for a more systematic exploration of other employment opportunities open to women graduates. Pioneering in this field was the Women's Educational and Industrial Union of Boston, which started the first "appointment bureau" for women in 1910 and published one year later a series of vocational bulletins on possible careers for them. The two of these on science (bacteriology and industrial chemistry) must rank as the earliest vocational guidance for women scientists.[29]

Continuing this tradition was the Intercollegiate Bureau of Occupations, which was set up in 1911 by the New York alumnae associations of nine women's colleges to help their graduates find suitable jobs and to do research on women's employment. After this bureau was absorbed by the federal government's new U.S. Employment Service in 1918, a second group of women philanthropists, presidents of women's colleges, and a few other male educators and officials (such as Samuel Capen of the American Council on Education and Vernon Kellogg of the National Research Council) formed the new Bureau of Vocational Information, also in New York City. For several years this bureau employed a director and an assistant director who put out several major studies of occupations (including those on statistics and chemistry mentioned earlier), and from 1922 until it ceased publication in 1926 it published a frequent *News Bulletin*.[30]

Several books on women's occupations (or "careers") also appeared in the 1920s. The most important was Catherine Filene's *Careers for Women,* which was published in 1920. The daughter of the remarkable Lincoln Filene, a prominent Boston merchant and liberal philanthropist who was known for his leadership in the fight for a minimum wage for women, she had become Mary Anderson's assistant at the new Women's Bureau in 1918. Seeing the need for a vocational handbook of short descriptions of careers open to women, Filene edited this anthology of short (four- or five-page) essays on 160 occupations, including 10 in the sciences. The book's value lay in the comprehensiveness of its coverage and in the honesty and authority of its authors, who were frequently the top women in their fields. Thus she asked Margaret Maltby of Barnard College to discuss careers in physics, Eleanora Bliss Knopf those in geology, Flora Wambaugh Patterson of the USDA in plant pathology, Helen Thompson Woolley in psychology, and Augusta Fox Bronner in clinical psychology.[31]

Many of these authors tried to be optimistic about women's future in their field, although the true test, their discussion of women's chances for advancement, reveals how difficult this was. For example, of physics, a field where women in 1920 continued to face such severe discrimination that almost all were still at the women's colleges, and in which she had had difficulty being promoted at Barnard College, Margaret Maltby gave this vacillating assessment:

> The opportunity for advancement in research depends entirely upon the character of the woman herself and her ability. There seems to be no prejudice against a woman, if she can do the work as well as or better than a man. It is difficult to be specific, for such opportunities have been open to women so few years and the cases are individual. A general notion has been prevalent that women have no interest or aptitude in fields requiring mechanical ability. But with the increasing use of automobiles and household mechanical or electrical devices women are acquiring familiarity with their construction and operation. Perhaps the conservative academic world is more imbued with the idea of women's limitations in this direction than the industries, for it has been difficult for women to get full professorships in the department of physics.[32]

By the late 1920s vocational guidance for women had become a branch of labor economics. It was no longer enough for well-meaning laywomen or even energetic staff members to collect and pass on their hints and clues, and in 1928 Catherine Filene Dodd (later Shouse) started the Institute of Women's Professional Relations at the Woman's College of North Carolina with a grant of $60,000. Its staff was to study systematically and to find ways to increase opportunities for women in the professions and other nontraditional areas. Under its energetic director, Dr. Chase Going Woodhouse, formerly an economist at the Bureau of Home Economics but now a professor of economics at the Woman's College of North Carolina, the institute issued many publications, most notably its bimonthly newsletter, *Women's Work and Education,* sponsored several volumes on the education and employment of college women (such as directories of fellowships open to women and Emilie Hutchinson's *Women and the Ph.D.* [1928]), and directed several symposia on women's employment during the depression. Woodhouse collected much economic information on expanding and contracting sectors of the economy, highlighted new and interesting jobs, and undertook more intensive studies of areas that offered opportunities to women. For example, during the 1930s her institute studied careers for women in interior decorating (including the new field of airplane interiors), photography, and the cosmetic industry. In 1938 the institute announced that in response to many queries it was going to make a special project of studying careers for women in science. Quickly its coverage of this area, which had been strong before, began to increase still further, and numerous articles appeared in *Women's Work and Education* on women radio engineers, botanists, bacteriologists, scientific illustrators, protozoologists, and other Civil Service positions. There can hardly have been many openings in these areas, but the institute's publications provide a measure of the job market for women scientists in the 1930s as well as of the atmosphere of desperation that prevailed. Since there were few jobs open to women in the standard or traditional fields, they were being urged by economists as well as by the women chemists mentioned earlier not to try to pioneer in hopelessly crowded and hostile fields like engineering, chemistry, medicine, or college teaching, but instead to keep alert for odd new areas tucked in the interstices of the employment world, and then scramble for them as best they could in the tight market.[33]

A common kind of advice in articles about careers that began to appear in some women's magazines, such as the *AAUW Journal* and especially the *Independent Woman* (published by the National Federation of Business and Professional Women) in the 1930s, was to try to turn a unique hobby or an unusual interest into a satisfying position. For scientists this generally meant either selling a traditional skill (like home economics) to a new or unusual employer (as an airline or a radio station) or else practicing an exotic specialty (herpetology, perhaps) at a traditional institution (such as a zoo or museum). Thus "Exploring the Museum Field" in the *Independent Woman* in 1933 described the work of several women at the American Museum of Natural History in New York City and encouraged readers to think that by starting as volunteers they too could rise to interesting positions. The magazines especially liked to publicize the kind of job where low-level positions could prove to be "opening wedges" to more important ones later on. This strategy had apparently paid off for several women employed at city zoos in the 1920s and 1930s, where responsible positions were still open to nonprofessionals who had a "way" with animals. Thus Belle Benchley had been a bookkeeper and animal-lover at the San Diego Zoological Gardens in 1924 when the financially strapped directors, apparently unable to find a professional curator, decided to let her run the zoo. She proved an exceptionally good choice, despite her lack of professional training or perhaps because of it, for she made the San Diego Zoo famous by rearranging the grounds so that the visitors rode on buses while the animals seemed to roam freely, a landmark in humane zoo design. By publicizing such ingenious success stories, the women's magazines of the 1930s encouraged women to capitalize on their special interests and private hobbies and build them into satisfying and profitable careers outside the confines of the corporate world.[34]

In general, then, the finding of jobs for women scientists in industry and business in the 1930s was a difficult task that required the best efforts of some of the shrewdest women available. A few women who had entered science in the 1920s held on to their jobs and had long careers of accomplishment, but their cases seem to have been unusual, and hundreds, perhaps thousands, of others were either unemployed or underemployed and felt their skills and talents going to waste. Though the number of women scientists employed in industry was up considerably over 1920, contemporary writings indicate that women were everywhere in marginal and precarious circumstances.

It was thus in the years after World War I that the industrial employment of women scientists had its real beginnings. Aware of their limited welcome, these women at first advocated a strategy of stringent self-discipline and cheerful overqualification as a way to reward or placate those employers who would hire them. But this tactic failed, and, rather than being praised for their hard work and promoted for their persistence, women scientists in industry, like those elsewhere, found themselves blamed for their lack of initiative and forced to make their careers in low-level and low-paying jobs. By the 1930s they were forced to fall back on the even more conservative strategy of advocating and accepting "women's work" as the chief alternative to unemployment.

10

DOUBLE STANDARDS AND UNDERRECOGNITION: TERRITORIAL AND HIERARCHICAL DISCRIMINATION

The expectation that good work and superior science will receive eventual recognition and reward, meted out according to proper professional standards, is by all accounts a strong belief among scientists and a powerful factor in their morale.[1] So essential and widespread has such recognition become in the twentieth century that most scientific societies (and even local sections of some national ones) make annual awards for a variety of professional services and accomplishments, and several wholly honorary societies devote considerable time to the selection of new members.[2] It is as if the rise of professionalism and the increasingly competitive modern scientific career have required a series of honorific benchmarks by which to identify the worthiest candidates and propel them upward.

Yet the women's experience in the 1920s and 1930s would indicate that the whole process is both more complex and less disinterested than this expectation implies. Women were entering science in large numbers and were doing important work both individually and on the major research projects at the time, and yet rarely did they win any awards from or hold any office in the major professional associations. This systematic underrecognition occurred in at least two ways: territorially, as in the low visibility given "women's work" in science and the little prestige accorded those fields in which women were prevalent; and especially hierarchically, as in the progressively greater underrepresentation of women in the higher ranks of all major scientific societies and among winners of all types of awards. Prize and selection committees repeatedly passed over even outstanding women as if they could not see them properly, were attributing their work to someone else, or were systematically discounting the importance of their work. Needless to say, it was the more talented and professionally conscious women scientists, those aspiring to be "Madame Curies," who suffered the most from this overall exclusion and deprivation, since they were the ones most eligible for such awards and most likely to be expecting them. But in a larger sense all women scientists suffered from the lowered expectations resulting from the nonrecognition of the top women. For both sorts of women, as their correspondence reveals, the issue of recognition, or, more precisely, the lack of it, became a sensitive one in the 1920s and 1930s. The women's reaction, which is discussed in chapter 11, is of some theoretical importance, for the sociology of science has little to say about the breakdown in morale that can occur when recognition is systematically denied to large groups of scientists. To a certain extent the phenomena belong in the realm of the sociopathology of science.

Part of the women's difficulty in gaining recognition in the 1920s and 1930s was a structural one that grew out of their subordinate position in the world of employment. Those women most likely to be doing prizeworthy work—the research associates at the universities and the government employees in research agencies—were caught in a double bind between two contradictory social roles. They were on the one hand permanently subordinate, helpful, almost invisible "associates" in the work of others, but on the other hand they were successful scientists who sought full professional recognition for their own (or their friends') discoveries and accomplishments and felt deprived when it was not forthcoming. Only a very few women were able to avoid this trap and win any kind of recognition from the major professional associations of the time. The circumstances behind these few "successes" suggest that in addition to outstanding scientific research, a woman needed many extrascientific assets to win awards. Chief among these additional factors was the enthusiastic backing of powerful and politically astute male colleagues, without whose support even the most meritorious work would go unrewarded. Thus to a far greater extent than is generally realized, the fate of the top women in a field says much about the attitudes and behavior of the more prominent men in that specialty.[3]

In science, as in any profession, there are a limited number of ways in which a group can honor formally its distinguished members: through fellowships, memberships, positions of leadership, prizes, and naming or dedicating objects or organisms in their honor or memory. A few women scientists achieved each of these honors (except the Nobel Prize) in the 1920s and 1930s, although, as described later, there were certain recurrent patterns in their victories: they won the lesser honors more often than the higher ones, and they did better in anthropology and perhaps anatomy than in any other field. Usually, however, their percentage of winners was only a fraction of their share of the eligible candidates.

The most common kind of recognition and the only kind many women scientists, especially the younger ones, were likely to receive in the 1920s and after was the postdoctoral fellowship. This was an area in which the Association of Collegiate Alumnae (later American Association of University Women) had pioneered when it awarded its first doctoral fellowships in the 1890s and its first postdoctoral one, the Alice Freeman Palmer Fellowship, in 1908. (It had also taken over the privately funded Sarah Berliner Research and Lecture Fellowship in 1919.) Until the 1920s these were among the very few postdoctoral fellowships for either sex in the world, and the distinguished recipients knew that they carried the pride and high hopes of thousands of AAUW members across the nation. In 1927 the AAUW ambitiously started its Million Dollar Fellowship Fund (which was finally completed twenty-six years later) to increase the number of its pre- and postdoctoral fellowships still further. During the 1920s and 1930s, when many other fellowships became available, the AAUW awards (approximately ten per year in the 1930s, about one-half of which went to scientists) were still highly coveted, since, as shown in what follows, relatively few other fellowships went to women. These awards also served a second purpose, since having one of the AAUW's endowed fellowships named for her was long one of the top honors an accomplished woman, scientist or not, could hope to receive.[4]

After World War I, however, several other large postdoctoral fellowship programs were established in response to the pervasive (and sexist) feeling, expressed in the *Bulletin of the American Association of University Professors* and elsewhere, that the academic man, especially the younger one, was overworked and underpaid at most colleges and that if he were ever to contribute anything to science or scholarship he needed subsidized advanced training and sabbatical leaves. The inadequacy of the traditional academic career for new postwar ambitions became standard rhetoric in the 1920s, as academic leaders and scientific spokesmen campaigned to upgrade university faculties.[5] Although the education and training of scientists was not yet seen as a responsibility of the federal government, it was a role superbly suited to the large new philanthropic foundations, which were in the 1920s seeking prestigious ways in which to disperse their vast fortunes. American scientists and academic men, whose genteel poverty was being amply documented at the time, formed an army of eager applicants for these millions in newfound patronage.

Preeminent among these giant philanthropic organizations that began to support postdoctoral fellowships after World War I was the Rockefeller Foundation of New York City. Working through several boards and committees, it underwrote postdoctoral fellowships so extensively that it changed the whole scale of philanthropy in this area. It created entire programs of fellowships and thus transformed an unheard-of luxury into a reality for thousands of scientists and scholars in this country and abroad. Yet the proportion of women among these fellows remained quite low, and only in the field of child development did more than a few of these new fellowships go to women.

The first such Rockefeller program and probably its best-known and most prestigious was that of the National Research Council Fellowships, established in 1919. At first limited to physics and chemistry, its coverage in later years was increased by the foundation to include the medical sciences (1922), anthropology, botany, psychology and zoology (1923), mathematics (1924), astronomy (1925), agriculture and forestry (1929), and, much later, geology (1937). These awards were highly publicized at the time and used by many administrators and spokesmen of science as indicators of the relative popularity of research programs at their universities. If the purpose of this new fellowship program was to recruit the "bright, young men" of the time into careers in research, the program certainly seems to have succeeded. Although hardly any data exist on the gender of the NRC's applicants, the little that is available indicates that male applicants were given roughly a two-to-one advantage over the women, as shown in table 10.1. Since an applicant for a postdoctoral fellowship had to hold or to have nearly completed a doctorate from an American institution, a population on which there is an abundance of data, one can compare those eligible for NRC fellowships with those actually supported. Although not conclusive, this comparison provides striking evidence of the general process of defeminization going on at even this low level of recognition—women, who constituted 13 percent of the U.S. doctorates in these science fields in this period, were only 5.4 percent of the NRC fellows. The funneling occurred in all fields except the tiny one of anthropology, where the women, who had earned 29.9 percent of the nation's doctorates from 1923 to 1938, received ten fellowships, or a striking 38.5 percent of the total. This was several times better than they did in the much larger field of psychology, where, despite earning 26.7 percent of the doctorates in the relevant years, they were restricted to just three, or 3.5 percent, of the NRC fellowships.[6] (This was just the first of several signs of the hostility faced by women psychologists in the 1920s and 1930s.)

TABLE 10.1. U.S. Doctorates and National Research Council Fellows, 1920–1938 (By Field and Sex)

	U.S. Doctorates*			NRC Fellows		
Field	Total	Women	Women's Percentage of Total	Total	Women	Women's Percentage of Total
Medicine	1,194	254	21.3	250	14	5.6
Chemistry	6,052	487	8.0	185	4	2.2
Physics	1,831	86	4.7	178	2	1.1
Zoology	2,503	395	15.8	129	14	10.9
Mathematics	954	132	13.8	110	4	3.6
Botany	1,098	219	19.9	86	5	5.8
Psychology	1,559	417	26.7	86	3	3.5
Agriculture (1929–38)	285	6	2.1	40	0	0
Anthropology	197	59	29.9	26	10	38.5
Forestry (1929–38)				8	0	0
Astronomy		(under physics)		5	1	20.0
Geology (1937–38)	123	4	3.3	1	0	0
Total or average	15,796	2,059	13.0	1,047	57	5.4

SOURCES: *National Research Council Fellowships, 1919–1938* (Washington, D.C., 1938): Lindsay Harmon and Herbert Soldz, comps., *Doctorate Production in United States Universities, 1920–1962* (Washington, D.C.: National Academy of Sciences-National Research Council Publication no. 1142, 1963), pp. 50–51.

*Doctoral data combine: (1) biochemistry and chemistry under chemistry; (2) anatomy, physiology, and other medical sciences under medical sciences; and (3) microbiology, zoology, and other biosciences under zoology.

More direct evidence of the systematic discrimination occurring at this level comes from a statement in an unpublished report of the NRC's own Medical Fellowships Board in 1926. It stated that between 1922 and 1926 "Only 30 of the 328 applicants were women [9.1 percent]—[but that they formed] less than 5 percent of the appointees [5 of 107, or 4.7 percent]." Whereas today these figures would be interpreted to mean that this fellowship board (on which no woman ever served) was biased against the female candidates, making it twice* as difficult for them to become fellows as it did for their male competitors, the 1926 report did not suggest such a possibility. Instead it seemed to imply, as did the prevailing mentality at the time, that the all-male committee was unbiased and that the female candidates were weak and undeserving of the precious fellowships. But the report admitted elsewhere that certain additional outside constraints had affected some of the board's selections. Since, for example, too many candidates wished to study neurology for the few spaces available at the one such program in the country, it had restricted the number of fellowships in that area to the size of that department. Nor had it granted any of these to women, since, as the report put it, "There is a justifiable hesitancy to appoint a woman whose eventual aim is Surgery, without definite assurance that there will be a place for her in university work."[7] Thus unlike the AAUW fellowships, which had been used since the 1890s to create that very leverage and pressure necessary to pry open new fields to women, the NRC fellowships of the 1920s and 1930s restricted women to those already open to them. Even the top women of the time were not allowed to use the NRC's prestigious awards to overcome the remaining barriers, and the NRC fellowships served more to reinforce existing segregation than to break it down.

An example has come to light of how these two-to-one odds worked in practice. Women applicants for such fellowships faced a deliberately negative reception from foundation officials, who were twice as skeptical of their abilities as they were of male applicants'. One Rockefeller Foundation official noted in his diary in 1934 that he had gladly explained the situation to a woman applicant (cytologist Mary Jane Guthrie of the University of Missouri): "Difficulties of a scientific career for a woman were discussed, and it was pointed out that while women are not officially excluded from our fellowship program, the burden of proof would be unusually heavy if such an appointment were to be made."[8] Thus despite the fact that a woman applicant might be a good scientist and be highly recommended by her colleagues or professors, she faced an extra "burden of proof" that doubled the odds against her being chosen. Foundation officials did not think of this as being at all discriminatory, but insisted such doubly high standards were necessary since women were, in general, considered weak applicants. The woman therefore had to run not only against the other applicants but also against the selection committee's stereotypes of most women's abilities. Needless to say, such deliberate discouragement must have achieved its purpose of deterring all but the most strongly motivated women of the time.

*That is, 5 of 30 women applicants won awards (16.7 percent), while 102 of 298 men did (34.2 percent).

Among the women NRC fellows who did beat these two-to-one odds and were in acceptable fields were the young Margaret Mead, heading off for fieldwork in Samoa, Cecilia Payne (later Gaposchkin), continuing her studies of stellar spectra at the Harvard Observatory, Barbara McClintock, moving to the California Institute of Technology to start her lifelong work on corn genetics, and physicist Jenny Rosenthal, studying quantum mechanics at The Johns Hopkins University. They formed an elite among the young women scientists of the time, but even so, it is perhaps indicative of these years and of women's position in science that most of them did not become professors and department chairmen at major institutions, like the male fellows, but remained research associates or were unemployed for most of their careers.[9]

Other Rockefeller Foundation programs established in the 1920s were the Social Science Research Council Fellowships (1925), which assisted many anthropologists and psychologists, the Laura Spelman Rockefeller Memorial Fellowships in child development (most of which went to women), the American Council of Learned Societies' fellowships and grants-in-aid to scholars in all fields, and the Rockefeller Foundation's own international fellowships, which not only sent many Americans abroad for further study but also brought many Europeans and Latin Americans to the United States and Europe. If these fellowships and grants had presented new opportunities to the scientists of the 1920s, they were basic to the survival of many during the 1930s. Several refugee scientists from Nazi Germany, including physicist Hertha Sponer and renowned mathematician Emmy Noether, both of Göttingen, found their Rockefeller awards indispensable passports to freedom in the 1930s. Many young American scientists, unable to find jobs during the depression, also considered their fellowships godsends, as their autobiographies attest.[10]

Anthropology presents an extreme case of this dependency on foundations and fellowships during the 1920s and 1930s. Since it was a small field and had few teaching positions available, most of its younger women did important work and built whole careers on little more than a series of temporary fellowships from the NRC and SSRC. In fact, there seems to have been a tendency, in this field at least, to give the fellowships to the women to "tide them over" while the few jobs available went to the men. Thus Hortense Powdermaker has described the ease with which she obtained both NRC and SSRC fellowships in the 1920s and 1930s, and Margaret Mead, shown in plate 32, has also reported Franz Boas's willingness to support her applications for grants in those years (although Edward Sapir did not think her physically strong enough for fieldwork far from home). Both women held temporary research positions between field trips, but strangely, despite their many publications, seem not to have been candidates for the few academic positions opening up in their field. (As a group anthropologists seem far more prone to talk of their adventures in the field than of those back home in academia; for example, Margaret Mead's article "My First Job" is about her arrival and early days in Samoa.)[11]

One other large postdoctoral fellowship program started in these years was that of the John Simon Guggenheim Memorial Foundation, established in 1925

PLATE 32. Anthropologist Margaret Mead, shown here with husband, Gregory Bateson, in Tambunam, New Guinea, in 1938, was perhaps the best-known woman scientist of her time. (From Margaret Mead, *Letters from the Field, 1925–1975* [New York: Harper & Row, 1977], p. 225.)

to offer stipends to creative persons in all fields under the age of forty, a restriction soon dropped. Although its secretary, Henry Moe, claimed in 1936 that the Guggenheim fellows were "appointed solely on the basis of the quality of their accomplishments past and prospective" with no restrictions or "thought of distributing" them by fields of knowledge, colleges or universities, geographic regions, or "any other extraneous, although interesting, factors," he then proceeded to present cumulative data that did reveal certain patterns. Women had been 23.7 percent of the applicants but only 13 percent of the fellows chosen. A far higher proportion of these 68 women (32, or 47.1 percent) were in the creative arts (dance, writing, painting, and photography) than were the male fellows (133 of 457, or 29.1 percent). Of the remaining 36 women only 5 were in the sciences, where they constituted just 3.5 percent of the total (144). Among this elite were the anthropologists Gladys Reichard and Ruth Bunzel, the mathematician Olive Hazlett, the zoologist Mary MacDougall of Agnes Scott College, and the geneticist Barbara McClintock. In 1936 the selection committee made the more venturesome choice of black anthropologist-turned-writer Zora Neale Hurston, who used the fellowship to study the religious cults of the Caribbean region.[12]

Women scientists were able to win a few of these fellowships, despite the strong bias of most committees, for a variety of reasons. First of all, as we have seen, many of them were now getting excellent undergraduate preparation and were being admitted to most of the major graduate programs of the time; secondly, some of them were doing outstanding work; and thirdly, some of their favorite graduate school professors, such as Lafayette B. Mendel or Ross Harrison of Yale or Frank Lillie of the University of Chicago, or other contacts (occasionally even a prominent woman like Florence Sabin) were on the fellowship selection committees and knew them personally. Thus, for example, biochemist Florence Seibert has recorded a particularly poignant episode: in 1934 Professor Lafayette B. Mendel, who had been her thesis adviser but who was at the time dying of diabetes, asked what last favor he could do for her. Knowing that he was on such a fellowship committee, she asked for a Guggenheim. Unfortunately he died before the committee met, but a few years later Florence Sabin of the Rockefeller Institute was on the committee. Eager to help highly qualified younger women scientists, she had no difficulty convincing the selection committee of the importance to future tuberculosis research of Seibert's traveling to Sweden to learn the latest protein-separation techniques. Seibert was just one of many women scientists chosen as Guggenheim fellows during Sabin's tenure on its selection committee.[13]

This kind of personal contact raises the question of how the gender of the committee members affected the choice of applicants. The relationship is complex. Perhaps because the Guggenheim Foundation dealt with the arts (or because Henry Moe or the trustees wanted it that way), its selection committee, unlike many others, long had at least one woman among its five or six members. In general, therefore, committees that were 17 to 20 percent female appointed fellows of whom 13 percent were women. There was a limit to this linkage, however, for when in 1937 and 1940 there were two women (Marjorie Hope Nicolson, a perennial member, and Florence Sabin) on the committee of five (40 percent), the percentage of female fellows actually dropped to about 10 percent! Thus the presence of certain women on the selection committee could be detrimental to female applicants. But if the percentage of women chosen did not increase in Florence Sabin's years on the committee, the number and percentage of them who were scientists did markedly. Of the thirty-five women fellows appointed from 1925 to 1935, when the woman on the committee was in the humanities, only four of the women fellows (11.4 percent) were in the sciences. From 1936 to 1946, when Sabin was the committeewoman, sixteen of sixty-one (26.2 percent, more than twice as high) were in science. Thus Sabin's presence did help women scientists become Guggenheim fellows, but chiefly at the expense of women in other fields, who had been aided earlier by Nicolson's position. This shifting of the applicant pool toward their own interests suggests that these women exerted their influence before the actual selections, chiefly, as in Sabin's contact with Seibert, in encouraging the top women in their area to apply. Certainly little could be done at the actual meeting, if the remark fellow committee member Louis B. Wright made about Florence Sabin is accurate: "Sharp and incisive in her judgments, she never once advocated the appointment of a fellow because she was a woman."[14] To have done so would have undercut her credibility and shortened her term on the committee. Sabin, therefore, as the statistics suggest, encouraged particular women before the meeting.

A second kind of recognition open to most women scientists in the 1920s was that of admission to the major professional societies in their fields, but even here, as more and more women joined, new barriers and forms of stratification arose almost as rapidly as others fell.[15] Many societies have at some point in their history tried either to restrict their membership or to mark off a select inner circle within it by creating a hierarchy of membership categories such as "fellows," "members," and, sometimes, "associate members" and "junior members." Although most women scientists seem to have approved, at least initially, of this kind of upgrading, feeling it a fair and convenient way to honor the leading members of the group, they learned too late that in practice it had led (consciously or unconsciously) to a general defeminization of the higher ranks. Although this happened in all fields and in a variety of types of societies in the late nineteenth and early twentieth centuries, the most blatant examples and later the strongest protest came in the field of psychology.

This process of creating separate levels of membership seems to have started as early as 1874 in the American Association for the Advancement of Science (AAAS), where, as discussed in chapter 4, it was introduced as a means not only of recognizing the achievers within the group, but also of excluding the "amateurs" from the elective offices. In practice this two-tiered kind of membership proved a way to keep the women, who were perceived as "mere amateurs," subordinate to the men, who, though also often lacking higher degrees, were believed to be "professionals." By 1900 the percentage of AAAS fellows who were women was just 4.2 percent (36 of 864), far lower than their 13.4 percent of the general membership (141 of 1050).[16] In general such "professionalization" led to a progressive defeminization of the higher ranks of an organization.

In later decades the process spread to other fields and societies. When the Geological Society of America was formed in 1888 all its charter members were considered fellows, but subsequent additions had to be elected to that status by nine-tenths of the ballots cast by the all-male membership and council. Though the society's male-authored history asserts that "the Society has never discriminated against women," others might disagree, since, as the same author later reported, by 1922 the society had elected 684 fellows, of whom only six (less than 1 percent) were women. (By contrast women constituted 5.1 percent of the geologists in the *American Men of Science* in 1921.) In 1946, when the GSA changed its membership rules, it still had only 24 women among its 1,053 living fellows (2.3 percent).[17]

Another example of this sexual stratification within a society occurred among the ornithologists in the period 1880 to 1930. When the American Ornithological Union was established in 1883, there were no women among its twenty-three founders, although several were becoming active in the field. Two years later, in 1885, Florence Merriam (later Bailey), a Smith College undergraduate and the sister and later wife of prominent government naturalists who was soon to be a leading ornithologist herself, was elected its first "lady associate." In 1901 she was raised to "member" for her numerous writings in the 1890s, but though nominated for fellow in 1912, she was not elected one until 1929. By that time, however, the AOU's long-standing refusal to elect women fellows was causing very angry private comment among its more critical women members. Although the group's male historian later asserted that "the Union has always welcomed ladies to membership on the same basis as men and has elected them to all classes of membership except Honorary Fellow,"[18] the outspoken Althea Sherman of National, Iowa, strongly disagreed. In 1924 she wrote her protégée, Margaret Morse Nice, who was just starting her distinguished career, "I have said and believe it, that no woman will ever be made a Fellow of AOU. In 1912 I was told Mrs. Bailey was nominated . . . but think of Bergtold made a Fellow and Mrs. Bailey not. No, man's nature must change before a woman is a Fellow." She added, more practically, "With the ability to pack the meetings and vote for their associates that are candidates for election, the Washington or the New York Museum men can secure the election of any candidate they may have up. On the whole I believe the condition of maintaining 'classes' is a mistake: there are always as many of the 'outs' just as worthy as the 'ins,' a state that will always exist I fear."[19] Sherman's protest spurred Nice to greater efforts, and eight years after Bailey's election she too was chosen a fellow of the AOU. Both were wise choices, for it would be hard to imagine two more highly qualified ornithologists than Bailey and Nice. Among Bailey's numerous writings was the *Handbook of the Birds of the Western United States* (1902), a standard reference work that went through many editions and reprintings, and Nice developed the concept of "territoriality" to explain birds' nesting behavior, "the outstanding contribution of the present quarter century to ornithological thinking in America," as her honorary degree from Mount Holyoke College put it in 1955 (see plate 33). Because the two women did not hold a doctorate or any professional position (although Nice did edit *Bird Banding* for a number of years), they became an inspiration to numerous other women ornithologists. Nice, who did her important work at home while raising four children and was the only woman ever elected president of any of the three major ornithological societies in the United States (the Wilson Ornithological Club, in 1938) has even had a woman's ornithological society in Toronto named in her honor. In fact, in her case, child rearing seems to have contributed greatly to her scientific work. Originally trained in psychology, she wrote several early articles on her own children's behavior. When these failed to stimulate any response or encouragement, she turned to bird-watching, where by applying the same psychological techniques, she pioneered in the new field of animal behavior studies.[20]

The desire for stratification and some sense of quality control led to a similar defeminization even within the largely female field of nutrition and home economics. The American Home Economics Association (AHEA) had been formed in 1908 and had grown rapidly thereafter. Its membership was largely female and highly diverse, including high school teachers, extension workers, busi-

PLATE 33. Preeminent ornithologist Margaret Morse Nice was a pioneer in the study of animal behavior. She is shown here receiving an honorary degree in 1955 from her alma mater, Mount Holyoke College. It was one of the few awards she received. (Courtesy of Mount Holyoke College Library/Archives.)

nesswomen, and journalists as well as professors and researchers in the interdisciplinary field of nutrition. In 1940 the AHEA's leadership devised a plan to upgrade the society and its journal—henceforth "members" must hold at least a bachelor's degree (or the equivalent) in home economics; others could be "junior members" but were not allowed to vote. After a lengthy and "heated" debate, the proposal was adopted, but even this modicum of stratification proved so unpopular that it was revoked in 1946, when new leaders found a way to bridge the divisions among the society (mostly in the editing of the *Journal of Home Economics*).[21]

By then, however, the scientists in the field had long since created other forums for their professional work. In 1928 Professor Lafayette B. Mendel of Yale University, recognizing the long-felt need for a journal that would bring together the nation's far-flung nutrition researchers (from physiology and biochemistry as well as nutrition and home economics) in both government and academia, foster their communication, and recognize their professional achievements, organized a group to publish the *Journal of Nutrition*. In 1934 this same group established a new society, the American Institute of Nutrition, whose membership was to be selective and deliberately limited to "qualified investigators who have independently conducted and published meritorious original investigations in some phases of the chemistry and or physiology of nutrition." Although this stipulation may seem to have been sex blind, it deliberately excluded so many home economists who did not publish and research associates who did not work "independently" that the percentage of women in the AIN was just 22 percent (42 of 192), far below their 90 percent or higher membership in the AHEA. This low representation of women in a nutrition society, traditionally an area of "women's work," created such lasting bitterness toward the institute that thirty years later, in 1965, the obituary of Martha Trulson, associate professor of nutrition at the Harvard School of Public Health, stated (incorrectly) that she had been one of the very few women ever elected to the AIN. Yet within the institute the situation was apparently quite favorable, at least for those women who had been students of Mendel. One suspects that it was not a coincidence that the two women who were president of the group before 1950—Mary Swartz Rose of Teachers College, Columbia University, in 1938 and Icie Macy Hoobler of the Children's Fund of Michigan in 1944—were both prominent Mendel students, while the other notable women in the AIN, who were not Yale doctorates, such as Agnes Fay Morgan of the University of California and Lydia Roberts of the University of Chicago, never attained this honor.[22]

The situation was even more deliberately exclusive in the field of psychology, especially at the higher levels. This field, as mentioned in earlier chapters, had long had a relatively high percentage of women and was growing very rapidly in the 1920s and 1930s. About 20 percent of the psychologists listed in the *American Men of Science* in both 1921 and 1938 were women (see table 6.2), but the percentage grew even higher in these years in the American Psychological Association (APA), the main professional society in the field. Not only did the APA's membership grow more rapidly than did the number of psychologists in the *AMS*, increasing about fivefold in just fifteen years (from 457 in 1923 to 2,318 in 1938), but the number of women there grew even more impressively (or alarmingly, depending on one's point of view). Just 81 women had belonged to the APA in 1923 (18 percent of the total), but this figure had jumped eightfold, to 687, or 29.6 percent in 1938—and had reached a peak of 34 percent in 1928.[23]

This trend must have concerned at least some of the leaders of the association, who began in 1925 to discuss introducing some kind of quality control over the membership. Although most—eighteen—of the twenty-five leaders polled favored the creation of a second tier of membership for fellows, they did so for a variety of reasons, none of which showed any concern that the move might affect women differently from men. Mary Calkins agreed with the majority that it would raise standards but feared that it would make the society "undemocratic." Five of the leaders opposed the change, saying it would add little while creating jealousies and invidious distinctions, as between researchers and teachers. Margaret Washburn was among the two "doubtful" respondents, since she was worried that the admission standards might be lowered for the "members" or nonfellows and was "not clear on advantages to be secured." Designed to keep the numbers of fellows relatively small, the new category was introduced unobtrusively in 1926—all current APA "members" were to become fellows, but new additions to this status would have to be elected by the (usually all-male) council. But as elsewhere, this seemingly innocent reform, which should have rewarded both women and men equally, was later found to have restricted the women's position within the society almost as effectively as a deliberate quota system would have. Thus though the women's percentage of the membership had jumped from 18 percent to 30 percent between 1923 and 1938, their percentage of the fellows had remained almost constant, rising from 18 percent in 1923 to just 19 percent in 1938! The new category had in effect been used to restrict the women to their former proportion of the society and had nullified their tremendous growth there. Though few if any women seem to have noticed this tendency at the time, the large discrepancy was cited in the early 1940s as prime evidence of the strong and deliberate discrimination against women within the American Psychological Association.[24]

A more striking example of the prevailing misogyny and deliberate discrimination against even the top women in psychology in these years is evident in the curious history of a more select society in the field. In 1904 Professor Edward Titchener of Cornell University formed a group around himself which met annually and which soon called itself a scientific society, the Society of Experimental Psychologists. Since Titchener's group discussed and criticized the latest researches in the field and developed and exchanged ideas about further work, participation in it was considered a very profitable experience as well as a great honor. Membership in the group was highly prized, but, as one participant has written, "From the first, the group—especially Titchener—was opposed to the presence of women at its meetings." Other members occasionally admitted that they were not so strongly opposed to the women's coming and even thought that some might have something to contribute, but dared not oppose Titchener. The women, however, knowing of the group's existence and of the professional value of attending its meetings, made various attempts to secure an invitation.[25]

The most persistent proponent of admitting women was the outspoken veteran Christine Ladd-Franklin, who tried a variety of tactics over the years. (She apparently so frightened the men—if one can believe Edwin G. Boring's distorted later account—that they feared she would take over and so resisted her efforts all the more strongly.) Ladd-Franklin's anger justifiably reached a fever pitch in 1914 when she learned that the "Experimentalists" were to meet at Columbia University, where she was a lecturer, and discuss her specialty, color theory, without inviting her. Her first reaction to this insult was to write Titchener an angry letter, which reveals the nebulous extralegal standing of proper scientific behavior in the Progressive Era. Thus although she asserted that his group was not merely a private club but a scientific society that should be open to all scientists who had something to contribute, she had no way to enforce this and in the end could appeal only to his personal good manners for an invitation: "Have your smokers separate if you like (tho I for one always smoke when I am in fashionable society) but a scientific meeting (however personal) is a necessarily public affair and . . . it is not open to you to leave out a class of fellow-workers without extra discourtesy. Suppose you were to exclude Jews or Japanese because they were personally disagreeable to you? Would you not be considered very petty + bonné? ~~Such things can't be done as private personal matters~~. Do correct this error of yours at once." When Titchener did not respond, Ladd-Franklin sent him a second letter in which she reiterated her criticisms. After discussing some items in logic (another of her specialties), the fiery Ladd-Franklin went on:

> There is certainly a crying need for all the psychologists who have any logic in them to pull well together and to put up a good fight against all the irrational cranks! . . . Is this then a good time, my dear Professor Titchener, for you to hold to the mediaeval attitude of not admitting me to your coming psychological conference in New York—at my very door? So unconscientious, so immoral—worse than that—so unscientific! (I have a good article under way on the subject for the *Nation*.) Both the Psychological and the Philosophical Associations have long since admitted women to their smokers and everything. (I smoke—I should be very unfashionable if I didn't. No separation of the sexes is necessary any more on this ground.) It is only this acute-thinking and discussing little organization of yours (which seems to be so sadly dominated by *you!*) which still holds out. So mediaeval!—such an indignity!—well meant, I know—you have told me so—but such a mistaken kindness! Do quickly repent! And you need me! I particularly want to discuss for you at this meeting the present vagaries of Watson, Dunlap, and Rand and Feree—(Watson doubly).

This second approach was also unsuccessful, so Ladd-Franklin tried another tactic suggested by fellow psychologist (though nonexperimentalist) Mary Calkins of Wellesley. She had recommended her own favorite tactic for attending male smokers: Ladd-Franklin should ask James McKeen Cattell, the meeting's host, to invite her as his guest. Ladd-Franklin did, Cattell complied, and though she was later considered to have "invaded the masculine sessions," none of the men were quite so rude as to force her to leave the discussion of color theories, on which she was the international authority. Yet even this "victory" did not keep any doors open, since Titchener never knowingly allowed another woman into a meeting.[26]

After his death in 1928, however, the group reorganized itself, and a year later it elected its first two women: the noncontroversial Margaret Washburn of Vassar and June Downey of the University of Wyoming. (Unfortunately Ladd-Franklin was not offered even an honorary membership. Although aged eighty-one, she was still vigorous and had recently published a volume of her collected papers. She would have been delighted at the recognition, however belated; after all, she had just gotten her long-overdue Johns Hopkins doctorate in 1926.) Downey and Washburn were both outstanding psychologists, and more importantly here, they were personally far less controversial than Ladd-Franklin. Downey at age fifty-four was an authority on "handedness" and handwriting, but she probably never attended a meeting of the "Experimentalists," since she rarely left the University of Wyoming, even for meetings, and died just three years after her election. Washburn, the author of several books on animal intelligence, a past president of the APA, and at the time coeditor of the *American Journal of Psychology,* was even more eminent and professionally active. Although her exclusion from Titchener's club for twenty-five years must have stung deeply, she apparently never protested; perhaps, as she had been a student of his, she was resigned to his ways. (Later, when asked to write an obituary of him, she refused, saying she had never felt close to him.) Not generally known for either her feminism or her sense of humor, she nevertheless described to Karl Dallenbach, her coeditor and apparently closest friend, the mock astonishment with which some of the men greeted her election: "When Woodworth heard that I was permitted to come to the Ex meeting, signs of shock and suffering appeared on his usually placid face, and he warned me that I would have to sit with my feet elevated on the table before me." Dallenbach, who had had his own difficulties with Titchener but was now his successor at Cornell, replied in kind: "We are counting on your attendance at the meetings of the Experimentalists. There will be no segregation. . . . The inclusion of women marks a great occasion. T's ashes will smolder from the thought of it." If Titchener's ashes were undisturbed by the women's election in 1929, they must surely have smoldered two years later when Washburn invited the Experimentalists to meet at Vassar, an occasion that marked the high point of her long and distinguished career, since it coincided with her own election to the National Academy of Sciences. Despite this progress from 1929 to 1931, there were still limits, however, on how far the Experimentalists' welcome extended, since they elected no other women, not even noncontroversial ones, during the 1930s.[27]

This over thirty years of deliberate exclusion of women by prominent psychologists was obviously just one part of a widespread campaign at the APA and elsewhere to demean women psychologists and deflect them from academic careers of distinction. Yet the practice was widely tolerated and, despite Ladd-Franklin's appeal to good taste in 1914, was not sufficiently politicized for any of the male Experimentalists to have resigned in protest, although several (like Cattell) personally disagreed with prevailing policy and practice.[28] In fact, if one is to believe Boring's account of the Experimentalists' activities, they took great delight in thwarting the women's attempts to enter the group, sounding more like adolescent fraternity boys than mature professionals. Perhaps they were afraid that the entrance of so many women into the field might "tip" it into a kind of subprofessional social work. In any case, one suspects that the greatest impact of this exclusion was on the younger women in psychology, who were entering the field in very large numbers and constituting about one-third of the APA, but who were being channeled into clinical positions and discouraged from taking up academic careers, let alone trying to join select professional circles. The result of years of this strongly negative selection was that hardly any of this large generation of women psychologists became eminent experimentalists, and, as described in chapter 7, only two women were considered to succeed Margaret

Washburn of Vassar in 1937. This whole "weeding-out" must be compared with the experience of the women anthropologists, who were apparently welcomed into their field, given certain minor professional encouragements (such as fellowships for fieldwork), and accordingly flourished—up to a point. Although they too could find few academic positions in the 1930s, it was not for a seeming lack of suitably qualified candidates, as occurred in psychology.

Despite these indignities and professional slights, deliberate and otherwise, a few women scientists did manage to achieve positions of leadership and honor within their professional societies in the 1920s and 1930s, and a very few were even elected to presidencies of national organizations. These few are listed in table 10.2, which cannot be considered complete, but which does reflect a systematic search of the major national professional associations. The table indicates that by the 1920s and 1930s, when the number of women in the societies had become sizeable, a few outstanding women had been elected to presidencies. These women presidents (twenty-two women, as Parsons and Patch each headed two organizations) still represent a very small percentage of the leading officers of approximately forty organizations over forty years, from 1900 to 1940. Only

TABLE 10.2. Female Presidents of National Scientific Societies to 1940 (Home Economics Omitted)

Year	Scientific Society	President
1905	American Psychological Association	*Mary Calkins
	American Folklore Society	*Alice Fletcher
1913	American Nature Study Society	*Anna Botsford Comstock
1918	Child Hygiene Association	Sara J. Baker
1921	American Psychological Association	Margaret Washburn
	Association of American Geographers	*Ellen Semple
	National Vocational Guidance Association	Helen T. Woolley
1923	American Ethnological Association	Elsie Clews Parsons
1924	American Association of Anatomists	*Florence Sabin
	American Folklore Society	Louise Pound
1926	National Council of Teachers of Mathematics	Elizabeth M. Gugle
1927	American Fisheries Society	*Emmeline Moore
	American Nature Study Society	Bertha Cady
1928	Society of American Bacteriologists	*Alice Evans
1929	Botanical Society of America	*Margaret Ferguson
1932	American Folklore Society	Martha Beckwith
	American Association of Variable Star Observers	Harriet Bigelow
1936	Entomological Society of America	*Edith Patch
	National Council of Geography Teachers	Alison Aitchison
1937	American Nature Study Society	Edith Patch
1938	American Institute of Nutrition	*Mary Swartz Rose
	Wilson Ornithological Club	*Margaret Morse Nice
1939	American Speech and Hearing Association	Sara Stinchfield Hawk
1940	American Anthropological Association	*Elsie Clews Parsons

*Known to be first female president.

three societies—the American Psychological Association, the American Nature Study Society, and the American Folklore Society (financed to a large extent by Elsie Clews Parsons)—elected more than one woman president in all these years. Although the women presidents might today be cynically considered "tokens," that term would not accurately reflect their role then, which was more like that of a "favorite daughter" or "sister": they were women who were so outstanding in their field and who had been so highly regarded for decades that they might just be chivalrously accorded the honor. Dorothy Stimson said modestly about her election to the presidency of the History of Science Society (established in 1924) in the 1950s that, since she was the only one of the founders left who had not been president, the others felt they could safely entrust it to her for one term.[29]

It is also important to note that few of these twenty-two women presidents of national scientific organizations listed in table 10.2 held professorships at major universities. Although their contributions to their fields were sufficiently impressive to justify election to high office, such work had not led to prestigious jobs, which for women were, as described in chapter 7, only obtainable through local favoritism or patronage. The positions these twenty-two women held at the time of their presidencies varied: five held professorships at women's colleges, four were either unemployed or working intermittently, four were in government, two in the public schools, and one (Bertha Cady) was a naturalist for the Girl Scouts. The remaining six were at coeducational colleges or universities in some capacity: Rose and Aitchison were professors at teachers colleges; Hawks was a lecturer at the University of Southern California; Comstock was a professor of nature study at Cornell; Sabin was a professor of anatomy at The Johns Hopkins Medical School; and Pound was a professor of English at the University of Nebraska. Although it is hard to know to which men such a diverse group of women might be compared, one suspects that the male presidents of these scientific societies would have been professors at the major graduate schools more often than were these few women.

The omissions from table 10.2 are also interesting. No woman was elected president of the American Chemical Society, the Geological Society of America, the American Physical Society, the American Astronomical Society, the American Mathematical Society, the American Ornithological Union, the American Public Health Association, or the American Society of Zoologists, to name only the largest and most prominent in the major fields. The lack of female leaders in these years is even more noticeable in those specialized professional groups where women made up a particularly high percentage of the society and/or were making particularly important contributions. For example, although several women played a major role in the establishment of the Society of Economic Paleontologists and Mineralogists in 1926 and 1927, none held office for decades. Nor, despite the important work of the many women plant pathologists in this country, was there a female president of the American Phytopathological Society until the 1950s. Likewise, there has never been a woman president of the American Society of Biological Chemists, and the Paleontological Society, established in 1909, did not elect its first woman president until forty years later, despite the availability of several women who were acknowledged to be outstanding. Nor were there any women presidents of the American Association of Applied Psychology, despite the facts that Leta Stetter Hollingworth and Gertrude Hildreth, both of Teachers College, were active in its founding and that by 1941 34 percent of its members and 32 percent of its fellows were women. The reason, Edwin Boring suggested, for this lack of women leaders was that women members, like the men, preferred male leadership, but there is no evidence that nominating committees ever gave them a choice. In fact there is a hint in the minutes of the early AAAP that one reason for its formation may have been to head off several (women) clinical psychologists who were beginning to form their own national group. In any case, by the 1940s critics would see the women's underrepresentation in leadership positions in even applied psychological organizations as still more evidence of the pervasive discrimination against women in that field.[30]

Although there would seem to be a host of social factors involved in the election of a society's president (and the whole subject remains unexplored by sociologists), the inclusions and omissions from table 10.2 suggest that there is a certain critical barrier that arises against a woman president once a society reaches a particular size. For example, when the American Psychological Association was a relatively small group (under 400 members), in the years from 1893 to 1921, it had two women presidents.[31] But once the society had grown beyond this size, certain barriers—both perceptual, such as "stereotypes" and "roles," and political, such as factions and cliques—arose and sharply limited the chance of any other woman's holding the office, despite a rapid rise in the percentage of women in the society. It may be that the job itself changed from one of leading an informal business meeting among friends once a year to the more "public" one of chairing large sessions and becoming a major spokesperson for the field. In a small and intimate professional group where the leaders had known each other for years and in a sense had grown up together, the women might be considered almost equal and their abilities noted, but in a larger group elements of power and domination might so "masculinize" the job description that even highly productive women would be overlooked and expected to take a modest or subordinate position (perhaps as a secretary or assistant editor), while a male department chairman took over the presidency. Such size factors might also help explain why the women were apparently more successful in holding the top job in the biological and anthropological societies—these fields, for reasons apparently endemic to the subject matter, were often subdivided into a great many smaller and more welcoming societies, such as those for bacteriology, anatomy, fisheries, entomology, folklore, ethnology, and others. But other sciences, especially the physical sciences, did not fractionate into so many small units, and there the women, who were less numerous anyway, were rarely elected to office, and even then only in the less prestigious teachers' and amateurs' groups.[32]

This tendency to overlook the women in large groups would also help explain why there was only one woman (a nonscientist) president before 1940 of any of the several supradisciplinary professional organizations, such as the American Association for the Advancement of Science, the American Association of University Professors, Phi Beta Kappa, and Sigma Xi, although several women held lesser offices in each. All four societies had a federal structure whereby most of the national officials were not elected directly but were chosen from among the leading figures in the major subunits. This helped the women, who were active in many sections and chapters, to percolate upward and hold some offices to which they might not have been elected directly. Several women thus held positions of responsibility in most of these organizations as committee chairmen, councillors, secretaries, and occasionally vice-presidents. Thus although no woman was president of the AAAS until 1969, three were listed as co-vice-presidents long before then, since this was an honor regularly accorded section chairmen. Nevertheless, the three women vice-presidents all chaired the same AAAS section, that of anthropology and psychology, where, as mentioned in chapter 4, they had been prominent since the 1880s: Alice Fletcher in 1896, Lillien Martin in 1915, and Margaret Washburn, its psychology portion, in 1926.[33]

Women had also been active in the AAUP since its founding in 1915. Almost all of them came from the women's colleges, which maintained active chapters of the association. Starting in 1917 there was a steady stream of women councillors, and after 1921, of committee chairmen as well. In 1925 psychologist Mary Calkins capped her long career at Wellesley College by serving as vice-president of the AAUP.[34] Phi Beta Kappa, which had opened its first chapter at a woman's college (Vassar) in 1899 and had twenty-one such units by 1940, elected its first woman president in 1940 (Marjorie Hope Nicholson, professor of English Literature and dean of Smith College).[35] By contrast Sigma Xi was inexplicably slow in establishing chapters at the women's colleges and apparently had no women officers at any but the local level until after World War II. Although the society had been established at Cornell in 1886 and had elected five women in 1888 (including Anna Botsford Comstock, the Cornell entomologist and pioneer in the nature study movement), it did not establish its first chapter at a woman's college (Smith) until 1935. The historian of the association has written that there was so little interest in science at the early women's colleges that they could not meet the association's requirement of sponsorship by at least ten faculty members (of either sex). But this seems hardly plausible, given the active interest in science at Wellesley, Vassar, and Mount Holyoke in the late nineteenth century and the presence of at least the necessary number of science faculty there by 1921, if not earlier. (Table 1.3 counts only women faculty members listed in the *American Men of Science;* it thus omits the men and other unlisted women there.) One suspects that if the Cornelia Clapps, Sarah Whitings, Mary Whitneys, and Emma Perry Carrs had wanted a chapter of Sigma Xi they would have gotten one. Why they did not remains a mystery.[36]

The nation's most prestigious scientific society, the National Academy of Sciences, finally acknowledged the contribution of women to science and elected its first two women in 1925 and 1931. The pressure for this seems to have come from some liberal men in the group, the indisputably outstanding accomplishments of a few women by the 1920s, and such unbridled favoritism in the academy's sections that even "third-rate men" were being selected.[37] There is a story here, however, for a certain amount of politics was involved in the election of the academy's first woman. As described in chapter 5, Madame Curie had made a much-publicized tour of the United States in May 1921, during which she received many honorary degrees and a gram of radium from President Harding at the White House. Her visit to Washington, D.C., also raised certain problems of protocol for the male leaders of the National Academy, whose headquarters was in the capital. They (George Ellery Hale, C. G. Abbott, R. A. Millikan, A. A. Noyes, and Charles D. Walcott) wondered whether to give her a medal or, as they were doing for Albert Einstein who was also touring the United States that spring, elect her a foreign associate. The official correspondence at the academy's archives indicates that both these ideas were raised and rejected, not only because of the imminence of her visit but also because any such action would require a vote of the membership, which was dispersed all over the country and would be unlikely to agree anyway. (Einstein's visit had, by contrast, coincided with the academy's annual meeting.) But Curie's visit had raised the issue of women's membership in the academy as early as 1921.[38]

Then in 1923 Raymond Pearl, a biostatistician at The Johns Hopkins School of Hygiene and Public Health, whose wife, Maud DeWitt Pearl, was a biologist and who had long bemoaned the poor quality of most of the men elected to the academy, corresponded with E. B. Wilson of the Harvard School of Public Health on the question of electing a woman, in general, and the fitness of astronomer Annie Jump Cannon, in particular. Both agreed that her scientific accomplishments were more than sufficient for the honor, but Pearl, a eugenicist, said he could not vote for her because she was deaf. It was hard enough, he joked, to run the academy meetings with the misfits already there without adding any more "physical defectives"![39] But it was not up to Pearl or Wilson to nominate her. That was the task of the astronomy section of the academy, which never did put her name forward (not one woman astronomer was elected to the academy until 1978). This sort of treatment in the 1920s may have been one reason that Annie Jump Cannon found it necessary to start a separate prize for women astronomers in the 1930s (see chapter 11).

Some of the medical scientists within the academy were more amenable to the idea of electing a woman, however, and in 1923 the physiology and pathology section nominated Florence Sabin of The Johns Hopkins Medical School, a prominent researcher and one of the many Welch protégés there (see chapter 5). Sabin was not elected, which was a common occurrence in a first attempt. Two years later, after serving as the first woman president of the American Association of Anatomists in 1924, Sabin was renominated for the academy and won, though not without a battle. Edwin Conklin of Princeton attended a part of the election meeting, and as he described the situation to his friend Ross Harrison at Yale, one needed strong political friends as well as scientific accomplishments to be elected to the National Academy: "I don't know whether they succeeded in getting in Miss Sabine [*sic*] or not, but the evidence was that the Hopkins crowd were knifing everybody who stood above her in the preference list, in the hope of combining on her at the end. Maybe I am severe in my judgment but that is the way I sized it up."[40] But even Sabin's election does not seem to have hastened the addition of other women appreciably. Despite the election of ten to twenty new members per year, it was not until 1931 that the academy elected its second woman, the psychologists' favorite, Margaret Washburn, and not until 1944 that cytogeneticist Barbara McClintock became number three.

In general, the choice of these three women from the many who might have been considered eligible tells more about the men in these groups and their attitudes and politics than it does about the women elected, whose work was outstanding but hardly any more so (to judge by their professional obituaries) than that of the many others not chosen: Ruth Benedict, Mary Swartz Rose, Annie Jump Cannon, Edith Patch, Margaret Ferguson, or the numerous other women mentioned in this chapter. The key difference for some women seems to have been less in their work and more in the men around them, who had to be both politically powerful and sufficiently convinced of their merit to be able to overcome others' reluctance to have them elected. It was not enough for the men to be mildly impressed with their work; they had to be willing to fight hard on the women's behalf. That kind of commitment by key, busy people seems generally (in these cases at least) to date back to friendships made decades before, probably in graduate school, when the men and women had all been students together under the same great mentor—like the medical students of William H. Welch at The Johns Hopkins, the psychologists from Cornell and Columbia, and the geneticists from Cal Tech. Years later, when the young men had become full professors at major institutions and were themselves members of the academy, they might think it "only fair" for their old classmates to join them. They might also have felt guilty at certain mistreatments such women had received in the past—like Sabin's having been passed over for her department chairmanship at Johns Hopkins in 1919 or Washburn's exclusion from the Experimentalists for twenty-five years. That such slights were not easily forgotten is clear from one of the letters of congratulation that Sabin received upon her election to the academy in 1925. A colleague at The Hopkins wrote her, "Please let me say how happy I am to hear of your latest recognition—and how disgusted I am that the great men of Hopkins appear to be so unaware of what everyone else in the world sees so clearly."[41] Election to the National Academy of Sciences was thus a special gift for a very few women from their closest professional siblings who were eager to reward them years later for a lifetime of distinguished researches. But for women outside this charmed circle of influential "big brothers," these higher honors were, regardless of the quality of their work, still an impossible dream.[42]

(Worth mentioning here briefly is the curious case of Florence Bascom, the Bryn Mawr geologist who thought she was a member of the National Academy and listed herself as such in the 1938 *American Men of Science* but was not. She had been elected to a three-year term on the National Research Council's committee on geology and geophysics in 1935 and apparently confused election to the council [partly because of the flowery letter announcing her appointment] with election to the academy. She was in her dotage by the late 1930s and reportedly less than clear-headed, but perhaps her confusion was justified, and it is revealing, for as the matriarch of American geology who had trained numerous women geologists, had been a fellow of the Geological Society of America for decades, and had held several offices in the AAAS and the AAUP, she probably felt she deserved election, and many colleagues would have thought she was right.)[43]

But not all doors once opened to women stayed that way, as was shown by the American Academy of Arts and Sciences in Boston. That honorary group had been quite daring in 1848 when, as mentioned in chapter 4, it had elected Maria Mitchell its first woman member, but since then it had adamantly refused to add any others. Even a polling of the membership in 1920, which revealed a majority favored adding qualified women, did not convince the Boston-based council to authorize their election. This prohibition remained in effect until 1943, when the council finally changed its policy and approved the election of five women, including such local luminaries as psychologist Augusta Fox Bronner and Radcliffe College president Ada Louise Comstock. It added Margaret Mead in 1948 and botanist Katherine Esau in 1949, long before their elections to the National Academy in the 1970s.[44]

Other forms of recognition available to women scientists in the 1920s and 1930s were the many prizes and honors granted by the numerous professional societies of the time. The *Handbook of Scientific and Technical Awards* (1956), edited by Margaret Firth, lists a few women among the thousands of winners of these hundreds of awards granted between 1900 and 1952. Almost all of the women so honored were the first to be so. Again, the sociology of prizes and awards—who gets them, who gives them, and why—is a relatively new area of the sociology of science. A recent study has made a beginning, however, by pointing out that within a field, there is a rough hierarchy of the prestige value or desirability of various awards. Thus to appreciate fully the fact that Florence Merriam Bailey won the Brewster Medal of the American Ornithological Union in 1931, Annie Jump Cannon the Draper Medal of the National Academy of Sciences also in 1931, or Louise Boyd the Cullom Gold Medal of the American Geographical Society in 1938 and were the first women to do so, one would have to know the particular field and its potential awards and winners rather well. One suspects, however, not only that these early women winners were highly deserving of their awards, but also that they should have received them years earlier. Thus Gertrude Rand and her husband, Clarence Ferree, had pioneered in the field of physiological optics and indoor lighting starting in the 1910s. When she became the first woman to receive the gold medal from the Illuminating Engineering Society in 1963, long after her husband had died and she had retired, she said that the award would have meant a lot more if it had been granted years earlier. Most scientists, however, were publicly grateful for any prize whenever it was offered.[45]

There were two awards whose value was apparent to all in the 1920s and 1930s: a "star" in the *American Men of Science,* and the supreme accolade, the Nobel Prize. Although there is a large amount of data on the "starred" scientists in the *AMS,* these awards were not a completely accurate indicator of eminence in science in the 1920s and 1930s. In fact, this system of recognition began to break down in the late 1930s, when it became riddled with an accumulation of internal injustices. Its two chief difficulties were (1) a failure to grant stars to some persons in new or applied fields and (2) an unwillingness to take stars away from older scientists whose work no longer merited them. The first restriction meant that much research in the new, rapidly growing, and moderately feminized fields of biochemistry and microbiology was not recognized with stars; nor was work in the applied areas of nutrition, plant pathology, or engineering officially considered eligible. Even as eminent a microbiologist as Selman Waksman, the discoverer of streptomycin, never got a star in the *AMS,* although he won a Nobel Prize in medicine and was elected to the National Academy of Sciences (by the botany section).[46]

Yet Waksman was unusual, for most other important men in hybrid or applied fields managed to get stars from neighboring, often parent, fields despite the official ban. For example, the prominent microbiologists G. W. McCoy, W. T. Sedgwick, E. O. Jordan, Theobald Smith, and Augustus Wadsworth were all awarded stars by the pathologists; nutritionists W. O. Atwater, F. G. Benedict, and H. C. Sherman got theirs from the chemists; and E. V. McCollum and L. B. Mendel received stars from the physiologists. For women, however, no such friendly stretching of the boundaries took place. The NIH bacteriologists Bengtson, Evans, and Branham remained starless, as did the nutritionists Rose, Hoobler, Morgan, Roberts, and Parsons and all the women plant pathologists. Since these scientists seem in retrospect as worthy as many who were included, and since they did win other honors, the implementation of the *AMS* rule about eligibility for stars affected women unduly.[47]

If this first restriction on the *AMS* stars led to the omission of many important women, the system's second weakness, the retention of stars long after their recipients' best work had been done, so overloaded it with past winners that considerable confusion resulted as to what being awarded a star really meant. Although most readers of *Science* seem to have been aware that a star in the *American Men of Science* had originally (1906) signified inclusion in the top 1,000 scientists in the nation, few seem to have been aware that this honor ceased almost immediately afterwards to mean quite this. They did not realize how protective Cattell was being with the egos of his former winners and assumed all too readily that continued starring meant retention in the top 1,000, an honor that seemed to increase in value as the size of the directories grew from about 4,000 entrants in 1906 to 34,000 in 1943. As early as 1910, however, Cattell had made the fateful decision to let those former winners who were no longer ranked in the top 1,000 retain their stars as well as to award new ones to those others who now deserved them, too. By the third edition of 1921, the number of persons starred in his directory was over 1,600—and all of them continued to think that they were in the current top 1,000. Thus Alice Eastwood, a long-lived botanist at the California Academy of Sciences in San Francisco, was starred in the first edition of 1906 at the age of forty-seven and in every subsequent edition through the seventh in 1943, by which time she was well over eighty, the field had changed tremendously, and her actual standing was probably diminished. Yet she and her biographers proudly, though mistakenly, interpreted her retention of the star to mean that she was still one of the top people in her field. They never suspected what is now evident from Cattell's unpublished papers (such as the list entitled "[Persons] No Longer Starred in 3rd Edition")—that she had been dropped by 1921 and probably as early as 1910.[48] This dual system was apparently unsuspected and went undetected until the late 1930s, when detailed studies of the starred persons by geographer Stephen S. Visher began to alert scientists to the many injustices and inconsistencies in the system. Before long the stars had received such negative comment in *Science* and elsewhere that Cattell's son and successor as editor of the *AMS* decided to discontinue the practice after the seventh edition (1943).[49] Nevertheless, the scientists of the 1920s and 1930s were unaware of Cattell's double standard and, thinking that a star in the *American Men of Science* was as valid an indicator of scientific merit as any (and better than some others, such as membership in the National Academy), they scrutinized each list of additions carefully. But for the reasons outlined here these lists should be interpreted cautiously—inclusion was certainly a great honor, but exclusion was not necessarily a dishonor, since it could be the result of several factors, many of them weaknesses in Cattell's system.

Although counting stars and using them for comparative purposes can be seen to be fraught with danger, stars were thought at the time to be valid measures of how women were faring in science. When, as described in chapter 4, Cattell had revealed in 1910 that the number of women in the top 1,000 had dropped from 19 to 18, this had caused great consternation among the feminists of the day. But in 1921, when Cattell reported privately to Christine Ladd-Franklin that the number was down to 15, he chose not to publish that embarrassing statistic. Nor would anyone have suspected it, since about twenty-five women are starred in the third edition, and by 1921 Cattell had shifted his frequent statistical reports on his starred persons away from the characteristics of the entire group to just those of the roughly two hundred and fifty new additions. From the women's point of view, this approach stressed the positive—a few women were being added—and made no mention of the distressing fact that perhaps just as many others were being dropped or had died. But even these additions were too few to have constituted major inroads on the top 1,000: the women constituted 4 of the 356 additions in 1921 (1.1 percent); 6 of 256 in 1927 (2.3 percent); 3 of 250 in 1933 (1.2 percent); 10 of 251 in 1938 (4 percent, their best showing); and 4 of 255 in 1943 (1.2 percent). Thus, despite the fact that several new women were being added, the women's overall representation in the select top 1,000 probably remained about 20 (or 2 percent) throughout this period. In fact, the actual number remained so low that it did not matter much when, for example, Alice Evans of the U.S. Public Health Service made the understandable mistake of assuming that all the starred women in the fourth edition (in 1927 about 30) were still in the top 1,000. Even her 50 percent error boosted the women to just 3 percent, a figure she still found deplorable.[50]

But if one follows Cattell's lead and talks only of the twenty-seven women who were given new stars in the years 1921 to 1943, it becomes clear, as shown in table 10.3, that they were not distributed uniformly across all twelve of his fields of science—nine (one-third of the total) were in zoology; three each (another one-third) were in botany, psychology, and astronomy; and there were one or two each in five other fields. Significantly, there were no women starred in all these years in chemistry, pathology, or physiology. (Not even Gerty Cori, later awarded a Nobel Prize for her work on glycogen metabolism, was considered worthy of a star in the *American Men of Science* in the 1930s, in itself an excellent cause for caution.) When the distribution of the starred women is compared with that of the total stars given in these years, as also shown in table 10.3, the women are found to have done best in anthropology, where they won 6.9 percent of the stars, and in anatomy, where their share was 5.9 percent. When this rather good showing in two of the smallest fields of the time (though still below their percentages in each field, as shown in the 1938 *AMS* [table 6.2]) is compared with their very poor showing in chemistry and physics, the two

TABLE 10.3. Women Starred in the *American Men of Science*, 1921–1943 (By Field, Number, and as a Percentage of the Total)

Field	Total Starred	Women Starred	Women's Percentage of Total
Anatomy	34	2	5.9
Anthropology	29	2	6.9
Astronomy	75	3	4.0
Botany	133	3	2.2
Chemistry	240	0	0
Geology	138	2	1.4
Mathematics	110	2	1.8
Pathology	92	0	0
Physics	197	1	0.5
Physiology	62	0	0
Psychology	69	3	4.3
Zoology	190	9	4.7
Total or average	1,369	27	2.0

SOURCE: Stephen S. Visher, *Scientists Starred in "American Men of Science," 1903–1943* (Baltimore: The Johns Hopkins University Press, 1947), pp. 113 and 148–49 (addition errors corrected).

largest fields (for stars anyway), it suggests that, as theorized earlier in connection with association presidencies, perhaps the size of a field is an important factor in how often women win awards there. To a certain extent the larger the field, the less good the women's chances for visibility and so for honors. The chief exceptions here would be physiology, where the women won no stars despite the relatively small size of the field, and zoology and botany, where they won several despite the large size of the field (as measured in the total number of stars given). Otherwise most fields seem to follow this pattern.

Another interesting characteristic of these twenty-seven new women "stars" in the *AMS* in the 1920s and 1930s that should be mentioned is the fact that eleven of them (40.7 percent) were or had been married, at a time when only 26.4 percent of all the women in the *AMS* were married and when only six of the twenty-two (or 27.3 percent) female presidents of professional societies listed in table 10.2 whose marital status is known were married. Apparently a star was one honor regularly awarded to those highly motivated married women who remained active scientists. More often they were overlooked, as they were in elections in scientific societies, probably because so many were married to top men in their own fields, whose position eclipsed and subsumed theirs. Thus it is interesting to note that of these eleven starred married women, nine were married to scientists, usually but not always in the same field; six of these husbands were also starred in the *AMS* (Sally Hughes-Schrader [Mrs. Franz] and Ethel Brown Harvey [Mrs. E. Newton] in zoology, Eleanora Bliss Knopf [Mrs. Adolph] in geology, Margaret Reed Lewis [Mrs. Warren] in anatomy, Anna Pell-Wheeler [Mrs. Alexander] in mathematics, and Ruth Benedict [divorced from Stanley Rossiter Benedict] in anthropology). Three other married women (the astrophysicists Charlotte Moore Sitterly [Mrs. Bancroft] and Cecilia Payne-Gaposchkin [Mrs. Sergei] and psychologist Helen Thompson Woolley [divorced from Paul Woolley]) were starred, although their husbands were not, a most unusual phenomenon.[51] A full list of the fifty-two women who were ever starred in the *AMS*, together with what is known of their marital status, is presented in table 10.4.

Although no American women won a Nobel Prize in science during the 1920s and 1930s, at least four got rather close. In 1925 a Swedish mathematician wrote Henrietta Leavitt of the Harvard College Observatory that he wished to nominate her for the Nobel Prize in physics, since her discovery of the link between a star's luminosity or brightness and its distance from earth had opened up much modern astrophysics. Unfortunately, unbeknownst to the Swede, Leavitt had died four years earlier.[52] Also in 1925 Gladys Dick and her husband were nominated for the Nobel Prize in medicine for their developing the "Dick test" for scarlet fever, but that particular prize was not awarded that year.[53] In 1934 pathologist Frieda Robscheit-Robbins, who had performed part of George Hoyt Whipple's prize-winning work on iron metabolism, had looked on unrewarded when, as mentioned in chapter 7, he shared the award with two men; and Gerty Cori and her husband Carl were busy in the 1930s with the work on glycogen that would bring them (and Bernardo Houssay of Argentina) the prize in 1947.[54]

TABLE 10.4. Women Starred in the *American Men of Science*, 1903–1943 (By Field, Edition First Starred, Name, and Marital Status)

Field	Edition First Starred	Name	Marital Status	Husband
Anatomy	1	Lydia Adams DeWitt	Married; separated	Alton D. DeWitt
	1	Florence Sabin	Single	
	6	Margaret Reed Lewis	Married	*Warren H. Lewis
	7	Elizabeth Crosby	Single	
Anthropology	1	Alice Fletcher	Single	
	4	Elsie Clews Parsons	Married	Herbert Parsons
	5	Ruth Fulton Benedict	Married; separated	*Stanley Rossiter Benedict
Astronomy	1	Williamina P. Fleming	Married; separated	James Orr Fleming
	3	Annie Jump Cannon	Single	
	4	Cecilia Payne-Gaposchkin	Married	Sergei Gapsochkin
	6	Charlotte Moore Sitterly	Married	Bancroft Sitterly
Botany	1	Elizabeth G. Britton	Married	*Nathaniel Lord Britton
	1	Alice Eastwood	Single	
	2	Margaret Ferguson	Single	
	4	M. Agnes Meara Chase	Married; widowed	William I. Chase
	6	Sophia H. Eckerson	Single	
	7	Barbara McClintock	Single	
Chemistry	1	Ellen Swallow Richards	Married	Robert H. Richards
	2	Mary E. Pennington	Single	
Geology	1	Florence Bascom	Single	
	6	Julia Gardner	Single	
	6	Eleanora Bliss Knopf	Married	*Adolph Knopf
Mathematics	1	Charlotte A. Scott	Single	
	3	Anna Johnson Pell Wheeler	Married; widowed; remarried; widowed	*Alexander Pell; Arthur Wheeler
	4	Olive Hazlett	Single	
Physics	1	Margaret Maltby	Single	
	7	Katharine B. Blodgett	Single	
Psychology	1	Mary W. Calkins	Single	
	1	Christine Ladd-Franklin	Married	*Fabian Franklin
	1	Margaret Washburn	Single	
	2	Ethel Puffer Howes	Married	Benjamin Howes
	2	Lillien Martin	Single	
	3	Helen Thompson Woolley	Married; divorced	Paul G. Woolley
	4	June Downey	Single	
	6	Florence Goodenough	Single	
Zoology	1	Cornelia Clapp	Single	
	1	Katharine Foot	Single	

(*continued*)

TABLE 10.4.—*Continued*

Field	Edition First Starred	Name	Marital Status	Husband
Zoology	1	Helen Dean King	Single	
(*continued*)	1	Agnes Claypole Moody	Married	Robert O. Moody
	1	Margaret L. Nickerson	Married	*Winfield Scott Nickerson
	1	Florence Peebles	Single	
	2	Susanna Phelps Gage	Married	*Simon H. Gage
	2	Nettie Stevens	Single	
	3	Mary Jane Rathbun	Single	
	4	E. Eleanor Carothers	Single	
	5	Libbie H. Hyman	Single	
	5	Ann Haven Morgan	Single	
	6	A. Elizabeth Adams	Single	
	6	Mary Jane Guthrie	Single	
	6	Ethel Browne Harvey	Married	*E. Newton Harvey
	6	Sally Hughes-Schrader	Married	*Franz Schrader
	7	Hope Hibbard	Single	

SOURCES: *American Men of Science,* 1st ed. (1906); 2d ed. (1910); 3d ed. (1921); 4th ed. (1927); 5th ed. (1933); 6th ed. (1938); and 7th ed. (1944). Stephen S. Visher, *Scientists Starred in the "American Men of Science," 1903–1943* (Baltimore: The Johns Hopkins University Press, 1947), pp. 148–49; Edward T. James, Janet Wilson James, and Paul S. Boyer, eds., *Notable American Women, 1607–1950,* 3 vols. (Cambridge: Harvard University Press, 1971); and Barbara Sicherman and Carol Hurd Green, eds., *Notable American Women: The Modern Period* (Cambridge: Harvard University Press, 1980).

*Indicates husband starred also.

The absence of an American female Nobelist in the 1920s and 1930s is, as with the *AMS* stars, a weak indicator of the quality of the women's work in science in these years. Again, it would certainly have been a great honor if one had, but the fact that no woman became a laureate may have been the result of several factors, some of them built into the selection system. The most serious problem here was the fact that Nobel prizes were given in only three fields of science: physics, chemistry, and medicine/physiology, which were, curiously, those very same fields in which women had made their weakest showing (for whatever reason) in winning "stars" in the *American Men of Science* in these same years. Unfortunately for the women, Nobel prizes are not awarded for work, however outstanding, in those other scientific fields in which they were more prevalent (like nutrition or psychology) or more successful at winning awards (like anthropology, anatomy, and zoology). One thus begins to wonder if there is some link here: why, for whatever unarticulated reason, did Alfred Nobel pick the most professionalized, laboratory-oriented, Germanic, male-dominated, "macho" fields of science for his awards, those very fields that had been and were remaining the least receptive to women, and ignore the others in which they were then or have since become more prominent?[55] In doing so he served to minimize the women's accomplishments territorially as well as hierarchically and thus to perpetrate the illusion that there were very few women in the whole of "science." Anthropologist Margaret Mead, who despite a wide popular reputation suffered from this professional discrimination (or "asymmetry," as she called it euphemistically), observed this phenomenon in many cultures in the 1920s and 1930s and mentioned it frequently in her writings, perhaps most succinctly in *Male and Female* in 1949: "In every known society, the male's need for achievement can be recognized. Men may cook, or weave, or dress dolls or hunt hummingbirds, but if such activities are appropriate occupations of men, then the whole society, men and women alike, votes them important. When the same occupations are performed by women, they are regarded as less important."[56]

Thus although sociologists talk about the "reward system" of science, as if it functioned fairly and well, the women's experience in science before 1940 clearly shows that for them it did not. Not only was good work by women very often not rewarded, but most women were weeded out long before they might have been considered eligible for the top honors or most prestigious positions. Such systematic defeminization occurred in all areas, but especially in the largest fields and societies and at the highest levels. It was particularly blatant in honorary groups, especially the two major academies of science, where even the most highly qualified women of their generation were not taken realistically as candidates until the supply of not only first-rate men but also admittedly second-rate ones had been exhausted. Only when the society was faced with electing third-rate men did its committees begin to rethink their admissions policies. Even then the election of women required more than the usual amount of political maneuvering by friends and supporters within the society. Thus the growing numbers of talented, trained, and accomplished women scientists in the 1920s and 1930s (7 percent of the total in the 1938 *American Men of Science*) found themselves restricted to the bare fringes of the top honors (2 percent or less) meted out by the main (male-dominated) "reward system." Had it been the overt purpose of professional organizations to weed women out, the process could hardly have done a more thorough job than that accomplished in most fields in these years.

In fact, one begins to suspect, as Mead suggested, that the whole reward or prestige system, which arose so rapidly in the first half of the twentieth century, may have been designed to fulfill other, less lofty and benevolent, purposes than the favorable concepts of "rewards" and "recognition" would suggest. Like the doctorates (see chapter 2) and the "higher standards" of the "professional" societies (in chapter 4), the new prestige system also served to weed out (indirectly, through "objective" criteria like credentials) those whose presence was for whatever real and unarticulated reason not desired in the inner circles. This process not only created hierarchies or stratification where none had existed before, but also empowered one kind of scientist to determine where the others should stand. Just how this authority was created, wielded, and accepted needs to be explored more fully, even though it will be extraordinarily difficult to document the intentions behind the process, since the principal figures were usually silent about, if even aware of, their own motives. Meanwhile the lack of much of this "real" recognition spurred some women's groups and individuals to create some other supplementary or "compensatory" clubs and awards that did fit their needs.

11

WOMEN'S CLUBS AND PRIZES: COMPENSATORY RECOGNITION

Because the main reward system of science was remote from the careers and realistic expectations of all but a very few women scientists in the 1920s and 1930s, many women either withdrew from the process or created their own separate groups and prizes to supplement their otherwise unacknowledged existence in science.[1] Although largely unnoticed by male scientists and those women who were not involved personally, this system of "compensatory recognition" played a moderately important part in the professional lives of thousands of rank-and-file women scientists of the 1920s and 1930s. These organizations offered women teaching in the schools of home economics, the women's colleges, and other small colleges, women working in government laboratories, industrial offices, or libraries, and women who were at home or unemployed a certain sense of belonging and acceptance, encouragement and psychological support, and a chance to be active in some role, including the leadership positions denied them in male dominated societies. So many of the supporters of these organizations and prizes were loyal to them for decades that most are still in existence today.

Although separate from the main (men's) groups, these women's societies were rarely radical or even "feminist" in their orientation. They were essentially conservative and nonconfrontational, as were most of the women's groups in other professions in the 1920s and 1930s. Rather than attacking the status quo or the male establishment that had excluded them, the members of these clubs accepted the separate spheres and worked to make the best of the segregation.[2] In fact, they grew to enjoy their separate clubs so much that, although Elizabeth Adams, formerly of Smith College and an early analyst of the professional woman worker and her problems, predicted in 1921 that these groups would be just temporary organizations ("a stage in the professional evolution of women"), they were surprisingly long-lived.[3] Even when the chance arose to merge or federate in some way with the men's groups in the field, these women's organizations usually refused. They treasured the atmosphere and independence of their own groups, and sensed, correctly no doubt, that they would control less than one-half of any joint society. Nor did more than a handful of these trained women leaders ever move on to hold major offices in the larger, main-line professional associations in their field.

Yet the purpose of these groups and prizes was not merely social or therapeutic. Over the years a few groups took on a modest political role within their field or parent professional organization. Usually the women's goal was the limited one of gaining some recognition or at least acknowledgment of their presence from their male colleagues—some indication that women were a part of the field and that not all of the women at meetings were the wives of scientists. These groups thus did not challenge the basic patterns of discriminatory employment that underlay their marginal status, but chipped gingerly away at only the most flagrant kinds of professional exclusion; they petitioned for admission to certain public health schools, requested admission to dinners at all-male clubs, suggested more women be added to a society's standing committees, and arranged for favorable publicity of women's achievements. These gentle reminders that women should be included in professional activities were typical of the cautious politics that followed the collapse of the more confrontational feminist tactics of the early 1920s. Yet it was also symptomatic of the times that even these limited and at best merely palliative objectives were controversial among the women, many of whom feared that such tactics would upset powerful men and thus backfire upon all women scientists. To many, preserving the harmony had become more important than fighting for greater equality.

The ten women's scientific societies of the 1920s and 1930s discussed briefly here were predominantly of two types: the university-based "honor fraternities"* for undergraduates and graduates, and the independent or professionally affiliated "para-scientific" women's groups in specific fields. Undergraduate and graduate "honor societies" and "honor fraternities" began to proliferate at the nation's land-grant universities around 1900, when it became clear to the faculty there that not all the students were putting academic or intellectual pursuits foremost. In an attempt to stem this tide of antiintellectualism, many universities established "honor" groups (with a deliberately restrictive membership) in order to recognize and reward those who did achieve academic excellence and thus to encourage others to emulate them.[4] Some of these societies, like Phi Beta Kappa and Sigma Xi, admitted both sexes and covered several fields, but others had all-male memberships and were restricted to particular fields, such as engineering, pharmacy, or journalism. Recognizing the value of these groups, the stronger women faculty at these institutions (almost always in home economics) felt it necessary to create additional honor societies for the women students and put considerable energy into sustaining them.

The stated purpose of these "honor fraternities" was usually the development of "leadership skills" and a "professional attitude," which probably meant responsibility, independence, and a positive outlook. Although such training and faculty attention may have proven helpful to all members, the contacts made and guidance received must have been especially useful to those students embarking

*These women eschewed the word *sorority*, apparently for its frivolous connotations.

on careers in home economics. Most of the burden of this formal and informal socialization necessarily fell on the faculty who served as mentors and role models to the younger women by presiding at local and national meetings, by describing their own careers in the society's bulletin, or by writing inspiring essays on such topics as "What Home Economics Means to Me" or "Reflections on Omicron." Because these groups were not challenging any male domain, their goals hardly constituted radical feminism. Yet they were clearly committed to excellence within the realm of "women's work"—to improving the training of future home economists and dignifying their role in society.

The first such women's honor fraternity in home economics was Phi Upsilon Omicron, established in 1909 at the University of Minnesota. By 1938 it had twenty-five active chapters, several alumnae ones, and a semiannual journal.[5] But it was Omicron Nu, established at Michigan State University in 1912 by Maude Gilchrist, a former botanist but now dean of the college of home economics, that was the most powerful group in the 1920s and 1930s. By 1940 it had thirty-two chapters, mostly at the large midwestern land-grant institutions, a membership of about 10,000, and a quarterly bulletin.[6] Its activities must have been similar to those Agnes Fay Morgan arranged for Alpha Nu, an independent group she established at the University of California in 1915. There the students did useful service projects around the school of home economics, such as running the annual spring picnic, fixing toys for the nursery school, occasionally setting up an exhibit, planning a course curriculum, or putting out an alumnae directory.[7] Through such extracurricular activities, the top women students, who were probably planning to be wives and schoolteachers, could develop their confidence and character. Although one suspects that the more scientific among them might have been better off taking another laboratory course, these home economics clubs provided the kind of supportive atmosphere or subculture a woman student would probably not find in the science departments on campus.

In addition to these three primarily undergraduate "honor fraternities," there were others that either emphasized their alumnae chapters or limited their membership to graduate students. The goals of these groups stretched beyond the character development of undergraduates to the larger and more political role of "mutual advancement" and "higher professional standards." The primary activity of these graduate fraternities for women was still at the local level, where, through an annual round of get-acquainted teas in the fall, monthly talks by invited speakers, and awards banquets in the spring, they helped nearby women scientists meet and support each other and participate in meaningful projects outside the main professional societies.

In addition to this round of local activities, most of these graduate fraternities also developed various national projects, which over the years served to build unity and sustain morale. Some established a newsletter, and a few even supported a clubhouse for the members' convenience. Many raised money both for scholarships for younger women just entering science and for grants-in-aid for researchers who lacked adequate support, and then took great pride in following their protégées' subsequent achievements. The largest groups (those with thousands of members) had national officers and devoted great energy to holding a national convention every few years. In addition, almost all of these fraternities elected a pantheon of particularly prominent women scientists as honorary members. Madame Curie was a perennial favorite, but other successful and already greatly honored women, such as Florence Sabin, Florence Seibert, Alice Evans, and Ruth Benedict shared this role. Although these women were themselves rarely active in these organizations, their acceptance of the honor was an inspiration to those who were members—a reminder that a few women at least could scale the heights, and, perhaps equally important, a reassurance that once there they had not forgotten or abandoned those less favored.

The first of these groups was Iota Sigma Pi, a national honor society for women chemists, which grew out of chapters established at the University of California at Berkeley in 1900 and (by Agnes Fay Morgan) at the University of Washington in 1911. By 1922 Iota Sigma Pi had eleven other chapters (named for the elements in the periodic table rather than letters of the Greek alphabet) with 585 members, and by 1937 it had twenty-one chapters with over 2,000 members. Its alumnae were particularly active, and, if the four purposes of "Oxygen," the chapter at the University of Washington, are a valid indication, the fraternity's founders may have had women's postgraduate needs specifically in mind. The chapter's purposes were, according to Agnes Fay Morgan, not only "(1) To promote interest and enthusiasm among women students in chemistry;" but also, "(2) To foster mutual advancement in academic, business and social life; (3) To stimulate personal accomplishment in chemical fields; (4) To be, by our earnest endeavor, an example of practical efficiency among women workers."[8]

In the 1930s Iota's national leadership began to forge out a modest political role for itself within the profession by supporting a women's luncheon or breakfast at the annual meeting of the American Chemical Society, and in the late 1930s its secretary, famed nutritionist Icie Macy Hoobler of Detroit (who had helped establish chapters at the University of Colorado in 1918 and Yale University in 1920), ran a placement service for women chemists that was reportedly busy and successful. At the same time, the leaders of Iota Sigma Pi rejected an offer to merge with Phi Lambda Upsilon, an honor society for male chemists, on the grounds that they wished to "safeguard the separate existence of Iota Sigma Pi." The association has continued to grow and is still active today.[9]

The broadest in its membership and most visible in its actions of these graduate "fraternities for women" was Sigma Delta Epsilon, in some ways the women's equivalent of the American Association for the Advancement of Science (AAAS). Sigma Delta Epsilon was founded at Cornell University in May 1921, the same month that Madame Curie visited the United States, but the society's historian considers the catalyst to its formation to have been certain local, though common, problems facing graduate women at Cornell.

> The life of a woman graduate student at Cornell University in the early 1920s was anything but enviable. Not only did the graduate student not fit in with the undergraduate activities, but she was forced to live off campus with isolated families. Thus there was little social life, little opportunity for exchange of ideas so vital to wholesome living. One

> of the women graduate students, Adele Lewis Grant, recognized the great good the men graduate students derived from their scientific fraternity, Gamma Alpha, and decided that a similar fraternity for women would be equally fine. With the help of a few collaborators in the field, she succeeded in rounding up most of the isolated women in research in the various departments at an informal picnic held on May 14, 1921. There were twelve persons present.[10]

These twelve charter members then rented a house near campus, where they roomed together. As they moved on to other universities, they opened new chapters, which by 1940 numbered fourteen with 2,200 members.

From its start Sigma Delta Epsilon took a strong interest in both the recognition awarded women scientists and some affiliation with the AAAS. Its leaders devoted considerable effort over the years to finding out who the achieving women scientists in this nation and abroad were and in selecting them for honorary memberships. The members of SDE also contributed to the Ellen Richards Research Prize (see the following), sent a delegate to the annual meeting of its selection committee, and spoke up for the interests of young women scientists when the committee reformulated the prize to go to older women. In addition, Sigma Delta Epsilon held a luncheon at the annual meeting of the AAAS starting in 1922. Shortly thereafter it petitioned to become an official AAAS affiliate, but this request was not granted until 1936, when the society was finally deemed to have a sufficiently large membership. In the 1970s it changed its name to Graduate Women in Science and was still the only women's group affiliated with the AAAS. It was proud to have enough members who were AAAS fellows to allow it to send two delegates to the annual meeting of the council.[11]

A third and initially more ambitious but in the end unsuccessful honor fraternity for women was Kappa Mu Sigma, a society for graduate students in chemistry, founded at the University of Chicago in 1920. Section two of its constitution reveals that its members were eager to play an even more active role in the sexual politics of science than Iota or Sigma and were venturing toward the kind of "upgrading" and demand for "standards" that was becoming common in the 1920s: "The aim of the Society shall be to raise the standards of professional chemistry among women by insisting on the importance of complete (graduate) chemical training for a professional career. This is to be accomplished by making active membership signify graduate preparation for a career in chemistry. By union of effort along such lines, we hope to promote both professional and social cooperation among women in the chemical profession."[12] Its founders (some of whom may have been from Mount Holyoke College) evidently shared the view of Professor Emma Perry Carr and other women chemists of the early 1920s that top-notch training, strong motivation, and stoic persistence were the only means to the full acceptance of women in the chemical profession (see chapter 9).

Unlike the groups discussed previously, however, Kappa Mu Sigma went beyond this standard self-improvement and self-discipline tactic in the late 1920s to protest at least one continuing kind of discrimination against women in chemistry. In 1927 the Columbia University chapter took on the New York City section of the American Chemical Society for its continued exclusion of women members from its dinners, which were held at the men-only Chemists Club. Kappa firebrand Louise Giblin sent a questionnaire to the forty-five women members of the section whose current addresses were known asking if they wished to have women admitted to the dinners and, further, if they were admitted, whether they would attend. Twenty-five responded immediately, agreeing enthusiastically that it was time that the men-only policy was changed. Although Giblin then reported the results of her survey, supported with many quotations from the questionnaires, to the male leaders of the section, it is unclear whether she achieved her objective.[13] Nor is anything known about the group's subsequent activities. Although it seemed to be flourishing in 1927, with a membership of 180 in two chapters at Columbia and the University of Chicago, many of whom went on to major academic positions in biochemistry and nutrition,[14] the fraternity vanished. It may be significant, though with the dearth of information one cannot be sure, that the one even slightly confrontational fraternity in the 1920s dropped out of sight shortly after taking on the American Chemical Society.

These six fraternities represented one kind of scientific society for women in the 1920s and 1930s; a second kind organized itself around a particular field or professional association and might therefore be termed a "parascientific" society. Although limited to specialists in one field, like some of the graduate fraternities, these groups dispensed with the chapter structure and prescribed rituals of pins, initiations, "grand councils," and the rest that meant so much to the fraternities and belied their campus origin. Nor did any of the three groups considered here run scientific meetings, publish a professional journal, or try to use their restricted membership to control a specialty (child welfare, for example), as had the all-male Society of Experimental Psychologists discussed in chapter 10. Instead, the chief aim of these "parascientific" organizations, whose members were marginal in their fields, was the usual one of helping other women adapt to the whole exclusionary employment and reward systems as best they could.

Once again the first step in this direction was primarily social—holding a separate women's breakfast or luncheon at the annual meetings of the major scientific associations. This might help to build professional camaraderie among the growing numbers of women, especially young ones, entering the field. But what happened next varied greatly from field to field, generally in response to the enthusiasm and political astuteness of the leaders and the extent of the members' goals. For some, as in geography, a vigorous separate society was the result; for others, as in public health, a weaker women's group barely outlasted its leader and founder; and for still others, as in chemistry, a muted and cautious women's committee was deemed suitable. In still other fields, most noticeably geology, the chance to form a women's group led to little more than acute embarrassment all around.

The earliest of these "parascientific" societies was the Association of Women in Public Health, which was formed in June 1920 by about one hundred women to "develop a sense of comradeship" among the many women entering the field after World War I (as discussed in chapter 8).[15] One of the group's founders later recalled that she had urged formation of the group when she saw that so many of "our women roamed like lost sheep with no opportunity for contact with other women of similar interests" at the national public health meetings.[16] In 1921 Gertrude Seymour of the U.S. Public Health Service in New York City, the association's first president, started its *Bulletin,* which she continued to publish at irregular intervals for five years, and in 1922 the group held its first formal luncheon at the annual meeting of the American Public Health Association. If this group was moderately successful at introducing women in public health to each other, it was much less so in its second initial purpose, that of trying "to raise their professional status." Its chief project in this direction was a series of attempts in the mid-1920s to induce the major schools of public health to award women degrees. When these petitions proved unsuccessful, the group voted in 1928 to give up such political goals and to limit itself in the future to social functions. The association was still holding its luncheons as late as 1943, but had disbanded by 1947.[17]

A second such group that has by contrast been successful and meaningful to its members is the quasi-professional Society of Women Geographers. It was established in 1925 by a hardy group of women travelers, geographers, and explorers who had been rebuffed by the all-male Explorers' Club of New York City. Led by Harriet Adams, Elsie Grosvenor Bell, and Blair Niles, the women immediately formed their own separate group, which has gone on quite contentedly ever since. Membership is international and, cutting across several specialties, includes anthropologists, geologists, naturalists, and independent travelers who delight in telling each other of past and future travels. One of the group's traditions is the carrying of its flag to distant places, as to the North Pole (by Marie Peary Stafford), New Guinea (Margaret Mead), under the ocean in a bathysphere (Gloria Hollister), and around the world (or almost, by Amelia Earhart on her final flight). The Women Geographers have a national headquarters in Washington, D.C., and informal area groups (rather than formal chapters) in larger cities across the nation. Although the SWG has met on occasion in New York City with the Explorers' Club, its members seem to enjoy greatly their independent status. They have held teas at meetings of the Association of American Geographers, the main professional society in that field, but were not concerned with reforming that group and securing the election of women officers. As one member put it in 1950, on the society's twenty-fifth anniversary, "The purpose was not grandiose, it was not to promote scientific knowledge of geography, but, more modestly, it was to be a society for the interchange of experiences. As Blair Niles has said, 'To my mind even more important was the comradeship which the organization would give to women inevitably lonely because of the unusual nature of their interests and ambitions.' "[18]

The growing number of women members of the American Chemical Society in the 1920s led to the formation of a women's committee within that association in 1927. Yet even that modest step was taken very cautiously for fear of inviting criticism. Although there had been only five women in the ACS before 1900, so many joined in the early 1920s that by 1924 the number had jumped to almost 500, 60 percent of whom had joined since 1921.[19] Still only 3.2 percent of the ACS membership in 1924, this was a large enough group for informal women's luncheons to start in the mid-1920s. At one in 1926 Glenola Behling Rose, a chemist at the Du Pont Company in Delaware since 1918, was chosen to start some more formal kind of women's organization. She and a committee of five decided not to form a separate association (perhaps like that of the women geographers), whose purpose might be misinterpreted, but just to establish a standing committee within the ACS. After some discussion on how to present this idea to the all-male governing council, Rose's committee settled for the "altruistic" phrase "Women's Service Committee," whose purpose would be to increase women's membership in the ACS. Although greatly relieved when the ACS accepted this motion, the committee's members had few ideas on how to proceed. One suggested that their future plans be "not too aggressive—just enough to keep the ball rolling."[20] Accordingly they devoted most of their energies to setting up luncheons at the ACS's semiannual meetings. Otherwise their sole accomplishment before 1935 was taking a poll of the ACS section leaders to ascertain what role women were playing in local affairs and then to suggest that more women be added to ACS committees. As the depression dragged on and women chemists (including Chairman Rose) were laid off or could not find jobs, the committee was so overwhelmed with requests for advice that, though its members felt helpless, it held symposia and published articles offering the conservative vocational guidance discussed in chapter 9. "Women's work" was the women's committee's only prescription for survival.

Yet the activities of the chemists' Women's Service Committee looked quite daring when compared with those of the women in the Geological Society of America, who as late as 1937 did not want to form a women's group of any sort. The opportunity arose that year at the fiftieth anniversary of the GSA, when a special luncheon for women geologists was held to honor the seventy-five-year-old Florence Bascom. The occasion turned out to be an acutely uncomfortable affair, largely because Bascom was displeased with the all-female attendance. (She, who had joined the Society of Women Geographers only reluctantly in 1930, remarked scornfully at the luncheon that women geologists had little more in common than did blue-eyed geologists.) Not even the several women journalists present could unearth much support for a women's group in geology.[21] Thus, rather than thinking that they had anything to gain, either socially or politically, by meeting together, the women geologists wished to deny that they had, aside from an acquaintance with Bascom (who had trained many of them), any reason for meeting. Similarly derisive views must have prevailed in the many other fields that did not have a women's group in the 1920s and 1930s.

Yet the lack of a women's organization in a science was not necessarily a sign that all was harmonious there. In fact, the women in such fields could be in great psychological danger. It is perhaps significant that the two best-known mental breakdowns among women scientists in these years took place among women who were outside even the modest protective subculture that a women's organization could provide. Both paleontologist Winifred Goldring, who recovered well, and mathematician Olive Hazlett, who never did, were in fields that lacked such organizations, though they might not have joined them had they been available. Isolated and moderately successful but with aspirations of full equality, they denied the potential psychological dangers in their situation and preferred to "make it" on their own. Then when the obstacles they encountered proved too much for them, they were left unprotected.

This topic, the mental health of women scientists, raises the larger question, which the histories of these organizations generally fail to address and which would have been interesting to pursue, of just how helpful these clubs were to their members or what membership meant to them. Many women found these groups, especially the honor fraternities, sufficiently meaningful and rewarding to merit their belonging for decades. Untold others, however, never joined or belonged only briefly. It is unclear what effect these groups had in alleviating, individually or en masse, what must have been major morale problems, such as isolation and depression, before they snowballed into nervous breakdowns and (possibly) suicides. Little is known of this side of the women's (or men's) professional life, but one suspects that there was some correlation between membership in such a group and mental health. The joiners were at least aware of the deficiencies of their situation and were already taking steps either to adjust or to protest them as best they could. The others, the nonjoiners, were probably in the greater danger, but they refused to join such groups, typically scorning them as second-class organizations to which only weak women would belong.

In addition to the many women's groups that appeared in various forms in the 1920s and 1930s, a second kind of "compensatory recognition" developed in these years—separate women's prizes in science. In fact this tactic was the women's most sophisticated one of the time, because it required a group or individual to move beyond mere sociability, controversial and awkward as even that could be, and onto the semipolitical realm of creating and awarding recognition. Although such prizes might seem harmless enough, since ostensibly they merely rewarded those women who were, despite a lifetime of accomplishments, not being recognized by their main professional associations, they served some larger political purposes as well. By making the top women in the field more visible, these prizes increased the pressure on the main organizations to notice them and acknowledge their presence and contributions. There was also an element of psychological warfare in such awards, for even if the men ignored the import of the women's prize and continued to insist publicly that there were no women of consequence in the field, the growing list of women winners would make it increasingly difficult for subsequent skeptics to repeat the error. In addition, these prizes and their attendant publicity offered the women's supporters a discreet way to present their accomplishments, which were still often minimized both within and outside science, in a positive light to the broader nonscientific public. An appreciation of the value of such public relations or image-building in science was one lasting lesson from Madame Curie's visit to the United States in 1921.

To a large extent the women's prizes of the 1930s were outgrowths of the feminists' earlier strategy of fellowships or "creative philanthropy." The longest-lived prize for a woman in science was the $1,000 Ellen Richards Prize, given since 1901 by the Association to Aid Women in Science (formerly the Naples Table Association, discussed in chapter 2) for the best piece of experimental work submitted by a woman in science. Although the association had awarded this prize several times by 1921, it had often been dissatisfied with all its candidates and had withheld the prize on the grounds that none of the works submitted was strong enough.[22] Then in 1921 its members were so caught up in the "Curie mania" that they increased their award to $2,000 and made a special presentation to Madame Curie on her triumphal tour. This experience increased the association's smoldering dissatisfaction with its usual candidates, recent doctorates. Why should it continue to honor such beginners for one piece of good work, difficult as that was to assess, when it could be rewarding the best of the older women, many of whom were now ending long careers in science and were still unrecognized by their professional colleagues? In 1925 the association undertook a comprehensive survey of its own past performance. Rather than reveling in its own high standards when it found that it had awarded its prize only six of the fourteen times that it had been offered, or that only 3.3 percent of its 184 candidates had been winners, the group turned upon its former applicants, blaming them for submitting such undistinguished work.[23]

The association then voted to do what it wanted in any case—upgrade the award into a woman's Nobel Prize. It based its new rules intentionally but loosely on those of the Nobel Foundation in Stockholm: candidates must have made important contributions to science and be nominated by responsible third parties, such as, for example, their country's association of university women. The first award under these new rules went in 1928 jointly to physicist Lise Meitner of the University of Berlin and chemist Pauline Ramart-Lucas of the University of Paris. When the leaders of Sigma Delta Epsilon protested on behalf of the disqualified younger women that the prize, which had been awarded so erratically before, was now no longer going to them at all, the association did not change its thinking.[24] It did respond to some other criticism, however. Chastised by many for not awarding the prize to American astronomer Annie Jump Cannon, who had received more nominations than any other candidate, the association broadened its definition of "laboratory" or "experimental" science to include astronomical classification, and in 1932 split its second award between Cannon and zoologist Helen Dean King of the Wistar Institute (whom they had rejected twice before for their earlier award).

And yet, just as the association seemed to have established a highly visible role for itself in awarding an important international prize, its remaining members decided that the time had come to dissolve the organization. Many of the original members had died, others were over seventy years old and in ill health, and financial supporters had been withdrawing as the group's purpose had wa-

PLATE 34. Shown in this rare 1928 photo of the Naples Table Association, which long aided women in biological research, are, left to right, front: Louise Lilley Howe, Mrs. Samuel F. Clarke, Virginia Gildersleeve (dean of Barnard), Dr. Lillian Welsh, and Florence R. Sabin. Rear: Mary F. Calkins, Mary W. Woolley (president of Mount Holyoke), Ellen F. Pendleton (president of Wellesley), Mrs. John W. Blodgett, H. Jean Crawford, and R. Louise Fitch (dean of Cornell). (New York *Herald Tribune*, 28 April 1928, p. 18.)

vered (see plate 34). Nor had the leaders made much attempt to take in those younger women scientists who were eager to help advance women's careers in science, such as Agnes Fay Morgan, Icie Macy Hoobler, Florence Seibert, Alice Evans, or perhaps some of the new NRC fellows. Instead, the association stuck with its traditional friends, the deans of the women's colleges, including Virginia Gildersleeve of Barnard, Ada Comstock of Radcliffe, and Marjorie Hope Nicolson of Smith, even when these women were at best only lukewarm feminists, had little enthusiasm for the project, and had far less contact with prominent women scientists than had their predecessors. Finally, in 1932, the association voted to disband, though rather than admitting their personal weariness, waning feminism, or failure to seek out suitable successors, the members claimed that women's progress in science had so improved of late that the prize was no longer necessary: "*Whereas,* the objects for which the Association has worked for thirty-five years have been achieved, since women are given opportunities to engage in Scientific Research on an equality with men, and to gain recognition for their achievements, be it *Resolved,* that this Association cease to exist after the adjournment of this meeting."[25]

Even though this pioneer association had thus pronounced women's problems in science solved in 1932, Annie Jump Cannon, the recipient of one-half of its final prize, was not so sure that full recognition had arrived for women in astronomy. A year later she revived the tradition in an important new format, when she contributed her share of the last Ellen Richards Prize ($1,000) to the council of the American Astronomical Society for a triennial award for distinguished contributions to astronomy by a woman of any nationality. Perhaps because she had never won an award from the AAS or been elected its president (she was treasurer from 1912 to 1919), she wanted more recognition for younger women. The administration of this novel award apparently presented no difficulties to the AAS, which, starting with Cecilia Payne-Gaposchkin in 1934 and Charlotte Moore Sitterly in 1937, awarded the Annie Jump Cannon Prize every three years until the 1970s.[26]

Even more illustrative of the sexual politics behind the new women's prizes of the 1930s was the establishment in 1935 of the Francis P. Garvan Medal for a distinguished woman chemist, in time the best-known prize for a woman in science. By 1935 the number of women in the American Chemical Society, as mentioned earlier, had been increasing for over a decade, and a Women's Service Committee, with primarily social functions, had been established. Matters might have rested there for some time to come had not in 1935 Francis P. Garvan, a wealthy patent lawyer who directed the Chemical Foundation, been overheard joking to a friend that there were still no women chemists. Some committee members immediately challenged him upon this. He admitted his mistake and then, agreeing that others probably shared his misconception, came to their meeting and offered to publicize their achievements by endowing a prize for outstanding work by a woman chemist. The ACS did not entrust the award selection to the Women's Service Committee, however, although that committee's suggestions were sought, but set up a separate committee of both men and women to formulate the rules for the new prize. Some members suggested that it be called the "Madame Curie Medal," which would have been appropriate for a woman chemist's award, but the society settled upon a gold Francis P. Garvan Medal and $1,000 (later raised to $2,000). This prize has since honored a long series of women chemists and biochemists, starting with Emma Perry Carr of Mount Holyoke College in 1937, Mary E. Pennington in 1940, Florence B. Seibert in 1942, and Icie Macy Hoobler in 1946.[27] (Carr's selection in 1936 was doubly ironic—she was on the selection committee herself at the time and voted for someone else; and ten years earlier she had recommended to the fledgling Women's Service Committee that there was no need to set up awards for women; the important thing was for them to be doing good work in chemistry.)[28] One might well speculate whether these talented women or others in chemistry might have won any of the numerous other prizes granted annually by the ACS if the Garvan Medal had not existed and served as a kind of political safety valve. It seems unlikely, since women were rarely included on the ACS program until the 1940s (and seldom even then),[29] and no woman won any other ACS prize until 1967 (and then for high school teaching rather than research). The Garvan Medal thus performed an important, though necessarily compensatory, function in recognizing and publicizing women's contributions to chemistry. Without it, the ACS would not have been recognizing any women at all.

In addition to the Ellen Richards Prize, the Annie Jump Cannon Prize, and the Francis P. Garvan Medal, women scientists were eligible for a variety of other

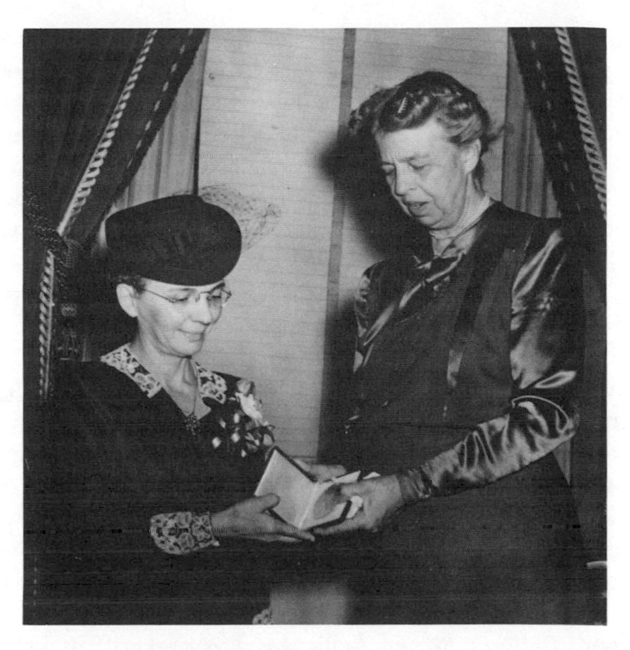

PLATE 35. Biochemist Florence Seibert of the Phipps Institute of the University of Pennsylvania won many awards for her outstanding work on tuberculin. Here she is being presented with the National Achievement Award by Eleanor Roosevelt at the White House in 1944. Such awards helped to publicize women's accomplishments in science. (Courtesy of Florence Seibert.)

more general awards. Two of the best publicized of these were the "Woman of the Year" Award from the *Ladies' Home Journal* and the National Achievement Award, which was established in 1930 by Maud Wood Park, a former suffragist who now sought new goals for women, and was administered by a committee headed by Eleanor Roosevelt and Beatrice Hinkle, a New York psychoanalyst. Women in the medical sciences or others whose work was widely known by the public seem to have had an advantage in the selection for these general awards, since among the early winners of the National Achievement Award were Florence Sabin, Alice Hamilton, Florence Seibert (plate 35), and Margaret Mead. (Other recipients were actress Katherine Cornell, suffragist Carrie Chapman Catt, and politician Madame Chiang Kai Shek.)[30] In 1943 the AAUW established another such prize for women with its annual Achievement Award, which is still one of the highest honors regularly given to an American woman scientist or scholar. Florence Seibert, Katharine B. Blodgett, Ruth Benedict, and Barbara McClintock were among the early winners.[31] In addition, many of the women's colleges granted honorary degrees to women scientists. Winifred Goldring, who did not have a doctorate, was especially excited about the honorary doctorate of science she received from Russell Sage College in 1937, the same year that Mount Holyoke College honored several women scientists at its centennial:

botanist Margaret Ferguson, mathematician Anna Pell Wheeler, and nutritionist and college president Katherine Blunt. Margaret Ferguson's letter of acceptance to President Mary Woolley is typical of the genre: it professes great humility and desire to do even more in the future, and in this case, also reveals the strong hold that even rival women's colleges still maintained on some women scientists in the late 1930s: "It is with a deep sense of unworthiness that I accept so signal an honor. There is no other College in America whose honorary degree I should cherish as that of Mount Holyoke's, with her devotion to truth and her high standards of living and of scholarship. This honor will ever be a stimulus to greater endeavor in my chosen field of science."[32]

In pondering the significance of these general awards, one wonders who won them, who noticed, and why were they given? Since most of the winners of these general prizes were the visible women who had already won other awards in their own field, the purpose of these prizes was not to inspire a young winner to greater efforts, as might a fellowship, though most winners, like Margaret Ferguson, promised to redouble their efforts to be worthy of the award. Even if such women felt discouraged in their own way, they were the women scientists least in need of encouragement. Anyone, however, could use the extra money that often accompanied such awards. Nor were these necessarily trifling sums. Florence Sabin's *Pictorial Review* Achievement Award in 1930 for the American woman who "has made the most distinctive contribution to American life," for example, was accompanied by a check for $5,000, a princely sum at the time.[33] Some of the few women who won most of these general awards greatly enjoyed the many degrees and honors that came their way, though typically they later protested that they had had more than their share of prizes. Others, like Florence Sabin, however, apparently tired of the commotion associated with winning too many nonscientific awards and had a large correspondence turning them down and suggesting the names of younger, less visible women, like Florence Seibert, who would appreciate the recognition much more.[34] From Sabin's point of view (and she was after all until 1931 the only woman member of the National Academy of Sciences), receiving most awards was a lot of extra bother: she would have to drop her experiments, buy a new dress, and perhaps write a talk, all of which she would rather not do. Typically she would rather spend the time in her laboratory.

But who was the audience for these awards? Who was supposed to notice and care? In the case of the Cannon Prize or the Garvan Medal, the audience was deliberately the other, mostly male, members of the society granting the prize. But this can hardly have been the case for those awards established by magazines or other organizations of nonscientists. Their audience was the general public, particularly women, a fact that suggests that the basic purpose behind these many awards was the quasi-political one of publicizing favorably both the organization giving the award and its cause, as, here, the achievements of women. Such a tactic might impress skeptics of women's proven abilities and achievements in science, inform young women and their parents and teachers of the careers and opportunities now available to women in science, and perhaps even inspire some younger women to emulate the winner and study science. In fact, this indirect approach of using favorable publicity to modify traditional stereotypes was fast becoming a basic technique in the newly emerging field of public relations, which, whether they were aware of it or not, was what many of the women's organizations were beginning to engage in during the 1920s and 1930s.

In the 1920s the new field of "public relations" emerged from traditional fund-raising (including that by celebrities like Madame Curie) and moved on to the more subtle art of shaping "public opinion" on such topics as political candidates and quasi-political issues. An effective way to do this was to manipulate the media into blurring the former distinction between "news" and "advertising." One technique involved what one expert called the "overt act" but that others have since termed the "pseudo-event": the staging of one's own "news," which then received much free publicity for one's self or cause. This meant, for example, that a candidate for mayor of a large city might arrange for representatives of one of its large ethnic groups, as perhaps Irish-Americans, to present him with a "brotherhood" award. This presentation would receive free more favorable and more credible publicity than he could buy or attract with his own campaign funds or speeches. In this way the candidate might reach those people who might otherwise have dismissed or overlooked him, induce them to think well of him, and perhaps even influence them to vote for him. Similarly, but on a slightly higher plane, such award-granting ceremonies could help to educate the public to appreciate or accept a certain new view, as that anti-Semitism was no longer respectable or that a college education was desirable. It was a technique that interest groups from universities to labor unions to professional associations and beyond were learning to use in the 1920s.[35] No one was going to make major political advances in this way (the confrontational approach had failed), but the tactic might break down traditional stereotypes and broaden public opinion of what were, for example, now acceptable activities for women. But ingenious as this approach was, women's groups would not have been staging such separate ceremonies (and might not have existed at all) if their members had been getting adequate recognition from the traditional organizations in their fields. Such "compensatory recognition" was, unfortunately, necessary if there was to be any recognition of women at all.

Thus, several women's organizations and a few women's prizes emerged in the 1920s and 1930s to help remedy what were felt to be serious deficiencies in most women's careers in science. On a local, personal level there were needs for companionship and psychological support through what were in all likelihood demoralizing circumstances. On a national, more political plane, the women scientists began to want some acknowledgment or appreciation of their presence and some recognition or publicity for their accomplishments. When the main-line professional organizations did not respond to these needs, some women (and lawyer Francis P. Garvan) created alternatives. Few of these women's groups aspired to more than sociability or "mutual advancement" in the 1920s and 1930s, although the three women's prizes of the period did imply some mild political protest. These gentle reminders that there were women in science after all were the women's strongest political protest after the collapse of their more angry, confrontational tactics of the early 1920s.

Yet it is debatable how effective even these clubs and prizes could be. Although they made it more comfortable to be a woman in certain fields or professional societies, they certainly could not make up for all the exclusion and underepresentation from the main organizations described in chapter 10. Nor did they make it any easier for women scientists either to interact with the men in their field[36] or to educate them to any noticeable extent as to the women's contributions to science. These organizations thus were at best only partial palliatives to a segregated system, and the women remained marginal in their fields. Unable to change science or many scientists, they could only try to make their small enclave within it a bit more humane.

CONCLUSION

By 1940, women had been participating in American science for over a century, though in a variety of carefully circumscribed and camouflaged ways. Even this limited involvement had required the struggles and strategies of a great many committed women and men. The essential structure of women's place in American science had been set by 1910; thereafter sexual segregation not only persisted but even spread into other newly emerging areas of science. Protest movements were unable to change this pattern in any significant way, and war and depression only strengthened it.

At first, before Emma Willard's pioneering work in the 1820s, the lack of a secondary education for women seemed like the major stumbling block to their equality, and the hope was that once women were educated, they would be accepted as men's intellectual equals. Yet the rapid development of secondary and then higher education for women in the nineteenth century came only as the result of a shrewd political and intellectual compromise with the prevailing antifeminism: women might be educated, critics acquiesced, but only if it was for motherhood, their basic role in American society. This rationale was so swiftly and widely accepted that by midcentury America led the world in the amount of education it offered its women, and by the 1880s and 1890s educated women, less constrained by motherhood than formerly, began to encroach upon men's former monopoly of the nation's intellectual life. To head off this impending feminization of science, new barriers and restrictions, unnecessary earlier, quickly arose.

Among these new barriers were the introduction and then the requirement of a doctoral degree for faculty positions at universities. Since these degrees had long been restricted to men (by both custom and practice in the United States and by law as well in Germany), women gained them only after decades of struggle marked by their distinctive strategies of patient "infiltration" of friendly departments and of skillful use of "coercive philanthropy" elsewhere. Claiming that since women were now men's equals in intellectual matters it was "only fair" that they be given the same degrees as men, the women's leaders had by 1910 convinced most (but not all) university presidents to award women doctorates. Henceforth the number of such degrees going to women continued to grow, and the lack of doctoral training ceased, in most fields at least, to be the stumbling block it had been earlier.

Yet even these coveted degrees did not open full careers in science to women, largely because at the same time that the universities were showing themselves to be liberal educators, they were also proving to be highly discriminatory employers. Although the claim of sexual equality and "fairness" had proven quite effective in opening the graduate schools to women, these liberal notions did not win faculty positions for them. Accordingly, advocates of women scientists' employment had to fall back on the more conservative notion of their "special skills" and "unique talents" for positions that they termed "women's work." Jobs could be considered "womanly" if they were segregated either hierarchically (as in assisting men in tedious, anonymous, and low-paying tasks in scientific institutions like observatories, museums, and laboratories) or territorially, as by working on such "feminine" subject matter as home economics, botany, or child psychology. So widespread and tenacious was the acceptance of such jobs as "womanly" that few American women scientists, even those with doctorates, would ever be allowed to do anything else.

Also in the 1880s and 1890s, as women gained both doctorates and employment in science, another barrier, that of "professionalism" or "higher standards," arose, because many men interested in upgrading science into a full profession were afraid that the presence of women would lower their "prestige," a new concept linked to masculinity. Accordingly, they restricted in a variety of ways which women, if any, would be allowed to enter their new professional societies. For example, they limited them to lower-level ("associate") memberships, required higher qualifications (such as doctorates) of them, restricted them to only the most formal and public of the group's activities, or simply excluded them altogether from their organizations. The women's only recourse was to establish their own separate organizations, but even the largest of these, the Woman's National Science Club of Washington, D.C., could not fully substitute for the main professional organizations of the time.

Although acceptance of such patterns of "womanly" involvement in science had at first seemed the only way to convince a hostile public that women could indeed "do" science, in time it became clear that this partial and segregated acceptance had not proven the "entering wedge" to a broader range of employment and activities for which the women of the 1890s had hoped. Around 1910, as the limits to "women's work" seemed more and more unjust and as the larger "women's movement" began to gather political strength and to point to still other societal inequities, some women scientists, angry about their limited welcome in science, began to organize to protest more forcefully their exclusion. Yet the tactics and leverage available to such reformers were insufficient to compel institutions to change. For example, the women's chief tactic in academia between 1910 and 1924 was to prepare statistical reports on their unequal status. Although these may have twinged some consciences and induced a few men to hire or promote an occasional "exception," they led to no major changes in employment practices. The pattern of sex-segregated faculties was so firmly entrenched and educational institutions so autonomous that even organized pressure groups could not force them to change. In the 1920s and 1930s, most of academia, including even the women's colleges, was fixed instead on maximizing its status or prestige; hiring women, even "brilliant" ones, was seen as detrimental to this. Those women who were hired were restricted to marginal roles, and even they were endangered when in the late 1930s the coming of the "tenure track" began to eliminate those very long term instructorships and lifetime assistant professorships that constituted many women's niche in the hierarchy.

In government, there was a great expansion of women's employment at the federal, state, and local levels in the years after World War I, but again mostly in segregated ways. Although many women's groups had politicked for and gotten passed seemingly key Civil Service reforms (including equal pay) in the years 1919–24, these were rapidly undercut, and most of the expansion in women scientists' employment in government in the 1920s and 1930s was in the "feminine" fields of home economics, botany, microbiology, statistics, and clinical psychology.

Women scientists in industry before 1940 were so few that they knew that they were on their own. Unable (and unwilling) to mount any campaigns to challenge or change the major patterns of women's employment, they developed the tactic of alerting other women to the prevailing attitudes and advising them on how to adapt to them. At first they counseled others to be aware of sexual stereotypes so that they might overcome them and, by dint of excellent training and credentials, hard work, personal modesty, and stoicism, become "exceptions." When the depression reduced and shifted women's employment prospects in science, the best advice became the opposite—rather than try to defy sexual stereotypes, one should now accept them and enter such old and new kinds of "women's work" as that of the "chemical secretary," home economist, bacteriologist, or, as seemed likely only briefly, the "cosmetic chemist." Advocating such jobs, however, was, as the advisers admitted privately, more of a survival tactic in difficult times than a strategy for lasting change or reform.

Thus, by 1940, women in all areas of science had reached an impasse. They could be educated to the doctoral level but would encounter great restrictions on their employment. Political realities were such that the most that they could do was help each other adjust and accept prevailing patterns. One way to do this was by forming separate women's groups, several of which appeared in the 1920s and 1930s. Although the stated purpose of such groups was usually sociability or psychological self-help rather than any overt confrontation, some did move into the modest political role of "mutual advancement" within their field or put pressure upon their professional society for more recognition for women. The most daring and sophisticated approach of these difficult years, coinciding with the rise of public relations in science, was that of awarding a separate "women's prize" to a particularly accomplished scientist as a way to put some pressure on the main men's group to recognize some women's contributions or at least acknowledge their presence in the field. Needless to say, these groups were only a mild palliative for the women and the prizes a gentle prod on the men's consciousness; they were not designed or able to change the established structure of scientific employment in a sexist society. For that ever to happen, there would have to be other, more basic changes in American society, values, and government.

ABBREVIATIONS

AAUP	American Association of University Professors
AAUW	American Association of University Women
ACA	Association of Collegiate Alumnae
AMS	*A Biographical Dictionary of American Men of Science*. Edited by J. McKeen Cattell. Garrison, N.Y.: Science Press, 1906, and subsequent editions.
APS	American Philosophical Society
BAAUP	*Bulletin of the American Association of University Professors*
CBY	*Current Biography Yearbook*
GPO	Government Printing Office
HUA	Harvard University Archives
JAAUW	*Journal of the American Association of University Women*
JACA	*Journal of the Association of Collegiate Alumnae*
NAS	National Academy of Sciences
NAW	*Notable American Women*. Edited by Edward T. James, Janet Wilson James, and Paul S. Boyer. 3 vols. Cambridge: Harvard University Press, 1971.
NAW:MP	*Notable American Women: The Modern Period*. Edited by Barbara Sicherman and Carol Hurd Green. Cambridge: Harvard University Press, 1980.
NCAB	*The National Cyclopedia of American Biography*. vols. 1–53. New York: James T. White, 1898–1972. vol. 54. Clinton, N.J.: James T. White, 1973.
NYT	*New York Times*
PACA	*Publications of the Association of Collegiate Alumnae*
SLRC	Schlesinger Library, Radcliffe College
WB*B*	Women's Bureau *Bulletin*. 293 vols. Washington, D.C.: GPO, 1919-67.
WWE	*Women's Work and Education*

NOTES

CHAPTER 1

1. Linda K. Kerber, *Women of the Republic: Intellect and Ideology in Revolutionary America* (Chapel Hill: University of North Carolina Press, 1980), and idem, "The Republican Mother: Women and the Enlightenment—An American Perspective," *American Quarterly* 28(1976): 187–205; Mary Beth Norton, *Liberty's Daughters: The Revolutionary Experience of American Women, 1750–1800* (Boston: Little, Brown, 1980).

2. Ann D. Gordon, "The Young Ladies Academy of Philadelphia," in *Women of America: A History*, ed. Carol Ruth Berkin and Mary Beth Norton (Boston: Houghton Mifflin, 1979), pp. 68–91; Kerber, *Women of the Republic*, pp. 210–12, 221, and 228. In 1787 the prominent Philadelphia physician Benjamin Rush delivered a famous address at this school, in which he advocated a new kind of no-frills education for Republican women (Benjamin Rush, "Thoughts upon Female Education, Accommodated to the Present State of Society, Manners, and Government in the United States of America," reprinted in *Essays on Education in the Early Republic*, ed. Frederick Rudolph, John Harvard Library [Cambridge: Harvard University Press, Belknap Press, 1965], pp. 26–40).

3. Linda Kerber, "Daughters of Columbia: Educating Women for the Republic, 1787–1805," in *The Hofstadter Aegis: A Memorial*, ed. Stanley Elkins and Eric McKittrick (New York: Alfred A. Knopf, 1974), pp. 36–59; Eleanor Flexner, *Mary Wollstonecroft: A Biography* (New York: Coward, McCann & Geoghegan, 1972); [Benjamin Silliman], *Letters of Shahcoolen, A Hindu Philosopher, Residing in Philadelphia; To His Friend, El Hassan, An Inhabitant of Delhi*, with introduction by Ben Harris McClary (1802, reprint ed., Gainesville, Fla.: Scholars Facsimiles and Reprints, 1962).

4. "Colden" in *NAW;* Anna Murray Vail, "Jane Colden: An Early New York Botanist," *Torreya* 7 (1907): 21–34; James Britten, "Jane Colden and the Flora of New York," *Journal of Botany* 33 (1895): 12–15; Jane Colden, *Botanic Manuscript*, Harold W. Rickett and Elizabeth C. Hall, eds. (Garden Clubs of Orange and Dutchess Counties, N.Y., 1963); other early women agriculturalists and horticulturalists are discussed in Joan Hoff Wilson, "Dancing Dogs of the Colonial Period: Women Scientists," *Early American Literature* 7 (1973): 225–35, and "Eliza Lucas Pinckney" in *NAW*.

5. Advertisements for Charles W. Peale's lectures and museum [Philadelphia], *Aurora General Advertiser*, 26 Sept., 3, 8, 16, 23 Oct., and 6 Nov. 1799, and *Poulson's American Daily Advertiser*, also of Philadelphia, 2 Jan. 1801. (I thank Toby Appel of the Charles Wilson Peale Papers, Smithsonian Institution, for sending copies of these items.) See also Toby Appel, "Science, Popular Culture, and Profit: Peale's Philadelphia Museum," *Journal of the Society for the Bibliography of Natural History* 9 (1980): 624. Charles Coleman Sellers, *Mr. Peale's Museum: Charles Willson Peale and the First Popular Museum of Natural Science* (New York: W. W. Norton, 1980), however, does not mention the innovation. For later examples, see Wyndham D. Miles, "Public Lectures on Chemistry in the U.S.," *Ambix* 15 (1968): 129–53; John C. Greene, "Science and the Public in the Age of Jefferson," *Isis* 49 (1958): 20 and 21; Margaret W. Rossiter, "Benjamin Silliman and the Lowell Institute: The Popularization of Science in Nineteenth-Century America," *New England Quarterly* 44 (1971): 612; Carl Bode, *The American Lyceum* (New York: Oxford University Press, 1956). The *Dictionary Catalog of the Manuscript Division [of the New York Public Library]* (Boston: G. K. Hall, 1967) claims that the Maria Trumbull Silliman Church Papers mention the attendance of women at

Professor Silliman's Lowell Lectures in Boston. I was not able to find this reference, but Silliman did say in a letter to her father of 26 March 1842 that she was so interested in a geology book he had recently sent her, if she were a man she would have wanted to become a geologist.

6. Ann B. Shteir of York University, who is preparing a full biography of Wakefield, has examined eighteenth-century women's interest in botany in " 'With Bliss Botanic': Women and Plant Sexuality" (Paper presented at session on "Women and Gardens in Theory and Practice," Annual Meeting of the American Society for Eighteenth-Century Studies, San Francisco, April 1980) and "Women and Plants, A Fruitful Topic," *Atlantis*, in press, which also includes several 19th-century Canadian women botanists. (I thank Sally Kohlstedt for telling me of her work and Professor Shteir for sending copies.)

7. Editions are taken from the Library of Congress's *National Union Catalog of Pre-1956 Imprints*. Astronomy also had a certain vogue with Montgomery Robert Bartlett, who wrote *Young Ladies' Astronomy* (Utica, N.Y.: Colwell & Wilson, 1825), and Denison Olmsted, who wrote *Letters on Astronomy: Addressed to a Lady* (Boston: Marsh, Capen, Lyon & Webb, 1840). See also Greene, "Science and the Public," pp. 16 and 17–18; Wyndham D. Miles, "Books on Chemistry Printed in the United States, 1755–1900: A Study of Their Origin," *Library Chronicle* 18 (1951–52): 51–62; idem, "America's First Chemistry Syllabus-and-Course for Girls," *School Science and Mathematics* 58 (1958): 111–18; Marcet and Wakefield in *The Dictionary of National Biography from the Earliest Times to 1900*, Sir Leslie Stephen and Sir Sidney Lee, eds. (1893, 21 vols; reprint ed., Oxford University Press by Humphrey Milford, London, 1921–22), 12: 1007–8, and 20: 455–56; Eva Armstrong, "Jane Marcet and Her 'Conversations on Chemistry,'" *Journal of Chemical Education* 15 (1938): 53–57; G. E. Fussell, "Some Lady Botanists of the Nineteenth Century: Five: Jane Marcet," *Gardners' Chronicle* 130 (1951): 238; K. R. Webb, "Conversations on Chemistry, Mrs. Jane Marcet (1769–1859)," *Chemistry and Industry*, no. 38 (1958), p. 1225. See also John K. Crellin, "Mrs. Marcet's 'Conversations on Chemistry,'" *Journal of Chemical Education* 56 (1979): 459–60. Only Shteir, "With Bliss Botanic" and "Women and Plants," examines how these women's textbooks differed in content from other scientific books of the time. Emanuel D. Rudolph, "How It Developed that Botany Was the Science Thought Most Suitable for Victorian Young Ladies," *Children's Literature* 2 (1973): 92–97, discusses some themes that are treated in chapter 3. (I thank Ravenna Helson for a reference to this article.) See also Harriet C. Rogers, "Books in Medicine, Botany and Chemistry Printed in the American Colonies and the United States before 1801" (Master's thesis, Columbia University, 1932), and Caroline John Garnsey, "Ladies' Magazines to 1850," *New York Public Library Bulletin* 58 (1954): 74–88.

8. Anne Firor Scott, "The Ever Widening Circle: The Diffusion of Feminist Values from the Troy Female Seminary, 1822–1872," *History of Education Quarterly* 19 (1979): 6, is based on Emma Willard's *Address to the Public, Particularly to the Members of the Legislature of New York, Proposing a Plan for Improving Female Education,* 2d ed. (Middlebury, Vt., 1819; reprint ed., Middlebury, Vt.: Middlebury College, 1918), pp. 16 and 19. Alma Lutz, *Emma Willard: Pioneer Educator of American Women* (Boston: Beacon Press, 1964), is still useful. *Emma Willard and Her Pupils, or Fifty Years of Troy Female Seminary, 1822–1872* (New York: Mrs. Russell Sage, 1898) has biographies of several thousand alumnae. Willystine Goodsell, ed., *Pioneers of Women's Education in the United States* (New York: McGraw-Hill, 1931), reprints Willard's *Address* on pp. 43–82. See also "Willard" in *NAW*. For more on other reformers of the time, see Kathryn Kish Sklar, *Catharine Beecher: A Study in American Domesticity* (New Haven: Yale University Press, 1973), and Joan N. Burstyn, "Catharine Beecher and the Education of American Women," *New England Quarterly* 47 (1974): 386–403. Two other works describe the prevailing beliefs about women's incapacities that these reformers were both using and trying to overcome: Barbara Welter's "Anti-Intellectualism and the American Woman, 1800–1860," *Mid-America* 48 (1966): 258–70, and Nancy Cott's *The Bonds of Womanhood: "Woman's Sphere" in New England, 1780–1835* (New Haven: Yale University Press, 1977), chap. 3, "Education."

9. Richard M. Bernard and Mavis A. Vinoskis, "The Female School Teacher in Ante-Bellum Massachusetts," *Journal of Social History* 10 (1977): 332–45. See also Thomas Woody, *A History of Women's Education in the United States,* 2 vols. (New York: Science Press, 1929), 1: chap. 10, for a full account of the feminization of schoolteaching.

10. Scott, "Ever Widening Circle," p. 8; Lutz, *Emma Willard,* p. 58.

11. Willard, *Address,* pp. 20–22; Lutz, *Emma Willard,* pp. 27, 35, 47, and 92; Scott, "Ever Widening Circle," passim. Scott considers Willard's curriculum fully equivalent and perhaps even more rigorous than those of male colleges of the time, whose presidents often deplored the low level of learning among applicants (pp. 8 and 22-23 n. 10). She and Patricia Hummer have a larger work on Willard under way.

12. Scott, "Ever Widening Circle," passim; less explicit in Lutz, *Emma Willard.*

13. Lutz, *Emma Willard,* pp. 54–55, and Ethel M. McAllister, *Amos Eaton: Scientist and Educator, 1776–1842* (Philadelphia: University of Pennsylvania Press, 1941), pp. 258–62, 353–55, and 483-91. At one point Eaton sought to admit women directly to Rensselaer.

14. "Phelps" in *NAW;* Emma L. Bolzau, *Almira Hart Lincoln Phelps: Her Life and Work* (Philadelphia: University of Pennsylvania Press, 1936), passim, p. 221; Mary Elvira Weeks and F. B. Dains, "Mrs. A. H. Lincoln Phelps and Her Services to Chemical Education," *Journal of Chemical Education* 14 (1937): 53–57; Wyndham D. Miles, ed., *American Chemists and Chemical Engineers* (Washington, D.C.: American Chemical Society, 1976), p. 237. Hannah Mary Bouvier's correspondence about her *Familiar Astronomy or Introduction to the Study of the Heavens* (1856) is in the John Bouvier Collection, Huntington Library, Pasadena, California; Martha Ward and Dorothy A. Marquardt, *Authors of Books for Young People* (New York: Scarecrow Press, 1964; supplement, 1967), lists the numerous women science writers of more recent times.

15. Almira H. Lincoln (Phelps), *Familiar Lectures on Botany* (Hartford: H. and F. J. Huntington, 1839); Mrs. Lincoln Phelps, *The Educator, or Hours with My Pupils* (New York: A. S. Barnes, 1872).

16. Kathryn Kish Sklar, "The Founding of Mount Holyoke College," in *Women of America,* ed. Berkin and Norton, pp. 177-201. Elizabeth Alden Green, *Mary Lyon and Mount Holyoke* (Hanover, N.H.: University Press of New England, 1979); "Lyon" in *NAW;* Marion Lansing, ed., *Mary Lyon through Her Letters* (Boston: Book, 1937), esp. chap. 9, "Science with Professor Eaton"; Emma Perry Carr, "The Department of Chemistry: Historical Sketch," *Mount Holyoke Alumnae Quarterly* 2 (1918): 159–61; Arthur C. Cole, *A Hundred Years of Mount Holyoke College* (New Haven: Yale University Press, 1940), chaps. 2–4; Beth B. Gilchrist, *The Life of Mary Lyon* (Boston: Houghton Mifflin, 1910); and Edward Hitchcock, *The Power of Christian Benevolence Illustrated in the Life and Labors of Mary Lyon,* new ed. (New York: American Tract Society, 1858); David F. Allmendinger, Jr., "Mount Holyoke Students Encounter the Need for Life-Planning, 1837–1850," *History of Education Quarterly* 19 (1979): 39–41.

17. Deborah Jean Warner, "Science Education for Women in Antebellum America," *Isis* 69 (1978): 58–67; Thomas Woody, *History of Women's Education,* 1: chaps. 8 and 9, remains the fullest account of these academies and seminaries (see esp. table I, p. 419, which records the great increase in the amount of science in their curricula from 1749 to 1871). See also Keith Melder, "Mask of Oppression: The Female Seminary Movement in the United States," *New York History* 55 (1974): 261–79, and Thomas C. Johnson, *Scientific Interests in the Old South* (New York: D. Appleton, 1936), chap. 4, "Sweet Southern Girls," which discusses the amount of science taught at Southern female academies in this period. Data are taken from *Report of the Commissioner of Education for 1872* (Washington, D.C.: GPO, 1873), table 9, pp. 797–801.

18. *Report of the Commissioner of Education for 1873* (Washington, D.C.: GPO, 1874), table 7, pp. 650–62 (summaries reported on p. liv differ).

19. The best general history of women's higher education is still Mabel Newcomer, *A Century of Higher Education for American Women* (New York: Harper & Bros., 1959). Two recent programmatic statements are Patricia Albjerg Graham, "So Much to Do: Guides for Historical Research on Women in Higher Education," *Teachers College Record* 76 (1975): 421–29, and Jill K. Conway, "Perspectives on the History of Women's Education in the U.S.," *History of Education Quarterly* 14 (1974): 1-12. The history of coeducation is beginning to be written and reveals that some colleges that opened with lofty ideals of treating men and women equally often vacillated on this promise and gave women subordinate status after all. Ronald W. Hogeland, "Coeducation of the Sexes at Oberlin College," *Journal of Social History* 6 (1972-73): 160–76; Charlotte Williams Conable, *Women at Cornell: The Myth of Equal Education* (Ithaca: Cornell University Press, 1977); and Patricia Foster Haines, "Men, Women and Coeducation: Perspectives from Cornell University, 1865–1900" (Paper presented at Third Berkshire Conference on the History of Women, Bryn Mawr College, June 1976);

Dorothy Gies McGuigan, *A Dangerous Experiment: One Hundred Years of Women at the University of Michigan* (Ann Arbor: Center for Continuing Education of Women, 1970); Walter A. Donnelly et al., eds., *The University of Michigan: An Encyclopedic Survey*, 4 vols. (Ann Arbor: University of Michigan Press, 1942–58), 4: 1783–85. Much older, but a classic on coeducation at the University of Wisconsin, is Helen R. Olin, *The Women of a State University: An Illustration of the Working of Coeducation in the Middle West* (New York: G. P. Putnam's Sons, 1909). The trickle of women onto coed faculties is discussed in Lucille Addison Pollard, *Women on College and University Faculties: A Historical Survey and a Study of Their Present Academic Status* (New York: Arno Press, 1977), pp. 108–54.

20. The history of the women's colleges is now also undergoing a revival. See Elaine Kendall, *"Peculiar Institutions": An Informal History of the Seven Sister Colleges* (New York: Putnam, 1976), and the unfortunately titled Liva Baker, *I'm Radcliffe! Fly Me!* (New York: Macmillan, 1976). (I have omitted Radcliffe here, since it had no separate faculty and thus seems more a part of the history of coeducation than of separate women's colleges.) Most of the newer scholarly work focuses on founders' attitudes, presidential styles, and alumnae careers. See Roberta Wein, "Women's Colleges and Domesticity, 1875–1918," *History of Education Quarterly* 14 (1974): 31–47; Roberta (Wein?) Frankfort, *Collegiate Women: Domesticity and Career in Turn-of-the-Century America* (New York: New York University Press, 1977); and Sarah H. Gordon, "Smith College Students: The First Ten Classes, 1879–1888," *History of Education Quarterly* 15 (1975): 147–67. Older studies stressed the curriculum, such as Newcomer, *Century of Higher Education,* chap. 5; list of "Textbooks Mentioned by Women's College Catalogs Since 1850" in Woody's monumental *History of Women's Education,* 2: app. 1; Mabel L. Robinson, *The Curriculum of the Woman's College,* U.S. Bureau of Education Bulletin no. 6 (1918), for Vassar, Wellesley, Radcliffe, Barnard, and Mount Holyoke; and Elizabeth Barber Young, *A Study of the Curricula of Seven Selected Women's Colleges of the Southern States,* Teachers College Contributions to Education no. 511 (1932), pp. 136–52. All overlook the faculty, who are stressed here and in Pollard, *Women on College and University Faculties,* pp. 69–108.

21. *Report of the Commissioner of Education, 1887–88* (Washington, D.C.: GPO, 1889), p. 595 (table 44, pt. 2). The value of Vassar's scientific apparatus constituted one-quarter of all that ($216,000) at 207 women's institutions in the entire nation in 1887 (p. 585, table 43). For doctorates, see M. Carey Thomas, *Education of Women,* published as vol. 7 of *Monographs on Education in the United States*,3–1. Nicholas Murray Butler, Department of Education for the United States Commission to the Paris Exposition of 1900 (Albany: J. B. Lyon, 1900), 1: 19–20.

22. A Mrs. Mary Griffiths of New York City published two articles on vision in 1834 and a third on haloes in 1840, but nothing is known of this aspiring physicist (Mrs. Mary Griffiths, "Observations on the Vision of the Retina," *Philosophical Magazine* 4 [1834]: 43–46; idem, "Observations on the Spectra of the Eye and the Seat of Vision," *Philosophical Magazine* 5 [1834]: 192–96; and idem, "On the Halo or Fringe Which Surrounds All Bodies," *American Journal of Science* 38 [1840]: 22–32). I thank John Greene for calling her to my attention and Clark Elliott for his efforts to identify her.

23. Helen Wright, *Sweeper in the Sky: The Life of Maria Mitchell, First Woman Astronomer in America* (New York: Macmillan, 1949); Dorothy J. Keller, "Maria Mitchell, An Early Woman Academician" (Ed. D. diss., University of Rochester, 1974); Stephen Catlett, "An Exhibit on Women Members of the American Philosophical Society, April 1981," mimeographed; seven publications are listed in the *Royal Society Catalog* for Mitchell. Mitchell's salary situation at Vassar was complicated by her getting free board and room at the college observatory and by her retaining, initially anyway, her Coast Survey computer work (though farming it out to someone else). Despite this, she and Dr. Avery felt that their salaries were still far too low ($800 versus the men's $2,500), and after several years of protest, they succeeded in getting them equalized in 1871. This policy did not extent to Mitchell's successor, Whitney, however, to whom a later president explained that salaries had to be unequal, since not everyone should (or could) get a raise just because one faculty member had a more lucrative outside offer (Mitchell to Rufus Babcock, n.d. [1865?], "Confidential"; numerous letters to and from President Raymond, 1871–72; J. M. Taylor to M. Whitney, 6 May 1890, all in Maria Mitchell Papers, Vassar College Archives).

24. Mary Whitney to James Monroe Taylor, 25 Jan. 1913, cited in James Monroe Taylor, *Before Vassar Opened* (Boston: Houghton Mifflin, 1914), p. 273; Frances Fisher Wood, "Sketch of Maria Mitchell," in *What America Owes to Women,* ed. Lydia Hoyt Farmer (Buffalo: Charles Wells Moulton, 1893), chap. 27. (I thank Mary Ellen Bowden for a copy of this handsome volume.) "Ladd-Franklin," "Richards," and "Whitney" in *NAW*; Christine Ladd-Franklin's diaries for 1860-72 are in the Vassar College Archives; Ellen Swallow to Maria Mitchell, 29 Nov. 1872, Maria Mitchell Papers, Vassar College Archives. Ellen Semple (Vassar 1882), the famed geographer, was probably also a student of Mitchell (*NAW*). Data on the Vassar graduates of 1867 to 1896 are presented in Frances M. Abbott, "Three Decades of [Vassar] College Women," *Popular Science Monthly* 65 (1904): 350-59.

25. Edward H. Clarke, *Sex in Education: Or, A Fair Chance for the Girls* (Boston: J. R. Osgood, 1873). Julia Ward Howe, ed., *Sex and Education: A Reply* (Boston: Roberts Bros., 1874), was one of several collections of rebuttals. Mary Roth Walsh, *"Doctors Wanted, No Women Need Apply": Sexual Barriers in the Medical Profession, 1835-1975* (New Haven: Yale University Press, 1977), pp. 120-25, places the controversy in its medical context (women gynecologists were threatening the professional standing of male practitioners). Clarke himself had suffered a breakdown from overstudy (*Dictionary of American Biography*, ed. Dumas Malone [New York: Charles Scribner's Sons, 1924], 8: 213).

26. Janice Law Trecker, "Sex, Science and Education," *American Quarterly* 26 (1974): 353-66; Marie Tedesco, "Science and Feminism: Conceptions of Female Intelligence and Their Effect on American Feminism, 1859-1920" (Ph.D. diss., Georgia State University, 1978); Susan S. Mosedale, "Science Corrupted: Victorian Biologists Consider 'The Woman Question,'" *Journal of the History of Biology* 11 (1978): 1-55. For the many articles on this topic in the *Popular Science Monthly,* see Frederik A. Fernald, comp., *Index to the Popular Science Monthly, from 1872 to 1892* (New York: D. Appleton, 1893), pp. 262-64.

27. Sally Gregory Kohlstedt, "Maria Mitchell: The Advancement of Women in Science," *New England Quarterly* 51 (1978): 39-63; Keller, "Maria Mitchell," chap. 5; *Historical Account of the Association for the Advancement of Women, 1873-1893* (Dedham, Mass.: Transcript Steam Job Print, 1893).

28. Professor Maria Mitchell, "Address of the President," in *Papers Read at the Third Congress of Women, Syracuse, October 1875* (Chicago: Fergus Printing, 1875), p. 5. Opinions as to the propriety of women's speaking in public changed rapidly during the mid nineteenth century. The first to break with tradition were female lecturers on physiology and hygiene, temperance (Virginia Penny, *The Employments of Women: A Cyclopedia of Women's Work* [Boston: Walker, Wise, 1863], pp. 18-19), and abolition. Although Emma Willard had had men read her speeches and abhorred the thought of female politicians, her student Elizabeth Cady Stanton broke the ban in the 1840s by speaking out publicly against slavery and for women's rights. One address in 1850 electrified Susan B. Anthony into a career in politics (*NAW*). Although still not respectable enough for the initial Vassar College curriculum, public speaking by women increased in favor in the 1870s. It was long a major barrier, however, to women's full participation in scientific societies (see chapter 4).

29. This composite biography is based on a reading of faculty files and obituaries of over one hundred women faculty members at these six women's colleges: Wellesley College Archives, Smith College Archives, Mount Holyoke College Archives, Vassar College Archives, Barnard College Archives, and Goucher College Archives. I am greatly indebted to helpful archivists at these institutions and at Bryn Mawr College.

30. For example, in the 1870s Susan Bowen of Mount Holyoke married ichthyologist David Starr Jordan, the president of Indiana University and later of Stanford University. Though she died young, he remained a good friend of several other women scientists, including Cornelia Clapp. Similarly, zoologist Emily Nunn of Wellesley College married Charles O. Whitman, later of the University of Chicago, in the 1880s. They had met one summer at Woods Hole, which her wealthy father later helped Whitman endow (see chapter 4).

31. Harriet Brooks to Dean Laura Gill, 18 July 1906, Departmental Correspondence 1906/1908, File 41, Barnard College Archives. (I thank Patricia Ballou for bringing this episode to my attention.)

32. Dean Laura Gill to Miss Brooks, 23 July 1906, Barnard College Archives. As dean she must

have known that Elsie Clews Parsons, who lectured at Barnard from 1902 to 1905, was married. (*NAW;* I thank Rosalind Rosenberg for bringing this to my attention.) Perhaps Gill had fired Parsons as well, or perhaps some problems with her led to a change in previously tolerant policies. Since Brooks's fiancé was not on the faculty, the issue was not one of antinepotism.

33. Margaret Maltby to Miss Gill, 24 July 1906, Barnard College Archives.

34. "J. J. Thomson informed me that she [Brooks] was the best woman researcher, next to Mrs. Sidgwick, he had at the Cavendish. She has already a good knowledge of experimental work in ionization of gases and radioactivity. She has as strong a claim as any probable candidate for the position [at McGill]," cited in John L. Heilbron, "Physics at McGill in Rutherford's Time," in *Rutherford and Physics at the Turn of the Century,* ed. Mario Burge and William R. Shea (New York: Dawson and Science History Publications, 1979), pp. 57–58 n. (Heilbron thanks Lawrence Badash for this reference and I thank them both.) Brooks is also mentioned in A. S. Eve, *Rutherford* (New York: Macmillan, 1939), pp. 74, 81, 94, 185, and 231.

35. Laura Gill to Miss Maltby, 30 July 1906, Barnard College Archives.

36. The month of strain must have helped bring on the broken engagement. Brooks's parting shot to Dean Gill that "Although it might, with justice, be the general policy, particular circumstances might cause its enforcement to be a great injustice" (Harriet Brooks to Miss Gill, 13 Sept. 1906, Barnard College Archives), shows that she had glimpsed what can be called the "strategy of exceptions," that is, when faced with an intransigeant administrator or an inflexible rule, one should seek to become an "exception." This allows discreet compromise—the dean can feel that she was enforcing her rule prudently while allowing the protester to get her way "just this once." (Maltby's comment that Brooks was unlike other women would have justified a decision of this sort here.) Allowing such "exceptions" has the additional advantage of "setting a precedent" for other persons. In time the cumulative effect of many "exceptions" can be a necessary change in the rules, a prospect unthinkable at the outset. Of course, if one can find an administrator who has the authority and is willing to make an exception, the battle is already half won. Here, in Brooks's case, Gill was unequal to the task.

37. Grace Langford in *AMS,* 3d ed. (1921), p. 398. By 1914 Brooks had become Mrs. Pitcher and had three children. She was still interested in physics but not able to do any (Eve, *Rutherford,* p. 231).

38. M. Carey Thomas was proud of having overcome the prejudice of hiring unmarried men at a woman's college (Thomas, *Education of Women,* 1: 24 n. 1); Ian Shine and Sylvia Wrobel, *Thomas Hunt Morgan: Pioneer of Genetics* (Lexington: University of Kentucky Press, 1976), pp. 43–45; "Rand," in *NAW: MP*; "Gertrude Rand," *NCAB*, vol. G (1946), p. 480; and "Clarence Ferree," *NCAB* 33 (1947): 96–97. Morgan married thirteen years after coming to Bryn Mawr and Ferree eleven.

39. The spartan economies of the women's colleges are rarely mentioned in their largely sentimental histories. Exceptions are Cole, *Hundred Years,* and Jean Glasscock, "The Development of Wellesley's Financial Resources," in *Wellesley College, 1875–1975: A Century of Women,* ed. Jean Glasscock et al. (Wellesley, Mass.: Wellesley College, 1975), pp. 370–86.

40. Emily Gregory to Bessie Helmer, 5 May 1893; Emily Gregory to Miss Weed, 22 June 1893 and n.d.; Emily Gregory to Emily James Smith, 27 Jan. 1896, all in Barnard College Archives. Wealthy geologists Ida Ogilvie and Carlotta Maury also accepted only token salaries from Barnard for their teaching services.

41. Horsford is mentioned frequently in Jean Glasscock et al., *Wellesley College*, esp. pp. 18–21, and in Sarah F. Whiting's "History of the Physics Department at Wellesley College from 1878 to 1912," S. F. Whiting Papers, Wellesley College Archives. He also left the college a sizeable bequest.

42. Margaret Ferguson to Sophie Hart, 18 May 1932, Ferguson Papers, Wellesley College Archives.

43. "Whiting" in *NAW*.

44. Curtis J. Smith, "Charlotte Haywood," *Biological Bulletin* 143 (1972): 16–17. Harriet Zuckerman has talked of the importance of such master-protégée relationships in the careers of Nobel laureates in *Scientific Elite: Nobel Laureates in the United States* (New York: Free Press, 1977),

chap. 4; and Rosabeth Kanter has found this "homosexual reproduction" important in successful careers in corporate life: *Men and Women of the Corporation* (New York: Basic Books, 1977), p. 63.

45. "Carr" in *NAW:MP;* "Emma Perry Carr," *CBY, 1959*, pp. 55–57; Emma Perry Carr, "One Hundred Years of Science at Mount Holyoke College," *Mount Holyoke Alumnae Quarterly* 20 (1937): 135–38; the sciences at Wellesley are discussed in Dorothy W. Weeks and Harriet B. Creighton, "Early Years in the Sciences," *Wellesley Alumnae Magazine* (Winter 1975): 28, and idem, "Pioneer Professors," *Wellesley Alumnae Magazine* (Winter 1975): 28–29; also in "Science and the Ideals of the Founder," in the proceedings of a science conference on "Energy" held at Wellesley College, 16–18 March 1949, pp. 7-10, Wellesley College Archives; C. Stuart Gayer, "Wellesley College and the Development of Botanical Education in America," *Science* 67 (1928): 171–78; "Whiting" in *NAW*, and clipping of obituary from *Boston Herald* (1927) in Faculty File, Wellesley College Archives; Dorothy Weeks, comp., "Biographical Notes on Miss Louise Sherwood McDowell," Jan. 1945, Faculty File, Wellesley College Archives; Caroline E. Furness, "Mary W. Whitney," *Popular Astronomy* 30 (1922): 597-608, and *Popular Astronomy* 31 (1923): 25–35; Maud W. Makemson, "Caroline Ellen Furness," *Publications of the Astronomical Society of the Pacific* 48 (1936): 97-100; "Maud W. Makemson," *CBY, 1941*, pp. 552–54; "Henry W. Albers," *American Men and Women of Science*, 13th ed. (1976): 1:38.

46. For example, "Destruction by Fire of Scientific Laboratories [at Mount Holyoke]," *Science* 43 (1918): 40–41.

47. Emma Paddock Telford, "The American College for Girls at Constantinople," *New England Magazine* 18 (1898): 10–20; Eveline A. Thomson, "Constantinople College: The American College for Girls at Constantinople," *JACA* 9 (1916): 244–49; Meta Glass, "The American College for Girls—Istanbul," *JAAUW* 45 (1952): 95–96; "Patrick" in *NAW*; Louise McCoy North, "The Women's Christian College, Madras, India," *JACA* 11 (1917–18): 149–53; "Women's Christian College, Madras," in Madras (City) University, *History of Higher Education in South India*, 2 vols. (Madras: Associated Printers, 1957), 2.86–88.

48. Obituary of Ellen Hayes, *Wellesley Magazine* 15 (Feb. 1931): 151–52, and Louise Brown, *Ellen Hayes: Trail Blazer* (n. p. [c. 1931]). Hayes was a socialist, possibly a Communist, and was arrested for protesting the Sacco-Vanzetti decision in 1927. See also Ellen Hayes, *Letters to a College Girl* (Boston: G. H. Ellis Co., 1909), for her views. Clipping about Emma Byrd from *Boston Globe*, 7 May 1906, in Emma Byrd Faculty File, Smith College Archives; "Miss Mary E. Byrd's Resignation," *Popular Astronomy* 14 (1906): 447–48; Louise Barber Hoblit, "Mary E. Byrd," *Popular Astronomy* 42 (1934): 496–98.

49. Mary W. Calkins of Wellesley deplored the policy of the *American Journal of Psychology*, which would not publish student papers (Calkins to James McKeen Cattell, 30 July 1894, Cattell Papers, Manuscript Division, Library of Congress).

50. M. Carey Thomas to Florence Bascom [spring 1906], Florence Bascom Papers, Sophia Smith Collection, Smith College Library. Bascom had skipped previous commencements. Thomas had tolerated Bascom's absence in previous years but was particularly upset with her this year since Bascom had successfully demanded a $500 raise (to $2,500) and a full professorship, which Thomas had said at first was impossible (Thomas to Bascom, 16 Apr. 1907, Bascom Papers).

51. Katharine Luomala, "Martha Warren Beckwith, A Commemorative Essay," *Journal of American Folklore* 75 (1962): 341–53; Janet Lewis Zullo, "Annie Montague Alexander: Her Work in Paleontology," *Journal of the West* 8 (1969): 183–99; Stephen G. Brush, "Nettie M. Stevens and the Discovery of the Sex Determination by Chromosomes," *Isis* 69 (1978): 170; "Stevens" in *NAW*.

52. Numerous obituaries; William Henry Welch, "Contribution of Bryn Mawr College to the Higher Education of Women," *Science* 56 (1922): 1–8. Baker, *I'm Radcliffe!* also blames the women's colleges for not pushing their faculty and graduates to achieve more. "The Confessions of a Woman Professor," *Independent* 55 (1903): 954–58, complains of the frequent put-downs encountered by women professors of the time.

53. Ferguson, *NAW: MP;* Sophie C. Hart, "Margaret Clay Ferguson," *Wellesley Magazine* 16 (1932): 409–10; Marian E. Hubbard, "The Plight of Our Zoology Department," *Wellesley Alumnae Magazine* 12 (1928): 123–30.

54. Cornelia Meigs, *What Makes A College? A History of Bryn Mawr* (New York: Macmillan,

1956), pp. 50 and 216; Emily Gregory was there from 1886 to 1888, before moving on to the University of Pennsylvania and later Barnard College. She described her problems at Bryn Mawr in a lengthy letter to William G. Farlow (2 Jan. 1887), now in the Farlow Letterbooks, Farlow Herbarium, Harvard University. As one of the first American women with a doctorate in botany, she had been hired to start a department along German lines. But the college had also hired E. B. Wilson to start a similar program in zoology. When there were not enough students to sustain both, he won out and formed a combined program in "biology."

Gregory has been underestimated, probably because she was too specialized too soon for the existing job market for women scientists and then died within a few years of setting up a program at Barnard. Because she passed away before 1906 she was never in the *American Men of Science,* but she should have been in *Notable American Women.* Her letters show her to have been one of the more vigorous women scientists of her time. (The only major obituary is Elizabeth G. Britton, "Emily L. Gregory," *Bulletin of the Torrey Botanical Club* 24 [1897]: 221–28. A fact sheet, "Biography of Emily Louisa Gregory," in the Barnard College Archives is also informative.)

The problems of Bryn Mawr's geology department can be traced in the Florence Bascom Papers in the Sophia Smith Collection at Smith College. Since Bascom was still reeling from the death of her father that autumn, the rescue effort was led by alumna and graduate student Eleanora Bliss (later Knopf) who aroused alumnae and trustee support, as shown in the Tasker Bliss Papers Manuscript Division, Library of Congress. (I thank Lucy Fisher West and Michele Aldrich for bringing these collections to my attention.) "Knopf" in *NAW: MP.* I thank Maria Luisa Crawford for sending me a copy of Isabel Fothergill Smith's *Stone Lady: A Memoir of Florence Bascom* (Bryn Mawr, Pa.: Bryn Mawr College, 1981), which is based on the Bascom Papers at Smith.

55. "Reichard" in *NAW: MP;* see also note 12 to chapter 6 herein.

56. "Charlotte Scott" in *NAW* and "Anna Pell Wheeler" in *NAW: MP;* see also Beale W. Cockey, "Mathematics at Goucher, 1888–1979," mimeographed (Baltimore: Goucher College Mathematics Department, 1979).

57. "Sarah Gibson Blanding," *CBY, 1946,* pp. 55–57; "Katharine Blunt," *CBY, 1946,* pp. 58–59, and in *NAW: MP.*

58. Some of the stronger departments kept close track of the subsequent career of their graduates; Margaret Ferguson to Sophie Hart, 18 May 1932 (Ferguson Papers, Wellesley College Archives), lists the accomplishments of 18 of "our girls"; *Mount Holyoke Alumnae in Medicine and Science* (n.d. [1925?]) lists jobs held by 114 alumnae in several fields. Ann Miller, ed., *A College in Dispersion: Women of Bryn Mawr, 1896–1975* (Boulder, Colo.: Westview Press, 1976), has an abundance of data on its graduates. The earliest data in M. Elizabeth Tidball and Vera Kistiakowsky, "Baccalaureate Origins of American Scientists and Scholars [Doctorates]," *Science* 193 (1976): 646–52, are for the 1910s.

59. "Preface," James McKeen Cattell, *A Biographical Dictionary of American Men of Science* (Garrison, N.Y.: Science Press, 1906); idem, "A Biographical Index of the Men of Science of the United States," *Science* 16 (1902): 746–47; idem, "A Biographical Directory of American Men of Science," *Nation* 80 (1905): 457; see also p. 358 n. 3.

60. Harrison White has analyzed how such "vacancy chains" operate in a closed labor market, such as that of the women physicists in 1920 (Harrison White, *Chains of Opportunity: System Models of Mobility in Organizations* [Cambridge: Harvard University Press, 1970]).

61. Thus Nellie Goldthwaite, assistant professor of household science at the University of Illinois, returned to Mount Holyoke in 1912 for the college's seventy-fifth anniversary and collected a scrapbook full of clippings of the occasion. (She had been head of Holyoke's chemistry department from 1897 to 1905.) (Nellie Goldthwaite Papers, University of Illinois Archives.)

62. Thomas, *Education of Women,* 1: 18–22. Her criticisms of Mount Holyoke led to a major change in hiring policy under new president, Mary Woolley, after 1901 (Elaine Kendall, *"Peculiar Institutions,"* p. 174). See Ella Keats Whiting, "The Faculty," in *Wellesley College,* ed. Glasscock et al., pp. 87–113, for a most perceptive and critical discussion of male faculty at the women's colleges. They would seem to be an interesting group that merits further study. The proportion they formed of the faculty (see the annual *Reports of the Commissioner of Education* for data) and the attitude of the women and administrators toward them varied greatly among the individual colleges

and over time. Some men were there because, as they would admit, it was the only job they could get at the time, and they left as soon as possible for a university, sometimes after distinguishing themselves with important researches—as E. B. Wilson and T. H. Morgan did at Bryn Mawr in the 1890s—and marrying a student or alumna. (Thus the proportion of men on these faculties might be a good barometer of their fortunes elsewhere.) The women on these faculties had almost no opportunity to leave, regardless of the quality of their work (Welch, "Contribution of Bryn Mawr College to the Higher Education of Women," pp. 1-8; "Washburn" in *NAW*). Other men, however, stayed at the women's colleges for decades. Apparently they liked these colleges and their rôle there, or else had no opportunity to leave. One wonders, though, if they weren't subject to the "advice" or taunts of other men urging them to avoid such colleges—M. J. Zigler, who taught psychology at Wellesley in the 1920s, considered it a "handicap" to be at such a school and was apparently anxious to leave (E. B. Titchener to M. J. Zigler, 7 Jan. 1925 and 1 May 1925, E. B. Titchener Papers, Cornell University Archives). See also E. B. Titchener to Mary Calkins, 17 Apr. 1924, where he had only compliments about Zigler and Wellesley. See also note 25 to chapter 7 herein. William Longley of Goucher College spent his summers at an all-male research laboratory (Tortugas Laboratory run by the Carnegie Institution) with professors from other women's colleges (Lucile Moore Burns, "Dr. William Hardy Longley," *Goucher Alumnae Quarterly* 15 [1937]: 9-10).

CHAPTER 2

1. The movement is not even mentioned in either of the two standard accounts of the rise of graduate education in America: Frederick Rudolph, *The American College and University: A History* (New York: Vintage Books, 1962), and Laurence R. Veysey, *The Emergence of the American University* (Chicago: University of Chicago Press, 1965). Thomas Woody mentions it briefly in his *A History of Women's Education in the United States*, 2 vols. (New York: Science Press, 1929), 2: 333-40. Events at particular universities are described in Hugh Hawkins's excellent *Pioneer: A History of The Johns Hopkins University, 1874-1889* (Ithaca: Cornell University Press, 1960), chap. 14, "The Uninvited," and at the University of Wisconsin in Helin Olin, *The Women of a State University: An Illustration of the Working of Coeducation in the Middle West* (New York: G. P. Putnam's Sons, 1909), chap. 12.

2. There is no history of the Bryn Mawr Graduate School, although Cornelia Meigs, *What Makes a College? The History of Bryn Mawr* (New York: Macmillan, 1956) mentions its problems occasionally. See also Eleanor Bliss, "Bryn Mawr Studies Its Ph.D.s," *JAAUW* 48 (1954): 14-16; Ann Miller, ed., *A College in Dispersion: Women of Bryn Mawr, 1896-1975* (Boulder, Colo.: Westview Press, 1976), esp. tables 123 and 124; and "The Academic Committee's Report on the Bryn Mawr Graduate School," *Bryn Mawr Alumnae Bulletin* 7 (1927): 3-36 (I thank Gertrude Reed for sending me a copy). Now that the M. Carey Thomas Papers are available, at least on microfilm, a full history seems feasible.

3. "Address by M. Carey Thomas," in *Addresses Delivered at the Opening of the Graduate Department of Women [at the University of Pennsylvania] on Wednesday, May 4th, 1892* (Philadelphia: Allen, Lane & Scott's Printing House, 1892), pp. 17-19; M. Carey Thomas, "Present Tendencies in Women's College and University Education," *PACA*, series 3, no. 17 (Feb. 1908), pp. 58-62; Mary Calkins, "The Relation of College Teaching to Research," *JACA* 4 (1911): 78-80; "Washburn" in *NAW;* Margaret Washburn to Robert Yerkes, 5 July 1910, Robert A. Yerkes Papers, Yale University Library; and Margaret Washburn to Christine Ladd-Franklin, 20 Feb. 1914, Christine Ladd-Franklin Papers, Special Collections, Butler Library, Columbia University. None of the women's colleges seems to have justified a graduate school by stressing what might be characterized "womanly" subjects as a way of increasing the size of its pool of applicants.

4. "Richards" in *NAW* and Robert Richards, *Robert Hallowell Richards: His Mark* (Boston: Little, Brown, 1936), p. 153. The first woman applicant to the Harvard Medical School in 1847 and again in 1850 was rejected not on the grounds that she was unqualified (she was highly qualified, with previous medical training) or because the school was crowded or lacked proper facilities, conditions that might improve over time. The standard reason given was the evasive and quasi-political one that it was "inexpedient" to admit women, a position that was largely unassailable. Mary Roth Walsh,

"Doctors Wanted, No Women Need Apply": Sexual Barriers in the Medical Profession, 1835–1975 (New Haven: Yale University Press, 1977), pp. 28–32, 164–66. MIT did not award its first Ph.D. until 1907, and its first to a woman was awarded in 1922 (Elizabeth Gatewood in chemistry). (Agnes L. Rogers, "Preliminary Report of the Committee on Fellowships of the American Association of University Women," *JAAUW* 17, no. 1 [Jan.-Mar. 1924]: 18).

5. "Richards" in *NAW*, and Caroline L. Hunt, *The Life of Ellen H. Richards* (Boston: Whitcomb and Barrows, 1912), p. 88.

6. Hawkins, *Pioneer*, chap. 14; "Ladd-Franklin" in *NAW*. See also note 48 to this chapter.

7. Item 27: Class notes for 1867 (12 Jan. 1868), Maria Mitchell Memorabilia, microfilm, American Philosophical Society, Philadelphia (I thank Sally Gregory Kohlstedt for this reference); "Whitney" and "Whiting" in *NAW;* Sarah F. Whiting, "History of the Physics Department at Wellesley College from 1878 to 1912," manuscript, 1912, in Wellesley College Archives. Edward Pickering was an early supporter of scientific education for women ([Edward C. Pickering], "Education," *Atlantic Monthly* 33 [1874]: 760–64), and a hearty supporter of the young Association of Collegiate Alumnae (Edward Pickering to Marion Talbot, 18 Dec. 1882, Marion Talbot Papers, Special Collections, Regenstein Library, University of Chicago). Genth: "Women in the University," *Philadelphia Public Ledger and Daily Transcript,* 25 July 1877, clipping in Rachel Bodley Scrapbook, p. 179, Rachel Bodley Papers, Archives and Special Collections, (Women's) Medical College of Pennsylvania.

8. Hugh Hawkins, *Pioneer,* chap. 14; "Thomas" and "Ladd-Franklin" in *NAW*. Sylvester was familiar with some of Ladd's previous publications. For more on The Johns Hopkins University, see notes, 46–48 and 50 to this chapter.

9. Walter Crosby Eells, "Earned Doctorates for Women in the Nineteenth Century," *BAAUP* 42 (1956): 647 and 651.

10. W. Freeman Galpin, *Syracuse University: The Pioneer Days* (Syracuse: Syracuse University Press, 1952), chap. 14; for mention of Winchell's later support of early women doctorates at the University of Michigan, see Lewis B. Kellum, "The Museum of Paleontology," in *The University of Michigan: An Encyclopedic Survey*, 4 vols., ed. Walter A. Donnelly et al. (Ann Arbor: University of Michigan Press, 1942–58), 4: 1488; Wooster's graduate school was later seen as an "unfortunate" addition (Lucy Lilian Notestein, *Wooster of the Middle West,* 2 vols. [New Haven: Yale University Press, 1937], 1: 89–90).

11. See Margaret Farrand Thorp, *Smith Grants Radcliffe's First Ph.D.* (to Kate Morris [later Cone] in 1882) (Northampton, Mass.: Smith College, 1965).

12. Gulliver: Eells, "Earned Doctorates," p. 646; Hooker: Eells, "Earned Doctorates," p. 651, and *AMS*, 3d ed. (1921), p. 328; "Clapp" in *NAW*; Crow: Galpin, *Syracuse University*, p. 217, and John S. Nollen, *Grinnell College* (Iowa City: State Historical Society of Iowa, 1953), p. 87; "White" in *NAW;* Slosson: Eells, "Earned Doctorates," p. 651, and *AMS,* 3d ed. (1921), p. 632; Cook: Eells, "Earned Doctorates," p. 651, and *AMS,* 5th ed. (1933), p. 225.

13. Jane M. Bancroft, "Occupations and Professions for College-Bred Women," *Education* 5 (1885): 486–95; Martha Foote Crow, "The Status of Foreign Collegiate Education for Women," *PACA,* series 2, no. 37 (1891), and, from Zurich, "Women in European Universities," *Nation* 54 (1892): 247.

14. Notestein, *Wooster,* 1: 90.

15. *Nation* 49 (1889): 426–27 and 446–47; *Nation* 54 (1892): 247; *Nation* 57 (1893): 483–84; *Nation* 58 (1894): 116–17, 137, 151–52, 154, 193, 212; *Nation* 59 (1894): 232–33, 247–48, 268; *Nation* 64 (1897): 223–34, and 262. (One may also see Christine Ladd-Franklin's hand here, since both she and her husband were frequent contributors to the *Nation*.) Helene Lange, *Higher Education of Women in Europe,* trans. L. R. Klemm (New York: D. Appleton, 1890); "A Profession for Women," *Popular Science Monthly* 38 (1890-91): 701–2.

16. Thomas Woody, *History of Women's Education,* 2: 333–40. There is some indication that the doctorate given by the University of Pennsylvania to a woman physician in 1880 may have been to avoid giving her a bachelor's degree (and thus condoning undergraduate coeduation). Her degree was later reduced to a B.S. (Martin Meyerson and Dilys Pegler Winegrad, *Gladly Learn and Gladly Teach: Franklin and His Heirs at the University of Pensylvania, 1740–1976* [Philadelphia: Univer-

sity of Pennsylvania Press, 1978], p. 122). See also the peculiarities surrounding Mary Pennington's doctorate at Pennsylvania in 1895: Barbara Heggie, "Profiles, Ice Woman," *New Yorker* 17 (1941): 23–24. The best discussion I have seen of the complex situation at Columbia and Barnard is in Rosalind Rosenberg, *Beyond Separate Spheres: Intellectual Roots of Modern Feminism* (New Haven: Yale University Press, 1982), chap. 4.

17. Arthur T. Hadley, "The Admission of Women as Graduate Students at Yale," *Educational Review* 3 (1892): 486; see also Timothy Dwight, "Education for Women at Yale," *Forum* 13 (1892): 451–63.

18. *Report of the President of Yale University for the Year Ending December 31, 1891*, p. 25 (Yale University Archives). See also Yale University Corporation Records, Book 1876–1900 (reel 6), 3 Mar. 1892, p. 297, and "Women in Post Graduate Courses," *Yale Alumni Weekly* 1, no. 23 (1892): 3, both at Yale University Archives. I thank Patricia Bodak Stark and Judith Schiff for assistance.

19. Both Dwights are in the *Dictionary of American Biography*, ed. Dumas Malone (New York: Charles Scribner's Sons, 1930), 15: 573–78. The Dwight Family Papers at the Yale University Archives have no material on this episode; Edgar S. Furniss, *The Graduate School of Yale: A Brief History* (New Haven: Printed for the Yale Graduate School by Carl Purington Rollins Printing-office, 1965), pp. 72–73; and Morris Hadley, *Arthur Twining Hadley* (New Haven: Yale University Press, 1948), pp. 68–70, 79, and 160–63.

20. Marion Talbot to Millicent Shinn, 27 Aug. 1892, Millicent Shinn Papers, Bancroft Library, University of California, Berkeley. (She also reported that women had won three out of five Yale fellowships.) Marion Talbot, *History of the Chicago Association of Collegiate Alumnae, 1888–1917* (Chicago: private, 1920), p. 6. The Association of Collegiate Alumnae also put out a bulletin on the opening of Yale, "Notes on Graduate Instruction, Yale University," *PACA*, series 2, no. 37-2 (1892).

21. Woody, *History of Women's Education*, 2: 336; Richard J. Storr, *Harper's University: The Beginning* (Chicago: University of Chicago Press, 1966), p. 109; "Palmer" in *NAW*.

22. Eells, "Earned Doctorates," p. 646, and "Clapp" in *NAW*.

23. Eells, "Earned Doctorates," p. 648.

24. Thus at the University of Chicago, Charles O. Whitman, professor of zoology, had 6 women among his 41 Ph.D.s ("Biographical Sketch of Charles O. Whitman," *Journal of Morphology* 22 [1911]: xlvi–xlvii); John Merle Coulter, professor of botany, had 25 women among his 82 doctorates (*A Record of the Doctors in Botany of the University of Chicago, 1897–1916, Presented to John Merle Coulter by the Doctors in Botany* [Chicago: private, 1916]); and Leonard E. Dickson, professor of mathematics, had at least 16 women among his Ph.D.s (Raymond Clare Archibald, *A Semicentennial History of the American Mathematical Society, 1888–1938*, published as American Mathematical Society, *Semicentennial Publications*, 2 vols. [New York: American Mathematical Society, 1938], 1: 105. [Margaret T. Corwin], ed., *Alumnae, Graduate School, Yale University, 1894–1920* (New Haven: Yale University Press, 1920) lists its 117 women doctorates to date. Introductory statements by department chairmen show how welcome they were in some departments.

Later data on the institutions that awarded five or more doctorates to women in the first three editions of the *American Men of Science* (1906, 1910, and 1921), broken down by field, are presented in Margaret W. Rossiter, "Women Scientists in America before 1920," *American Scientist* 62 (1974): 319.

25. Women schoolteachers flocked to New York University's School of Pedagogy, because, as graduates of normal colleges, they were prohibited from entering its graduate school. Because the School of Pedagogy was so feminized, its administration organized a Woman's Advisory Council of wealthy New Yorkers who covered the school's annual deficits for many years. One member, Helen Gould, the daughter of financier Jay Gould, gave NYU more than $2 million in contributions. Joshua L. Chamberlain, ed., *New York University: Its History, Influence, Equipment and Characteristics* (Boston: R. Herndon, 1901 and 1903), 1: 240–42; and Theodore F. Jones, ed., *New York University, 1832–1932* (New York: New York University Press, 1933), chap. 14.

26. G. Stanley Hall of Clark University later claimed that it had been open to women since its founding in 1887 (*Life and Confessions of a Psychologist* [New York: D. Appleton, 1924], p. 318),

but M. Carey Thomas, who kept close watch over how women were faring at the various graduate schools, stated in 1900 that Clark was closed to them. She thought this particularly blatant discrimination, since Clark's principal field was pedagogy, which would otherwise have attracted many women there (M. Carey Thomas, *Education of Women*, published as vol. 7 of *Monographs on Education in the United States*, ed. Nicholas Murray Butler, Department of Education for the United States Commission to the Paris Exposition of 1900 [Albany: J. B. Lyon, 1900], pp. 349–50 n. 3). Clark enrolled its first woman the next year (Louis N. Wilson, comp., *List of Degrees Granted at Clark University and Clark College, 1889–1920*, Publications of the Clark University Library 6, no. 3 [Dec. 1920], p. 11).

27. ["Women Students at Leipzig"], *Atlantic Monthly* 44 (1879): 788–91; in the early 1870s Erminnie A. P. Smith "studied crystallography and German literature at Strassburg and Heidelberg and completed a two-year mineralogy course at the School of Mines in Freiburg" (*NAW*). Mary Hegeler (later Carus) also attended lectures at the Freiburg mining school in the 1880s (David Eugene Smith, "Mary Hegeler Carus, 1861–1936," *American Mathematical Monthly* 44 [1937]: 280). Emily L. Gregory earned her doctorate at Zurich in 1886, and famed geographer Ellen Semple was allowed to attend Friedrich Ratzel's lectures at Leipzig in 1891 and 1892, but not to enroll (*NAW*). For the endless refusals that awaited women applicants to German universities in the 1880s, see Nathan H. Dole, "Biographical Sketch," in Helen Abbott Michael, *Studies in Plant and Organic Chemistry and Literary Papers* (Cambridge, Mass.: Riverside Press, 1907), pp. 26–87. On Zurich, see Flora Bridges, "Coeducation in Swiss Universities," *Popular Science Monthly* 38 (1890–91): 524–30; and Ernst Gagliardi, Hans Nabholz, and Jean Strohl, *Die Universität Zürich, 1833–1933, und Ihre Verläufer* (Zurich: Verlag der Erziehungsdirektion, 1938), pt. 5, chap. 9. Helen D. Webster, "Our Debt to Zurich," in *The World's Congress of Representative Women*, 2 vols., ed. May Wright Sewall (Chicago: Rand, McNally, 1894), 2: 269–99, is superficial. Russian students, apparently including women, often went abroad for graduate training in these years (Alexander Vucinich, *Science in Russian Culture, 1861–1917* [Stanford: Stanford University Press, 1970]; and Jan M. Meijer, *Knowledge and Revolution: The Russian Colony in Zurich, 1870–1873* [Assen, Netherlands: Van Gorcum, 1955]).

28. Christine Ladd-Franklin, "Report of Committee on Endowment of Fellowship," *PACA*, series 2, no. 7 (1888); "Fellowship Fund," *PACA*, no. 11 (1889); "The European Fellowship," *PACA*, no. 24 (1890); Christine Ladd-Franklin, "The Usefulness of Fellowships," *PACA*, no. 31 (1890). Members of the committee were: Christine Ladd-Franklin, Ellen Richards, Alice Freeman Palmer, Anna Botsford Comstock, Kate Stephens, Mary Sheldon Barnes, and Heloise Edwina Hersey. There is some correspondence about the committee in 1889 and 1890 from Christine Ladd-Franklin and Marion Talbot in the Kate Stephens Collection, University Archives, Spencer Research Library, University of Kansas. Ladd-Franklin discussed the issue of whether fellows could be engaged or married with Kate Stephens on 27 Mar. 1890. A list of fellows appears in Margaret E. Maltby, comp., *History of the Fellowships Awarded by the American Association of University Women, 1888–1929, with the Vitas of the Fellows* (Washington, D.C.: AAUW, [1929]). They are also discussed in Marion Talbot and Lois K. M. Rosenberry, *The History of the American Association of University Women, 1881–1931* (Boston: Houghton Mifflin, 1931), chap. 10, and Marion Talbot, "Mrs. Richard's Relation to the Association of Collegiate Alumnae," *JACA* 5 (1912): 302–4.

Bryn Mawr College started its own European fellowship in 1891 ("Academic Committee's Report on the Bryn Mawr Graduate School," pp. 7 and 8 n), and the Women's Education Association of Boston another in 1892 (Katharine P. Loring, "A Review of Fifty-Seven Years' Work," *Fifty-Seventh and Final Annual Report of the Women's Education Association for the Year Ending January 17, 1929*, p. 8). See also the list of "Fellows of the Woman's Education Association," *PACA*, series 3, no. 8 (supplement; 1903–4), pp. 14–15.

29. The only general history of the ACA is that by Talbot and Rosenberry, *History of the American Association of University Women*. See also its own series of *Publications* and Roberta Frankfort, *Collegiate Women: Domesticity and Career in Turn-of-the-Century America* (New York: New York University Press, 1977), chap. 6.

30. Quoted in Crow, "Status of Foreign Collegiate Education," p. 7. See also Vera Brittain, *The*

Women at Oxford: A Fragment of History (New York: Macmillan, 1960), chap. 4, "Unofficially Present (1880-1890)"; and Rita McWilliams-Tullberg, "Women and Degrees at Cambridge University, 1862-1897," in Martha Vicinus, ed., *A Widening Sphere: Changing Roles of Victorian Women* (Bloomington: Indiana University Press, 1977), pp. 117-45.

31. Bessie Bradwell Helmer to Mrs. Hearst, 1 May 1894, Phoebe Apperson Hearst Papers, Bancroft Library, University of California, Berkeley. I thank Gayle Gullett Escobar for bringing this collection to my attention. Helmer was known for her strong support for the ACA fellowships (Marion Talbot, *History of the Chicago Association of Collegiate Alumnae*, pp. 3-4 and 5). Her papers have not been found.

32. Maltby, comp., *History of the Fellowships*, pp. 13-14.

33. Felix Klein to "Hochgeehrte Frau Professor!," 15 May 1892, Christine Ladd-Franklin Papers, Special Collections, Butler Library, Columbia University. I thank Ian Dengler for help in transcribing and translating this letter. See also "Ladd-Franklin" in *NAW*.

34. "Hyde" and "Maltby" in *NAW;* Margaret Maltby, *History of the Fellowships*, pp. 14-16 (but error in year of Winston's degree; cf. *AMS*, 3d ed. [1921], p. 505); Margaret Maltby to Ida Hyde, 21 Sept. 1929, Ida Hyde Papers, AAUW Archives; E. C. Scott Barr, "Margaret Eliza Maltby," *American Journal of Physics* 28 (1960): 474-75; Caroline Newson Beshers (daughter of Mary Winston Newson) to author, 11 June 1974; see also Margaret E. Maltby, "A Few Points of Comparison between German and American Universities," *PACA*, series 2, no. 62 (1896), and H. S. White, comp., "A Brief Account of the Congress on Mathematics, Held at Chicago in August, 1893," in *Mathematical Papers Read at the International Mathematical Congress, Held in Connection with the World's Columbian Exposition, Chicago, 1893*, ed. Committee of the Congress, E. Hastings Moore et al. (New York: Macmillan, 1896), pp. vii-xii.

35. Ida Hyde, "Before Women Were Human Beings: Adventures of an American Fellow in German Universities of the '90s," *JAAUW* 31 (1938): 226-36; Kate Stephens to Ida Hyde, 27 Sept. 1929, Kate Stephens Collection, University of Kansas Archives; "She Opened German Universities to Women," *Kansas City Star*, 13 Apr. 1902, clipping in Ida Hyde Papers; comments by Sally Gregory Kohlstedt, session on "Pioneers in Science, 1880-1910," Berkshire Conference on Women in History, Mount Holyoke College, 25 Aug. 1978; two papers by Gertraude Wittig describe the complex legal and administrative situation Hyde encountered in Germany: "Hyde's 1896 Doctorate: A Lesson from the History of Women in Science" (Paper delivered at the National Women's Studies Association Annual Meeting, Bloomington, Indiana, May 1980), and her "Hyde's 1896 German Doctorate: Opening Science to Women" (Paper delivered at the Berkshire Conference on Women's History, Vassar College, June 1981).

36. Bessie Bradwell Helmer to Mrs. Hearst, 20 Sept. 1894, and Marion Talbot to Mrs. Phoebe Hearst, 3 Nov. 1894, both in Phoebe Apperson Hearst Papers, Bancroft Library, Berkeley, California. See also note 31 to this chapter. For more on women's experiences at German universities in the 1890s, see Alice Hamilton, "Edith and Alice Hamilton, Students in Germany," *Atlantic Monthly* 215 (Mar. 1965): 129-32, and the Ethel Puffer Howes Papers in the Morgan-Puffer Family Papers, SLRC. An attempt by the ACA to prepare a full list of women with foreign (as well as American) doctorates apparently failed (Martha Foote Crow, "[Request for Assistance]," *PACA*, series 2, no. 59 [1896]).

37. Laura D. Gill, "Report of the Committee upon the Establishment of a Council for Foreign University Work," *PACA*, series 2, no. 59-2 (1896); "Council to Accredit Women for Advanced Work in Foreign Universities," *PACA*, series 3, no. 1 (1898), pp. 97-98, and *PACA* no. 4 (1901), pp. 82, 92-93; "Report of the Internal Committee of the Council to Accredit Women for Advanced Work in Foreign Universities," *PACA*, series 3, no. 5 (1902), pp. 69-72; Helen T. Woolley, "Report of the Committee on Foreign Universities," *PACA*, series 4, no. 1 (1911), pp. 30-32, discusses eligibility of foreign degree recipients for ACA membership.

38. Isabel Maddison, *Handbook of British, Continental, and Canadian Universities with Special Mention of the Courses Open to Women* (New York: Macmillan, 1896); an endorsement appears in *PACA*, series 3, no. 1 (1898), pp. 102-3.

39. Little has been written on the entrance of German women into the universities. Good starting places are Richard J. Evans, *The Feminist Movement in Germany, 1894-1933* (Beverly Hills: Sage

Publications, 1976), pp. 17–21, 72–73, 176–77, and 187; George Bernstein and Lottelore Bernstein, "Attitudes toward Women's Education in Germany, 1870–1914," *International Journal of Women's Studies* 2 (1979): 473–88; E. Dühring, *Der Weg zur höheren Berufsbildung der Frauen* (Leipzig: Fues's Verlag, 1877); Arthur Kirchhoff, *Die Akademische Frau, Gutachten hervorragender Universitätsprofessoren, Frauenlehrer und Schriftsteller über die Befähigung der Frau zum wissenschaftlichen Studium und Berufe herausgegeben* (Berlin: H. Steinitz, 1897); [Frances Graham French], "The Status of Woman from the Educational and Industrial Standpoint," in *Report of the [U.S.] Commissioner of Education for 1897–98,* 2 vols., 1: 631–72, esp. 637–44; and, for background, see Helene Lange, *Lebenserinnerungen* (Berlin: F. A. Herbig, 1921), which ought to be translated into English.

The American women's influence was quite direct in one case. Alice Hamilton, for example, stayed with Dr. and Mrs. Edinger when in Germany in 1896. Their daughter Tilly, born a year later, earned a doctorate at Frankfurt University in 1921 and subsequently became a prominent German-American vertebrate paleontologist (Alice Hamilton to Tilly Edinger, 2 May 1964, Tilly Edinger Papers, Museum of Comparative Zoology Archives, Harvard University).

40. Christine Ladd-Franklin to Marion Talbot, 8 Dec. 1896, Marion Talbot Papers, Special Collections, Regenstein Library, University of Chicago.

41. William James to Christine Ladd-Franklin, 3 Mar. 1892, Christine Ladd-Franklin Papers, Special Collections, Butler Library, Columbia University.

42. The full story is in Laurel Furumoto, "Mary Whiton Calkins (1863–1930), Fourteenth President of the American Psychological Association," *Journal of the History of the Behavioral Sciences* 15 (1979): 346–56, which uses much Calkins correspondence still in private hands. See also "Calkins" in *NAW; In Memoriam, Mary Whiton Calkins, 1863–1930* (Boston: Merrymount Press, printed by D. B. Updike, 1931); and Bruce Kuklick, *The Rise of American Philosophy, Cambridge, Massachusetts, 1860–1930* (New Haven: Yale University Press, 1977), app. 4, "Women Philosophers at Harvard"; Hugo Münsterberg to Miss Calkins, 20 May 1902, Hugo Münsterberg Collection, Department of Rare Books and Manuscripts, Boston Public Library, Boston, Mass.; and Margaret Münsterberg, *Hugo Münsterberg: His Life and Work* (New York: D. Appleton, 1922), p. 76. Kate Morris had left Harvard in 1881 after her petition to its corporation requesting she be admitted to candidacy for a doctorate was denied (Thorp, *Radcliffe's First Ph.D.,* p. 24).

Although Münsterberg was quite liberal for his time and place in supporting Calkins and Puffer for Harvard doctorates, he was otherwise quite conservative about women's roles and strongly opposed the increasing feminization of culture, including schoolteaching (Kuklick, *Rise of American Philosophy,* pp. 211–12, and Matthew Hale, Jr., *Human Science and Social Order: Hugo Münsterberg and the Origins of Applied Psychology* [Philadelphia: Temple University Press, 1980], pp. 61–64). Interestingly, one reason that he had come to the United States in the 1890s was that his wife wanted their daughters to take advantage of the educational opportunities here that were not yet available in Germany (Phyllis Keller, *States of Belonging: German-American Intellectuals and the First World War* [Cambridge: Harvard University Press, 1979], p. 29).

43. Puffer: Josiah Royce to Miss Puffer, 23 May 1898, and "Report of the Committee on Honors and Higher Degrees of the Division of Philosophy in Harvard University," both in Ethel D. Puffer Faculty File, Smith College Archives.

44. Unidentified clipping, in Ethel D. Puffer Faculty File, Smith College Archives.

45. "Radcliffe Day, Degrees Conferred on 100 Graduates," *Boston Evening Transcript,* 26 June 1902, clipping in Morgan-Puffer Family Papers, SLRC; Agnes Irwin, "Report of the Dean [of the Radcliffe Graduate School]," in *Annual Report of the President and Treasurer of Radcliffe College, 1901–1902* (Cambridge: [Radcliffe College], 1902), pp. 13 and 21–22; Walter Crosby Eells, "Earned Doctorates," p. 660; there is apparently no history of the Radcliffe Graduate School, although one may be forthcoming now that the college has hired an archivist. The volume by the Radcliffe College Committee on Graduate Education for Women, *Graduate Education for Women: The Radcliffe Ph.D.* (Cambridge: Harvard University Press, 1956), is not historical.

46. Kathryn Jacob, "How Johns Hopkins Protected Women from 'The Rougher Influences,'" *The Johns Hopkins Magazine* 27 (Mar. 1976): 4–5, 7; Hugh Hawkins, *Pioneer,* chap. 14; John C. French, *A History of the University Founded by Johns Hopkins* (Baltimore: The Johns Hopkins

University Press, 1946), p. 147; "Ladd-Franklin" in *NAW*. D. C. Gilman's "Memoranda Submitted to the Trustees [on the Education of Women]," 5 Nov. 1877, lists of "Applications Refused," reports of action taken on applications accepted, and correspondence are in three folders on the "Admission of Women" in Special Collections and Manuscripts, Eisenhower Library, The Johns Hopkins University.

47. "Bascom" in *NAW;* George D. Williams to Florence Bascom, 16 Apr. 1891; quotation from G. H. Williams to Professor Edward Orton, 5 Jan. 1893; D. C. Gilman to Miss Bascom, 26 May 1893; Rebekah W. Griffin to "My dear Florence," n.d. [spring 1893]; D. C. Gilman to Miss Bascom, 5 July 1893; numerous clippings from New York and Baltimore newspapers; all in Florence Bascom Papers, Sophia Smith Collection, Smith College. See also W. S. Bayley, "Contributions of Hopkins to Geology," and Florence Bascom, "Fifty Years of Progress in Petrography and Petrology," both in *Fifty Years' Progress in Geology, 1876–1926,* ed. Edward Bennett Matthews, The Johns Hopkins University Studies in Geology, no. 8 (Baltimore: The Johns Hopkins University Press, 1926). Both of Bascom's parents were suffragists: Emma C. Bascom, "Reports of Vice-Presidents," *Papers and Reports Read before the Association for the Advancement of Women at its Annual Congress, . . . Buffalo, N.Y., October 1881* (Boston: Alfred Mudge & Sons, 1881), pp. 17–20, and obituaries in Florence Bascom Papers; "Woman Suffrage, Meeting of the Wisconsin Advocates of the Movement Address by John Bascom, President of State University," n.d., copy in Florence Bascom Papers. Although Walter Eells counted Constance Pessels, Ph.D. 1894, as a woman, Hawkins cited evidence he was male (Hawkins, *Pioneer*, p. 266 n. 21).

48. Christine Ladd-Franklin to Marion Talbot, 8 Dec. 1896, Marion Talbot Papers, Special Collections, Regenstein Library, University of Chicago, discusses the imminent opening of Hopkins to women. Daniel C. Gilman, "The Future of American Colleges and Universities," *Atlantic Monthly* 78 (1896): 176; Henry A. Rowland to editor of the *Nation,* 27 Sept. 1896, and Gilman's request that he not publish it (2 Oct. 1896), in folder on "Admission of Women," Special Collections and Manuscripts, Eisenhower Library, The Johns Hopkins University, Baltimore. Additional material on Ladd-Franklin's degree and A. O. Lovejoy's criticisms are in the University Archives, Eisenhower Library, The Johns Hopkins University, and in the Christine Ladd-Franklin Papers, Special Collections, Butler Library, Columbia University.

49. Christian A. Ruckmick to Professor R. M. Yerkes, 14 Mar. and 25 June 1927; and Christian A. Ruckmick to President Abbott L. Lowell, Robert A. Yerkes Papers, Yale University Library.

50. "Garrett" in *NAW;* Edith Finch, *Carey Thomas of Bryn Mawr* (New York: Harper & Bros., 1947), pp. 197–202; Alan M. Chesney, *The Johns Hopkins Hospital and the Johns Hopkins University School of Medicine: A Chronicle,* 2 vols. (Baltimore: The Johns Hopkins University Press, 1943 and 1958); Simon Flexner and James Thomas Flexner, *William Henry Welch and the Heroic Age of American Medicine* (New York: Viking Press, 1941), pp. 215–30; William H. Welch to F. P. Mall, 7 Nov. 1891; William H. Welch to M. Carey Thomas, 16 Feb. 1893, and William H. Welch to Harvey Cushing, 8 Aug. 1922, all in William H. Welch Papers, The Alan Mason Chesney Medical Archives of The Johns Hopkins Medical Institutions. "Sabin," in *NAW*: *MP*. Another famous student, Gertrude Stein, 1901, never completed her degree.

51. Reuben Peterson, "The Department of Obstetrics and Gynecology," in *The University of Michigan: An Encyclopedic Survey,* 4 vols., ed. Wilfred B. Shaw (Ann Arbor: University of Michigan Press, 1951), 3: 866–86.

52. Edward P. Cheyney, *History of the University of Pennsylvania, 1740–1940* (Philadelphia: University of Pennsylvania Press, 1940), pp. 303–9; Eells, "Earned Doctorates," p. 647; notes 3 and 16 to this chapter; and Karl G. Miller, "Daughters of [the University of] Pennsylvania," *General Magazine & Historical Chronicle, University of Pennsylvania* 39 (1937): 405–20.

53. Jan Butin Sloan, "The Founding of the Naples Table Association for Promoting Scientific Research by Women, 1897," *Signs* 4 (1978): 208–16. Christiane Groeben to Mrs. Elizabeth ten Houten, 19 May 1977, in Ida Hyde Papers, AAUW Archives, encloses a "List of American Women who worked at the Naples Zoological Station on the 'American Women's Table' (1898–1933)." For more on the NTA, see its own and ACA publications, such as *Association for Maintaining the American Woman's Table at the Zoological Station at Naples and for Promoting Scientific Research by Women, 1898–1903* (n.p., 1903) in the Ellen Richards Papers, MIT Archives; annual reports by

the ACA's delegate to the NTA in *PACA*, series 3, no. 10 (Jan. 1905), pp. 91-93; no. 13 (Feb. 1906), pp. 71-72; no. 14 (1907), pp. 94-95; no. 17 (Feb. 1908), pp. 154-55; no. 18 (Dec. 1908), pp. 123-24; no. 20 (Feb. 1910), pp. 36-37; *JAAUW* 4 (1911): 13-14; 5 (1912): 224-25; advertisements for the table are also scattered through the ACA's journal. Occasional reports by the Smith College delegate appeared in the *Smith College Monthly:* Olive Rumsey, "The Naples Table and Research Association," vol. 10 (May 1903), pp. 538-40; Elizabeth L. Clarke, "Sixth Award of the $1000 Research Prize," vol. 11 (May 1905), pp. 517-18; idem, "The Naples Table Association," vol. 13 (June 1906), pp. 600-601; idem, "The Naples Table Association," vol. 15 (June 1908), pp. 595-96; and idem, "Annual Report of the Naples Table Association," vol. 16 (June 1909), pp. 605-6. There is much correspondence about the NTA in the Ida Hyde Papers, including Anton Dohrn to Miss Hyde, 30 Apr. 1897, about the establishment of the American ladies' table.

Emily Nunn (of Wellesley College, though later Mrs. Charles O. Whitman) visited the station as early as 1883 and described it in an article, "The Naples Zoological Station," *Science* 1 (1883): 479-81 and 507-10 (the first article by a woman in that journal). Alice Upton Pearmain described it much later for ACA members in "The Zoological Station at Naples," *PACA*, series 3, no. 14 (Feb. 1907), pp. 1-13. Christiane Groeben, *The Naples Zoological Station at the Time of Anton Dohrn, Exhibition and Catalog* (Paris: Goethe Institute, 1975), contains a bibliography of articles about the station.

54. H. Jean Crawford, "The Association to Aid Scientific Research by Women," *Science*, n.s. 76 (1932): 492-93. The association also paid about $1,200 for the publication costs of Margaret Ferguson's dissertation on pine seeds in 1903, although she had come in third (behind Florence Sabin), and it gave Frances Wick $1,000 in 1923 and 1924 for her researches in physics, although these items are not mentioned in any of the association's reports (note attached to association circular, Margaret Ferguson Papers, Wellesley College Archives, and "Frances Wick" in Maltby, *History of the Fellowships*, p. 80). For more on the association, see chapter 11.

55. Christine Ladd-Franklin, "Endowed Professorships for Women," *PACA*, series 3 (Feb. 1904), p. 61. See also her "Report of the Committee on the Endowment of Fellowships," *PACA*, series 3, no. 17 (Feb. 1908), pp. 114 and 143-46, where she admitted defeat. Earlier correspondence on the same themes exists in Kate Holladay Claghorn to Christine Ladd-Franklin, 14 and 15 Oct. 1898 and 26 Mar. 1899; Elizabeth M. Howe to C. L. F., 28 May 1903; Florence Cushing to C. L. F., 21 June 1903; Warner Fite to C. L. F., 28 May 1905; and undated fragment C. L. F. to ______, all in Christine Ladd-Franklin Papers, Special Collections, Butler Library, Columbia University.

Starting with its founding in 1876 The Johns Hopkins University had offered fellowships to some of its graduate students, some of whom already held Ph.D.s from German or other American universities (J. C. French, *A History of the University Founded by Johns Hopkins* [Baltimore: The Johns Hopkins University Press, 1946], pp. 40-41).

56. Christine Ladd-Franklin, "Endowed Professorships for Women," p. 61.

57. "Berliner Research Fellowship," *JACA*, 12 (Apr. 1919): 133; and Maltby, comp., *History of the Fellowships*, pp. 8 and 66-84. It was Maltby who was particularly critical of Ladd-Franklin's ways, as in Margaret Maltby to Florence Sabin, 28 Jan. 1917, Florence Sabin Papers, APS, saying she had "no sympathy whatever with this attempt to offer the bait of a lecturer's salary in the shape of a fellowship for the privilege of saying she gave lectures in a certain university. I think it is unethical as well as unwise." There is also a folder of C. L. F.'s correspondence about the Berliner Fellowship in 1911-14 in the University Archives, Eisenhower Library, The Johns Hopkins University.

58. The General Federation of Women's Clubs made at least two attempts to start a Rhodes Fellowship for American Women, in 1909 and 1912 (Laura D. Gill to Ira Remsen, 6 Oct. 1909, and reply, 12 Oct. 1909, in folder 167, University Archives, Eisenhower Library, The Johns Hopkins University, and Edith [Abbott?] to Marion Talbot, [15 Oct. 1912], Marion Talbot Papers, Special Collections, University of Chicago Library). See also note 28 to this chapter.

CHAPTER 3

1. (Rev.) Phebe A. Hanaford, *Daughters of America, or Women of the Century* (Augusta, Maine: True, 1882), p. 270. Chapter 9 is the best description available of the women scientists of the 1870s, but

it omits some important ones, as, most noticeably, Ellen Swallow Richards of the Woman's Laboratory at MIT; Mary E. Murtfeldt, "Woman and Science," *Moore's Rural New Yorker* 27 (4 Jan. 1873): 19; Maria Mitchell, "Address of the President," *Papers Read at the Third Congress of Women, Syracuse, October 1875* (Chicago: Fergus Printing, 1875), pp. 1-7; Graceanna Lewis, "Science for Women," *Third Congress of Women, Syracuse*, pp. 63-73; [Edward C. Pickering], "Education," *Atlantic Monthly* 33 [1874]: 760-64.

2. In 1876 Christine Ladd-Franklin, a secondary school teacher at the time, wrote the president of Vassar College about his recent commencement address, which had displeased her greatly. He seemed to expect that the graduates would all be returning home and did not even suggest that some might become teachers or possibly try to study medicine or the law. She asked rhetorically, "Doesn't Vassar recognize honest work?" (Christine Ladd to President John H. Raymond, 7 July 1876, Christine Ladd-Franklin Papers, Special Collections, Columbia University); Maria Mitchell had used the same theme a few months before in her address to the Association for the Advancement of Women (Maria Mitchell, "The Need for Women in Science," *Papers Read at the Fourth Congress of Women, Philadelphia, 1876* [Washington, D.C.: Todd Bros., 1877], which suggested [p. 9] that women students would make good assistants in observatories and museums). Decades later a retrospective survey revealed that several Vassar graduates of the 1870s had eventually held jobs (Frances M. Abbott, "Three Decades of [Vassar] College Women," *Popular Science Monthly* 65 [1904]: 350-59).

3. This theme is discussed in Daniel T. Rodgers, *The Work Ethic in Industrial America, 1850-1920* (Chicago: University of Chicago Press, 1978), chap. 7, "Idle Womanhood"; and Roberta Frankfort, *Collegiate Woman: Domesticity and Career in Turn-of-the-Century America* (New York: New York University Press, 1977), chap. 6. See also Jane M. Bancroft, "Occupations and Professions for College-Bred Women," *Education* 5 (1885): 486-95; Charles F. Thwing, "What Becomes of College Women?" *North American Review* 161 (1895): 546-53; Alice M. Gordon, "The After-Careers of University-Educated Women," *Nineteenth Century* 37 (1895): 955-60; "Compensation in Certain Occupations of Women Who Have Received College or Other Special Training," *PACA*, series 2, no. 56 (1896), also published separately in the Massachusetts Bureau of Statistics of Labor (Boston: Wright & Potter, 1896); M. Carey Thomas, *Education of Women*, published as vol. 7 in *Monographs on Education in the United States*, ed. Nicholas Murray Butler, Department of Education for the United States Commission to the Paris Exposition of 1900 (Albany: J. B. Lyon, 1900), 1: 37; and Frances M. Abbott, "Three Decades."

4. Mary W. Whitney, "Scientific Study and Work for Women," *Education* 3 (1882): 58-69.

5. For example, Mrs. Emily Crawford, "Journalism as a Profession for Women," *Contemporary Review* 64 (1893): 362-71, and Margaret E. Sangster, "Editorship as a Profession for Women," *Forum* 20 (1895-96): 445-55.

6. The best accounts of the women astronomers at Harvard (and a few from elsewhere) are Bessie Z. Jones and Lyle Boyd, *The Harvard College Observatory: The First Four Directorships, 1839-1919* (Cambridge: Harvard University Press, 1971), chap. 11, "A Field For Women," with full bibliographic notes, and Pamela E. Mack, "Women in Astronomy in the United States, 1875-1920" (B. A. honors thesis, Harvard University, 1977). Anna Winlock had worked at the HCO in the 1870s. See also entries for "Cannon," "Fleming," and "Leavitt" in *NAW*; "Maury" in *NAW: MP*; Annie J. Cannon, "Williamina Paton Fleming," *Science* 33 (1911): 987-88, and *Astrophysical Journal* 34 (1911): 314-17; Edward C. Pickering, *In Memoriam: Williamina Paton Fleming* (Cambridge, Mass., 1911) in HUA, attests that she "occupied one of the most important positions in the Observatory." Helen Buss Mitchell, "Henrietta Swan Leavitt and Cepheid Variables," *Physics Teacher* 14 (1976): 162-67. (Leavitt was "extremely deaf" and Cannon less so.)

7. Women astronomers received an inordinate amount of attention in popular journals in these years; see, for example, E. LaGrange, "Women in Astronomy," *Popular Science Monthly* 28 (1885-86): 534-37; Esther Singleton, "Women as Astronomers," *Chautauquan* 14 (1891): 209-12 and 340-42; Helen Leah Reed, "Women's Work at the Harvard Observatory," *New England Magazine*, n.s., 6 (1892): 165-76; Herman S. Davis, "Women Astronomers," *Popular Astronomy* 6 (1898): 129-38, 211-20, and 220-28, which is in part a review of Alphonse Rebière's *Les Femmes dans la Science*, 2d ed. (Paris: Nony, 1897); so too is Dorothea Klumpke, "La Femme dans l'Astronomie," *Astronomie* 13 (1899): 162-70 and 206-15; Edward S. Holden, "On the Choice of a

Profession, II: Science," *Cosmopolitan Magazine* 24 (1898): 543–49 (which hypothesizes that women will never find happiness in science); and Anne P. McKenney, "What Women Have Done for Astronomy in the United States," *Popular Astronomy* 12 (1904): 171–82 (which contains errors and copies from Fleming [see note 8 to this chapter] in part); [Mabel L. Todd], "Women Astronomers at Harvard," *Harvard Alumni Magazine* 29 (1927): 420–22; and most recently, Anne Gordon, "Williamina Fleming: 'Women's Work' at the Harvard Observatory," *Women's Studies Newsletter* 6, no. 2 (spring 1978): 24–27. (I thank Sally Gregory Kohlstedt for a copy.)

8. Fleming, "A Field for 'Woman's Work' in Astronomy," *Astronomy and Astrophysics* 12 (1893): 688–89. See also Donald E. Osterbrock, "First World Astronomy Meeting in America," *Sky and Telescope* 56 (Sept. 1978): 183, which reports that Fleming was absent and did not read this paper herself.

9. Pamela Mack, "Women in Astronomy," chap. 4.

10. Ibid.; Deborah Jean Warner, "Women Astronomers," *Natural History* 88 (May 1979): 12, 14, 16, 20, 22, 24, and 26.

11. Renate Britenthal and Claudia Konz, "Beyond Kinder, Küche, Kirche: Weimar Women in Politics and Work," in *Liberating Women's History: Theoretical and Critical Essays*, ed. Bernice A. Carroll (Urbana: University of Illinois Press, 1976), p. 318.

12. Although most of the women astronomical assistants rejoiced in being allowed to do such interesting work, Antonia C. Maury was too independent to be such a Pollyanna and rankled at the close supervision and lack of recognition accorded her work, which did contradict some of Director Pickering's own classifications. She left the observatory in 1896 upon completion of a major volume on stellar spectra and did not return until 1918, when Pickering was seventy-two (he died a year later). (*NAW: MP;* Jones and Boyd, *Harvard College Observatory,* pp. 395–400.)

13. Thomas E. Drake, *A Scientific Outpost: The First Half-Century of the Nantucket Maria Mitchell Association* (Nantucket, Mass.: Nantucket Maria Mitchell Association, 1968). The association also raised $12,000 in 1916 to endow a fellowship at Harvard for a woman in astronomy. But, for reasons that are not entirely clear, the Harvard Corporation refused to administer this new fund as it did all its others and insisted that Professor Pickering would have to do it himself. He did, and the fellowship has apparently been in use ever since, but it is debatable how much it freed women graduate students from the "woman's work" at the observatory, since in 1917 it was offering just $500 in exchange for 1,500 hours of work on the photographic plates. Rather than freeing a woman for her studies, the new fellowship seemed to perpetrate the old pattern. (Jones and Boyd, *Harvard College Observatory,* 411–13, 480–81 nn. 25, 27; Edward C. Pickering to R. H. Baker, 12 Apr. 1917, Laws Observatory Papers, 1877–1954, University of Missouri, Western Historical Manuscripts Collection, University of Missouri. I wish to thank Nancy Lankford for bringing this set of letters to my attention.)

14. "Journal of Williamina Paton Fleming, Curator of Astronomical Photographs, Harvard College Observatory," HUA. Archivist Clark Elliott informed me that it was part of a larger, university-wide, historical project of collecting job descriptions that were to be opened in 1950.

15. Ibid., entry for 18 Apr. 1900. Only Henrietta S. Leavitt was able to get a raise from Pickering. After her return to Wisconsin in 1900 for a "family crisis," she stayed there until 1902, when Pickering enticed her back with a raise. She was apparently worth this extra sum, since in 1908 she discovered the relationship between a star's luminosity and its period of rotation (and so its distance from the earth), an achievement that opened up new areas of astrophysics and earned her a posthumous nomination for the Nobel Prize (Jones and Boyd, *Harvard College Observatory,* pp. 367–70, 400, 402, and 479 n. 15).

16. The earliest women museum workers found so far were those at the Boston Society of Natural History in the late 1860s. See Sally Gregory Kohlstedt, "The Nineteenth-Century Amateur Tradition: The Case of the Boston Society of Natural History," in *Science and Its Public: The Changing Relationship,* ed. Gerald Holton and William A. Blanpied (Dordrecht, Holland: D. Reidel, 1976), pp. 183–84. For statistics on natural history museum employees, see *Report of the Commissioner of Education for the Year 1873* (Washington, D.C.: GPO, 1874), table 17, pp. 764–73, and *Report of the Commissioner of Education for the Year 1874* (Washington, D.C.: GPO, 1875), table 16, pp. 794–801. The Museum of Comparative Zoology did not list any women employees in 1874.

17. Jeanne E. Remington, "Katharine Jeannette Bush: Peabody's Mysterious Zoologist," *Discovery* 12, no. 3 (1977): 3–8. I thank Patricia Bodak Stark for this reference. A. E. Verrill to Mary Jane Ruthbun, 15 Dec. 1907, Mary Jane Rathbun Papers, Smithsonian Institution Archives, contains an apology for his very low salary to Miss Bush.

18. "Rathbun" in *NAW;* Mary Jane Rathbun Papers, Smithsonian Institution Archives. There are also 135 M. J. Rathbun letters in the Museum Collection at the Archives of the Museum of Comparative Zoology, Harvard University. See also Lucile McCain, "Mary Jane Rathbun," *Science* 97 (1943): 435–36, and Waldo L. Schmitt, "Obituaries," *Journal of the Washington Academy of Sciences* 33 (1943): 351–52.

19. "Fletcher," "Nuttall," and "Smith" in *NAW;* see also Ralph Dexter, "Guess Who's Not Coming to Dinner: Frederic Ward Putnam and the Support of Women in Anthropology," *History of Anthropology Newsletter* 5, no. 2 (1978): 5-6. Stevenson in Martin Meyerson and Dilys Pegler Winegrad, *Gladly Learn and Gladly Teach: Franklin and His Heirs at the University of Pennsylvania, 1740–1976* (Philadelphia: University of Pennsylvania Press, 1978), chap. 10; "Hearst" in *NAW,* and Phoebe Apperson Hearst Papers, Bancroft Library, University of California at Berkeley.

20. "Brandegee" in *NAW;* and "Eastwood" in *NAW: MP;* Alice Eastwood Papers, Library, California Academy of Sciences. (I thank archivist Margaret Campbell for allowing me to see these papers.) See also Susanna Bryant Dakin, *Perennial Adventure: A Tribute to Alice Eastwood, 1859–1953* (San Francisco: California Academy of Sciences, 1954), and Carol Green Wilson, *Alice Eastwood's Wonderland: The Adventures of a Botanist* (San Francisco: California Academy of Sciences, 1955).

21. There was also what can be called a "health" or "protection" rationale for keeping the women *out* of certain areas. Thus even the relatively liberal Pickering reasoned in 1885 that because it was too cold for the women to use the telescope on winter nights, he could not let them use it any night of the year. He did not want them to get all the colds and coughs that the men seemed to get from their night viewing. Nor did he think it unfair that this prohibition cut women off from participating in the whole field of observational astronomy, where there was, apparently, more glory (and higher pay) than in daytime computing and classifying in the back rooms. This ban on women using the major telescopes of the nation, presumably for their own good health, reportedly spread, persisted, and was rigidly enforced at most major observatories until recent times (women could and did use the smaller telescopes at the women's colleges), thus greatly limiting not only the role women played on a scientific team but also the subject matter of astronomical research that women could do (Jones and Boyd, *Harvard College Observatory*, p. 188).

22. For example, G. Stanley Hall thought that child study "should be preeminently the woman's science" (G. Stanley Hall, "Practical Child Study" [1894], cited in Dorothy Ross, *G. Stanley Hall, The Psychologist as Prophet* [Chicago: University of Chicago Press, 1972], p. 260).

23. The history of scientific librarianship remains to be written. Librarianship was already highly feminized in the United States by the 1890s, when the major scientific libraries began to hire full-time librarians and thus transferred into science a sex-typed occupational hybrid. Mary Day of the Gray Herbarium at Harvard is an example of a type of scientific worker which must have been common, though now largely lost from view. A former schoolteacher and public librarian, she was hired by the herbarium in 1893 because, though she claimed to know no botany, she was still better qualified than the other applicants. She not only did library work but also checked bibliographical references, proofread manuscripts, prepared indexes, and collected statistics on the library and the herbarium collections. From 1909 to 1915, when the herbarium was rebuilt, "Miss Day showed much executive capacity," to cite her main obituary. She also had a wide botanical correspondence, remembered everybody, and at the time of her death in 1924 had completed the massive *Card Index of New Genera, Species, and Varieties of American Plants*, consisting of about 170,000 cards. Although "this great work was merely incidental in Miss Day's routine," it was soon deemed "indispensable" to botanists. Despite all this, Day's salary, like Mrs. Fleming's at the observatory, remained meager (B. L. Robinson, "Miss Day," *Rhodora* 26 [1924]: 41–47; Dee Garrison, *Apostles of Culture: The Public Librarian and American Society, 1876–1920* [New York: Free Press, 1979]).

24. Michael Spence, *Market Signaling: Information Transfer in Hiring and Related Screening Processes* (Cambridge: Harvard University Press, 1974). I thank Jeffrey Escoffier for this reference.

25. J. F. A. Adams, "Is Botany a Suitable Study for Young Men?," *Science* 9 (1887): 117-18 (I thank Sally Kohlestedt for this reference), and Emmanuel D. Rudolph, "How It Developed that Botany Was the Science Thought Most Suitable for Victorian Young Ladies," *Children's Literature* 2 (1973): 92-97. Although the Garden Club of America was not formally established until 1913, so many women of the 1880s and 1890s were gardeners or "horticulturalists" as well as "botanists" that the men became alarmed. The Mira Lloyd Dock Papers in the Manuscript Division of the Library of Congress contain several items of correspondence about the sex typing under way in botanical fields. (For example, William Frear to Myra [*sic*] Dock, 23 May 1899, informs her that courses in horticulture at Pennsylvania State College are open to women, though none has attended, and that though women could become horticulturalists and floriculturalists at state experiment stations, again none had yet done so. F. E. Olmsted of the U.S. Bureau of Forestry wrote her on 5 Nov. 1902 that "forestry is a man's work," admitting, though, that her lectures on conservation had been helpful to the movement.)

26. Andrew Denny Rodgers, III, *Erwin Frink Smith: A Story of North American Plant Pathology,* published as *Memoirs of the American Philosophical Society* 31 (1952): passim, esp. pp. 174, 211, 379-80, 432, 481-82, 651. See also obituaries, as of Nellie Brown, *Washington Evening Star,* 15 Sept. 1956 (I thank Paul Lentz of the USDA for this), and Della E. Watkins, *Washington Evening Star,* 12 Apr. 1977; Alice L. Robert and Johns G. Moseman, "Charlotte Elliott, 1883-1974," *Phytopathology* 66 (1976): 237; and Gladys Baker, "Women in the United States Department of Agriculture," *Agricultural History* 50 (1976): 190-201. None of these women is mentioned in John M. Coulter, "The Botanical Work of the Government," *Botanical Gazette* 20 (1895): 264-68, which describes the work of twenty-seven male scientists at the USDA.

27. Margaret W. Rossiter, "Women Scientists in America before 1920," *American Scientist* 62 (1974): 316 (data based on first three editions of *AMS*). Flora Wambaugh Patterson was hired by another part of the BPI in 1896. See chapter 8 herein.

28. Office of Public Affairs, U.S. Civil Service Commission, *Biography of an Ideal: A History of the Federal Civil Service* (Washington, D.C.: GPO, 1973), pp. 161-62; "Barton" in *NAW;* R. McClelland to Hon. Alex DeWitt, 27 Sept. 1855, Letters Sent, vol. I. Patents and Miscellaneous Division, Records of the Office of the Secretary of the Interior, RG 48, reprinted as "Official Objections to Employing Women Clerks in the Patent Office, 1855," in *Clio Was a Woman, Studies in the History of American Women,* ed. Mabel E. Deutrich and Virginia C. Purdy (Washington, D.C.: Howard University Press, 1980), p. 2. See also Carroll Pursell, "The Cover Design: Women Inventors in America," *Technology and Culture* 22 (1981): 545-49.

29. "Smith," "Stevenson," "Fletcher," and "Hearst" in *NAW;* there are many letters from Matilda Stevenson in the Incoming Correspondence of the Bureau of American Ethnology, 1890-1908, in National Anthropological Archives, Smithsonian Institution; *Organization and Historical Sketch of the Women's Anthropological Society of America* (Washington, D.C.: The Society, 1889); see also chapter 4 herein; "Women in Science: Mrs. Nuttall Believes that a Promising Field in Archaeology is Open to Them," *New York Tribune,* 30 June 1900 (account of recent meeting of American Association for the Advancement of Science), clipping in Alice Fletcher Papers, National Anthropological Archives, Smithsonian Institution; Franz Boas to Zelia Nuttall, 18 May 1901, in Phoebe Apperson Hearst Papers, Bancroft Library, University of California, Berkeley. See also George W. Stocking, "Franz Boas and the Founding of the American Anthropological Association," *American Anthropologist* 62 (1960): 1-17.

30. Although most of the women scientists in the United States employed before 1910 taught at the women's colleges, and many persons must have thought this was clearly one kind of "women's work," few said so, since many men taught there as well, often in the highest-ranking positions (Rossiter, "Women Scientists," p. 318).

31. Mitchell, "Address of the President," pp. 4-5.

32. Octavia W. Bates, "Women in Colleges," *Papers Read before the Association for the Advancement of Women, Nineteenth Woman's Congress, Grand Rapids, Michigan, October 1891* (Syracuse: C. W. Bardeen, 1892), pp. 19-21. Bates later became a lawyer (Dorothy Gies McGuigan, *A Dangerous Experiment: One Hundred Years of Women at the University of Michigan* [Ann Arbor: Center for Continuing Education of Women, 1970], p. 77). A year earlier, in 1890, Emily Wheeler of

the Chicago branch of the ACA had presented a paper on "Women in College Instruction," which claimed that state colleges and universities were discriminating against women faculty, but her report was apparently not published (Marion Talbot, *History of the Chicago Association of Collegiate Alumnae, 1888-1917* [Chicago: private, 1920], pp. 4-5).

33. Morris Bishop, *A History of Cornell* (Ithaca: Cornell University Press, 1962), p. 380; see also pp. 337-38, 379-81, and 388-89; see also Flora Rose, "Forty Years of Home Economics at Cornell University," in *A Growing College: Home Economics at Cornell University* (Ithaca: Cornell University Press, 1969), pp. 19 and 34-38. "Comstock" is in *NAW*.

34. Charlotte Williams Conable, *Women at Cornell: The Myth of Equal Education* (Ithaca: Cornell University Press, 1977), pp. 126-30.

35. "Rose" in *NAW;* Juanita Archibald Eagles, Orrea Florence Pye, and Clara Mae Taylor, *Mary Swartz Rose, 1874-1941: Pioneer in Nutrition* (New York: Teachers College Press, 1979); Elmer V. McCollum, *A History of Nutrition: The Sequence Ideas in Nutrition Investigations* (Boston: Houghton Mifflin, 1957); Margaret W. Rossiter, *The Emergence of Agricultural Science: Justus Liebig and the Americans, 1840-1880* (New Haven: Yale University Press, 1975); Edward C. Kirkland, "'Scientific Eating': New Englanders Prepare and Promote a Reform, 1873-1907," *Proceedings of the Massachusetts Historical Society* 86 (1974): 28-52.

36. Catherine Beecher, *Treatise on Domestic Economy for the Use of Young Ladies at Home and at School* (Boston: Marsh, Capen, Lyon and Webb, 1840); Kathryn Kish Sklar, *Catherine Beecher: A Study in American Domesticity* (New Haven: Yale University Press, 1973), chap. 11; Joan N. Burstyn, "Catherine Beecher and the Education of American Women," *New England Quarterly* 47 (1974): 386-403. Linda Marie Fritschner, "Women's Work and Women's Education: The Case of Home Economics, 1870-1920," *Sociology of Work and Occupations* 4 (1977): 209-234, is suggestive.

37. Dee Garrison, *Apostles of Culture;* and idem, "The Tender Technicians: The Feminization of Public Librarianship, 1876-1905," *Journal of Social History* 6 (1972-73): 131-59; James Leiby, *A History of Social Welfare and Social Work in the United States, 1815-1972* (New York: Columbia University Press, 1978); Roy Lubove, *The Professional Altruist: The Emergence of Social Work as a Career, 1880-1930* (Cambridge: Harvard University Press, 1965); Mrs. Frances Fisher Wood, "The Scientific Training of Mothers," *Papers Read before the Association for the Advancement of Women, Eighteenth Woman's Congress, Toronto, Canada, October 1890* (Fall River, Mass.: J. H. Franklin, 1891), pp. 39-48.

38. Eugene W. Hilgard, "Progress in Agriculture by Education and Government Aid," *Atlantic Monthly* 49 (1882): 531-41 and 651-61; Alfred C. True, *A History of Agricultural Education in the United States, 1785-1925* (Washington, D.C.: GPO, 1929), pp. 267-72; [Walton C. John, ed.] *Land Grant College Education, 1910-1920. Part V: Home Economics,* U.S. Bureau of Education, Bulletin 1925, no. 29.

39. Caroline L. Hunt, *The Life of Ellen H. Richards* (Boston: Whitcomb and Barrows, 1912), p. 91.

40. Ellen Richards's 1880 address on "Woman's Work in the Laboratory" to the Association for the Advancement of Women is reported in Mary Clemmer, "Women in the Laboratory," *Pacific Rural Press* 21 (8 Jan. 1881): 22. (The AAW's usual annual volume of "papers read" was not published for 1880.) The best short biography of Richards is that by Janet Wilson James in *NAW;* the standard one that by Hunt, *Life;* tributes by contemporaries include: Marion Talbot, "Mrs. Richards' Relation to the Association of Collegiate Alumnae," *JACA* 5 (1912): 302-4, and Isabel Bevier, "Mrs. Richards' Relations to the Home Economics Movement," manuscript in the Isabel Bevier Papers, University of Illinois Archives. A provocative recent biography is Robert Clarke, *Ellen Swallow: The Woman Who Founded Ecology* (Chicago: Follett Publishing, 1973). The early days of the Woman's Laboratory are discussed in Ellen Richards to Edward Atkinson, July 1878, included as an appendix to his "Report to the Committee on Subscriptions, 1882" (mimeographed) in MIT Archives. (I thank Helen Slotkin and Deborah Cozort for bringing this to my attention.)

41. Ellen Richards, "The Relation of College Women to Progress in Domestic Science," *PACA*, series 2, no. 27 (1890). (Some of her notes for this talk are in the Ellen Richards Papers, Sophia Smith Collection, Smith College Library, Northampton, Mass.) Besides the Richards Collections at

MIT and Smith, the best collection of her letters is her correspondence with Edward Atkinson, 1889-1900, in his papers at the Massachusetts Historical Society. There are also Ellen Richards Papers at the American Home Economics Association in Washington, D.C., in the Vassar College Archives, in the William Barton Rogers Papers at the MIT Archives, and at the MIT Special Collections. A published edition of her letters would be a great step forward. See also Sally Gregory Kohlstedt, "Single-Sex Education and Leadership: The Early Years of Simmons College," in *Women and Educational Leadership: A Reader,* ed. Sari Knopp Biklen and Marilyn B. Brannigan (Lexington: Lexington, 1980), pp. 93-112.

42. Emma Seifrit Weigley, "It Might Have Been Euthenics: The Lake Placid Conferences and the Home Economics Movement," *American Quarterly* 26 (1974): 79-96, analyzes the conferences' published proceedings. Isabel Bevier and Susannah Usher, *The Home Economics Movement* (Boston: Whitcomb & Barrows, 1912), describes the roots of the movement. Bevier expanded this in her *Home Economics in Education* (Philadelphia: J. B. Lippincott, 1924).

43. "Bevier," "Marlatt," and "Rose" in *NAW;* Lita Bane, *The Story of Isabel Bevier* (Peoria, Ill.: Charles A. Bennet, 1955); Richard G. Morris, *Fields of Rich Toil: The Development of the University of Illinois College of Agriculture* (Urbana: University of Illinois Press, 1970), pp. 176-204; Isabel Bevier, "Chapters from the Lives of Leaders: How I Came to Take Up Home Economics Work," *The Home Economist* 6 (1928): 117, 136, and 140. The Isabel Bevier Papers are at the University of Illinois Archives. Eugene Davenport, "Home Economics at Illinois," *Journal of Home Economics* 13 (1921): 337-41; Ida Hyde, "The Beginnings of the Science as a Development of Intellectual Personality," manuscript (1910) in Ida Hyde Papers, University of Kansas Archives, Spencer Research Library, esp. pp. 8, 9, 12. For more on early vocational guidance for women, see chapter 9 herein.

44. "E. Mosher" and "Welsh" in *NAW;* "C. Mosher" in *AMS,* 2d ed. (1910), and Kathryn Allamong Jacob, "Clelia Duel Mosher," *Johns Hopkins Magazine* (June 1979), pp. 8-16; William W. Guth, "Impressions of Lillian Welsh," *Goucher Alumnae Quarterly* 4 (Dec. 1924): 1-10, and Florence Sabin, "Dr. Lillian Welsh," *Goucher Alumnae Quarterly* 9 (Feb. 1931): 3-5; and "Dr. Mary Sherwood," *Goucher Alumnae Quarterly* 13 (July 1935): 9-13; *A Tribute to Lillian Welsh* (Baltimore: Goucher College, 1938); Lillian Welsh, "The Significance of Medicine as a Profession for Women," *Bulletin of the Woman's Medical College of Pennsylvania* 74 (1924): 3-9, is partly autobiographical; see also Thomas A. Storey, "The Status of Hygiene Programs in Institutions of Higher Education in the United States," *Stanford University Publications, University Series, Medical Sciences* 2, no. 1 (1927), and James F. Rogers, *Instruction in Hygiene in Institutions of Higher Education,* U.S. Office of Education, Bulletin no. 7 (1936); Willystine Goodsell, *The Education of Women: Its Social Background and Its Problems* (New York: Macmillan, 1923), chap. 9; Dorothy S. Ainsworth, *The History of Physical Education in the United States* (New York: A. S. Barnes, 1942); Arthur Weston, *The Making of American Physical Education* (New York: Appleton-Century-Crofts, 1962); Deobold B. Van Dalen et al., *A World History of Physical Education* (Englewood Cliffs, N.J.: Prentice-Hall, 1953), chaps. 23 and 24; Fred E. Leonard and George B. Affleck, *A Guide to the History of Physical Education*, 3d rev. ed. (Philadelphia: Lea and Febiger, 1947), chap. 23, esp. pp. 279, 289, and 342-43.

45. "Talbot," "E. Mosher," and "Washburn" in *NAW;* Breed: Thomas D. Clark, *Indiana University, Midwestern Pioneer,* 2 vols. (Bloomington: Indiana University Press, 1970, 1973), 2: 26-27 (which incorrectly reports her Bryn Mawr doctorate as being from Heidelberg); I wish to thank Dolores Lahrman of the Indiana University Archives for copies of Breed's correspondence with President Joseph Swain about her title and the faculty status of the dean in 1901; "Mitchell" in *NAW: MP* and Lucy Sprague Mitchell, *Two Lives: The Story of Wesley Clair Mitchell and Myself* (New York: Simon and Schuster, 1953), pp. 192-93; Gates: Mary Louise Filbey, "The Early History of the Deans of Women, University of Illinois, 1897-1923," typescript, 1969 (which ought to be published), p. 90, in Filbey Family Papers, University of Illinois Archives; a few letters from Gates's earlier, happier, days are in the Edwin G. Conklin Papers, Special Collections, Princeton University Library, and the Ernest Rutherford Papers, Cambridge University Library (on microfilm at Office for the History of Science and Technology, University of California, Berkeley. I thank J. L. Heilbron for bringing them to my attention). The early assumption that women professors would be very

solicitous about the personal needs of their female students is evident (humorously so) in Mrs. Caroline A. Soule, "A Collegiate Education for Women, and the Necessity of a Woman-Professor in the Mixed College," *Papers and Letters Presented at the First Woman's Congress of the Association for the Advancement of Women, New York, October 1873* (New York: Mrs. William Ballard, Printer, 1874), pp. 60-67.

46. W. H. P. Faunce to Marion Talbot, 14 Feb. 1900, Marion Talbot Papers, Special Collections, University of Chicago Library.

47. Gertrude S. Martin, "The Position of the Dean of Women," *JACA* 4, series 4 (1911): 65-77; the issue of faculty status bothered Martin particularly, since Cornell considered her only an "adviser" to the women students, not even a "dean." (The administration did not want to admit to the segregation that a deanship would imply.) In 1916 Martin resigned in protest, and the ACA passed a resolution of protest on her behalf ("Condensed Minutes of the Council Meeting, Held in Chicago, April 22, 1916," *JACA* 9, no. 3 [May 1916]: 85; J. Schurman to Miss Talbot, 26 Apr. 1916, and her reply, 3 May 1916, both in Marion Talbot Papers, Special Collections, University of Chicago Library); also Marion Talbot, *More Than Lore: Reminiscences of Marion Talbot* (Chicago: University of Chicago Press, 1936), esp. chap. 9; Lulu Holmes, *A History of the Position of Dean of Women in a Selected Group of Co-Educational Colleges and Universities in the United States,* published as Teachers College, Columbia University, Contributions to Education, no. 767 (1939); other Contributions to Education publications are: Sarah M. Sturtevant et al., *Trends in Student Personnel Work,* no. 787 (1940), and Jane Louise Jones, *A Personnel Study of Women Deans in Colleges and Universities*, no. 326 (1928), plus several others in the same series, as well as Lois Kimball Mathews, *The Dean of Women* (Boston: Houghton Mifflin, 1915), and Mildred Bunce Sayre, "Half a Century, An Historical Analysis of the National Association of Deans of Women, 1900-1950" (Ed.D. diss., Teachers College, Columbia University, 1950).

48. For example, Ellen Hayes, "Women in Scientific Research," *Science* 32 (1910): 864-66; and Marion Talbot, "Eminence of Women in Science," *Science* 32 (1910): 866.

CHAPTER 4

1. Cited in Paul Monroe, *Founding of American Public School System* (New York: Macmillan, 1940), 1: 449. See also Barbara Welter, "Anti-Intellectualism and the American Woman, 1800-1860," *Mid-America* 48 (1966): 258-70.

2. Mrs. Percival to John Torrey, n.d. [1837?], John Torrey Papers, New York Botanical Gardens Archives (hereafter NYBGA).

3. John W. Harshbarger, *The Botanists of Philadelphia and Their Work* (Philadelphia: T. C. Davis & Son, 1899), p. 300-301 (cited in Sally Gregory Kohlstedt, "In from the Periphery: American Women in Science, 1830-1880," *Signs* 4 [1978]: 90). Treat seems to have been of two minds over the issue of modesty and recognition. Although always asserting her reluctance to speak out, she did not hesitate to correct Asa Gray and Charles Darwin on matters regarding her specialty, the insectivorous plants, and she published many popular articles and books (Harshbarger, pp. 298-302). A review of her *Home Studies in Nature* (1885) appeared in *NYT*, 19 Apr. 1885 (clipping in Museum Collection, Museums of Comparative Zoology Archives, Harvard University). Treat and her work deserve a more complete study.

4. Kohlstedt, "In from the Periphery," pp. 83-85. Erminnie A. P. Smith was very active in the Aesthetic Society of Jersey City, New Jersey, from 1876 until her death in 1886 (*NAW: Echoes of the Aesthetic Society of Jersey City* [New York: Thompson & Moreau, 1882]; *In Memoriam: Mrs. Erminnie A. Smith* [Boston: Lee & Shepard, 1890]). Harshbarger, *Botanists of Philadelphia,* describes the male leaders in the Botanical Society of Pennsylvania in the text (pp. 22-23) and lists the largely female membership from 1897 to 1899 in the appendix (pp. 417-21); Cora Clarke of Boston mentioned her little club in a letter to Dr. Hermann Hagen in 1883 (Cora Clarke to Dr. Hagen, 30 Sept. 1883, Museum Collection, Museum of Comparative Zoology Archives) and reported frequently between 1890 and 1906 to Elizabeth Britton about her unnamed Boston botanical group, which may have been affiliated with the New England Woman's Club. In 1903 she wrote, "It's much

nicer studying together than alone" (Clarke to Britton, 2 Feb. 1903. Elizabeth Britton Papers, NYBGA). See also "Cora H. Clarke," *Psyche* 23 (1916): 94, and "Miss Clarke is Winning Fame as 'Plant Doctor,' " *Boston Sunday Post,* Jan. 1910(?), p. 40, clipping in Museum Collection, Museum of Comparative Zoology Archives.

In his book *Rockdale* (New York: Alfred A. Knopf, 1978), about life in the Mid-Atlantic region south of Philadelphia before the Civil War, Anthony F. C. Wallace discusses the women's intellectual life. Often graduates of Philadelphia academies and well trained in botany and natural history, they discussed local flora and fauna in their correspondence. At one point Sophie Du Pont proposed to a friend that they prepare a naturalist's journal of the region and then publish it anonymously. Even this upset her friend, who recorded in her diary, "Oh dear! I was shocked at the bare idea of any words of *mine* in any way appearing in print!" (p. 29).

5. Harry B. Weiss and Grace M. Ziegler, *Thomas Say: Early American Naturalist* (Springfield, Ill.: Charles C. Thomas, 1931), chap. 16, on Lucy Say. Graceanna Lewis was the academy's ninth woman member when she was elected (after an initial refusal) in 1870 (Deborah Jean Warner, *Graceanna Lewis: Scientist and Humanitarian* [Washington, D.C.: Smithsonian Institution Press, 1979], pp. 96–98, and Harshbarger, *Botanists of Philadelphia*, p. 235). Several other women joined in the 1870s, as Lewis reported to Maria Mitchell in the latter's survey of women scientists in 1880 (Graceanna Lewis's reply, Maria Mitchell Microfilms, reel 6, item no. 47 [1880], APS). See also Anthony F. C. Wallace, *Rockdale,* p. 275, for mention of a Mme. Fretageot, who was on the fringes of the academy in the 1820s.

6. "Mitchell" and "Phelps" in *NAW;* Sally Gregory Kohlstedt, *The Formation of the American Scientific Community: The Association for the Advancement of Science, 1848–1860* (Urbana: University of Illinois Press, 1976), p. 103.

7. Robert C. Miller, "The California Academy of Sciences and the Early History of Science in the West," *California Historical Society Quarterly* 21 (1942): 368; "List of Members of the California Academy of Natural Sciences, January 1863," *Proceedings of the California Academy of Natural Sciences* 2 (1858–62): v–vii. Botanist Mary L. K. C. Brandegee, who joined in 1879, was added to the staff in the 1880s.

8. *Proceedings of the American Association for the Advancement of Science, 1866–80*; regarding fellows see particularly the volumes for *1873*, pp. xvii and 400–403; *1874*, pp. xli–iv, 150–51, and "Constitution," article 4, n.p.; unidentified clipping of 1880 meeting, Alice Fletcher Papers, box 5, National Anthropological Archives, Smithsonian Institution. Historian and writer Martha Lamb kept some memorabilia from the AAAS meetings of 1880 and 1881 and diaries of her experiences at those of 1882 and 1891 (Martha Lamb Papers, Forbes Library, Northampton, Mass.).

9. *Proceedings of the Boston Society of Natural History, 1875–77,* and Sally Gregory Kohlstedt, "The Nineteenth-Century Amateur Tradition: The Case of the Boston Society of Natural History," in *Science and Its Public: The Changing Relationship,* ed. Gerald Holton and William A. Blanpied (Dordrecht, Holland: D. Reidel, 1976), pp. 184–86 and 190 n. 64.

10. T. S. Palmer, "Brief History of the American Ornithologists' Union," in *Fifty Years' Progress of American Ornithology, 1883–1933,* ed. Frank M. Chapman and T. S. Palmer (Lancaster, Pa., 1933), p. 10; Paul H. Oehser, "In Memoriam: Florence Merriam Bailey," *Auk* 69 (1952): 26.

11. "1774–Centennial of Chemistry–1874," *American Chemist* 5 (1874): 35–40; "Bodley" in *NAW; Papers Read at the Memorial Hour Commemorative of the Late Professor Rachel L. Bodley, M.D.* ([Philadelphia: Women's Medical College of Philadelphia], 1888); Gulielma Fell Alsop, "Rachel Bodley, 1831–1888," *Journal of the American Medical Women's Association* 4 (1949): 534–36; Harshbarger, *Botanists of Philadelphia,* pp. 283–85, discusses her contributions to that field; an obituary of Bodley by Caroline Dall in Caroline Dall Papers, SLRC; Mary L. Willard, "Pioneer Women in Chemistry," mimeographed (Pennsylvania State University, 1940), p. 7; Rachel Bodley Papers, Archives and Special Collections on Women in Medicine, (Women's) Medical College of Pennsylvania; "Shattuck" in *NAW*; clipping from *New York World,* 1 Aug. 1874, p. 1, in Lydia Shattuck Papers, Mount Holyoke College Archives. None of these women is mentioned in Herman Skolnik and Kenneth M. Reese, eds., *A Century of Chemistry: The Role of Chemists and the American Chemical Society* (Washington, D.C.: American Chemical Society, 1976), pp. 4–7.

12. "Proceedings" and "List of Officers, Honorary Members, Members, and Associates of the American Chemical Society," in *Journal of the American Chemical Society,* 1879–95, passim; *The*

Misogynist Dinner of the American Chemical Society, Boston, August 27, 1880, Photophonically Reported [by Henry Morton] (New York: Russell Bros., 1880). Professor T. Sterry Hunt of MIT took a prominent role in the dinner's toasts and songs. The episode is not mentioned in a report of the meeting (F. W. Clarke, "The Subsection of Chemistry at Boston," *American Chemical Journal* 2 [1880–81]: 274–79), which did list two women (Helena Stallo and Rachel Lloyd) as coauthors of two papers presented.

13. Cited in Kohlstedt, "In from the Periphery," p. 82.

14. Ralph W. Dexter, "The Salem Meeting of the American Association for the Advancement of Science (1869)," *Essex Institute Historical Collections* 93 (1957): 263; *Proceedings of the American Association for the Advancement of Science, 1881–84;* "Smith" and "Fletcher" in *NAW;* see also Nancy Oestreich Lurie, "Women in Early Anthropology," in *Pioneers of American Anthropology*, ed. June Helm [MacNeish] (Seattle: University of Washington Press, 1966), pp. 42–43. The only biography of Putnam (Alfred M. Tozzer, "Frederic Ward Putnam, 1839–1915," *Biographical Memoirs of the National Academy of Sciences* 16 [1936]: 125–53), gives no clue as to the source of his feminism, but Joan Mark suggests his first wife, who died about 1878 (letter to author, 3 June 1980). See also Ralph Dexter, "Guess Who's Not Coming to Dinner: Frederic Ward Putnam and the Support of Women in Anthropology," *History of Anthropology Newsletter* 5 (1978): 5–6.

15. The Women's Anthropological Society of America is discussed in Nancy O. Lurie, "Women in Early Anthropology," pp. 29–81, and most recently in Joan Cindy Amatniek, "The Women's Anthropological Society of America: A Dual Role—Scientific Society and Woman's Club" (undergraduate honors thesis, Harvard University, 1979). (I thank Joy Harvey for this reference.) Older items about it include *The Organization and the Constitution of the Women's Anthropological Society, Founded 1885* (Washington, D.C. [1885]); Anita Newcomb McGee, "The Women's Anthropological Society of America," *Science* 13 (1889): 240–42; and Susan A. Mendenhall, "Statistical Sketch of the Woman's Anthropological Society," in its *Proceedings of the One-Hundredth Meeting, January 28, 1893* (Washington, D.C.: Gibson Bros., 1893), pp. 4–9; the Library of Congress has several of the WASA's minor publications. See also D. S. Lamb, "The Story of the Anthropological Society of Washington," *American Anthropologist* 8 (1906): 564–79. The WASA is also mentioned briefly in Jennie C. Croly, *The History of the Woman's Club Movement in America* (New York: Henry G. Allen, 1898), pp. 341–42; George M. Kober, *The History and the Development of the Housing Movement in the City of Washington, D.C., 1897–1927* (Washington, D.C., Washington Sanitary Housing Companies, 1927), pp. 10–11; and J. Kirkpatrick Flack, *Desideratum in Washington: The Intellectual Community in the Capital City, 1870–1900* (Cambridge, Mass.: Schenckman Publishing, 1975), pp. 128–31.

16. Anita Newcomb McGee to Mrs. Stevenson, 18 Nov. 1891, Matilda Stevenson Papers, National Anthropological Archives, Smithsonian Institution, informs Stevenson of her election to the ASW ("Hurrah! You are a member of the Anthropological Society."); Alice Fletcher mentioned the WASA a few times in ambivalent terms (5 Aug. 1887 and 26 Jan. and 20 Mar. 1890) and her election to the presidency of the ASW with bemusement (10 and 21 Jan. 1903) in her correspondence with Frederic W. Putnam (Frederic W. Putnam Papers, HUA). The Alice C. Fletcher Papers at the National Anthropological Archives are less revealing. The origins of the American Anthropological Association are described in Franz Boas, "The Founding of a National Anthropological Society," *Science* 15 (1902): 804–9; [W. J. McGee], "The American Anthropological Association," *American Anthropologist* 5 (1903): 178–92, which includes a membership list, and George W. Stocking, Jr., "Franz Boas and the Founding of the American Anthropological Association," *American Anthropologist* 62 (1960): 1–17, but none mentions the women's group as having played any role. "Parsons" in *NAW*. Hamilton Cravens also has some perceptive comments on Boas in *The Triumph of Evolution: American Scientists and the Heredity-Environment Controversy, 1900–1941* (Philadelphia: University of Pennsylvania Press, 1978), pp. 102–3.

17. *Records of the Society of Naturalists of the Eastern United States* vol. 1, pt. 1 (for 1883), p. 24.

18. Ibid, p. 5, Constitution, art. 2, sec. 1.

19. Ibid., "By-Laws," p. 9.

20. *Records of the American Society of Naturalists*, vol. 1, pt. 4 (for 1886), p. 112, and pt. 10 (for 1892), p. 281. There were by 1892 also two American women Ph.D.s in botany, Alice Carter Cook

(1888) and Henrietta Hooker (1889), both from Syracuse University, but neither was elected to the ASN. Cornelia Clapp, Ph.D. in zoology from Syracuse, never joined.

21. Only 31 of 112 (or 27.7 percent) of the original fellows of the GSA held the doctorate (Ph.D. or Sc.D.), but many of these were older men, including James Hall and James Dwight Dana, who dated from the predoctoral era. Yet of the 50 persons elected in May 1889 (which included Mary E. Holmes) only 11 (22 percent) held the doctorate, and of the 11 elected in August 1894 (including Florence Bascom) only 3 held doctorates (27.3 percent) ("List of Officers and Fellows of the Geological Society of America," *Bulletin of the Geological Society of America* 1 [1890]: 579–86, and "Proceedings of the Sixth Summer Meeting, Held at Brooklyn, New York, August 14 and 15, 1894," *Bulletin of the Geological Society of America* 6 [1894]: 2). Bascom's professor G. H. Williams of The Johns Hopkins University had been uncertain whether to put her name forward, fearing that she would be upset if she was not elected (H. L. Fairchild to Professor Van Hise, 20 Jan. 1894, Florence Bascom Papers, Sophia Smith Collection, Smith College).

22. "List of Members, Revised and Corrected, January 1891," *Torrey Botanical Club Bulletin* 18 (1891): i–viii. The main historical articles about the early club studiously avoid mention of its feminization, however (Edward S. Burgess, "The Work of the Torrey Botanical Club," *Torrey Botanical Club Bulletin* 27 [1900]: 552–8, and John Hendley Barnhart, "Historical Sketch of the Torrey Botanical Club," *Torrey Botanical Club Memoirs* 17 [1917]: 12–21); Harold W. Rickett, "The Torrey Botanical Club, A Retrospect," *Garden Journal* 17 (1967): 144–50; and Douglas Sloan, "Science in New York City, 1867–1907," *Isis* 71 (1980): 44–46.

23. William T. Davis, "Annie Trumbull Slosson," *Journal of the New York Entomological Society* 34 (1926): 361–65; in the same journal: A. T. Slosson, "A Few Reminiscences," 23 (1915): 85–91; idem, "A Few Memories, II," 25 (1917): 93–97; idem. "Reminiscences of the Early Days of the New York Entomological Society," 26 (1918): 134–37; and Charles W. Leng, "History of the New York Entomological Society, 1893–1918," 26 (1918): 129–33; Annie Morrill Smith, "The Early History of *The Bryologist* and the William Sullivant Moss Society," *Bryologist* 20 (1917): 1–8; and letters in Elizabeth G. Britton Papers, NYBGA; and F. W. H., "Mary Louise Duncan Putnam," *American Anthropologist* 5 (1903): 173–74.

24. Sloan, "Science in New York City," discusses the whole move to oust amateurs in the 1880s and 1890s.

25. Elizabeth Britton to Annie M. Smith, 12 Dec. 1906, Elizabeth Britton Papers, NYBGA; by 1913 she had still not attended a BSA dinner (Elizabeth G. Britton to William G. Farlow, 28 Apr. 1913, Farlow Letterbooks, Farlow Herbarium, Harvard University, where she referred to his comment about gossip at BSA dinners in William G. Farlow, "The Change from the Old to the New Botany in the United States," *Science* 37 [1913]: 84); "Britton" in *NAW*; Marshall A. Howe, "Elizabeth Gertrude Britton," *Journal of the New York Botanic Garden* 35 (1934): 97–103; John Hendley Barnhart, "The Published Work of Elizabeth Gertrude Britton," *Bulletin of the Torrey Botanical Club* 62 (1935): 1–13, lists 346 items. Oswald Tippo, "The Early History of the Botanical Society of America," *American Journal of Botany* 43 (1956): 53. The Botanical Society of America's archives are at the Humanities Research Library, University of Texas, Austin. The BSA's principal founder was Charles Reid Barnes of the University of Wisconsin, later of the University of Chicago, whom Clara Cummings and Susan Hallowell of Wellesley found particularly conceited and chauvinistic (Clara Cummings to Mrs. Britton, 7 Oct. 1890), an assessment with which Britton agreed, as her note at the end of his letter to her of 13 Feb. 1892 reveals (all in Elizabeth G. Britton Papers, NYBGA). The few published notes about the founding of the BSA are in Barnes's own *Botanical Gazette:* "Proceedings of the Botanical Club of the AAAS," 17 (1892): 289; "Proceedings of the Botanical Club, AAAS, Madison Meeting," 19 (1894): 388; and "Editorial," 22 (1897): 171–72.

26. J. F. A. Adams, "Is Botany a Suitable Study for Young Men?" *Science* 9 (1887): 117–18. See also chapters 1 and 3 about the feminization of botany.

27. Joan N. Burstyn, "Early Women in Education: The Role of the Anderson School of Natural History," *Boston University Journal of Education* 159, no. 3 (Aug. 1977): 50–64; David Starr Jordan, *The Days of a Man*, 2 vols. (Yonkers-on-Hudson, N.Y.: World Book, 1922), 1: 106–19, 132, 165, and 185; "E. Agassiz" in *NAW*.

28. Ralph W. Dexter, "The Annisquam Sea-Side Laboratory of Alpheus Hyatt, Predecessor of

the Marine Biological Laboratory at Woods Hole, 1880–1886," in *Oceanography: The Past*, ed. Mary Sears and Daniel Merriman (New York: Springer-Verlag, 1980), pp. 94–100. (I thank Professor Dexter for sending me a preprint.) Frank R. Lillie, *The Woods Hole Marine Biological Laboratory* (Chicago: University of Chicago Press, 1944), pp. 34–46, 204–8, and 252–55. See also Edwin Grant Conklin, "Early Days at Woods Hole," *American Scientist* 56 (1968): 112–20; Cornelia M. Clapp, "Some Recollections of the First Summer at Woods Hole, Mass., 1888," typescript in Clapp Papers, Mount Holyoke College Archives. The multifarious activities of the Women's Education Association of Boston are described in Katharine P. Loring, "A Review of Fifty-Seven Years' Work," *Fifty-Seventh and Final Annual Report of the Women's Education Association for the Year Ending January 17, 1929*, pp. 5–18, but the association needs a full history. Frances L. Goudy, special collections librarian at Vassar College, has been good enough to send me several obituaries of the ubiquitous but elusive Florence Cushing. Much of the March 1928 issue of the *Vassar Quarterly* (13, no. 2) was devoted to her memory. Maria Mitchell reportedly nominated her for president of Vassar in 1886, and others hailed her as "one of the builders of opportunity in our generation."

29. Frank Lillie, *Woods Hole*, passim; Henry F. Osborn et al., "Marine Biological Laboratory: A Statement to the Corporation from the Trustees," *Science*, n.s. 6 (1897): 475–76; [Samuel F. Clarke, Edward G. Gardiner, and J. Playfair McMurrich], *A Reply to the Statement of the Former Trustees of the Marine Biological Laboratory* (Boston: Alfred Mudge & Son, 1897); Edwin G. Conklin, "Cornelia Maria Clapp and the Marine Biological Laboratory," 1949, typescript in Clapp Papers, Mount Holyoke College Archives; Frank R. Lillie, "Cornelia Maria Clapp at Woods Hole," *Mount Holyoke Alumnae Quarterly* 19 (May 1935): 3–4; "Harvey" in *NAW: MP;* and E. G. Butler, "Ethel Browne Harvey," *Biological Bulletin* 133 (1967): 10. Recent general articles on the MBL include Luther J. Carter, "Woods Hole: Summer Mecca for Marine Biology," *Science* 157 (1967): 1288–92, and Detlev W. Bronk, "Marine Biological Laboratory: Origins and Patrons," *Science* 189 (1975): 613–17. Whitman apparently knew millionaires C. R. Crane and Otho Sprague of Chicago, and his wife, Emily Nunn Whitman, formerly a professor at Wellesley College, was the daughter of the wealthy L. L. Nunn, an engineer and pioneer in high-voltage transmission of alternating current from Telluride, Colorado, who later became an MBL trustee and benefactor. For mention of some of L. L. Nunn's other Telluride charities, see George Martin, *Madame Secretary: Frances Perkins* (Boston: Houghton Mifflin, 1976), p. 555 n. 9.

30. Emily Ray Gregory to Elizabeth G. Britton, 29 Oct. 1898, enclosing "Circular of the Committee of the National Science Club on the Endowment of Tables at Biological Stations," in Elizabeth G. Britton Papers, NYBGA. The Naples Table Association also raised money for a table at Woods Hole after 1901 (see chapter 2).

31. "Calkins," "Ladd-Franklin," and "Washburn" in *NAW;* "Proceedings of the American Psychological Association, Third Annual Meeting . . . 1894," *American Journal of Psychology* 6 (1895): 617; James McKeen Cattell to Christine Ladd-Franklin, 7 Nov. and 11 Dec. 1894, in Christine Ladd-Franklin Papers, Special Collections, Butler Library, Columbia University; "Proceedings of the American Psychological Association, Fifth Annual Meeting, 1896," *Psychological Review* 4 (1897): 108; Hinman in *NCAB* 26 (1957): 269–70; "Proceedings of the American Psychological Association, Thirteenth Annual Meeting, 1904," *Psychological Bulletin* 2 (1905): 37–38; G. Stanley Hall, "A Reminiscence," *American Journal of Psychology* 28 (1917): 299; Wayne Dennis and Edwin G. Boring, "The Founding of the APA," *American Psychologist* 7 (1952): 95–97.

32. "Scott" in *NAW;* Ruth Gentry in Will E. Edington, "Biographical Sketches of Indiana Scientists IV," *Proceedings of the Indiana Academy of Sciences for 1967*, pp. 36–37; Raymond Clare Archibald, *A Semicentennial History of the American Mathematical Society, 1888–1938* (New York: American Mathematical Society, 1938), pp. 97 and 106; see also its p. 6.

33. K. K. Darrow, "The Names of Those Who Met on 20 May 1899 to Organize the American Physical Society," *APS Bulletin* 24, no. 5 (16 June 1949): 34; Ernest Merritt, "Early Days of the Physical Society," *Review of Scientific Instruments* 5 (1934): 148; "Maltby" and "Whiting," in *AMS*, 2d ed. (1910).

34. *Proceedings of the Sixth Annual Session of the Association of American Anatomists* (Washington, D.C.: Beresford, Printer, 1894), p. 5; both "Moodys," "De Witt," "Gage," and "Sabin" in *AMS*, 2d ed. (1910); "Sabin" in *NAW: MP;* "DeWitt" in *NAW;* and Pearl Bliss Cox,

"Pioneer Women in Medicine, Michigan XV," *Medical Woman's Journal* (Apr. 1949), p. 51, repeated the next month, p. 49; Simon H. Gage, "Susanna Phelps Gage, Ph.B.," *Journal of Comparative Neurology* 27 (1916): 5–18; list of members in *American Journal of Anatomy* 3 (1904): xxi–xxviii. Nicholas A. Michels, "The American Association of Anatomists: A Tribute and Brief History," *Anatomical Record* 122 (1955): 679–714, adds little.

35. W. H. Howell, "The American Physiological Society during Its First Twenty-Five Years," in *History of the American Physiological Society, Semicentennial, 1887–1937* (Baltimore, 1938), pp. 70–71 (in 1912 Hyde completed most of the requirements for an M.D. from the Rush Medical College in Chicago); "Hyde" in *NAW;* Eliza H. Root, "Frances Emily White, M.D.," *Woman's Medical Journal* 14 (1904): 97–99; White also wrote several articles on "the woman question" for the *Popular Science Monthly;* Wallace O. Fenn, *History of the American Physiological Society: The Third Quarter-Century, 1937–1962* (Washington, D.C.: American Physiological Society, 1963), pp. 65–67.

36. Preston E. James and G. J. Martin, *The Association of American Geographers: The First Seventy-Five Years, 1904–1979* (n.p., Association of American Geographers, 1978), pp. 31, 36–38, 47, and 50. "Semple" in *NAW*.

37. "Proceedings of the Montreal Meeting, September 1879," *Transactions of the American Institute of Mining Engineers* 8 (1879–80): 135 (I thank Deborah Cozort for this date); Stoiber: Clarke C. Spence, *Mining Engineers and the American West* (New Haven: Yale University Press, 1970), pp. 6–7; Nora Stanton Blatch de Forest (Barney) in *NAW:MP;* Suzy Fisher, "Nora Stanton Barney, First U.S. Woman CE, Dies at 87," *Civil Engineer* 41 (Apr. 1971): 87; and *Father of Radio: The Autobiography of Lee de Forest* (Chicago: Wilcox & Follett, 1950), pp. 222–25, 232, 236, 243–46, 248, 251–53, for comments on the advantages and disadvantages of marrying someone in the same field; Margaret Ingels, "Petticoats and Slide Rules," *Midwest Engineer* 5 (1952): 2–4 and 10–16.

38. Quoted in Caroline E. Furness, "Mary W. Whitney," *Popular Astronomy* 31 (1923): 31; Whitney had earlier been rejected by the German but largely international Astronomische Gesellschaft on the basis of her sex (p. 28). See also Joel Stebbins, "The American Astronomical Society, 1897–1947," *Popular Astronomy* 55 (1947): 404–13, and, for the story of the founding, Richard Berendzen, "Origins of the American Astronomical Society," *Physics Today* 27 (1974): 32–39.

39. The AAS also met at the Maria Mitchell Observatory on Nantucket Island in 1936 (*Publications of the American Astronomical Society*, vols. 4, 5, and 10).

40. Robert Sobel, *They Satisfy: The Cigarette in American Life* (Garden City, N.Y.: Anchor Press/Doubleday, 1978), especially pp. 61, 87, 92, and 95–106. Alcoholic beverages may also have been drunk at "smokers."

41. Sarah F. Whiting, "History of the Department of Physics at Wellesley College, 1878–1912," manuscript, 1912, Sarah F. Whiting Papers, Wellesley College Archives. She then summed up her long career: "For many years I was almost alone in college work in this line [physics,] meeting the somewhat nerve-wearing experiences of constantly being in places where a woman was not expected to be, and doing what women had not to that time conventionally done."

42. Mary Whiton Calkins to James McKeen Cattell, 17 Jan. 1911, James McKeen Cattell Papers, Manuscript Division, Library of Congress.

43. Mary Whiton Calkins to James McKeen Cattell, 12 Mar. 1911, James McKeen Cattell Papers, Manuscript Division, Library of Congress.

44. Cited in Allen D. Bushong, "Some Aspects of the Membership and Meetings of the Association of American Geographers before 1949," *Professional Geographer* 26 (1974): 437.

45. Ibid.

46. Reginald A. Daly, "William Morris Davis, 1850–1934," *National Academy of Sciences Biographical Memoirs* 23 (1945): 264 and 279–80.

47. See note 26 to chapter 10.

48. Slosson, "Reminiscences of the Early Days of the New York Entomological Society," p. 134.

49. Mary L. Davis, "Sketch of the 'Eistophos': Early Days," in Eistophos Club Program for 1927–28, pp. 1–7; Maeme T. Ault, *Eistophos Science Club History, 1927–1950* (Washington, D.C.,

1950); *Proceedings of the National Science Club, 1898-99,* all in Rare Book Room, Library of Congress. Members of the AAAS "Ladies' Reception Committee" for 1891 are listed in the *Proceedings of the American Association for the Advancement of Science, 1891,* pp. xiv and 458. There are a few letters from Laura Osborne Talbott in the "Admission of Women" file, Special Collections, Milton Eisenhower Library, The Johns Hopkins University, and in the Assistant Secretary's correspondence, U.S. National Museum, 1860-1918 series (RU 189, Box 128, folder 5). (I thank Sally Gregory Kohlstedt for bringing this folder to my attention.)

50. Mary Murtfeldt, "Report on [the] Present Status of American Women in Entomology," *Proceedings of the National Science Club* 3 (Apr. 1897): 11-14; she had been crippled by polio in her youth and must therefore been one of the earliest handicapped women scientists (obituaries in *Journal of Economic Entomology* 6 [1913]: 288-89, and *Entomological News* 24 [1913]: 241-42). Rosa Smith Eigenmann, "Women in Science," *Proceedings of the National Science Club* 1 (Jan. 1895): 13-17; she was an early ichthyologist and friend of David Starr Jordan who never recovered from a mental depression she suffered after a move to San Diego and the birth of defective children ("Eigenmann" in *NAW*). The club is also mentioned in Oehser, "Florence Merriam Bailey," p. 21, and Croly, *History of the Woman's Club,* pp. 348-49. For further comment on the fictitious dangers of praising women unduly, see Edward S. Holden to Mrs. Stanhope Sams, of Atlanta, Georgia, 9 Mar. 1895, in Outgoing Correspondence, Lick Observatory Archives, University of California, Santa Cruz. (I thank Arthur Norberg for a copy of this interesting letter.)

51. Elizabeth G. Britton, "Report of the Chairman of the Division of Bryophyta," *Proceedings of the National Science Club* 3 (Apr. 1897): 11. Several letters in the Elizabeth G. Britton Papers, NYBGA, relate to the National Science Club; on William Sullivant Moss Society, see note 23 to this chapter; naturalists of both sexes became quite well organized in Britain at the turn of the century; see David Elliston Allen, *The Naturalist in Britain: A Social History* (London: Allen Lane [Penguin Books], 1976), which also describes women's long-standing role in British natural history.

52. Rossiter Johnson, ed., *A History of the World's Columbian Exposition, Held in Chicago in 1893,* 4 vols. (New York: D. Appleton, 1897), is the main history of the Fair. Moses P. Handy, ed., *The Official Directory of the World's Columbian Exposition, 1893* (Chicago: W. B. Conkey, 1893), contains a history of the Board of Lady Managers. For Susan B. Anthony's involvement, see Ida Husted Harper, *The Life and Work of Susan B. Anthony,* 3 vols. (Indianapolis: Bowen-Merrill, 1898), 2: chap. 41, and Anna Howard Shaw, *The Story of a Pioneer* (New York: Harper & Bros., 1915), p. 261. See also Virginia C. Meredith, "The Columbian Exposition," *Papers Read before the Association for the Advancement of Women, Nineteenth Women's Congress, Grand Rapids, Michigan, October 1891* (Syracuse, N.Y.: C. W. Bardeen, 1892), pp. 104-10; Jeanne M. Weiman, "A Temple to Women's Genius: The Woman's Building of 1893," *Chicago History* 6 (Spring 1977): 28-33.

For some discussion of women scientists at the 1876 exposition, see Deborah Jean Warner, "The Women's Pavilion," in *1876: A Centennial Exhibition,* ed. Robert C. Post (Washington, D.C.: National Museum of History and Technology, 1976), pp. 165-73. Sarah Whiting's visit is mentioned in Bessie Z. Jones and Lyle Boyd, *The Harvard College Observatory: The First Four Directorships, 1839-1919* (Cambridge: Harvard University Press, 1971), p. 406, and some women exhibitors are discussed in Phebe (sic) A. Hanaford, *Daughters of America, or Women of the Century* (Augusta, Maine: True, 1882), chap. 9.

53. The exposition's exhibits are described in Handy, *Official Directory,* and Johnson, *History of the World's Columbian Exposition,* vol. 3. The Chicago Historical Society also has some manuscript material, such as the "Report of Committee on Scientific Exhibit, Woman's Building, World's Columbian Exposition, Chicago 1893," in Miscellaneous Committee Reports, Board of Lady Managers, World's Columbian Exposition Papers, Chicago Historical Society. I thank Sally Gregory Kohlstedt for bringing this report to my attention. Bertha H. Palmer to Professor Putnam, 27 Feb. 1894, Frederic W. Putnam Papers, HUA, thanks him for featuring Zelia Nuttall's contribution to his Department of Ethnology exhibits.

54. George H. Johnson, "The World's Congress Auxiliary of the Columbian Exposition," *Science* 22 (1893): 117; May Wright Sewall, ed., *The World's Congress of Representative Women,* 2 vols. (Chicago: Rand, McNally, 1894). Johnson, *History of the World's Columbian Exposition,*

vol. 4, is a history of most of the congresses with summaries of many of the speeches and a bibliography of published proceedings (pp. 497-508).

55. Ellen Richards, "Efficiency of a System of Ventilation as Shown by the Amount of CO_2," *Journal of the American Chemical Society* 15 (1893): 572-74; Hortensia M. Black, "To the Rescue of Birds," *Papers Presented to the World's Congress on Ornithology*, ed. Mrs. E. Irene Rood under the direction of Elliott Coues (Chicago: Charles H. Sergel, 1896), pp. 171-78; Mrs. M. Fleming, "A Field for 'Woman's Work' in Astronomy," *Astronomy and Astrophysics* 12 (1893): 683-89; and Dorothea Klumpke, "The Bureau of Measurements of the Paris Observatory," *Astronomy and Astrophysics* 12 (1893): 783-88; for biographical material on Dorothea Klumpke Roberts and other members of her talented San Francisco family, see *NCAB*, 31, pp. 403-6; papers by Nuttall, Fletcher, M. C. Stevenson, S. Y. Stevenson, and Englishwoman Mrs. M. French-Sheldon are in C. Staniland Wake, ed., *Memoirs of the International Congress on Anthropology* (Chicago: Schulte Publishing, 1894); W. H. Holmes "The World's Fair Congress of Anthropology," *American Anthropologist* 6 (1893): 423-34; comment on Nuttall in "The International Congress of Anthropology," *Science* 22 (1893): 150; Sara Y. Stevenson to Professor Putnam, June 1893, Putnam Papers, HUA, thanks him for encouraging her.

56. The World's Congress on Geology apparently did not publish its proceedings, but a copy of the program, which includes that of the Woman's Department, is in the Mary Louise Foster Faculty File, Smith College Archives, and the Frederic W. Putnam Papers, HUA.

57. Some kinds of sex segregation (but not all) persisted at later world's fairs and other international gatherings held in the United States. For example, women's and men's exhibits were judged together, not separately, at the next American world's fair, the Louisiana Purchase Exposition in St. Louis in 1904, but separate women's auxiliary congresses continued to be held, such as that at the Second Pan-American Scientific Congress in Washington, D.C., in 1916. Several American women scientists (including Alice Fletcher, Florence Bascom, Cornelia Clapp, Frances Densmore, Louise McDowell, Maud Slye, and others) participated. One lasting result of this particular congress was an ACA (later AAUW) fellowship for a Latin American woman to study in the United States (Mrs. Glen Levin Swiggett, *Report on the Women's Auxiliary Conference, Held in the City of Washington, U.S.A., in Connection with The Second Pan American Scientific Congress, December 28, 1915-January 7, 1916* [Washington, D.C.: GPO, 1916], and Margaret E. Maltby, comp., *History of the Fellowships Awarded by the American Association of University Women, 1888-1919, with the Vitas of the Fellows* [Washington, D.C.: AAUW (1929)], pp. 6-7, 58-61, and 99-100).

CHAPTER 5

1. The concept of "stereotyping" apparently entered the social sciences through the work of journalist Walter Lippmann, *Public Opinion* (New York: Harcourt, Brace, 1922). See also Margaret W. Rossiter, "History of the Word 'Stereotype'" (Paper delivered at the Annual Meeting of the Society for the History of Technology, Toronto, October 1980), and Hamilton Cravens, *The Triumph of Evolution: American Scientists and the Heredity-Environment Controversy, 1900-1941* (Philadelphia: University of Pennsylvania Press, 1978), pp. 123-53, which says there was no real sociology before 1920.

2. The classic source on the conservatism of the suffrage campaign is Aileen Kraditor, *The Ideas of the Woman Suffrage Movement, 1890-1920* (New York: Columbia University Press, 1965), which describes the increasingly racist and anti-immigrant tone of some of the final hard-fought campaigns.

3. "Woolley" in *NAW;* Helen Bradford Thompson, *Psychological Norms in Men and Women* (Chicago: University of Chicago Press, 1903); see also Viola Klein, *The Feminine Character: History of an Ideology* (1946; reprint ed., Urbana: University of Illinois Press, 1972), chap. 6; Rosalind Rosenberg, "In Search of Woman's Nature, 1850-1920," *Feminist Studies* 3(1975): 141-54; and at greater length in idem, *Beyond Separate Spheres: Intellectual Roots of Modern Feminism* (New Haven: Yale University Press, 1982), chap. 3.

4. Eleanor Flexner (niece of Simon and Helen Thomas Flexner), *Century of Struggle: The*

Woman's Rights Movement in the United States (Cambridge: Harvard University Press, 1959), is the outstanding history of the movement. "Phelps" and "Richards" in *NAW;* Mitchell: Mary W. Whitney, "Life and Work of Maria Mitchell, L.L.D.," *Papers Read before the Association for the Advancement of Women, Eighteenth Woman's Congress, Toronto, Canada, October 1890* (Fall River, Mass.: J. H. Franklin, 1891), p. 26; Lewis: Frances E. Willard and Mary A. Livermore, eds., *A Woman of the Century* (Buffalo: Charles W. Moulton, 1893), p. 462; ACA: "Editorial," *JACA* 10(1917): 601, and "Minutes of the Thirty-fourth General Meeting April 1917," *JACA* 10(1917): 660.

5. *Proceedings of the Thirty-Eighth Annual Convention of the National American Woman Suffrage Association, Held at Baltimore, Maryland, February 7–13, 1906* (Warren, Ohio: William Ritezel, n.d.), pp. 6, 48–49; the 1906 meeting is described in Ida Husted Harper, *The Life and Work of Susan B. Anthony,* 3 vols. (Indianapolis: Hollenbeck Press, 1908), 3: chap. 69 (I thank Gayle Gullett Escobar for this reference); Anna Howard Shaw's autobiography, *The Story of a Pioneer* (New York: Harper & Bros., 1915), pp. 221-26; also in Edith Finch, *Carey Thomas of Bryn Mawr* (New York: Harper & Bros., 1947), pp. 246–48; Lillian Welsh, *Reminiscences of Thirty Years in Baltimore* (Baltimore: Norman Remington, 1925), chap. 7; Anna Mary Wells, *Miss Marks and Miss Woolley* (Boston: Houghton Mifflin, 1978), pp. 88 and 94–95.

6. The addresses are published in *The College Evening of the Thirty-Eighth Annual Convention of the National American Women Suffrage Association, Held in Baltimore, February 8, 1906* (n.p.); Calkin's address got four stars and Thomas's five in Margaret Ladd Franklin, *The Case for Woman Suffrage; A Bibliography* (New York: National College Equal Suffrage League, 1913), p. 101.

7. William Albert Noyes and James Flack Norris, "Ira Remsen, 1846–1927," *National Academy of Sciences Biographical Memoirs* 14(1932): 238; Ira Remsen, "Original Research," *PACA,* series 3, no. 6(Feb. 1903), pp. 20-29, and mentioned in "Report of the Meeting of the Association of Collegiate Alumnae," *PACA,* series 3, no. 6 (Feb. 1903), pp. 76–77.

8. William H. Welch to sister (Emeline Welch Walcott), 25 Jan. and 10 Feb. 1906, quoted in Simon Flexner and James Thomas Flexner, *William Henry Welch and the Heroic Age of American Medicine* (New York: Viking Press, 1941), p. 231. Welch's desk calendar with quotations, 7–10 Feb. 1906, Welch Papers, The Alan Mason Chesney Medical Archives of The Johns Hopkins Medical Institutions; *Proceedings of the Thirty-Eighth Annual Convention of the National American Woman Suffrage Association,* p. 58; see also Victor O. Freeburg, ed., *William Henry Welch at Eighty* (New York: Milbank Memorial Fund, 1930), pp. 201-3.

9. Florence Sabin to Adelaide Nutting (dean of The Johns Hopkins School of Nursing), 15 Mar. 1909, quoted in Helen E. Marshall, *Mary Adelaide Nutting: Pioneer of Modern Nursing* (Baltimore: The Johns Hopkins University Press, 1972), p. 166; Nutting was not a suffragist, however; the stalwart antifeminists at Johns Hopkins, Gilman and Rowland, whom Christine Ladd-Franklin had faced in the 1880s, and W. K. Brooks, who later became a suffragist, had all died by 1908. Cf. n. 30.

10. There is no history of the National College Equal Suffrage League, but see "Maud Wood Park" in *NAW:MP*; Edith Finch, *Carey Thomas*, pp. 248–51, and Ethel Puffer Howes, "The National College Equal Suffrage League," *Smith Alumnae Quarterly* (Nov. 1920): 42–45; psychologist Howes, the first Radcliffe doctorate in science, was executive secretary to the NCESL and corresponded with M. Carey Thomas until 1915, when she "retired" to have a baby (obituary in *Smith Alumnae Quarterly* [Feb. 1951]: 93, and correspondence in Morgan-Puffer Family Papers, SLRC); Mount Holyoke College is mentioned in Wells, *Miss Marks and Miss Woolley,* p. 154.

11. For discussion of heightened interest in sex differences after 1910, compare Helen Thompson Woolley, "General Reviews and Summaries, The Psychology of Sex," *Psychological Bulletin* 11(1914): 353, with her "Review of the Recent Literature on the Psychology of Sex," *Psychological Bulletin* 7(1910): 335–47; Franklin P. Mall's study "On Several Anatomical Characters of the Human Brain Said to Vary According to Race and Sex, with Especial Reference to the Weight of the Frontal Lobe," *American Journal of Anatomy* 9(1909): 1–32; strong interest in Freud did not start until after 1915, however—see Nathan Hale, *Freud and the Americans: The Beginnings of Psychoanalysis in the United States, 1876–1917* (New York: Oxford University Press, 1971), p. 417; Christine Ladd-Franklin deplored the "absurdities" of Freudian doctrines in a letter to the editor of the *Nation* 103(1916): 373–74.

12. Maxine Seller, "G. Stanley Hall and Edward Thorndike on the Education of Women: Theory and Policy in the Progressive Era," *Educational Studies* (in press), is the best account of this thinking. (I thank Sally Gregory Kohlstedt for sending me a preprint.) This controversy, which is so important for women's history, is not mentioned in the major biographies of Hall and Thorndike (Dorothy Ross, *G. Stanley Hall: The Psychologist as Prophet* [Chicago: University of Chicago Press, 1972], though pp. 301–2 and 416 discuss related ideas, and Geraldine Joncich, *The Sane Positivist: A Biography of Edward L. Thorndike* [Middletown, Conn.: Wesleyan University Press, 1968], where pp. 341–42 touch on sex differences). See also G. Stanley Hall, *Adolescence*, 2 vols. (New York: D. Appleton, 1908), 2: chap. 17, "Adolescent Girls and Their Education" (I thank Dorothy Ross for this reference); Daniel J. Kevles, "Women, Catholics, and Jews in American Science, The Interwar Background" (Paper presented at Symposium on Women and Minority Groups in American Science and Engineering, California Institute of Technology, December 1971); Edward L. Thorndike, "Sex in Education," *Bookman* 23 (1906): 211–14, and Sylvia Kopald, "Where Are the Women Geniuses?," *Nation* 119 (1924): 619–22. See also the early Edward L. Thorndike, ed., *Heredity, Correlation, and Sex Differences in School Abilities*. Studies from the Department of Educational Psychology at Teachers College, Columbia University, published as *Columbia University Contributions to Philosophy, Psychology, and Education* 11, no. 2 (Feb. 1903). Challenges to variation were numerous before 1910 (Karl Pearson, "Variation in Man and Woman," in his *The Chances of Death* [London: E. Arnold, 1897]; "On the Laws of Inheritance in Man, II. On the Inheritance of the Mental and Moral Characters in Man," *Biometrika* 3, p. 2 [1904]: 131–90; and Raymond Pearl, "Relative Variability of Man and Woman," *Science* 19 [1904]: 73), but apparently were unconvincing to the conservatives.

13. James McKeen Cattell, "A Statistical Study of Eminent Men," *Popular Science Monthly* 62 (1903): 375. The previous issue had included an article critical of Pearson: Havelock Ellis, "Variation in Man and Woman," *Popular Science Monthly* 62(1903): 237–53.

14. James McKeen Cattell, "The School and the Family," *Popular Science Monthly* 74(1909): 92–93. (I thank Rosalind Rosenberg for this reference.) Contemporary protests against female schoolteachers are discussed in Thomas Woody, *A History of Women's Education in the United States* 2 vols. (New York: Science Press, 1929), 1:chap. 10, and the New York feminists' fight in 1914 and 1915 to allow women teachers who married to retain their jobs is described in June Sochen, *The New Woman: Feminism in Greenwich Village, 1910–1920* (New York: Quadrangle Books, 1972), pp. 52–57.

15. James McKeen Cattell, "A Further Statistical Study of American Men of Science," *Science* 32 (1910): 676, reprinted from Cattell, *AMS*, 2d ed. (1910), p. 584. In 1910 seven of the original nineteen women starred were dropped from the top 1,000, although they retained their stars in the directory (apparently Lydia DeWitt, Alice Eastwood, Katharine Foot. Margaret Maltby, Agnes Moody, Margaret Nickerson, and Florence Peebles), and six others were added (Margaret Ferguson, Susanna Gage, Ethel Howes, Lillien Martin, Mary Pennington, and Nettie Stevens). (Deduced from "Leading Men of Science in the United States in 1903 Arranged in the Order of Distinction in Each Science," in AMS, 5th ed. [1933], pp. 1269–78, and list, "No Longer Starred in 3rd Edition [1921]," in James McKeen Cattell Papers, box 62, Library of Congress.) For more on this covert "destarring" process, see chapter 10 herein.

16. See especially the 1905 ranking sheets of the astronomers and psychologists, the only field with many women, in box 62, James McKeen Cattell Papers, Manuscript Division, Library of Congress. The astronomers, all male, included anywhere from one (George Ellery Hale) to eleven (Edward C. Pickering) women astronomers on their lists. For example, the rankings of Williamina P. Fleming, the only woman on all lists, ranged from a low of 58 from George Comstock of the Washburn Observatory at the University of Wisconsin to a high of 11 from Pickering, her employer at Harvard, for an eventual average of 36. The psychologists were asked to select names from a list for the top five and the top fifty persons in their field. Only Christine Ladd-Franklin put any women in her top five (Calkins and Washburn), and most of the seven women whose rankings are in the collection included more women in the top fifty than did the men. Thus Ladd-Franklin had nine (and omitted herself, as everyone did, or she would have had ten), Mabel Fernald seven, June Downey seven, Helen Woolley six, Margaret Washburn and Mary Calkins five each, and Kate Gordon only

four. Of the major men submitting rankings, Boring listed six women in the top fifty, Cattell had five, and Lewis Terman and Edward Titchener just four (Hugh Münsterberg's was missing). In 1937, when the pathologists were asked to rank each other, Florence Sabin put two women in her top five, including Rebecca Lancefield, her colleague at the Rockefeller Institute, as number 1.

17. Ellen Hayes, "Women in Scientific Research," *Science* 32 (1910): 866; Marion Talbot, "Eminence of Women in Science," *Science* 32 (1910): 866; I thank Professor John Burke of UCLA for bringing these two items to my attention. Even before 1910 Hayes and Talbot had liked and respected Cattell a great deal and favored some of his stands on controversial issues, including the "woman question" (Ellen Hayes to J. M. Cattell, 4 Jan. 1900 and 3 Mar. 1904, and Marion Talbot to Ware Cattell, 24 Jan. 1944, all in J. M. Cattell Papers, Library of Congress).

18. "Women and Scientific Research," *Science* 32(1910): 919–20. One can see here the hands of Fabian Franklin and Christine Ladd-Franklin, since he was an editor of the *New York Post* at this time and she frequently sent clippings and postcards to her scientific friends.

19. "Women and Scientific Research," *Scientific American* 104(Jan. 21, 1911): 77; an editorial on Madame Curie's candidacy for the French Academy also appeared in the same issue, "Sex and Scientific Recognition," p. 58 (she was rejected). Beach: *NCAB* 17(1927): 327–28 and *Dictionary of American Biography*, 20 vols., ed. Allen Johnson and Dumas Malone (New York: Charles Scribner's Sons, 1928–36), 2:81; Munn: *Who's Who in America* 12(1922–23): 2273. See also "Orson Munn," *NCAB* 7(1897): 83.

20. Marion Talbot, *The Education of Women* (Chicago: University of Chicago Press, 1910), pp. 244–46; see also *More Than Lore, Reminiscences of Marion Talbot* (Chicago: University of Chicago Press, 1936); Rosalind Rosenberg, "The Academic Prism: The New View of American Women," in *Women of America: A History*, ed. Carol Ruth Berkin and Mary Beth Norton (Boston: Houghton Mifflin, 1979), pp. 318–38; idem, *Beyond Separate Spheres*, chap. 2; Mary Jo Deegan, "Women and Sociology, 1890–1930," *Journal of the History of Sociology* 1 (1979): 11–32.

21. Susan M. Kingsbury, "Committee on Academic Appointments," *JACA* 4(1911): 20–22; *NAW*. Kingsbury deserves a full biography.

22. C[harles] H. Handschin, "The Percentage of Women Teachers in State Colleges and Universities," *Science* 35(1912): 55–57.

23. Gertrude S. Martin, "The Position of the Dean of Women," *JACA* 4, series 4 (1911): 65–77.

24. "Proceedings of the Association," *JACA* 6, no. 1 (Jan. 1913): 28–29.

25. "General Meeting of the Association, April 12, 1919," *JACA* 12, no. 3(Apr. 1919): 178 and 181; "Council Meeting, April 3, 1919," *JACA* 12, no. 3 (Apr. 1919): 187–88; Ada Comstock, "Report of the President," *JACA* 14, nos. 5 and 6 (Feb.–Mar. 1921): 123–24; Gertrude S. Martin, "Report of the Executive Secretary," *JACA* 14, nos. 7 and 8(Apr.–May 1921): 166; "Meeting of the Council, March 30, 1921," *JACA* 14, nos. 7 and 8 (Apr.–May 1921): 190–93, and "Report of Committee on Recognition of Colleges and Universities," *JACA* 14, nos. 7 and 8 (Apr.–May 1921). 194–96; Dean F. Louise Nardin, "Report of the Committee on Recognition, April 7, 1922," *JAAUW* 15, no. 4(July 1922): 111–16; a copy of the Johns Hopkins University registrar's response to her questionnaire is in the "Admission of Women" file, Special Collections, Eisenhower Library, The Johns Hopkins University; F. Louise Nardin, "Additions to the List of Approved Colleges," *JAAUW* 17, no. 2(May 1924): 11–12.

26. "Minutes of the Thirty-Fourth General Meeting," *JACA* 10, no. 9(May 1917): 644 and 648; M. Carey Thomas, "Joint Conference of Trustees, Deans, and Professors," *JACA* no. 10(June 1918), pp. 686–88. See also chapter 7 herein.

27. Christine Ladd-Franklin to J. M. Cattell, 28(?) Apr. 1920, in Cattell Papers, Manuscript Division, Library of Congress. "Ladd-Franklin" in *NAW*; the full extent of Ladd-Franklin's propagandist activity cannot be known without a complete bibliography of her writings. The bibliography of Christine Ladd-Franklin's works in the Christine Ladd-Franklin Papers at Special Collections, Columbia University, lists only her scientific books and papers; a partial bibliography of her contributions to the *Nation* (signed articles only) appears in Daniel C. Haskell, comp., *The Nation, Volumes 1–105, New York, 1865–1917: Indexes of Titles and Contributions,* 2 vols. (New York: New York Public Library, 1953), 2:168. Most of the biographical material on Ladd-Franklin minimizes her

interest and involvement in the feminist cause: R. S. Woodworth, "Christine Ladd-Franklin," *Science* 71(1930): 307; and *NCAB* 26(1937): 422-23; but see *NAW*; Cora Sutton Castle, "A Statistical Study of Eminent Women," *Archives of Psychology* 4, no. 2 (1913); a shortened version appears in *Popular Science Monthly* 82(1913): 593-611, under the same title.

28. "Parsons" in *NAW*, and Franz Boas, "Elsie Clews Parsons," *Science* 95(1942): 89-90; Elsie Clews Parsons, "Feminism and Conventionality," *Annals of the American Academy of Political and Social Sciences* 56(1914): 47-53; idem, "Feminism and Sex Ethics," *International Journal of Ethics* 26(1916): 462-65; idem, "Anti-Suffragists and the War," *Scientific Monthly* 1(1915): 44-46; idem, *The Old-Fashioned Woman: Primitive Fancies about the Sex* (New York: G. P. Putnam's Sons, 1913), reviewed by ornithologist Elsa Naumberg in the *Survey* 30(1913): 437-38; a review by Janet James which suggests that Parsons was in close touch with feminists in Greenwich Village is in *American Historical Review* 80(1975): 516-17; the Elsie Clews Parsons Papers at the American Philosophical Society in Philadelphia do not cast any light on this relationship (see Gladys Reichard, "The Elsie Clews Parsons Collection," *Proceedings of the American Philosophical Society* 94[1950]: 308-9, and her obituary, "Elsie Clews Parsons," *Journal of American Folklore* 56[1943]: 45-56 [with full bibliography]). Rosalind Rosenberg, *Beyond Separate Spheres*, chap. 6, and Barbara Keating, "Elsie Clews Parsons: Her Work and Influence in Sociology," *Journal of the History of Sociology* 1(1978): 1-9.

29. H. J. Mozans [pseud.], *Woman in Science* (1913; reprint ed., Cambridge: MIT Press, 1974); Ralph E. Weber, *Notre Dame's John Zahm* (Notre Dame, Ind.: University of Notre Dame Press, 1961), pp. 193-94, and Michael J. Crowe, "Who Was H. J. Mozans?" *Isis* 68(1977): 111; Theodore Roosevelt, "Woman in Science," *Outlook* 106(1914): 93-95; the ACA's reviewer thought it inadequate (especially for the recent period): *JACA* 11(1917-18): 347-48; neither the *Nation* nor *Science* apparently reviewed the book, which is strange, since Christine Ladd-Franklin agreed to do it for *Science:* Christine Ladd-Franklin to J. M. Cattell, n.d., Cattell Papers, Library of Congress (box 113). Zahm later wrote another book, *Great Inspirers* (New York: D. Appleton, 1917), about four women who encouraged St. Jerome and Dante, which pleased the feminists much less. His papers are located in the Notre Dame University Archives, South Bend, Ind.

30. W. L. George, "What the Feminists Are Really Fighting For," *NYT*, 14 Dec. 1913, Magazine section (section 6), p. 6; George MacAdam, "Feminist Revolutionary Principle is Biological Bosh," *NYT*, 18 Jan. 1914, Magazine section (section 5), p. 2; "Indignant Feminists Reply to Professor Sedgwick," *NYT*, 15 Feb. 1914, Magazine section (section 5), p. 4; other strong feminists at Johns Hopkins at this time were: Florence Sabin, promoted to full professor of anatomy in the medical school in 1917, and the wives and daughters of Howell and Mall, especially Janet Howell Clark, later an associate professor of physiology at The Johns Hopkins School of Hygiene and Public Health, and Mrs. Mabel Mall, who had gone to medical school with Sabin. Professor Mall died in 1917 and protégée Sabin wrote his biography: Florence Rena Sabin, *Franklin Paine Mall: The Story of a Mind* (Baltimore: Johns Hopkins Press, 1934), but omitted his liberal views on women's abilities. W. K. Brooks, professor of zoology, who died in 1908, had begun to favor woman suffrage as early as 1896 ("Woman from the Standpoint of a Naturalist," *Forum* 22[1896]: 286-96). On stereotypes, see note 1 to this chapter.

For more on Sedgwick, see chapter 8 herein and E. O. Jordan, G. C. Whipple, and C.-E. A. Winslow, *A Pioneer of Public Health, William Thompson Sedgwick* (New Haven: Yale University Press, 1924), esp. pp. 50-51.

31. "Hollingworth" in *NAW;* Stephanie A. Shields, "Ms. Pilgrim's Progress: The Contributions of Leta Stetter Hollingworth to the Psychology of Women," *American Psychologist* 30(1975): 852-57; Ludy (*sic*) T. Benjamin, Jr., "The Pioneering Work of Leta Hollingworth in the Psychology of Women," *Nebraska History* 56(1975): 493-505; Harry L. Hollingworth, *Leta Stetter Hollingworth: A Biography* (Lincoln: University of Nebraska Press, 1943); her connections in Greenwich Village are discussed in Sochen, *New Woman*, pp. 82-83; Floyd Dell, "The Nature of Women," *Masses* 8(Jan. 1916): 16; Rheta Childe Dorr, *A Woman of Fifty* (New York: Funk & Wagnalls, 1924), chap. 16, "A Feminist Group in New York," and idem, "Is Woman Biologically Barred From Success?" *NYT Magazine*, 19 Sept. 1915, pp. 15-16; "Dorr" in *NAW;* Robert H. Lowie and Leta Stetter Hollingworth, "Science and Feminism," *Scientific Monthly* 2(1916): 277-84, and L. S.

Hollingworth to R. H. Lowie, 29 Apr. 1914, Robert H. Lowie Papers, Bancroft Library, University of California, Berkeley; Arthur I. Gates (her colleague for two decades), "Leta S. Hollingworth," *Science* 91(1940): 9-11; George Malcolm Stratton, "Feminism and Psychology," *Century* 92(1916): 420-26; Leta Stetter Hollingworth, "The Vocational Aptitudes of Women," in H. L. Hollingworth, *Vocational Psychology: Its Problems and Methods* (New York: D. Appleton, 1916); idem, "Social Devices for Impelling Women to Bear and Rear Children," *American Journal of Sociology* 22(1916): 19-29; idem, "Comparative Variability of the Sexes," in Leta Stetter Hollingworth, *Public Addresses* (Lancaster, Pa.: Science Press, 1940), pp. 11-16; "Norsworthy" in *DAB*.

32. For example, Leta Setter Hollingworth, "Sex Differences in Mental Traits," *Psychological Bulletin* 13(1916): 377-84, and idem, "Comparison of the Sexes in Mental Traits," *Psychological Bulletin* 15(1918): 427-32; Hugo Münsterberg of Harvard University lost much caste by publishing nonexperimental essays after 1910. (Bruce Kuklick, *The Rise of American Philosophy, Cambridge, Massachusetts, 1860-1930* [New Haven: Yale University Press, 1979], pp. 211-12.) Hamilton Cravens, *Triumph of Evolution*, omits the feminists, leaving Rosalind Rosenberg, *Beyond Separate Spheres*, the major study of this episode.

33. Washburn supported the idea of women's suffrage but was not active in any groups to obtain it, since she reportedly disapproved so strongly of any "separate spheres" for women that she would not even join the ACA (Polyxenie Kambouropoulou to Edwin Boring, 21 Sept. 1960, in Washburn folder, Edwin Boring Papers, HUA). H. L. Mencken, "Woman Suffrage," in *In Defense of Women* (New York: Alfred A. Knopf, 1922), p. 135; Helen Norton Stevens, *Memorial Biography of Adele M. Fielde* (New York: Fielde Memorial Committee, 1918), chap. 29; de Forest (later Barney) under "Barney" in: *NAW:MP*, and Suzy Fisher, "Nora Stanton Barney, First U.S. Woman CE, Dies at 87," *Civil Engineer* 41(Apr. 1971): 87. Martin: *Winning Equal Suffrage in California: Reports of Committees of the College Equal Suffrage League of Northern California in the Campaign of 1911* (n.p., 1913), p. 5; Miriam A. De Ford, *Psychologist Unretired: The Life Pattern of Lillien J. Martin* (Stanford: Stanford University Press, n.d.), p. 87, reports her incorrectly as president in 1911; Owens: "Women Voters to Mark 40th," clipping from *Centre Daily Times*, State College and Bellefonte, Pa., Aug. 25, 1960, in Helen Brewster Owens Collection, SLRC.

34. "Woolley" in *NAW;* "Hollingworth" in *NAW*, and H. L. Hollingworth, *Leta Stetter Hollingworth*, pp. 119-20; Sabin: Florence Sabin to Mabel Mall, 28 Dec. 1919, Florence Sabin Letters, The Alan Mason Chesney Medical Archives of The Johns Hopkins Medical Institutions, and Elinor Bluemel, *Florence Sabin. Colorado Woman of the Century* (Boulder: University of Colorado Press, 1959), p. 63; Welsh: Lillian Welsh, *Reminiscences of Thirty Years*, p. 112; Chase in *NAW:MP*, and *NYT*, 26 Sept. 1963, p. 35; Caroline Furness, "Mary W. Whitney," *Popular Astronomy* 31(1923): 35; Cunningham: *NCAB* 6(1929): 504.

35. The standard history of science during World War I is Robert M. Yerkes, *The New World of Science: Its Development during the War* (New York: Century, 1920). The only field to try open recruitment was chemistry, where the wartime shortages were desperate. (Charles L. Parsons, "War Service for Chemists," *Science* 46[1917]: 32-33, 107-8, 451-52; *Science* 47[1918]: 234-35, and *Science* 48[1918]: 377-86, which indicates that in October 1918, just before the war ended, a roster was being set up and a long list of openings organized.) See also "Women Electrical Engineers," *JACA* 11(1917-18): 61, which describes a training program at Kansas Agricultural College, and the entry for Euphemia Worthington in *AMS*, 3d ed. (1921), which indicates that she left the Wellesley College mathematics department in 1918 to work for the Gallaudet Aircraft Corporation. Margaret Culkin Banning, *Women for Defense* (New York: Duell, Sloan and Pearce, 1942), chap. 2, describes women's contributions to winning World War I in industry.

36. Bureau of Vocational Information, *Women in Chemistry: A Study of Professional Opportunities* (New York: Bureau of Vocational Information, 1922). For more on the BVI, see chapter 8 herein. The papers of the BVI at SLRC contain many clippings and comments about women's wartime contributions to industrial chemistry. See also [Emma Hirth], "Industrial Chemistry for Women," *Smith Alumnae Quarterly*, Feb. 1918, pp. 3-6; Margaret B. MacDonald, "A Service Call to Scientific College Women," *JACA* 11, no. 1 (1917-18): 359-61; "Industrial Chemistry Is Opening for Women Many Doors of Opportunity Heretofore Closed to Them," [New York?] *Evening Sun*, 5 Feb. 1918, clipping in BVI Papers; Dan J. Forrestal, *Faith, Hope, and $5,000: The Story of Monsanto*

(New York: Simon and Schuster, 1977), p. 36. *Industrial Chemistry as a Vocation for Women,* Women's Educational and Industrial Union Bulletin no. 8 (Boston, 1911), was far from encouraging. Alfred Gradenwitz, "Women Chemists in Wartime," *Scientific American* 116 (1917): 343, had many examples of women chemists in Germany, where they were utilized even more intensively than in the United States. An editorial two years later ("Women and the Labor Shortage," *Scientific American* 119 [1918]: 206) had many examples of how well women were filling in for men in Britain.

An examination of the careers of the forty-two women chemists in the third edition of the *American Men of Science* (1921) reveals that only one woman, Mary Louise Foster, later of Smith College, had worked for an industrial firm before 1916, but starting with Ruby Rivers Orcutt in 1916, five others went to work for chemical companies (the Hercules Powder Company, National Aniline and Chemical Company, General Chemical Company, Standard Aniline Products, and Hammersley Manufacturing Company) in the next two years. The actual numbers must have been, as for the women astronomers in chapter 3, far greater than this, since from later accounts of their poor work and lack of training, most must have been little more than college students or just high school graduates.

37. "The Woman Chemist Has Come to Stay," *Journal of Industrial and Engineering Chemistry* 11(1919): 183.

38. "Notes," *Journal of Industrial and Engineering Chemistry* 11(1919): 508. Elsewhere in this issue (for example, p. 255) there was evidence of ominous employment conditions in the chemical industry in 1919. Other articles and several letters to the editor urged returning veterans to seek advanced training in graduate school, since there would not be enough jobs available for all of them in industry immediately. Correspondence about this episode is in the Bureau of Vocational Information Papers at SLRC. Brady said the women realized that it was "only fair" to give the returning veterans their jobs back and so resigned (William Brady to Harriet Sisson Gillespie, 23 Apr. 1919), but one of the women who resigned said that they had been told that the company preferred male chemists and that many of the men hired were not veterans but college men, including undergraduates (Elisabeth Duncan to Miss Hirth, 24 Feb. 1920, both in BVI papers).

39. "McDowell" in *AMS*, 3d ed. (1921); Dorothy W. Weeks, "There Are Diversities of Gifts but the Same Spirit . . ." (L. S. McDowell), *Wellesley Alumnae Magazine* (Winter 1975), inside back cover, and press magazine releases in McDowell's Faculty File, Wellesley College Archives; Grace McDermut Mulligan, an early graduate of the Colorado School of Mines, worked as a draftsman at the NBS from 1904 to 1952 ("Women at NBS: Something of an Experiment," NBS *Standard* vol. 24, no. 15 [25 July 1979]: 4–5. I thank Walter W. Weinstein for sending me a copy.). Chemist Martha Austin Phelps, Ph.D., had worked at NBS as an "analyst" in 1908 and 1909 (*AMS*, 3d ed. [1921]). Wick: *NCAB* 34 (1948): 433; Lura Beam, *Bequest from a Life: A Biography of Louise Stevens Bryant* (Baltimore: Waverly Press, 1963), pp. 70–4; Parsons, "Anti-Suffragists and the War." Many younger women scientists got their start in government jobs in 1918, as discussed in chapter 8.

40. "[Press Release] From the Committee on Public Information, Division on Women's War Work, June 7, 1918," Bureau of Vocational Information Papers, SLRC. This press release also adds, "According to Major R. M. Yerkes of the Psychological Division, trained women can be used for the highly specialized work of handling the Army reports and may eventually be called upon to assist with work in special hospitals dealing with cases of reconstruction. He does not anticipate the appointment of women to the psychological staffs now 'measuring the minds' of the soldiers in the camps." "Fernalds" in *AMS*, 3d ed. (1921); obituary of Mabel Fernald, *NYT*, 10 Oct. 1952, p. 25; "Cobb" in *AMS*, 6th ed. (1938). For more on this project, see Daniel J. Kevles, "Testing the Army's Intelligence: Psychologists and the Military in World War I," *Journal of American History* 55(1968): 565–81; Franz Samuelson, "World War I Intelligence Testing and the Development of Psychology," *Journal of the History of the Behavioral Sciences* 13(1977): 274–82, and Cravens, *Triumph of Evolution*, pp. 83–86 and 227; as well as the project's final report: Robert M. Yerkes, ed., *Psychological Examining in the United States Army*, published as *Memoirs of the National Academy of Sciences* 15(1921), and idem, "Testing the Human Mind," *Atlantic Monthly* 131(1923): 358–70. Two other women psychologists contributed to the war effort in other ways: Gertrude Rand of Bryn Mawr College and her husband, Clarence Ferree, studied night vision for the United States Navy (his

obituary, *NCAB* 33[1947]: 96), and Eleanor Rowland Wembridge, formerly of Reed College, was a supervisor of physical therapists and rehabilitation workers ("reconstruction aides") in the Surgeon General's Office (*AMS*, 3d ed. [1921]).

41. "Strong" in *AMS*, 5th ed. (1933), and Anne Hard, "Friendly Impressions, Dr. Helen Strong," *The Woman Citizen* 12(June 1927): 12–13, 45 (which has many errors); "Semple" in *NAW*, and Charles C. Colby, "Ellen Churchill Semple," *Annals of Association of American Geographers* 23(1933): 235. "The Inquiry" is described in John Kirtland Wright, *Geography in the Making: The American Geographical Society, 1851–1951* (New York: American Geographical Society, 1952), pp. 199–202.

42. Those histories of the Food Administration written by Hoover's friends and former staff members minimize the role of the women; see William C. Mullendore, *History of the United States Food Administration, 1917–1919* (Stanford: Stanford University Press, 1941), chap. 5 and app. A (list of personnel), and Maxcy Robsen Dickson, *The Food Front in World War I* (Washington, D.C.: American Council on Public Affairs, 1944), which stresses the conservation effort, especially chap. 5, "She Also Serves." Elliott F. Rose, "Food Will Win the War: The Nutritional Sciences and the U.S. Food Administration, 1917–19" (Master's thesis, University of California at Davis, 1966), discusses the war's impetus to the popular knowledge of vitamins. "Blunt" in *NAW:MP;* "Pennington" *in NAW:MP*, and "Nunc Dimittis, Mary Engle Pennington," *Poultry Science* 32(1953): 363; "Rose" in *NAW*, and Juanita Archibald Eagles, Orrea Florence Pye, and Clara Mae Taylor, *Mary Swartz Rose, 1874–1941: Pioneer in Nutrition* (New York: Teachers College Press, 1979), pp. 87–90; "Bevier" in *NAW*, and "War Record of Isabel Bevier," manuscript in Isabel Bevier Papers, University of Illinois Archives, Urbana, Ill.

43. Emily Newell Blair, *The Woman's Committee, United States Council of National Defense: An Interpretative Report, 1917–1919* (Washington, D.C.: GPO, 1920), discusses the women's contributions (Shaw's quotation taken from p. 57). Ida Tarbell, who had in her youth wished to be a biologist, worked for the educational division of the Food Administration and discussed the suffragists' complaints in her *All in the Day's Work: An Autobiography* (New York: Macmillan, 1939), chap. 16 ("Women and War").

44. Helen Hayes Peffer, "[Isabel Bevier] Sixty-Eight Years Young," *Independent Woman* 8(1929): 487, 527–28, interviews Bevier, who insists she is not a feminist.

45. Grace A. Hubbard, "War and the Woman's College," *New Republic* 15(1918): 285–87, Parke Rexford Kolbe, *The Colleges in War Time and After* (New York: D. Appleton, 1919), chap. 11, "College Women and the War". Gabrielle Elliott, "'The College Woman's Plattsburg,'" *JACA* 11(1917–18): 498–501, and editorial, "A Training Camp for Nurses," *JACA* 11 (1917–18): 511–12; Frances Wentworth Cutter, "College Women in the Nation's Service," *JACA* 12(1918): 7–11; Edith K. Dunton, "The Smith College War Emergency Summer School," *JACA* 12(1918): 11–15; editorial, "War Service Training for Women College Students," *JACA* 12(1918): 16–22, which endorses a plan by the American Council on Education to train them to be nurses. "The War Work of the Association of Collegiate Alumnae," *JACA* 12(1919): 100–103, summarizes the group's volunteer activities.

Emma Perry Carr of Mount Holyoke took advantage of the wartime demand for chemists to secure trustee approval of some industrial support for research in her department (Emma P. Carr to Dr. Charles Herty, 20 Oct. and reply, 22 Oct. 1918, both in Folder 35, box 117, Charles Holmes Herty Papers, Special Collections, Robert W. Woodruff Library, Emory University. I thank archivist Diane Windham for sending copies).

46. Penny Martelet, "The Woman's Land Army, World War I," in *Clio Was a Woman: Studies in the History of American Women*, ed. Mabel E. Deutrich and Virginia C. Purdy (Washington, D.C.: Howard University Press, 1980), pp. 136–46. Ida Ogilvie, "Agriculture, Labor, and Woman," *Columbia University Quarterly* 20(1918): 293–300; [Ethel Puffer Howes], "History of the Woman's Land Army of America," manuscript in Morgan-Puffer Family Papers, SLRC; Elizabeth A. Wood, "Memorial to Ida Helen Ogilvie (1874–1963)," Geological Society of America, *Bulletin* 75(1964): P36–P37, and clippings at Barnard College Archives; obituary of Hilda Loines, *NYT*, 7 July 1969, p. 33; for some of the problems of the Land Army, see Virginia C. Gildersleeve, "Women Farm Workers," *New Republic* 12 (1917): 132–34; Flora Rose et al., *A Growing College: Home*

Economics at Cornell University (Ithaca: Cornell University Press, 1969), pp. 60–62; Arthur C. Cole, *A Hundred Years of Mount Holyoke College* (New Haven: Yale University Press, 1940), pp. 259–60.

47. Robert N. Manley, *Centennial History of the University of Nebraska*, 2 vols. (Lincoln: University of Nebraska Press, 1969), 1:214–23; see also Carol S. Gruber, *Mars and Minerva: World War I and the Uses of the Higher Learning in America* (Baton Rouge: Louisiana State University Press, 1975); Marion Talbot, *More Than Lore: Reminiscences of Marion Talbot* (Chicago: University of Chicago Press, 1936), chap. 12 ("The Great War"—and how it affected student activities); and "Parsons" in *NAW*. The women faculty at Wellesley College were a particularly radical group. Ellen Hayes was, as mentioned in chapter 1, a socialist and communist (*Wellesley Alumnae Magazine* 15 [Feb. 1931]: 151–52), Sarah Whiting a prohibitionist ("Professor Whiting of Wellesley College Strong Supporter of National Constitutional Prohibition," *Union Signal*, 13 Aug. 1914, pp. 3–4, clipping in Whiting File, Wellesley College Archives), Mary Calkins an advocate of human rights (Jean Dietz, "Human Rights Her Battle Cry," Boston *Globe*, 18 June 1962, clippings in Calkins Faculty File, Wellesley College Archives; and Edgar Sheffield Brightman, "Mary Whiton Calkins: Her Place in Philosophy," *Wellesley Alumnae Magazine* 14 [June 1930]: 309), and Emily Balch, professor of economics, such an outspoken pacifist that the trustees fired her in 1918 (she was vindicated when she won the Nobel Peace Prize in 1946 [*CBY, 1947*, pp. 32–34]). See also "Vida Scudder" in *NAW:MP*.

48. "Mosher," in *AMS*, 3d ed. (1921), and David Starr Jordan, *The Days of a Man*, 2 vols. (Yonkers-on-Hudson, N.Y.: World Book, 1922), 1:411 n; "Gardner," in *NAW; MP*, and Harry S. Ladd, "Memorial to Julia Anna Gradner, 1882–1960," *Proceedings of the Geological Society of America for 1960*, pp. 87 and 88. "Bingham" in *CBY, 1961*, p. 55. Explorer Delia Akeley Howe may also have been a nurse with the Red Cross (*NYT*, 23 May 1970, p. 23).

49. "Editorial," *JACA* 10, no. 1 (Sept. 1916): 42–44, and Caroline Humphrey, "Address of the President," *JACA* 10, no. 9 (May 1917): 579–85. The ACA's identity crisis from 1916 to 1922, so obvious in its *Journal*, is not discussed in the official history (Marion Talbot and Lois K. M. Rosenberry, *The History of the American Association of University Women, 1881–1931* [Boston: Houghton Mifflin, 1931]). At one point the ACA leadership broached the idea of allowing men to join, as some suffragist groups did ("Shall We Have a Men's Auxiliary?," *JACA* 11 [1917–18]: 235).

50. "Report of the Committee on New Policies, October 1919," Naples Table Association, in Florence Sabin Papers, APS. The NTA later changed its mind about Curie (file on "Association to Aid Scientific Research by Women," Mary Thaw Thompson Papers, Vassar College Archives).

51. The best description of the Curie visit is in Robert Reid, *Marie Curie* (New York: Saturday Review Press, 1974), chaps. 20 and 21. More analytical is Helena M. Pycior, "The American Use of Madame Curie as Model of the Working Mother, the Female Healer and the Female Scholar" (paper, 1981). Contemporary accounts in the *New York Times, Science,* and the *Scientific Monthly* (indexed in the *Reader's Guide*) also capture the atmosphere of the visit, as do Eve Curie's *Madame Curie,* trans. Vincent Sheean (Garden City, N.Y.: Doubleday, Doran, 1937), chap. 23, and Marie Curie, *Pierre Curie, with the Autobiographical Notes of Marie Curie*, trans. Charlotte and Vernon Kellogg, with an introduction by Mrs. William Brown Meloney (New York: Macmillan, 1923), chap. 4, "My Visit to America." The Meloney-Curie correspondence is in Special Collections, Columbia University. Also useful are correspondence relating to her visit at the Archives of the National Academy of Sciences, the Abbott Lawrence Lowell Papers at the HUA (used with permission of Robert Shenton, Secretary to the Corporation), and the Florence Sabin Papers, APS. Clippings of the visit are in the Bureau of Vocational Information Papers, SLRC. "Meloney" in *Who's Who in America*, 12(1922–23): 2153. Her mother, Marie Mattingly, had been a prominent Southern educator in the 1870s and 1880s (*NCAB* 39[1943]: 458). Cullis: "In Retrospect: IFUW Beginnings," *JAAUW* 43(Summer 1950): 247–48, and *CBY, 1943*, pp. 158–59.

52. The gram of radium took on a life of its own after the trip. Marie Curie and her daughter Irene both used it before World War II, when it was evacuated from Paris ("Marie Curie's Gram of Radium," *JAAUW* 34[1941]: 167). It was last accounted for in 1948 ("Marie Curie's Gram of Radium," *JAAUW* 41[1948]: 85–86). Such lavish American philanthropy spurred the French government to increase its support of Curie's work in the 1920s (Spencer Weart, *Scientists in Power*

[Cambridge: Harvard University Press, 1979], pp. 17-18).

Although Warren Harding had not supported women's suffrage as Ohio's Republican senator until very late (1920), as a presidential candidate he courted women voters. His 1920 campaign was one of the first to be controlled by public relations experts (Randolph C. Downes, *The Rise of Warren Gamaliel Harding, 1865-1920* [Columbus: Ohio State University Press, 1970], pp 208, 225, 294, 502-9, and chap. 20 ["Of Managers and Management"]). Once elected he supported much women's legislation and appointed several women to high office (Eugene P. Trani and David L. Wilson, *The Presidency of Warren G. Harding* [Lawrence: Regents Press of Kansas, 1977], pp. 104-6).

53. In fact, the radium, mesothorium, and other ores Madame Curie took back to France were valued at over $162,000. She also took almost $7,000 in cash awards given her by various groups. Even so, about $54,000 remained in the "Marie Curie Radium Fund" at a New York bank. The interest was to provide her with a supplemental income until her death. (Marie Meloney to Dr. James Angell, 25 July 1921, Presidential Papers, Yale University Archives. I thank Patricia Bodak Stark for locating this item and sending me a copy.) Upon Curie's death in 1934, her daughter Irene Joliot-Curie became the recipient of the annual stipend. Then from 1948 to 1950, the AAUW fought for control of the fund, which it has since 1951 used for fellowships ("The Marie Curie Radium Fund," *JAAUW* 44[1951]: 183-84).

54. Nor did prominent suffragist Carrie Chapman Catt in her "Helping Madame Curie to Help the World," *Woman Citizen* 5(12 Mar., 1921): 1062.

55. "The Curie Fund," *JACA* 4, nos. 5 and 6 (Feb. and Mar. 1921): 116-17, and letterheads of Charlotte Kellogg to Mrs. Chipman, 14 Apr. 1921, in Meloney-Curie Correspondence, Special Collections, Columbia University (and Alice Ford Parsons to Florence Sabin, 21 May 1921, in Florence Sabin Papers, APS). The AAUW had still another group, the New York-based "Committee to Welcome Madame Curie," to run the Carnegie Hall reception. Two of its forty-two members were scientists: the home economists Martha Van Rensselaer of Cornell University and Louise McDanell Browne (wife of Charles A. of the USDA). The main speaker at the Carnegie Hall ceremony, which attracted an audience of 3,500, was M. Carey Thomas, whose lengthy address on international peace overshadowed the others' remarks in the *New York Times*'s account (*NYT*, 19 May 1921, p. 15) but was criticized strongly by her biographer, who apparently attended (Finch, *Carey Thomas*, pp. 282-83).

56. Eve Curie, *Madame Curie*.

57. So long have historians embraced this ideology that it is only recently that the whole public relations phenomenon has begun to be studied critically. See Spencer Weart, *Scientists in Power*, passim; Ronald Tobey, *The American Ideology of National Science, 1919-1930* (Pittsburgh: University of Pittsburgh Press, 1971); and Erwin Chargaff, "Building the Tower of Babble," in his *Voices in the Labyrinth: Nature, Man, and Science* (New York: Seabury Press, 1977), especially pp. 57-58.

58. By contrast, Yale's honorary degree to Curie was its sixth to a woman. Jane Addams of Hull House had been its first in 1910 (Lottie G. Bishop to Ysabella Waters, 20 July 1922, Mary Adelaide Nutting Papers, Teachers College Archives, quoted in Marshall, *Mary Adelaide Nutting*, p. 276 n. 70). Yale, however, refused to give Albert Einstein an honorary degree in 1921 on the grounds that he was in the United States not as a scientist but as a Zionist (Ronald W. Clark, *Einstein: The Life and Times* [New York: Avon Books, 1972], p. 475). A comparison of the reception accorded these two highly publicized and politicized scientists, who toured the United States within a month of each other in April and May 1921, might be instructive.

59. A (male) professor at Smith reported to Christine Ladd-Franklin that the "average" students there were indifferent to Curie's visit: "Mme. Curie is to be here tomorrow; there will be a special Convocation, and the occasion will be a memorial one for the College. The average student, however, I imagine to be much more interested in the Prom which comes next week. These students seem much less interested in the pursuit of learning than in avoiding as far as possible its pursuit of them" (H. N. Gardiner to Christine Ladd-Franklin, 12 May 1921, Christine Ladd-Franklin Papers, Special Collections, Columbia University). Ladd-Franklin also wrote a letter to the editor of the *New York Times* about Curie's eponymous fame ("Mme Curie and the Curie," *NYT*, 4 June 1921, p. 12). Icie Macy, a 1920 Yale doctorate in physiological chemistry and later a prominent biochemist,

attended the Curie reception in New York (Margaret A. Cavanaugh, "The Contributions of Icie Macy Hoobler to Chemistry" [Paper delivered at Women in Chemistry Symposium held in conjunction with American Chemical Society meeting, Washington, D.C., Sept. 1979]).

60. Simon Flexner, "The Scientific Career for Women," *Scientific Monthly* 13 (1921): 97-105; "Wants Women in Science, Dr. Flexner Tells Bryn Mawr Door of Opportunity Is Open," *NYT*, 3 June 1921, p. 32. Flexner's appointment of Louise Pearce as his assistant in 1913 also attracted the notice of the *New York Times* ("Woman Aid to Flexner," 28 Aug. 1913, p. 6). A pathologist, she later became well known for her trip to the Belgian Congo in 1920 to test tryparsamide as a remedy for African sleeping sickness. She also served as president of the Woman's Medical College of Philadelphia from 1946 to 1951. (*NAW:MP*; Marion Fay, "Louise Pearce, 1885-1959," *Journal of Pathology and Bacteriology* 82[1961]: 542-51; J. D. Fruton, "Dr. Louise Pearce," *Nature* 184[1959]: 588-89; Florence Sabin to Anne Morgan, 23 June 1932, Sabin Papers, APS; Tom Rivers, *Reflections on a Life in Medicine and Science* [Cambridge: MIT Press, 1967], pp. 82-85; the archives of the renamed Medical College of Pennsylvania, Philadelphia, has a file of biographical materials on Pearce.) Flexner also hired Sabin for the Rockefeller Institute in 1924 and helped elect her to the National Academy of Sciences in 1925 (see chapter 10). Ida Rolf also worked at the Rockefeller Institute from 1916 until 1928. Then a biochemist (Ph.D., Columbia University, 1920), she later gained fame as an applied physiologist who developed the technique of "rolfing" (obituary, *NYT*, 21 Mar. 1979, p. D21, and Lisa Connolly, "Ida Rolf," *Human Behavior* 6[May 1977]: 17-23).

61. Editorial, "Scientific Careers for Women," *NYT*, 4 June 1921, p. 12.

62. William H. Welch, "Contribution of Bryn Mawr College to the Higher Education of Women," *Science* 56(1922): 1-8. Thomas suggested the topic to Welch beforehand (M. C. Thomas to William H. Welch, 11 Apr. and 18, 27 and 29 May 1922, all in Welch Papers, The Alan Mason Chesney Medical Archives of The Johns Hopkins Medical Institutions). Welch was glad to comply and avoid the more controversial topic of the low marriage rate of college graduates, which bothered him and his friend C. W. Eliot of Harvard much more than discriminatory employment in academia (Welch to C. W. Eliot, 25 Aug. 1922, and reply 30 Aug. 1922, and probably Welch to S. Flexner, 28 Aug. 1922 [missing], all in Welch Papers).

CHAPTER 6

1. Lindsey Harmon and Herbert Soldz, comps., *Doctorate Production in United States Universities, 1920-1962* (Washington, D.C.: National Academy of Sciences-National Research Council Publication 1142, 1963), with the women on pp. 49-53. Interpreting this data to document "withdrawal" are Jessie Bernard, *Academic Women* (Cleveland: World Publishing, 1964), and John B. Parrish, "Top Level of Training of Women in the United States, 1900-1960," *Journal of the National Association of Women Deans and Counselors* 25(1962): 67-73. Using it more critically are: Ann Fischer and Peggy Golde, "The Position of Women in Anthropology," *American Anthropologist* 70(1968): 337-44; and Frank Stricker, "Cookbooks and Lawbooks: The Hidden History of Career Women in Twentieth-Century America," *Journal of Social History* 10(1976): 6-7.

2. Harmon and Soldz, *Doctorate Production*, pp. 11 and 57.

3. James McKeen Cattell, "A Biographical Index of the Men of Science of the United States," *Science* 16(1902): 746, reprinted in *Popular Science Monthly* 62(1903): 185 and the *Nation* 80(1905): 457. See also "Preface" of James McKeen Cattell, *American Men of Science, A Biographical Directory* (New York: Science Press, 1906), p. v. Several studies based upon data from the *AMS* directories include Dean Brimhall, "Family Resemblances among American Men of Science," *American Naturalist* 56 (1922): 504-47, and *American Naturalist* 57 (1923): 74-88, 137-52, and 326-44; James McKeen Cattell, "Origins and Distribution of Scientific Men," *Science* 66(1927): 513-16; and "The Distribution of American Men of Science in 1927," *AMS*, 4th ed. (1927), pp. 1118-29; L. Pressey, "The Women Whose Names Appear in the *American Men of Science* for 1927," *School and Society* 29(1929): 96-100; Harvey C. Lehman and Paul A. Witty, "Scientific Eminence and Church Membership," *Scientific Monthly* 33(1931): 544-49; James McKeen Cattell, "The Distribution of American Men of Science in 1932," *Science* 77(1933): 264-70, reprinted in

AMS, 5th ed. (1933), pp. 1261–68; Edward L. Thorndike, "The Production, Retention and Attraction of American Men of Science," *Science* 92(1940): 137–41; and Alice Wupperman, "Women in the *American Men of Science*," *Journal of Chemical Education* 18(1941): 120–21 (has many errors).

4. Spencer Weart, "The Physics Business in America, 1919–1940," in *The Sciences in the American Context: New Perspectives*, ed. Nathan Reingold (Washington, D.C.: Smithsonian Institution Press, 1979), p. 335 n. 5. Marie Farnsworth reported she had found just 42 women chemists in the third edition of the *AMS* (1921), but that there were an overwhelming 481 women members of the American Chemical Society (3.2 percent) ("Women in Chemistry: A Statistical Study," *Industrial and Engineering Chemistry* 3[10 Sept. 1925]: 4).

Thus in addition to the "scientists" of the *AMS* directories and the "doctorates" of the NRC data one could designate two other groups, "chemists" and "government scientists," as important and overlapping, but largely separate, populations of scientists in these years, as shown here diagrammatically:

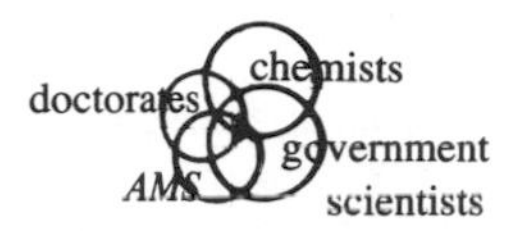

The "chemists" were the members of the American Chemical Society, many of whom neither held a doctorate nor met Cattell's criterion of contributing to research. The ACS, established in 1876, underwent a tremendous growth between 1890, when it had 250 members, and 1920, when it had over 15,000 (when the *AMS* had only 9,500 persons in all fields of science). (Charles L. Parsons, "The Activities of the American Chemical Society," in *A Half-Century of Chemistry in America, 1876–1926*, ed. Charles A. Browne [Easton, Pa.: American Chemical Society, 1926], p. 42.) The explanation is that about one-half of these "chemists" were in management or routine testing positions in industry, careers that were open to persons in few other fields. In the late 1940s, some critics could complain that the *AMS* had clearly underemphasized the importance of chemists in the United States. In a sense it had, but the "chemists" played many roles, and only about one-half of them contributed to "science." For data on government scientists, see chapter 8 herein.

5. U.S. Bureau of the Census, *Historical Statistics of the U.S., Colonial Times to 1957* (Washington, D.C.: GPO, 1960), p. 10.

6. This may be an artifact, since Joel Stebbins, "The American Astronomical Society, 1897–1947," *Popular Astronomy* 55(1949): 405, reports that the society grew from 370 members in 1922 to "about 625" in 1947. Steve Pyne claims that geology stagnated intellectually after the late 1920s but provides no statistical data to quantify the idea ("From the Grand Canyon to the Marianas Trench: The Earth Sciences after Darwin," in *Sciences in the American Context*, ed. Reingold, pp. 165–92.)

7. American Association of University Professors, Committee Y, *Depression, Recovery, and Higher Education* (New York: McGraw-Hill, 1937); Gardner Lindzey, ed., *History of Psychology in Autobiography*, 7 vols. (Englewood Cliffs, N.J.: Prentice-Hall, 1974), vol. 6, especially essays by Krech and Mowrer; A. L. Patterson, "Experiences in Crystallography, 1924 to Date," in *Fifty Years of X-Ray Diffraction*, ed. P. P. Ewald (Utrecht: A Osthoek's Uitgeversmij, 1962), p. 621. Lorenz J. Finison opens up important new ground with his "Unemployment, Politics, and the History of Organized Psychology," *American Psychologist* 31(1976): 747–55. Unemployment was apparently also a serious problem for chemists in New York City. See "Unemployment Among Chemists and Chemical Engineers," *Science* 76(1932): 10, and "Work of the Committee on Unemployment and Relief for Chemists and Chemical Engineers," *Science* 83(1936): 253–54.

8. It was not until the eighth edition of the *AMS* (1948) that entrants were asked their marital status and number of children, but even then many persons known from other sources to be married omitted this information. (There are also some amusing errors, as some husbands and wives differed on their marriage date or number of children.) Ching-Ju Ho, *Personnel Studies of Scientists in the United States*, Contributions to Education no. 298 (New York: Teachers College, Columbia University, 1928), p. 30; Jessica Peixotto, *Getting and Spending at the Professional Standard of Living* (New York: Macmillan, 1927), pp. 11 and 50, is based on 1922 data (her sample of 433 faculty

members included at least three women); Yandell Henderson and Maurice R. Davie, eds., *Incomes and Living Costs of a University Faculty* (New Haven: Yale University Press, 1928), chap. 3, table 1.

9. There is also evidence that Lise Meitner, the German physicist, might have come to America in 1938 to work on cancer research at a New York hospital rather than fleeing to Sweden (Esther Caukin Brunauer to Florence Sabin, 22 Apr. 1938, and reply, 29 Apr. 1938, in Florence Sabin Papers, APS; and Otto Hahn, *A Scientific Autobiography,* ed. and trans. Willy Ley [New York: Charles Scribner's Sons, 1966]).

10. Henry Menard, *Science: Growth and Change* (Cambridge: Harvard University Press, 1971); Ho, *Personnel Studies,* pp. 48, 51, and 58. A companion study to Ho's, Lycia Martin's "Personnel Study of Women Scientists" (Ph.D. diss., Teachers College, Columbia, 1928), is cited in Emilie J. Hutchinson, *Women and the Ph.D.,* Institute of Women's Professional Relations Bulletin no. 2 (Greensboro: North Carolina College for Women, 1929), p. 23 n. It was based on NRC data for 611 women, but may never have been completed.

11. These thoughts have been presented more fully in my "Sexual Segregation in the Sciences: Some Data and a Model," *Signs* 4 (1978): 146–51. Although the field of nutrition may seem to be a serious counterexample to this growth-induced feminization, since its percentage of women dropped by more than one-half from 1921 to 1938 despite a very large growth rate, its change is really a structural one due to the rise of a new commercial food industry in the 1930s; for example, General Mills and Quaker Oats hired many male "food chemists" or chemical engineers to develop efficient production methods. For lack of a better alternative this new field was classified as a part of "nutrition," although its practitioners were not in close contact with the women professors at the schools of home economics or even with the increasing numbers of women nutritionists in industry who worked in kitchens to develop recipes for consumers (see chapter 9 herein).

12. "Anthropology: Studied, Taught, and Practiced," *Barnard College Alumnae Monthly* 49(Nov. 1950): 4.

13. Carolyn Shaw Bell, "Women in Science: Definitions and Data for Economic Analysis," in *Successful Women in the Sciences: An Analysis of the Determinants,* ed. Ruth B. Kundsin, published as *Annals of the New York Academy of Sciences* 28(1973): 134–35.

14. Everyone who counts women, especially married ones, in the various editions of the *AMS* comes up with different totals. See n. 2 above and Mary Jo Huth, who found only 687 women in the 1927 edition, only 72 of whom were married ("A Comparative Study of Women Listed in the 1900 and 1950 Editions of *Who's Who in America*" [Master's thesis, Indiana University, 1951], pp. 17–18, cited by Peter Filene, *Him, Her, Self: Sex Roles in Modern America* [New York: Harcourt Brace Jovanovich, 1974], p. 305 n. 13.) The only discussion of couples of scientists that I have found for this or any period is Alma Smith Payne's *Partners in Science* (Cleveland, Ohio: World Publishing, 1968) for juveniles. Angelo Hall, *An Astronomer's Wife; The Biography of Angeline Hall* [wife of Asaph Hall] (Baltimore: Nunn, 1908), is one of several about nonscientist wives of scientists.

15. A good introduction to the literature and issues is Lois Scharf, *To Work and to Wed: Female Employment, Feminism, and the Great Depression* (Westport, Conn.: Greenwood Press, 1980). See also David W. Peters, *The Status of the Married Woman Teacher,* Contributions to Education no. 603 (New York: Teachers College, Columbia University, 1934). Walter Kotschnig, *Unemployment in the Learned Professions* (London: Oxford University Press, 1937), is internationally oriented and has relatively little on the American situation, but it does have a provocative thesis, which says that the increased enrollments of women in professional schools were a major cause of male unemployment in the professions; a guide to the numerous studies of unemployment among American women by the AAUW, other alumnae groups, the National Federation of Business and Professional Women, and the U.S. Women's Bureau is AAUW, comp., *Summaries of Studies on the Economic Status of Women*, WB Bulletin, U.S. Department of Labor, no. 134 (1935); Susan Slavin Schramm, "Section 213: Woman Overboard" (Paper presented at the Second Berkshire Conference on Women's History, Radcliffe College, 1974) (now in SLRC), discusses the effects and politics of the law.

16. Edward J. Power, *Catholic Higher Education in America: A History* (New York: Appleton-Century-Crofts, 1972), is a gold mine of information but is largely negative on the women's colleges, for example, blaming them for their failures rather than applauding their efforts; see

pp. 272–73 and chap. 2. So too, Mabel Newcomer, *A Century of Higher Education for American Women* (New York: Harper & Bros., 1959), p. 195, says that the curriculum of the Catholic women's colleges stressed teacher-training. Catholic University of America was established in the 1880s largely through the gift of a wealthy New York heiress, Mary G. Caldwell, but it did not admit women, even to its graduate school, until 1928 (Power, pp. 355–56 and 366–67). There is a need for a social and intellectual study of Catholic nuns; a good start is provided by Sister Cecile Agnes Forest, F.S.E., "The Religious Academic Woman: A Study of Adjustment to Multiple Roles" (Ph.D. diss., Fordham University, 1966). Three obituaries which indicate some had an active intellectual life are: Donald P. Costello, "Sister Florence Marie Scott," *Biological Bulletin* 133 (1967): 13–14 (she was a Columbia Ph.D., spent thirty-two summers at Woods Hole, and was elected a Trustee in 1964, one of the very few women given this honor); "Sister Amata (Rosalie McGlynn)," *Proceedings of the Indiana Academy of Sciences for 1938*, pp. 1–2; and "Mother Mary Verda (Margaret Dorsch)," *Proceedings of the Indiana Academy of Sciences for 1943*, pp. 12–13.

17. Robert H. Knapp and H. B. Goodrich, *The Origins of American Scientists* (Chicago: University of Chicago Press, 1952), p. 20 n and app. 2; M. Elizabeth Tidball and Vera Kistiakowsky, "Baccalaureate Origins of American Scientists and Scholars," *Science* 193(1976): 646–52. Although primarily interested in the religious backgrounds of American scientists, Kenneth R. Hardy also presents data on the most productive undergraduate colleges from 1920 to 1939 and from 1950 to 1961 in "Social Origins of American Scientists and Scholars," *Science* 185(1974): 501. Earl K. Wallace, "A Survey of Chemistry in Women's Colleges," *Journal of Chemical Education* 14(1937): 285–94, talks only of the curricula and not of personalities.

The relative importance of the women's colleges (and so of women faculty) for the education of later career women has been much discussed of late: M. Elizabeth Tidball, "Perspective on Academic Women and Affirmative Action," *Educational Record* 54(1973): 130–35; Mary J. Oates and Susan Williamson, "Women's College and Women Achievers," *Signs* 3(1978): 795–806; and M. Elizabeth Tidball, "Women's Colleges and Women Achievers Revisited," *Signs* 5(1980): 504–17.

18. "Melba Newell Phillips, Oersted Medalist for 1973," *American Journal of Physics* 42(1974): 357; K. P. Sopka, "Melba Phillips, Transcript of an Interview on a Tape Recorder, December 5, 1977," Niels Bohr Library, American Institute of Physics, New York City.

19. "Best Women [Scientists]," *Time* 21(20 Mar. 1933): 38. She protested privately that she had been misquoted (she had referred only to the difficulties of women anthropologists, not those of all women scientists). The correction was apparently never published in *Time*, although an editor promised it would be (Ruth Benedict to *Time*, 16 Mar. 1933, and I. van Meter to Ruth Benedict, 27 Mar. 1933). She also recommended omitting anthropology from a book on careers for women because "present opportunities [are] so pitifully inadequate—in terms of positions and funds for expeditions—that even gifted women have to be turned away" (Catherine Filene Shouse to Ruth Benedict, 23 Mar. 1933, and reply, 27 Mar. 1933). However, when Cora DuBois wrote a few months later that she expected to be unemployed soon, Benedict found some money for her (Cora DuBois to Ruth Benedict, 21 June 1933, and reply, 10 July 1933, all in Ruth Benedict Papers, Vassar College Library). See also Ruth Benedict, "Women and Anthropology," in *Report of the Commission on Education of the Woman's Centennial Congress in New York City, 1940*, reprinted in *Women's Work and Education* 11, no. 3(Oct. 1940): 11, and George Grant MacCurdy, "The Academic Teaching of Anthropology in Connection with Other Departments," *American Anthropologist* 21(1919): 49–60. Courses were often given in sociology departments.

20. John B. Daffin, "Why the Woman Student Does Not Elect Physics," *American Journal of Physics* 5(1937): 82–85, and Sister M. Ambrosia, "Teaching Physics to Women," *American Journal of Physics* 8(1940): 289–90. Oswald Blackwood limits his "Undergraduate Origins of American Physicists," *American Journal of Physics* 12(1944): 149–50, to males but does mention Mount Holyoke's R. D. Rusk in one table.

21. Harmon and Soldz, comps., *Doctorate Production*, pp. 19–26.

22. Garland E. Allen, *Thomas Hunt Morgan: The Man and His Science* (Princeton, N.J.: Princeton University Press, 1978), p. 163; Ian Shine and Sylvia Wrobel, *Thomas Hunt Morgan, Pioneer of Genetics* (Lexington: University of Kentucky Press, 1976), p. 105.

23. One article about Benedict attributes the large increase in the number and percentage of doctorates given to women by the Columbia anthropology department from 3 (of 12, or 25 percent) between 1901 and 1920 to 22 (of 51, or 43.1 percent) between 1921 and 1940 to her great activity there during those years rather than, as is more common, to Boas's receptivity (Virginia Wolf Briscoe, "Ruth Benedict, Anthropological Folklorist," *Journal of American Folklore* 92[1979]: 472). The Minnesota psychology department was also the first in the nation to award a doctorate to a black woman (Ruth Howard), in 1934 (Robert V. Guthrie, *Even the Rat Was White: A Historical View of Psychology* [New York: Harper & Row, 1976], p. 138).

24. So rare were women on even the subfaculty at Yale that it was news when one was appointed a "research assistant" in chemistry in 1925 ("Woman Joins Yale Staff, Elizabeth Gilman Gets Post Hitherto Held by Men," *NYT*, undated clipping in Bureau of Vocational Information Papers, box 21, SLRC).

25. Radcliffe College Committee on Graduate Education for Women, *Graduate Education for Women: The Radcliffe Ph.D.* (Cambridge: Harvard University Press, 1956), p. 18, table D, shows how very few women got doctorates from Harvard-Radcliffe from 1923 to 1952 despite relatively high female enrollments there.

An abundance of articles use doctoral data to go into more detail about particular fields than is possible here, such as Chester K. Wentworth, "American Doctorates in Geology," *Journal of Geology* 41(1933): 432–38; Louis L. Ray, "American Doctorates in Geology, 1931–1940," *Journal of Geology* 49(1941): 854–61; John F. Hall and J. Allan Cain, "Past and Present Provenances and Predilections of Pedagogical Ph.D.s," *Journal of Geological Education* 13(1965): 82–85; and R. G. D. Richardson, "The Ph.D. Degree and Mathematical Research," *American Mathematical Monthly* 43(1936): 199–215, which scoured mathematics journals and departments for a list of 1,286 doctorates (186, or 13.2 percent, of which went to women).

Other studies are: Emilie J. Hutchinson, "Women and the Ph.D.," *JAAUW* 22(Oct. 1928): 19–22, and idem, *Women and the Ph.D.*, which are based on questionnaires returned by women doctorates. The American Council on Education's *The Production of Doctorates in the Science, 1936–48* (Washington, D.C.: American Council on Education, 1951) has a bibliography of pre-1920 doctoral data, and several of the major universities have published directories of their doctorates arranged by departments.

CHAPTER 7

1. A. Caswell Ellis to Cora Beckwith, 27 Oct. 1919, in A. Caswell Ellis Papers, University of Texas Archives (Barker Texas History Center). He, too, would merit a biography, since the brief entry in *Who Was Who in America*, 6 vols. to date (Chicago: A. N. Marquis, 1950), 2:173, does not do justice to his far-ranging projects.

2. Clara F. McIntyre, "A Venture in Statistics," *JACA* 12, no. 1 (Oct. 1918): 1–7. This was written in response to the jocular "Pleasant Possibles in Lady Professors" by Elizabeth Hazelton Haight (Latin professor at Vassar College), published in a previous issue (*JACA* 11, no. 1[Sept. 1917]: 10–17.

3. Helen Sard Hughes, "The Academic Chance," *JACA* 12, no. 2(Jan. 1919): 79–82. ("My dear Miss———: You have been at a good deal of pains, I realize, to answer my questions and give me information about your work. I am sorry it all came to nothing. I ought not to have troubled you. The situation this year for instructors has been most unusual. We prefer to appoint a man. At one move of the kaleidoscope there seemed to be no men; and at the next change, a plenty. It was on one of the off days [for men] that I wrote you. The letters which Mr.———and others sent me were most cordial, and I hope that you will accept my best wishes for your scholarly success. Yours sincerely, ———.") Two years later, however, Hughes was advising women scholars to publish more ("College Women and Research Again," *JACA* 14, no. 4 [Jan. 1921]: 88–90).

4. "Opportunities and Salaries of Women in the Teaching Profession in Nebraska," *JACA* 13, nos. 5 and 6(Mar. and Apr. 1920): 10–13.

5. "The Nebraska Committee's Report," *JACA* 13, nos. 7 and 8(July and Aug. 1920): 1–4. See also questionnaires returned to AAUP Committee W in 1920 by Nebraska University professor Louise Pound and Chancellor S. Avery in A. Caswell Ellis Papers, University of Texas Archives. Marion Talbot was also part of a 1924 protest of the small number of women on the University of Chicago faculty (3 of 150) (*More Than Lore: Reminiscences of Marion Talbot* [Chicago: University of Chicago Press, 1936], chap. 7).

6. William H. Welch, "Contribution of Bryn Mawr College to the Higher Education of Women," *Science* 56(1922): 1–8, and chapter 5 herein.

7. A. Caswell Ellis to H. W. Tyler, n.d. [1924], A. Caswell Ellis Papers, University of Texas Archives. A. B. Wolfe, an economist formerly at the University of Texas but then at Ohio State University, joined the committee in 1924.

8. *BAAUP* 4(1918): 42; 5(1919): 74; 6(1920): 28; 7(1921): 43; "Preliminary Report of Committee W, on Status of Women in College and University Faculties," *BAAUP* 7, no. 6(1921): 21–32; see also "Committee W, Status of Women in Colleges and Universities," *BAAUP* 8, no. 2(1922): 161. A full article could be written on Committee W's activities and its internal divisions. Besides the Ellis Papers at the University of Texas Archives, correspondence relating to it include Gertrude S. Martin to Kate Stephens, 12 May 1917, and reply, 16 May 1917, in Kate Stephens Papers, University of Kansas Archives, Spencer Research Library; John M. Coulter to Marion Talbot, 16 Feb. 1918, in Marion Talbot Papers, Special Collections, University of Chicago Library; E. A. Birge to "My dear Florence," 13 Dec. 1920, A. Caswell Ellis to Dr. Bascom [3 Dec. 1921], and M. Carey Thomas to Professor Bascom, 20 Dec. 1921, all in Florence Bascom Papers, Smith College Archives; and H. W. Tyler to H. E. Bolton, 4 Mar. 1922, in H. E. Bolton Papers, Bancroft Library, University of California, Berkeley. Many of Bolton's 200 women graduate students at the University of California considered him an antifeminist, since he encouraged them to earn only a master's degree and not a doctorate (John Francis Bannon, *Herbert Eugene Bolton: The Historian and the Man, 1870–1953* [Tuscon: University of Arizona Press, 1978], pp. 106–7 and app.). Mark Aldrich has analyzed Willcox's naive (but typically "Progressive") beliefs about social statistics. He insisted that the numbers could be objective and thus separate from and superior to "mere opinion" or political assumptions, although his very categories and selection of data were riddled with political views (Mark Aldrich, "Progressive Economists and Scientific Racism: Walter Willcox and Black Americans, 1895–1910," *Phylon* 40[1979]: 1–14).

9. Marion O. Hawthorne, "Women as College Teachers," *Annals of the American Academy of Political and Social Science* 143 (1929): 153.

10. U.S. Office of Education, *Biennial Survey of Education in the U.S., 1956–58* (Washington, D.C.: GPO, 1958), chap. 4, p. 6, section 1, table 2.

11. "Second Report of Committee W on the Status of Women in College and University Faculties," *BAAUP* 10, no. 7 (1924): 65–73; the last notice of the committee's existence appears in *BAAUP* 13, no. 3 (April 1927): 284–85. Several responses to the questionnaire on publications and salaries are in the Ellis Papers at the University of Texas Archives. He later explained to an inquirer that the report had never been completed "on account mainly, I think, of the chairman's lack of time" (A. Caswell Ellis to Mrs. C. F. Martin, 24 April 1934, Ellis Papers).

12. Ella Lonn, "Academic Status of Women on University Faculties," *JAAUW* 17, no. 1 (1924): 5-11. A further study of college promotion and tenure practices was planned by the AAUW in 1924 but not funded (Frances Fenton Bernard, "Progress of the Education Program," *JAAUW* 17, no. 1 [1924]: 4, and her "Report of the Educational Secretary for the Period November 1923–April 17, 1924," *JAAUW* 17, no. 2 [1924]: 8). In 1926, however, the AAUW and the Bureau of Vocational Information of New York City did cosponsor Emilie Hutchinson's study of 1,025 women doctorates who earned their degrees before 1924 (Emilie J. Hutchinson, "Women and the Ph.D.," *JAAUW* 22 [Oct. 1928]: 19–22, and idem, *Women and the Ph.D.*, published as the Institute of Women's Professional Relations Bulletin no. 2 [Greensboro: North Carolina College for Women, 1929]).

Besides Lonn, there were two other major academic stateswomen of the 1920s. Marjorie Hope Nicholson, professor of English (and dean) at Smith College and, later, Columbia University, was also the first woman president of Phi Beta Kappa (1940) and the author of "Scholars and Ladies,"

Yale Review 19 (1930): 775–95. There is a small collection of her papers in the Smith College Archives, and an oral history memoir at Columbia University Oral History Office (see note 15 below). She merits a full biography, as does Louise Pound, professor of English at the University of Nebraska and first woman president of the Modern Language Association, who was the author of "The College Woman and Research," *JACA* 14, no. 2(Nov. 1920): 31–34, and is in *NAW:MP*.

13. Stanford psychologist Truman L. Kelley, formerly of Teachers College, suggested Leta Stetter Hollingworth be added to the committee in 1924 (Kelley to Ellis, 13 Sept. 1924, Ellis Papers, University of Texas Archives), but she was not. She may not have been a member of the AAUP, however, since this criterion had eliminated many other persons who were also suggested.

This kind of logical error, attributing to sex differences that which is only due to social practices, was quite common in the 1920s and 1930s. Rosabeth Moss Kanter has found that it pervaded the management literature of the time ("Women and the Structure of Organizations: Explorations in Theory and Behavior," in *Another Voice: Feminist Perspectives on Social Life and Social Science*, ed. Marcia Williams and Rosabeth Moss Kanter [New York: Octagon Books, 1976], pp. 34–74). More surprising than the presence of such fallacies was the lack of any criticism of them. One suspects that Christine Ladd-Franklin, an acute logician, would have detected the fallacy and written a scorching letter to the *New York Times*, the *Nation*, or *Science*. Though in her late seventies, she was still active and was probably kept informed of the committee's work by John Dewey, but apparently she had no public response.

14. Because women were not expected to become professors, observers rarely detected in them the early "promise" that could justify and open up the series of favorable opportunities necessary to fulfill the prophecy. The idea that young men would "develop" this "promise" if given the right opportunities dates back at least to the "advice literature" of the 1830s (Burton J. Bledstein, *The Culture of Professionalism: The Middle Class and the Development of Higher Education in America* [New York: W. W. Norton, 1976], p. 216). The embarrassment and anger that could result when a young man, once labeled "a comer," did not develop is evident in several letters of geneticist George Shull to Ralph Cleland (28 Mar. and 7 Apr. 1930) in the Ralph Cleland Papers, Lilly Library, Indiana University. (I thank Saundra Taylor for help in locating these items.)

15. President Nielsen of Smith deliberately hired men and promoted them more readily than women (John Wieler, "Interview with Professor Marjorie Nicolson," 1975, Oral History Memoir, Columbia University, pp. 273–74, 282, 333, and 354). Marjorie Nicolson was dean at the time and strongly endorsed the policy as the only way to upgrade the college.

16. Eunice Fuller Barnard, "Woman's Rise in Science," *NYT Magazine*, 27 Oct. 1935, pp. 9 and 18; Cornelia B. Meigs, *What Makes A College? A History of Bryn Mawr* (New York: Macmillan, 1956); Arthur C. Cole, *A Hundred Years of Mount Holyoke College* (New Haven: Yale University Press, 1940); Jean Glasscock, "The Buildings," in *Wellesley College, 1875–1975: A Century of Women*, ed. Jean Glasscock et al. (Wellesley, Mass.: Wellesley College, 1975), pp. 316–19; Samuel White Patterson, *Hunter College* (New York: Lantern Press, 1955), chaps. 7–9; "Katherine Blunt," *CBY, 1946*, pp. 57–59. The Carnegie Corporation of New York gave more than $3 million to over thirty women's colleges between 1934 and 1940 (Robert M. Lester, *A Thirty Year Catalog of Grants* [New York: Carnegie Corporation, 1942], pp. 51–64).

17. *Many a Good Crusade: Memoirs of Virginia Crocheron Gildersleeve* (New York: Macmillan, 1954), pp. 105–9. See also *NAW:MP*. The woman physicist may have been Agnes Townsend Wiebusch, who married in 1926 (*AMS*, 6th ed. [1938], and *AMS*, 8th ed. [1949]).

18. Elizabeth F. Genung, who taught bacteriology at Smith College starting in 1919, may have been the first to do so at a woman's college (Elinor V. Smith, "Obituary: Elizabeth F. Genung," *ASM* [*American Society for Microbiology*] *News*, Feb. 1975, p. 114 [I thank Ann Norberg for this item]). Elizabeth F. Genung, "Smith College's Part in the Field of Public Health," *Smith Alumnae Quarterly*, Nov. 1928, pp. 25–28, describes the program; and Elinor V. Smith to Miss Johnson, 22 Jan. 1954, Myra Sampson Papers, Smith College Archives, describes its more notable graduates. The beloved Edith Minot Twiss, an assistant professor of botany at Vassar College, inaugurated instruction in bacteriology there in 1920 ("Edith Minot Twiss," *Vassar Quarterly*, Nov. 1925, p. 25; clippings in Faculty File, Vassar College Archives). Ruth Wheeler of Goucher had been the only faculty member in home economics at a woman's college in the 1921 *AMS*, but she left soon

afterward for the University of Iowa; her "Home Economics in the Woman's College," *Journal of Home Economics* 11(1919): 375–80, reveals how lukewarm the welcome was. She later directed Vassar's program in physiology, nutrition, and "euthenics" (*AMS,* 6th ed. [1938]).

19. Margaret Washburn to Robert Yerkes, 24 Oct. 1931, Robert M. Yerkes Papers, Yale University Library. Florence Sabin also discussed the difficulties of doing research at a woman's college in a letter to Anne Morgan, 23 June 1932, Florence Sabin Papers, APS.

20. Marjorie O'Connell to Christine Ladd-Franklin, series of five letters, June 1920–December 1921, in Christine Ladd-Franklin Papers, Special Collections, Columbia University; Professor L. E. Dickson cited in Grace Keller to E. J. Townsend, 5 Mar. 1925, Personnel folder, Department of Mathematics, University of Illinois; see also Olive Hazlett to E. H. Moore, 16 May 1925, E. H. Moore Papers, Special Collections, University of Chicago Library. (Other letters in this folder recount the circumstances of her dismissal from Bryn Mawr College in 1918.) Neither woman contributed much to science after these decisions, however: O'Connell married and held a series of secretarial and administrative positions in New York and Washington ("Shearon" in *AMS,* 7th ed. [1944]), and Hazlett was so unhappy at Illinois that she had a series of mental breakdowns in the 1930s and 1940s from which she never recovered (she died in 1974). She was, as the chairman put it recently, "the most difficult personnel case in a hundred years in the Illinois math department" (Professor W. A. Ferguson to author, personal communication, 29 Mar. 1978). Obituary in *NYT,* 12 Mar. 1974, p. 40.

21. Cole, *Hundred Years,* pp. 326–27; Mount Holyoke College File at Rockefeller Foundation Archives; "Reichard" in *NAW:MP;* and Margaret Mead, "Commitment to Fieldwork," in *Gladys A. Reichard, 1893–1955* (New York: Barnard College, 1955), p. 22; the "Southwest Society" is identified in Margaret Mead, "Ruth Fulton Benedict, 1887–1948," *American Anthropologist* 51 (1949): 459.

22. Waldemar Kaempffert, "Women in Science: More Research Opportunity Held Needed in Colleges," *NYT,* 11 Apr. 1937, section 12, p. 6.

23. A. L. Patterson, "Experiences in Crystallography, 1924 to Date," in *Fifty Years of X-Ray Diffraction,* ed. P. P. Ewald (Utrecht: A. Oostoek's Uitgeversmij, 1962), pp. 621–22; Jane Dewey to H. P. Robertson, "April Fool [1936]" and other letters from her, 1936–40, in H. P. Robertson Papers, California Institute of Technology Archives, Pasadena. Dewey was apparently the daughter of philosopher John Dewey, about whom she wrote a brief biography (in *The Philosophy of John Dewey,* ed. Paul A. Schilpp [Evanston: Northwestern University Press, 1939]), and was the first woman postdoctoral fellow in physics at Princeton (Daniel Kevles, *The Physicists: The History of a Scientific Community in Modern America* [New York: Alfred A. Knopf, 1978], p. 207).

Bryn Mawr's men-only policy evidently allowed of certain exceptions, since in 1933 the college hired famed German mathematician Emmy Noether of Göttingen (with the help of the Rockefeller Foundation), but she died just two years later. (Edna E. Kramer, "Amalie Emmy Noether," *Dictionary of Scientific Biography,* 10 vols., ed. Charles Gillespie [New York: Charles Scribner's Sons, 1970–80], [1974], 10: 137–41; A. Dick, *Emmy Noether, 1882–1935* [Basel and Stuttgart: Birkhauser Verlag, 1970]; Hermann Weyl, "Emmy Noether," *Scripta Mathematica* 3 [1935]: 201–20; and correspondence in the Bryn Mawr College File, 1934–35, Rockefeller Foundation Archives, Tarrytown, New York.)

24. George P. Schmidt, *Douglass College: A History* (New Brunswick, N.J.: Rutgers University Press, 1968), pp. 95–96; John A. Small, "Torreya: Hettie Morse Chute," *Bulletin of the Torrey Botanical Club* 89(1962): 331–32.

25. W. H. Longley to Ross G. Harrison, 29 Jan. 1932, and reply, 30 Jan. 1932, in Ross G. Harrison Papers, Yale University Library. See also Bertram G. Smith to Harrison, 26 Apr. 1925, and reply, 6 May 1925, also in Harrison Papers, and the particularly strong and welcoming statement by Harrison in *Alumnae, Graduate School, Yale University, 1894–1920,* ed. Margaret T. Corwin (New Haven: Yale University Press, 1920), p. 61. Gairdner Moment got the job (*AMS,* 6th ed. [1938]). See also note 15 to this chapter.

26. Mary W. Whitney to Christine Ladd-Franklin, 26 Feb. 1905, Christine Ladd-Franklin Papers, Special Collections, Columbia University.

27. M. Carey Thomas, "The Future of Woman's Higher Education," *Mount Holyoke College:*

The Seventy-Fifth Anniversary (South Hadley, Mass: Mount Holyoke College, 1913), pp. 100-104, reprinted as "Marriage and the Woman Scholar," in *The Educated Woman in America: Selected Writings of Catharine Beecher, Margaret Fuller, and M. Carey Thomas,* ed. Barbara M. Cross, published as Classics in Education no. 25 (New York: Teachers College Press, 1965), pp. 170-75 (quotation from p. 171).

28. Gildersleeve, *Many a Good Crusade,* pp. 78-79, but see also p. 109. Physicist Margaret Maltby learned the hard way how reluctant some members of the Columbia faculty could be to promote women at Barnard. Although she had assumed that good work would be rewarded, she had finally asked a Mr. Brewster why she had not been promoted or received a salary increase in her twelve years at Barnard. He apparently knew nothing of her existence and cared even less, greatly stinging her feelings (Margaret Maltby to Miss Gildersleeve, n.d.). Dean Gildersleeve quickly secured a more favorable and appreciative reaction from other Columbia physicists and got Maltby an associate professorship (at age fifty-two) (George Pegram to Dean Gildersleeve, 9 Dec. 1912, both in Barnard College Archives. I thank Patricia Ballou for helping to locate these items).

29. H. N. Russell to Dr. H. N. MacCracken, 19 May 1936, Henry Norris Russell Papers, Princeton University Library. I thank Peggy Kidwell for bringing this item to my attention.

30. Correspondence is in Gleason and Peak folders of the Edwin G. Boring Papers, HUA. Peak later became the Kellogg Professor of Psychology at the University of Michigan, a chair reserved for a woman (*AMWS*, 12th ed. [1971], 8:1904-5). "Goodenough" is in *NAW:MP*; her main professional obituary unfortunately stresses her weak points, as her "devastating criticism" in seminars (Dale B. Harris, "Florence Goodenough," *Child Development* 30 [1959]: 305-6). The psychology departments at the women's colleges merit further study. They may have been tense, embattled places in the 1930s: they employed very few women and often refused to teach child psychology or child study (Herbert Moore, "Practical Aspects of Psychology in Women's Colleges," *Mount Holyoke Alumnae Quarterly* 20[1936]: 5-6).

See also Samuel W. Fernberger, "Academic Psychology as a Career for Women," *Psychological Bulletin* 36(1939):390-94.

31. Three presidential searches at the women's colleges in the late 1930s cast some additional light on this sensitive issue of preferential hiring and declining feminism on the campus. Evidently the sex of the new appointee was the first item settled upon, and then great steps were taken to find someone who fit. Wellesley College's trustees decided in 1935 that since they had always fared well under their female presidents they would seek another woman to succeed Ellen Fitz Pendleton. After an eighteen-month search they hired the thirty-six-year-old Mildred McAfee (later Horton), a Vassar graduate, Chicago M.A. in sociology, and dean of women at Oberlin College. They were apparently not upset at her youth and lack of experience and expressed great confidence in her ability to learn the job. She did, and was one of their great presidents. (Jean Glasscock, "The Selection of Wellesley's Presidents," in *Wellesley College*, ed. Glasscock et al., pp. 65-68, and "Mildred McAfee," *CBY, 1942*, pp. 539-42.)

But such faith in women was unusual, and even the trustees of Mount Holyoke College decided in 1936 to end their long line of women presidents (dating back to Mary Lyon a century before) and hire a man, for these reasons: "(1) We need some strong professors to replace the many soon to retire, and a man can get better ones, especially men. [The number of men on the faculty had been increasing under Woolley.] (2) A man with a nice wife would give social tone which the faculty has lacked. (3) A man can be a good pal with men outside (raising money . . .)." Although Frances Perkins, an alumna trustee (former chemistry major, in fact) and, as Secretary of Labor, the highest-ranking woman in the federal government, protested sharply, "I don't quite see the sense in conducting a college for 100 years if it and others like it don't produce at least one female capable of running such an institution," the trustees chose Roswell Ham, professor of English literature at Yale University, an inexperienced administrator and fundraiser but a family man. His selection set off one of the loudest brouhahas of the decade and was criticized in the *AAUW Journal* and by some faculty and alumnae in newspapers and the alumnae magazine but was approved by the *New York Times* (on the sexist grounds that the job was now too big for a woman). Recent research indicates that it may have been former president Mary Woolley's strong personal attachment to an unpopular woman professor that had turned the trustees toward a safe family man. (Anna Mary Wells, *Miss Marks and Miss*

Woolley [Boston: Houghton Mifflin, 1978]; George Martin, *Madame Secretary, Frances Perkins* [Boston: Houghton Mifflin, 1976], chap. 28, quotations from p. 370; *Mount Holyoke Alumnae Quarterly* 20[Aug. 1936], quotation from *NYT*, n.d., cited on p. 70; and Dorothy Kenyon, "The Presidency of Mount Holyoke College: Some Issues Involved in Choosing the Head of a Great Woman's College," *JAAUW* 30[1936]: 16-17, and "Editorial," *JAAUW* 30[1936]: 39. See also Jeannette Marks, *Life and Letters of Mary Emma Woolley* [Washington D.C.: Public Affairs Press, 1955]; other clippings and comments by Emma Perry Carr are in the Mount Holyoke College File, Rockefeller Foundation Archives.) In 1939, when Smith College continued its tradition of male presidents by appointing Herbert John Davis, professor of English at Cornell University, no one, not even the AAUW, complained, apparently because Smith College, unlike the others, had no tradition of female presidents ("The Presidency of Smith College," *JAAUW* 33[1939-40]: 94-95). Hope Nicolson, dean of the college, might have accepted had she been asked. Meyer F. Nimkoff and Arthur L. Wood, "Women's Place Academically," *Journal of Higher Education* 20(1949): 28-36, deplored the lack of female leadership at the women's colleges before and after 1940.

32. Lawrence Cremin et al., eds. *A History of Teachers College, Columbia University* (New York: Columbia University Press, 1954), p. 245, indicates that 36 percent of the Teachers College faculty was female in 1950.

33. Fred: Elizabeth M. O'Hern, "Elizabeth McCoy, Pioneer Microbiologist," *ASM* [*American Society for Microbiology*] *News* 42(1976): 531-35, and clippings, McCoy File, University of Wisconsin Archives; Diane Johnson, *Edwin Braun Fred: Scientist, Administrator, and Gentlemen* (Madison: University of Wisconsin Press, 1974), which has a list of his students. Bailey: see chapter 3. Mendel and Harrison: see chapter 6. Lillie: the Frank R. Lillie Papers at the Special Collections, University of Chicago Library, contain many enthusiastic letters of recommendation for his female students, but apparently he never hired any women faculty. Mall: Elinor Bluemel, *Florence Sabin: Colorado Woman of the Century* (Boulder: University of Colorado Press, 1959); P. D. McMaster and Michael Heidelberger, "Florence Rena Sabin, 1871-1953," *National Academy of Sciences Biographical Memoir* 34(1960): 217-19. Huber: Wilfrid B. Shaw, ed., *The University of Michigan: An Encyclopedic Survey*, 4 vols. (Ann Arbor: University of Michigan Press, 1951), 3:818-19, which also says of Crosby, "We of the University were late in recognizing that in our midst is one who ranks among the leading neurologists of the world." Van Wagenen: Solly Zuckerman, *From Apes to Warlords: The Autobiography of Solly Zuckerman* (London: Hamish Hamilton, 1978), pp. 73-74; *AMS*, 7th ed. (1944). Harriet Zuckerman has discussed the importance of such elite sponsorship in the careers of Nobel laureates, who were usually men, in *Scientific Elite: Nobel Laureates in the United States* (New York: Free Press, 1977), esp. chap. 4, "Masters and Apprentices in Science," and chap. 5, "Moving into the Scientific Elite."

34. Zuckerman, *From Apes to Warlords*, pp. 73-74; Earl B. McKinley to E. O. Jordan, 29 Sept. and 11 Oct. 1933, Edwin O. Jordan Papers, Special Collections, University of Chicago Library; "Verder" in *AMS*, 6th ed. [1938].

35. Flora W. Patterson, "The Plant Pathologist," in *Careers for Women*, ed. Catherine Filene (Boston: Houghton Mifflin Co., 1920), pp. 435 and 437.

36. "Downey" in *NAW*; and Richard Stephen Uhrbock, "June Etta Downey, 1875-1932," *Journal of General Psychology* 9(1933): 351-64; "Caldwells, George and Mary E.," in *AMS*, 6th and 8th eds. (1938 and 1949).

37. "Sabin" in *NAW:MP*; "Weed" and "Linton" in *AMS*, 6th ed. (1938). Esther S. Goldfrank, a graduate student and former department secretary, *Notes on an Undirected Life: As One Anthropologist Tells It*, published as Queens College [CUNY] Publications in Anthropology no. 3 (1977), pp. 110-11, says the Columbia department was split between "his" students and "hers" for many years.

38. "Hamilton" in *NAW:MP* and *AMS*, 6th ed. (1938); *CBY, 1946*, pp. 234-36; Madeleine P. Grant, *Alice Hamilton* (London: Abelard-Schuman, 1967), and *Journal of Occupational Medicine*, Feb. 1972, pp. 97-114. There is correspondence relating to her appointment at Harvard in the A. Lawrence Lowell Papers, HUA. Lowell agreed to recommend her appointment to the corporation but then denied to a critic that she had a faculty appointment, claiming it was a research position only. (A. Lawrence Lowell to Dr. Henry P. Walcott, 21 Dec. 1918, and A. Lawrence Lowell to H. H.

Moore, Esq., 4 Apr. 1919, both in A. Lawrence Lowell Papers). Florence Sabin to Marion Hines, 23 Apr. 1925, Florence Sabin Papers, APS.

39. Robin Wright, "Awards Document Full Life of Dr. Elizabeth Crosby," *Ann Arbor News*, 17 Jan. 1971, clipping in Crosby File, Bentley Historical Library, University of Michigan. Arlie Hochschild has discussed this phenomenon in "Making It: Marginality and Obstacles to Minority Consciousness," *New York Academy of Sciences* 208(1973): 179–84.

40. W. P. Few to Frank Blair Hanson, 4 June and 19 July 1936, Duke University File, Rockefeller Foundation Archives. James Franck worked hard in the spring of 1934 to get Sponer a job outside Germany—at Groningen and Oslo, and in the United States (James Franck Papers, Special Collections, University of Chicago Library, especially Coster and Sponer files). Franck and Sponer were married in 1946. They were both interviewed by T. S. Kuhn and Maria Mayer in 1962 for the Quantum Archives Project of the American Institute of Physics (Archives for the History of Quantum Physics, Office for the History of Science and Technology, University of California, Berkeley). Interview number six is her autobiography to the 1930s.

41. Robert A. Millikan to President W. P. Few, 24 June 1936, William Preston Few Papers, Correspondence File, Duke University Archives. (I thank archivist Mark C. Stauter for his expert help in locating this and other relevant documents.)

Edward U. Condon, "Reminiscences of a Life in and out of Quantum Mechanics," *International Journal of Quantum Chemistry*, Symposium no. 7(1973), p. 10, mentions Cal Tech's attitude toward women in the 1920s. (I thank Dr. Katherine R. Sopka for this reference.) Sponer is mentioned in Walter M. Elsasser, *Memoirs of a Physicist in the Atomic Age* (New York: Science History Publications, 1978), p. 44, and has an undated photograph facing p. 44. (I thank Pnina Abir-Am for this reference.)

42. Charles Edwards to Dean Alice Baldwin, 11 Jan. 1936, Correspondence File, William Hane Wannamaker Papers, Duke University Archives. Again I thank Mark Stauter, an exceptional archivist, for this item.

For more on Millikan's accomplishments and views, see Robert Kargon, "The Conservative Mode: Robert A. Millikan and the Twentieth-Century Revolution in Physics," *Isis* 68(1977): 509–26; *The Autobiography of Robert A. Millikan* (Englewood Cliffs, N.J.: Prentice-Hall, 1950); Ronald C. Tobey, *The American Ideology of National Science, 1919–1930* (Pittsburgh: University of Pittsburgh Press, 1971), and Kevles, *Physicists*. Kevles also discusses anti-Semitism in physics, pp. 211–15 and 278–79. Millikan's attitudes toward women seem to have been quite complex. He had been one of the scientists who welcomed Madame Curie to Washington, D.C., in May 1921, and in a 1937 nationwide radio address ("Woman's Century, Oberlin's Celebration of 100th Anniversary of Admission of Women, October 8, 1937," typescript in Robert A. Millikan Papers, California Institute of Technology Archives, Pasadena) he selected for praise the advance of women's education, improvements in women's dress, and higher wages (for domestic servants!), all of which he attributed to the spread of science. His wife had been a student of Marion Talbot at the University of Chicago (Marion Talbot Papers, University of Chicago Special Collections).

43. The "tenure-track" was very much a product of the depression. See Henry M. Wriston, "Academic Tenure," *American Scholar* 9(1940): 339–49, reprinted in *BAAUP* 27(1941): 337–47; idem, "Statement of Principles on Academic Freedom and Tenure," *BAAUP* 26(1940): 49–51; discussion continued in later volumes, such as *BAAUP* 27(1941): 40–46 and 625–26; James Bryant Conant, *My Several Lives: Memoirs of a Social Inventor* (New York: Harper & Row, 1970), chap. 14; Paul C. Reinert, S. J., *Faculty Tenure in Colleges and Universities from 1900 to 1940*, Saint Louis University Studies, Monograph Series, Social Sciences no. 1 (St. Louis: Saint Louis University Press, 1946), documents the increasing length of service of faculty members. None of his data are broken down by sex, however.

44. "Sarah R. Atsatt," in *University of California: In Memoriam* (Berkeley and Los Angeles: University of California Press, 1975), pp. 1–2; James V. Mink, "Chequered Career: Scotland to UCLA, Flora M. Scott," oral history transcript, 1976, UCLA, pp. 44–47 and 50; Olive Hazlett to Professor Carmichael, 4 July 1935, Personnel File, Mathematics Department, University of Illinois.

45. "Second Report of Committee W," pp. 72–73.

46. "Richards" and "Comstock" in *NAW* (Comstock became a full professor of nature study in

1920). Stowell: Shaw, ed., *University of Michigan,* 3:498 and 3:500. ("Her title did not do justice to her responsibilities and attainments" [p. 498].)

47. Elizabeth C. Sprague to Marion Talbot, 29 Apr. 1916, Marion Talbot Papers, University of Chicago Library, reports that Kansas had a state law against "the employment of two in one family." Professor Julius Stieglitz of the University of Chicago recommended Agnes Fay Morgan for a job at the University of Illinois in 1914, even though she was married, because, as he explained, her husband was sickly (Julius Stieglitz to W. A. Noyes, 8 May 1914, enclosed in W. A. Noyes to Dean K. C. Babcock, 19 May 1914, School of Arts and Sciences Files, Box 1, Chemistry Department, University of Illinois Archives. I thank Robert Kohler for calling my attention to this item). For a poignant statement of what it meant to her and her marriage to suffer from such roles, see the anonymous "Reflections of a Professor's Wife," *JACA* 14(Jan. 1921): 90-92.

48. Ethical: Mark Ingraham, "Ethics of Teaching," in *Encyclopedia of Education,* 10 vols. ed. Lee C. Deighton (New York: Macmillan and Free Press, 1971), 3:430. Political: Malcolm Moos and Francis E. Rourke, *The Campus and the State* (Baltimore: The Johns Hopkins University Press, 1959), pp. 148-49. Economic: Eleanor F. Dolan and Margaret P. Davis, "Anti-Nepotism Rules in American Colleges and Universities," *Educational Record* 41(1960): 285.

49. "Mayer" in *NAW:MP,* and Harold C. Urey, "Maria Goeppart Mayer (1906-1972)," *American Philosophical Society Yearbook, 1972,* p. 235; Robert G. Sachs stresses that though Hopkins's antinepotism rules were strict, members of the physics department allowed Mayer to use the library and gave her space to work (since she was a theoretical physicist, the space was minimal) ("Maria Goeppart Mayer, 1906-1972," *NAS Biographical Memoirs* 50[1979]: 314). "Cori" in *NAW:MP, AMS,* 8th ed. (1949), and Carl F. Cori, "The Call of Science," *Annual Review of Biochemistry* 39(1969): 13. Graham: *NCAB* 56(1975): 39-40, and Thomas B. Turner, *Heritage of Excellence: A History of The Johns Hopkins Medical Institutions, 1919-1947* (Baltimore: The Johns Hopkins University Press, 1974), pp. 452-53. Her papers at the Washington University Medical Center, St. Louis, Missouri, show her to have been a gracious and energetic scientist and hostess. In 1939 the AAUP opposed university rules that required one spouse to resign if two members of a faculty married ("Tenure Status of Teachers Who Intermarry," *BAAUP* 26[1940]: 278-79).

50. John Rodgers, "Memorial to Eleanora Bliss Knopf, 1883-1974," Geological Society of America, *Memorials* 6(1977): 2. See also *NAW:MP.*

51. "Opportunities and Salaries of Women in the Teaching Profession in Nebraska," pp. 10-13.

52. "Preliminary Report of Committee W," pp. 27-29. Although most discussions of academic salaries in the 1920s stressed how inadequate they were, Frederick E. Bolton's "College Teaching as a Career for Men," *School and Society* 21(1925): 213-17, took the opposite tack and sought "to show statistically that the pecuniary rewards of the really successful college teacher are much better than is generally supposed" (p. 213). He then described the many moonlighting opportunities a venturesome academic could engage in. Perhaps as the dean of the University of Washington School of Education he was upset at the small number of men earning higher degrees in education.

53. Amy Hewes, "Dependents of College Teachers," *Quarterly Publications of the American Statistical Association* 16(1919): 502-11.

54. Grace M. Morton and Marjorie R. Clark, "Income and Expenditures of Women Faculty Members in the University of Nebraska," *Journal of Home Economics* 22(1930): 653-56.

55. Quotation from C. E. Yarwood and Mabel Nebel, "Ruth Florence Allen, 1879-1963," *Phytopathology* 54(1964): 885; "Ruth Allen Award," *Phytopathology* 64(1974): 9. Olive Hazlett's salary was also reduced in 1933 from $4,000 to $3,500, but unlike those of the rest of the members of her department, her salary was not raised again after the depression. One result was that she retired on a pitiable pension (Personnel File, Mathematics Department, University of Illinois).

Child psychologists Nancy Bayley and Mary Cover Jones reported in oral histories years later that during the depression the hours of the staff members of the Institute for Child Welfare at the University of California at Berkeley were reduced by one-half and their salaries cut another 30 percent after that. Nevertheless, most members continued to work almost full time, since their project, a longitudinal study of child development, required annual collections of measurements. To paraphrase Bayley, "despite the Depression, the children kept on having birthdays!" (Milton J.

Senn, "Four Interviews [with Leaders in Child Development Movement]," Nancy Bayley [1969], p. 18, and Mary Cover Jones, p. 9.)

56. Turner, *Heritage of Excellence*, pp. 342–46. See also Janet Clark to President J. S. Ames, 16 Feb. 1935, and the curious [J. S. Ames] to Janet Howell Clark, 12 June 1935, where he did not realize she had resigned (Archives, The Johns Hopkins University). Sources differ on Baetjer's status, which is listed as "associate" in *AMS*, 6th ed. (1938).

57. Teachers College: in Mary Swartz Rose, "University Teaching of Nutrition and Dietetics in the United States," *Nutrition Abstracts and Reviews* 4(1935): 439–46; Juanita Archibald Eagles, Orrea Florence Pye, and Clara Mae Taylor, *Mary Swartz Rose, 1874–1941: Pioneer in Nutrition* (New York: Teachers College Press, 1979); "Rose" in *NAW*; Grace MacLeod, "Mary Swartz Rose, 1874-1941," *Journal of Home Economics* 33(1941):221-24, and Henry C. Sherman, "Mary Swartz Rose, 1874–1941," *Journal of Nutrition* 21(1941): 209–11; Clara Mae Taylor, "Grace MacLeod, A Biographical Sketch," *Journal of Nutrition* 95(May 1968): 3-7; and Cremin et al., *A History of Teachers College*, pp. 56–57.

University of Chicago: Marie Dye, *History of the Department of Home Economics, University of Chicago* (Chicago: University of Chicago Home Economics Alumni [*sic*] Association, 1972), in Special Collections, University of Chicago Library; "Roberts" in *NAW:MP*, Ethel A. Martin, "Lydia Jane Roberts," *Journal of the American Dietetic Association* 47(Aug. 1965): 127–28; and idem, "The Life Works of Lydia J. Roberts," *Journal of the American Dietetic Association* 49(Oct. 1966): 299–302.

University of California: "Morgan" in *NAW:MP;* Gladys A. Emerson, "Agnes Fay Morgan and Early Nutrition Discoveries in California," *Federation Proceedings* 36(1977): 1911–14; Ruth Okey, "Agnes Fay Morgan (1884–1968), A Biographical Sketch," *Journal of Nutrition* 104 (1974): 1103-7. Morgan's own "The History of Nutrition and Home Economics in the University of California, Berkeley, 1914-1962," mimeographed (1962), is an uncritical, almost a public relations, account of the program's success.

University of Wisconsin: Merle Curti and Vernon Carstensen, *The University of Wisconsin: A History, 1848-1925*, 2 vols. (Madison: University of Wisconsin Press, 1949), 2:404-6; "Helen Tracy Parsons, Transcript of Oral History," c. 1972, Office of Oral History, University of Wisconsin.

Kansas State: clippings and Glenda Odgers, "Margaret M. Justin," typescript, 1966, in Margaret Justin File, University Archives, Kansas State University.

Cornell: *A Growing College: Home Economics at Cornell University* (Ithaca: Cornell University Press, 1969). More general are Mary A. Grimes, "A Count of Home Economics Researchers," *Journal of Home Economics* 32(1940): 27-28; Gladys A. Branagan, *Home Economics Teacher-Training under Smith-Hughes Act, 1917 to 1927*, Teachers College Contribution to Education no. 350 (1929); and Marie Negri Carver, *Home Economics as an Academic Discipline: A Short History*, published as Center for the Study of Higher Education, College of Education, University of Arizona Topical Paper no. 15 (1979).

58. "Trade school": "Helen Tracy Parsons, Transcript of Oral History," pp. 49–50.

59. Clara M. Brown, "Appraisal of Trends in Home Economics Education Research," *Journal of Home Economics* 29(1937): 604.

60. Okey, "Agnes Fay Morgan," p. 1103, and Ruth Okey to author [June 1974]. The Agnes Fay Morgan Papers, Bancroft Library, University of California, Berkeley, add little, and her history of the school (see note 57 to this chapter) does not allow one to see the inner struggles. But see also Agnes Fay Morgan, "Undergraduate and Graduate Preparation for Home Economics Research," *Journal of Home Economics* 31 (1939): 685–91, which reportedly aroused "considerable discussion" among members of the home economics section of the Association of Land-Grant Colleges and Universities.

61. Mary S. Lyle, "Florence Fallgatter, Former AHEA President, Retires," *Journal of Home Economics* 50(1958): 659.

62. Nellie Kedzie Jones, "Abby L. Marlatt," *Journal of Home Economics* 35(Oct. 1943): 483-84; Lucy Rathbone, "Mary E. Gearing: Pioneer," *Journal of Home Economics* 39(1947): 5-6. A. F. Morgan's history of the Berkeley program (see note 57 to this chapter) fits this pattern and elaborates greatly on the research done.

63. Lafayette B. Mendel to Miss Macy (who had just left Berkeley), 28 Aug. 1923, Icie Macy

Hoobler Papers, Bentley Historical Library, University of Michigan; "Helen Tracy Parsons, Transcript of Oral History," pp. 38–39, 54–56, 65. (She was also very appreciative of Professor E. B. Hart in chemistry, though he would hire no women, and Dean Charles Schlichter, who scorned home economics, as well as Deans Conrad Elvehejm and E. B. Fred, who were supportive.) May S. Reynolds, oral history tape, University Archives and Oral History Project, University of Wisconsin, Madison (1977), tape 1 (side 1: markers 522, 648, 885, 959; side 2: entire). "Ruth Henderson, An Interview Conducted by Donna S. Taylor," typescript, University Archives and Oral History Project, University of Wisconsin, Madison, 1975, pp. 32–33. There is some evidence here that physical size had some importance in the women's reactions: Marlatt was large and at first frightened people, but they grew to like her enormously; Zuill was tiny but so forceful that few felt close to her. The whole subject is unexplored, but see Constantina Safilios-Rothschild, *Sex Role Socialization and Sex Discrimination: A Synthesis and Critique of the Literature* (Washington, D.C.: National Institute of Education, 1979), pp. 121–22.

64. Donald S. Napoli, "The Architects of Adjustment: The Practice and Professionalization of American Psychology, 1920–1945" (Ph.D. diss., University of California, Davis, 1975); Elizabeth Lomax, "The Laura Spelman Rockefeller Memorial: Some of Its Contributions to Early Research in Child Development," *Journal of the History of the Behavioral Sciences* 13(1977): 283–93; and idem, *Science and Patterns of Child Care* (San Francisco: W. H. Freeman, 1978); Hamilton Cravens, "The Laura Spelman Rockefeller Memorial, the Child Welfare Institutes, and the Creation of the Science of the Child, 1917–1940" (Paper presented at Annual Meeting of American Historical Association, New York, Dec. 1979).

Marion McPherson (Mrs. John A. Popplestone), personal communication, 12 Apr. 1977; Milton J. E. Senn, *Insights on the Child Development Movement in the United States,* published as Monographs of the Society for Research in Child Development, vol. 40, nos. 3–4 (Chicago: University of Chicago Press, 1975); John E. Anderson, "Child Development: An Historical Perspective," *Child Development* 27(1956): 181–96; various conference proceedings and publications by the Committee on Child Development of the National Research Council (Division of Anthropology and Psychology), 1925–27; "Mitchell" in *NAW:MP*, and Lucy Sprague Mitchell, *Two Lives: The Story of Wesley Clair Mitchell and Myself* (New York: Simon and Schuster, 1953), especially pp. 458–61; Cremin et al., *A History of Teachers College*, pp. 81–82; "Woolley" in *NAW*; Hamilton Cravens, "Inconstancy of the Intelligence Quotient: The Iowa Child Welfare Research Station and the Criticism of Hereditarian Mental Testing, 1917–1939" (Paper presented at Annual Meeting of the History of Science Society, Madison, Wisconsin, Oct. 1978); "Daniels" in *AMS,* 7th ed. (1944).

The Merrill-Palmer School, An Account of Its First Twenty Years, 1920–1940 (Detroit: Merrill-Palmer School, 1940); "Goodenough" in *NAW:MP,* and Dale B. Harris, "Florence Goodenough," *Child Development* 30(1959): 305–6; Albert J. Solnit, M.D., and Arthur Greenwald, "A Survey of Child Study at Yale," *Yale Alumni Magazine and Journal* 41, no. 6 (Feb. 1978): 21–28; "Frances Ilg" and "Louise Bates Ames" in *CBY, 1952*, pp. 299–301; "Katherine M. Wolf, In Memoriam," *Child Study* 35(Winter 1957): 2–3.

65. Wells: Biographical File at University Archives, Bryan Hall, Indiana University; and Thomas D. Clark, *Indiana University: Midwestern Pioneer,* 2 vols. (Bloomington: Indiana University Press, 1973), 2:156–57.

Mueller: Biographical File at University Archives, Bryan Hall, Indiana University, which includes a humorous autobiography in typescript (Apr. 1971). (I thank archivist Dolores Lahrman for bringing these items to my attention.) Detlev W. Bronk to President Alan Valentine, 19 May, 1938, Alan Valentine Papers, Special Collections, Rush Rhees Library, University of Rochester. Although no male scientists (that I have come across) became deans of women in these years, one woman mathematician was dean of men at Morehead State Teachers College in Kentucky from 1931 to 1936. Martha R. W. Blessing had been an assistant dean of women at Swarthmore, where her husband was head of the engineering department, but after his death she took the new position and was one of the first women to do so (obituary in *NYT,* 15 June 1954, p. 29). In 1938 the American College Personnel Association was established; the National Association of Deans of Women, which was founded in 1916 and long affiliated with the AAUW, started a journal in the late 1930s.

66. In talking of these "research associates" it is necessary to remember that there was often

confusion about their status and role in the 1920s and 1930s. The place of female assistants had expanded so greatly in these years as to separate into two different roles, though the public was hardly aware of it. Thus of the professional "research associates," almost all held doctorates and did supervised, collaborative, or even independent research, unlike the women "laboratory technicians" and "medical technologists," who were greatly romanticized in the 1930s and often confused in the periodicals of the time with women "scientists"; see, for example, "Science Opens Its Doors to Women," *Literary Digest* 123(10 July 1937): 17-18; Barnard, "Woman's Rise in Science"; Laura Z. Hobson, "Follies Girls to Scientist," *Independent Woman* 20(Oct. 1941): 297-98, 319; Clennie E. Bailey, "Silent Partners to Aesculapius," *Independent Woman* 15 (Mar. 1936): 77-78, 91; Eleanor Breitwieser, "Medical Technologist," in *Chemist at Work,* ed. Roy I. Grady and John W. Chittum (Easton, Pa.: *Journal of Chemical Education,* 1940), pp. 29-34, and Colette Corbett, "Chemistry and the Medical Technician," pp. 327-40 of *The Chemist at Work.*

Some women scientists, especially those who lacked doctorates, defied the distinction between technicians and research associates. Julia Tiffany Weld, an heiress whose father refused to let her attend college, was a research assistant at several New York medical schools and published more than fifty papers on medical and bacteriological topics alone and with her collaborators (including Dr. Hans Zinsser). (Obituary in *NYT,* 23 Nov. 1973, p. 34.) Similarly Katherine Mills Price, a "long time co-worker" of Alphonse Dochez, who directed research on the "common cold" at the Rockefeller Institute in the 1920s, portrayed her role in their work in terms that would baffle and fascinate a sociologist: "I worked for Dr. Dochez as a laboratory technician for over thirty years. We on his laboratory staff felt we were working *with* [her italics], as well as for, him. He encouraged us to consider ourselves his research assistants. Our names were included in publications; indeed, on many of Dr. Dochez's papers, the order of names frequently began with that of a younger colleague or a lesser member of the team. . . . We all realized we were privileged to work for a great man." (Michael Heidelberger, Yale Kneeland, Jr., and Katherine Mills Price, "Biographical Memoir of Alphonse Raymond Dochez," *NAS Biographical Memoirs* 42 [1971]: 35 and 37.)

67. Edith Wallace, M.A., T. H. Morgan's assistant, received $1,100 in 1920 and $1,800 in 1928 (Garland Allen, *Thomas Hunt Morgan: The Man and His Science* [Princeton; N.J.: Princeton University Press, 1978], p. 255); E. V. McCollum paid his research associate Elsa Orent, Sc.D., $2,750 in 1938 and 1939, the same as his male associate Harry Day, but she left in 1940 for the USDA's Bureau of Home Economics, where her salary was much higher (The Johns Hopkins University File, Rockefeller Foundation Archives Center). There is also some salary data in Herbert Evans's budgets in the University of California File at the Rockefeller Foundation Archives.

68. The funding of science in the 1920s and 1930s is a largely untapped subject. For starters, see Raymond B. Fosdick, *The Story of the Rockefeller Foundation* (New York: Harper & Brothers, 1952); Raymond M. Hughes, "Research in American Universities and Colleges," in National Resources Committee, Science Committee, *Research: A National Resource,* 3 vols. (Washington, D.C.: GPO, 1938-41), 1:167-93: Scott M. Cutlip, *Fund Raising in the United States: Its Role in America's Philanthropy* (New Brunswick, N.J.: Rutgers University Press, 1965), chap. 7 ("Cash for Colleges and Cathedrals"); Ronald Tobey, *American Ideology;* AAUP, Committee Y, *Depression, Recovery, and Higher Education* (New York: McGraw-Hill, 1937). More specialized are Spencer Weart, "The Physics Business in America, 1919-1940; A Statistical Reconnaissance," in *The Sciences in the American Context: New Perspectives,* ed. Nathan Reingold (Washington, D.C.: Smithsonian Institution Press, 1979), pp. 295-358; Virginia Cameron and Esmond R. Long, *Tuberculosis Medical Research, National Tuberculosis Association, 1904-1955* (New York: National Tuberculosis Association, 1959); Richard H. Shryock, *National Tuberculosis Association, 1904-1954* (New York: National Tuberculosis Association, 1957), pp. 127-35, 190-92, 210-17, 223-26, 249-55, chap. 14; Turner, *Heritage of Excellence,* p. 373; see also "Otho Sprague," *NCAB* 15 (1916): 244, and "Henry Phipps," *NCAB* 43(1961): 191-92.

69. A. T. Poffenberger, "Leta Stetter Hollingworth, 1886-1939," *American Journal of Psychology* 53(1940): 301.

70. "Sitterly" in *AMS,* 8th ed. (1948). See also Charlotte Moore-Sitterly, "Collaboration with Henry Norris Russell over the Years," in *In Memory of Henry Norris Russell*, ed. A. G. Davis Philip

and David H. DeVorkin, published as Dudley Observatory Reports no. 13 (Albany, N.Y.: Dudley Observatory, 1977), pp. 27–41. (I thank Peggy Kidwell for this reference.)

71. Turner, *Heritage of Excellence,* passim; Elmer V. McCollum, *From Kansas Farm Boy to Scientist* (Lawrence: University of Kansas Press, 1964), passim; *AMS*, 6th ed. (1938); "Helen Tracy Parsons, Transcript of Oral History," pp. 14–36; "Justina Hill" in *CBY, 1941*, pp. 384–85. Cf. n. 56.

72. Joseph H. Greenberg, "Melville Jean Herskovits," in *Biographical Memoirs of the National Academy of Sciences* 42(1971): 65. Frances Shapiro Herskovits did merit an obituary in the *New York Times,* however (8 May 1972, p. 40).

73. "Estimates of the contributions made by E. S. Clements to the accomplishments of Frederic E. Clements," Box 155, Frederic and Edith Clements Papers, University of Wyoming, Western History Collections; Edith S. Clements, *Adventures in Ecology* (New York: Hafner Publishing, 1960). A brief biography of her appears in *NCAB* G (1946): 352.

74. Edwin G. Boring, "Lewis Madison Terman, 1877–1956," *NAS Biographical Memoirs* 33(1959): 414–16; Lewis Terman and Melita Oden et al., *The Gifted Child Grows Up* (Stanford: Stanford University Press, 1947), pp. ix–xii; May V. Seagoe, *Terman and the Gifted* (Los Altos, Calif.: Kauffman, 1975). Julian C. Stanley, William C. George, and Cecilia H. Solano, *The Gifted and the Creative: A Fifty-Year Perspective* (Baltimore: The Johns Hopkins University Press, 1977), is largely about Terman's work and influence. George W. Corner, "Herbert McLean Evans, 1882–1971," *NAS Biographical Memoirs* 45(1974): 153–92; McCollum, *Kansas Farm Boy;* and Harry G. Day, "Elmer Vernon McCollum, 1879–1967," *NAS Biographical Memoirs* 45(1974): 263–335; George W. Corner, *George Hoyt Whipple and His Friends* (Philadelphia: J. B. Lippincott, 1963); "Helen Tracy Parsons, Transcript of Oral History," pp. 14–36; "Vincent du Vigneaud" in *CBY, 1956,* pp. 160–62. Beebe: Robert Henry Walker, *Natural Man: The Life of William Beebe* (Bloomington: Indiana University Press, 1975), chap. 8: "Female of the Species"; and Tim M. Berra, *William Beebe: An Annotated Bibliography* (Hamden, Conn.: Archon Books [Shoe String Press], 1977). There is a chapter on Jocelyn Crane in Edna Yost, *Women of Modern Science* (New York: Dodd, Mead, 1959), pp. 108–23. Another Beebe assistant, Ruth Rose, was the subject of a newspaper article, "Prefers Jungle Life to City Jumble Life," unidentified clipping [New York City?] [1925], p. 3, in Bureau of Vocational Information Papers, Zoology, SLRC. Helen Tee-Van, who wrote several articles mentioned in chapter 9 herein, was an illustrator for Beebe.

75. Anti-Semitism was enough of a problem in academia in these years to merit occasional mention. Libbie Hyman had intended to study botany at the University of Chicago but found enough anti-Semitism in that department to make a transfer to zoology necessary. Although Helen Tracy Parsons thought The Johns Hopkins "was essentially Jewish" when she was there in the early 1920s, Elmer V. McCollum complained to the Rockefeller Foundation officials that he could not get his assistant, Elsa Orent (later Keiles), a Russian Jewess, on the payroll at Hopkins in 1936 and again in 1938: "Both because she is a woman and because she is Jewish it will be difficult to place her elsewhere, but M[cCollum] says that she is the most useful and valuable research assistant he has ever had." (Horace W. Stunkard, "In Memoriam: Libbie Henrietta Hyman, 1888–1969," in *Biology of the Turbellaria*, ed. Nathan W. Riser and M. Patricia Moore [New York: McGraw-Hill, 1974], pp. ix–xxv; "Helen Tracy Parsons, Transcript of Oral History," p. 19; Frank Blair Hanson Diary, 7 Apr. 1936, and Warren Weaver Diary, 9 Dec. 1938, p. 183, both in The Johns Hopkins University File, Rockefeller Foundation Archive Center; Orent-Keiles left for a job at the Bureau of Human Nutrition and Home Economics of the USDA in 1940 [*AMS*, 8th ed., 1948].)

Two articles that discuss the achievements of Jewish women in science but not the discrimination they faced are Ida Welt, Ph.D., "The Jewish Woman in Science," *Hebrew Standard* 50, no. 11 (5 Apr. 1907): 4, and Morris Goldberg, "Women in the Realm of Science," *American Hebrew* 126 (3 Jan. 1930): 312.

76. The most common handicaps seem to have been deafness and crippling from polio suffered in youth. Henrietta P. Leavitt and Annie Jump Cannon, two of the most outstanding women astronomers of the early twentieth century, and Tilly Edinger, a well-known German vertebrate paleontologist who came to the United States in 1940, were all very hard of hearing, and Florence Seibert, highly regarded for her work on the biochemistry of tuberculin, was lame from polio. Lafayette B.

Mendel, her Yale professor, sent her, when she first sought an academic position, the well-meaning advice that with her handicap she would be better off as a research associate, apparently unaware of how well several other afflicted women were faring. Mary S. Case, for example, was a professor of psychology at Wellesley for forty years, and Marian Wesley Smith even did anthropological fieldwork despite crippling from polio. ("Cannon" and "Leavitt" in *NAW*; on Cannon's handicap, see Raymond Pearl to E. B. Wilson, 12 Mar. 1923, NAS Archives, Washington, D.C. [I thank Paul McClure for finding this outspoken item.] Siebert: Lafayette B. Mendel to Florence Seibert, 9 Aug. 1924, Florence Seibert Papers, APS, and her autobiography, *Pebbles on the Hill of a Scientist* [St. Petersburg, Fla.: private, 1968]. [Her sister was very close and even worked in the same laboratories.] Case: *NYT*, 2 Feb. 1953, p. 21; Smith: Frederica de Laguna, "Marian Wesley Smith, 1907-1961," *American Antiquity* 27[1962]: 567-69.)

77. Correspondence regarding Justine Garvey, 1966-70, in Daniel H. Campbell Papers, California Institute of Technology Archives, Pasadena.

78. Horace W. Stunkard, "In Memoriam: Libbie Henrietta Hyman, 1888-1969." An issue of the *Journal of Biological Psychology* (vol. 12, no. 1[Oct. 1970], pp. 1-23) is devoted to her memory. See also note 75 to this chapter.

79. Cecilia Payne to Henry Norris Russell, 11 Dec. 1930, Henry Norris Russell Papers, Princeton University Library. I thank Peggy Kidwell for telling me of this item.

80. Lewis Terman to Florence Goodenough, 8 Feb. 1927, Lewis Terman Papers, Stanford University Archives; Peter Olitsky Papers, APS, especially John R. Paul to Peter Olitsky, 10 June 1944. For more on Isabel Morgan (later Mountain), see Trudy Whitman, "On the Polio Trail," *Independent Woman* 27(Jan. 1948): 11 and 29; [Saul Benison, ed.], Tom Rivers, *Reflections on a Life in Medicine and Science* (Cambridge: MIT Press, 1967), pp. 407-11; John R. Paul, *A History of Poliomyelitis* (New Haven: Yale University Press, 1971), pp. 385-86; Ian Shine and Sylvia Wrobel, *Thomas Hunt Morgan: Pioneer of Genetics* (Lexington: University Press of Kentucky, 1976), especially chap. 7; "DeWitt" in *NAW;* Shaw, ed., *University of Michigan,* 2:817-18, and note 34 to chapter 4 herein. Merrill: Turner, *Heritage of Excellence,* pp. 362-63.

81. Jerry Gaston, *Originality and Competition in Science* (Chicago: University of Chicago Press, 1973), p. 172. A fine example of the "great man" approach to the history of science, which minimizes the contributions of associates, junior authors, and students, is Russell Chittenden, *The Development of Physiological Chemistry in the United States* (New York: Chemical Catalog, 1930). The contributions of at least fifty-five women biochemists are discussed or mentioned, but since almost all were either research associates or professors of home economics, their work is usually subsumed under that of someone else, whose inflated productivity is then considered even more remarkable.

82. Walker, *Natural Man,* chap. 8.

83. Corner, "Herbert McLean Evans," p. 176; Frank Blair Hanson's interview with Dr. Nellie Halliday, 23 Feb. 1937, in "Evans File" (University of California, Sex Research File), Rockefeller Foundation Archives; Florence Sabin to Marion Hines, 5 [?] Jan. 1925, Florence Sabin Papers, APS; my conversation with Elizabeth Scott, Berkeley, Calif., 16 May 1975.

84. Agnes Fay Morgan to Professor Evans, 13 Mar. 1930, in reply to Herbert Evans to Professor Agnes Fay Morgan, 11 Mar. 1930, both in Agnes Fay Morgan Papers, Bancroft Library, University of California, Berkeley.

85. Corner, *George Hoyt Whipple*, pp. 102-3, 179-85, 200-201, 208, 296-98; also numerous clippings in Whipple Scrapbooks at University of Rochester Medical School Library, especially Mildred Bond, "Along the Promenade," *Rochester Democrat and Chronicle,* 18 Nov. 1934; "A Nobel Christmas Gift, Dr. Whipple to Share Prize," *Rochester Evening Journal,* 26 Dec. 1934; "Dr. Whipple Shy at Honors, Makes Co-Worker Take Bow," *Rochester Democrat and Chronicle,* 16 Jan. 1935; see also *Rochester Times-Union,* 31 May 1955, for Robbins's retirement, and the same, 20 Dec. 1973, for her obituary (Mrs. Frieda Sprague). In an autobiography published in 1959 Whipple called Robbins "a loyal and able associate," surely an understatement (George Hoyt Whipple, M.D., "Autobiographical Sketch," *Perspectives in Biology and Medicine* 2[1959]: 265).

86. Orma F. Butler, "The Women's Research Club," in *University of Michigan*, ed. Shaw, 1:410-11; Malcolm H. Soule, "The Junior Research Club," in the same volume, pp. 407-10. In

1921 the members strongly rejected an offer to affiliate with the General Federation of Women's Clubs. The club's papers are at the Bentley Historical Library, University of Michigan. Archivist Mary Jo Pugh has written an unpublished history of the club to 1975 which is with this collection.

87. "A Speech Given by Marion Hall (Mrs. Harvey Hall) at the First Meeting of the Stanford Faculty Women's Club, November 3, 1964," and "Memorial Resolution, Hazel D. Hansen, 1899–1962," Faculty File, both at Stanford University Archives. UCLA: "Historical Sketch," in *Campus Directory and Handbook, 1949–1950: A Publication of the Faculty Women's Club of the University of California, Los Angeles* [1949], p. 4. Berkeley: Women's Faculty Club Papers, University of California, Berkeley, both in Archives, Bancroft Library, University of California, Berkeley. "Agnes Wells," *CBY, 1949,* pp. 632–34; Sara Browne Smith, "The [University of Michigan] Faculty Women's Club," in *University of Michigan,* ed. Shaw, 1:418–19 (their papers for 1921–71 are in the Bentley Historical Library on campus). Texas: the papers of the University of Texas Faculty Women's Clubs, 1922–27, are in the Barker Texas History Center, University of Texas. Columbia: Helen E. Marshall, *Mary Adelaide Nutting: Pioneer of Modern Nursing* (Baltimore: The Johns Hopkins University Press, 1972), p. 183.

88. Patricia L. Pilling to author, 25 Feb. 1975, regarding Icie Macy Hoobler; Edwin G. Boring to Helen Peak, 31 Jan. 1949, regarding Margaret Washburn; Washburn to Boring, 10 Apr. 1934, and reply, 12 Apr. 1934, all in Edwin G. Boring Papers, HUA.

CHAPTER 8

1. "Bascom" in *NAW*; she did not take the Civil Service examinations in geology until 1900 or become certified as an "assistant geologist" until 1901 (certificate, Department of the Interior, 1 July 1901, Florence Bascom Papers, Sophia Smith Collection, Smith College); "Pennington" in *NAW:MP;* Anne Hartwell, "Mary E. Pennington, Who Keeps Cold Storage Cold," *Woman's Journal* 15(Nov. 1930): 11 and 42–43; Anne Pierce, "Mary Engle Pennington: An Appreciation," *Chemical and Engineering News* 18(1940): 941–42; "Mary Engle Pennington: October 8, 1872–December 27, 1952," *Refrigerating Engineering* 61(Feb. 1953): 184; Alice C. Goff, *Women CAN Be Engineers* (Youngstown, Ohio: private, 1946) pp. 183–214.

2. Evans: quotation from "Autobiography" in Alice Evans Papers, Cornell University Archives and Regional History Office; *NAW:MP*; "Alice Evans," *CBY, 1943*, pp. 198–200, *NYT*, 7 Sept. 1975, p. 51; Paul de Kruif, "Before You Drink a Glass of Milk," *Ladies Home Journal* 46(Sept. 1929): 8–9, reprinted in idem, *Men against Death* (New York: Harcourt, Brace, 1932), chap. 5; Milton MacKaye, "Undulant Fever," *Ladies Home Journal* 61(Dec. 1944): 23 and 69–70; Ralph Chester Williams, M.D., *The United States Public Health Service, 1798–1950* (Washington, D.C.: Commissioned Officers Association of the U.S.P.H.S., 1951), pp. 245–46; and Eunice Fuller Barnard, "Women Microbe Hunters," *Independent Woman* 15(1936): 379, 396–97. Gerry: Lida W. McBeath, "A Woman of Forest Science: Eloise Gerry," *Journal of Forest History* 22(1978): 128–35; and Harriet M. Grace, "Blazing a Trail in Woodland Research," *International Altrusan* 21(Nov. 1944): 10–11 and 20. Other Forest Service women are discussed in "Women in the [Forest] Service," *American Forests* 61(Mar. 1955): 30–31; Henry Clepper, "Women in Conservation," *American Forests* 62(Dec. 1956): 20–22, 52–53; see also note 25 to this chapter, on C. Audrey Richards. Bengtson: Alice C. Evans, "Ida Albertina Bengtson," *Journal of the Washington Academy of Sciences* 43(1953): 238–40; Williams, *United States Public Health Service,* pp. 244–45; Mary Greiner Kelly, "If War Should Come, Humanity's Pain Will Be Eased by Woman Scientist's Discoveries," *Washington Post,* 17 May 1934, p. 13, clipping in Ida Bengston File, Historical Library National Library of Medicine, Bethesda, Md.

Other early women scientists in the federal government included Flora Wambaugh Patterson of the USDA's Bureau of Plant Industry, 1896 to 1923 (Vera K. Charles, "Mrs. Flora Wambaugh Patterson," *Mycologia* 21[1929]: 1–4, with much more personal material than her boss, Beverly T. Galloway provided in "Flora W. Patterson, 1847–1928," *Phytopathology* 18[1928]: 877–79); Alice Davis Dunbar at the USDA's Bureau of Chemistry, from 1909 until 1910, when she quit after her marriage to a fellow chemist there ("Report of the Committee on Necrology," *Journal of the*

Association of Official Agricultural Chemists 40[Feb. 1957]: 97); and Isabel Martin Lewis, a computer and in 1908 the first woman regular employee of the Nautical Almanac Office. She did not quit after her marriage in 1912 but merely reduced her job to part-time work and then resumed full time after the death of her husband in 1930 (Richard S. Westwood, "Isabel Martin Lewis," *Nature Magazine* 43[1955]: 100-101).

3. "Women Government Employees Protest," *JACA* 10, no. 6 (Feb. 1917): 432-33; [Lucille Foster McMillan], *Women in the Federal Service* (Washington, D.C.: U.S. Civil Service Commission, 1938), pp. 16-18. Anna Jespersen, "Memorial to Jewell Jeannette Glass (1888-1966)," *Proceedings of the Geological Society of America for 1966*, p. 225. But some other women who might have been chosen to become scientists never shook off the clerical role (if indeed they wished to). For instance, Milcey Zachary (later McGregor) volunteered as a navy "yeomanette" in April 1917 and immediately became an assistant in the Office of the Secretary of the Navy, and was assigned to attend the meetings of the Naval Consulting Board, which evaluated inventions for the military. After the war she was one of the few civilians retained by the navy, and as the only female at the Naval Research Laboratory, was put in charge of its elaborate filing system and the recruitment of other women to the office staff. Apparently there were no women scientists at the NRL until after World War II (A. Hoyt Taylor, *The First Twenty-Five Years of the Naval Research Laboratory* [Washington, D.C.: Navy Department, 1948], p. 5).

4. "Women and the Federal Government at the Biennial," *JACA* 10, no. 3(Nov. 1916): 211. The best general account of the women's involvement in this movement is J. Stanley Lemons, *The Woman Citizen: Social Feminism in the 1920s* (Urbana: University of Illinois Press, 1973), chap. 5. See also Paul P. Van Riper, *History of the United States Civil Service* (Evanston, Ill.: Row, Peterson, 1958), chap. 11.

5. Laura Dana Morgan, "A Report on the Status of Women in the Classified Civil Service of the United States Government in the District of Columbia," *JACA* 6(Apr. 1913): 88-94; "Women Government Employees Protest," pp. 432-33; and resolution passed in "[Minutes of the] General Meeting of the Association [April 13, 1917]"; Bertha M. Nienburg, *Women in the Government Service*, WB*B* no. 8 (1920), pp. 7 and 10-14; *Woman at Work: The Autobiography of Mary Anderson as Told to Mary N. Winslow* (Minneapolis: University of Minnesota Press, 1951), pp. 152-53; "Women in the Government Service," *JACA* 13, no. 3 (1920): xv-xvi. See also Gustavus A. Weber, *The Women's Bureau: Its History, Activities and Organization* (Baltimore: The Johns Hopkins University Press, 1923). In April 1920 President Woodrow Wilson appointed suffragist and feminist Helen H. Gardener of Washington, D.C., the first woman member of the Civil Service Commission. She had started a long career as a journalist and political activist with an 1888 essay on "Sex in Brain," a refutation of the antifeminist researches of the time (*NAW*).

6. "Veterans' preference" in federal employment started as early as 1865, when it was offered to disabled veterans with honorable discharges from the Union Army and Navy. Over the years this preference was reduced to the five- and ten-point rules adopted by the Civil Service Commission, and finally encoded in federal legislation in 1919. This was amended by various Executive Orders during the 1920s and 1930s, until reformulated in the Veterans Preference Act of 1944, which, as amended, is still in effect (*Veterans Preference*, Form 1481 [Washington, D.C.: U.S. Civil Service Commission, 1923], and *Veterans Preferred* [Washington, D.C.: U.S. Civil Service Commission, 1944], pp. 2-3). The fairness of the sex-preference rule was hotly discussed in the 1930s ("Progress Report on Thirty-five Years of Legislation," *Independent Woman* 33[1954]: 255 and 280; "Women in Public Service," *Women's Work and Education* 8, no. 4 [Dec. 1937]: 7; *Woman at Work*, p. 153), but it is debatable whether the chief alternative (two lists—one for men and another for women) would have helped or hurt the women scientists more. See also Lemons, *Woman Citizen*, p. 83 n. 52, for a full discussion.

7. Van Riper, *History*, pp. 296-304; Lemons, *Woman Citizen*, pp. 134-37. The progress of the bill can be followed in *JACA*. See the ACA's 1920 resolution, *JACA* 13, nos. 7 and 8 (1920): 36; Eunice R. Oberly, "An Employment Policy for Uncle Sam," *JACA* 14, no. 1 (Oct. 1920): 16-19; another resolution in 1922, *JACA* 15, no. 3 (Mar. 1922): 88-89; Laura Puffer Morgan, "Present Status of Reclassification Bill," *JACA* 17, no. 1 (Jan.-Mar. 1924): 16-17. See also *The Status of Women in the Government Service in 1925*, WB*B* no. 53 (1926).

8. [Rachel Fesler Nyswander and Janet M. Hooks], *Employment of Women in the Federal Government, 1923-1939,* WB*B* no. 182 (1941), pp. 8, 17, 30, and 51.

9. Malcolm L. Smith and Kathryn R. Wright, "Occupations and Salaries in Federal Employment," *Monthly Labor Review* 52(1941): table 1, p. 69. These categories were introduced in the 1923 Reclassification Act, which had deliberately ranked many women's jobs as "clerical." There is no way of determining how large a distortion this introduces into the data, but it may well have been sizeable, especially among female statisticians.

10. [Nyswander and Hooks], *Employment of Women*, p. 19. This bulletin also analyzed the results of the 1939 Civil Service examinations by sex (pp. 11-15). It found that 44 percent of the 463,400 men who took the examinations passed, but only 40.1 percent of the 62,560 women did. This discrepancy was even greater among those taking the exams for the scientific, technical, and professional positions, where 25.4 percent of the 40,200 men passed, but only 18.8 percent of the 5,455 women did. The reasons for the differences were not clear, but the report suggested several: (1) the "veterans preference" gave the men the advantage in passing most of the examinations (jobs above a certain level were exempt from it, however); (2) most of the professional examinations substituted relevant education and experience for a written test and thus may have favored those with practical experience in the field (usually men), although the women's more frequent higher degrees (see chapter 6 herein) should have helped them here; and (3) many of the women had had an inferior education or lacked the administrative experience necessary for promotion to advanced positions. Although it is unclear whether the examiners knew the sex of the candidate when grading the tests, the men passed more often in "masculine" fields. Thus none of the 8 women candidates passed the examination for junior engineer, although 2,325 of the 11,910 men did (19.5 percent); nor did any of the 6 women applicants for "junior agronomist" pass, although 316 of the 744 men did (42.5 percent). On the other hand, the women had only a slight edge in the home economics exams, where 23 of the 1,327 women passed tests for several such positions (1.7 percent), but only 3 of the 283 men did (1.1 percent).

11. *Status of Women in the Government Service in 1925*, pp. 10-13. *Women in Chemistry,* a private study (see chapter 9 herein), found fourteen women chemists in four federal agencies in 1921, including eight at the Treasury (pp. 178-89).

12. Smith and Wright, "Occupation and Salaries," table 8, p. 83.

13. Data on starred men of 1938 are included in Raymond M. Hughes, "Research in American Universities and Colleges," *Research: A National Resource, Report of the Science Committee to the National Resources Committee,* 3 vols. (Washington, D.C.: GPO, 1938), 1:191.

14. The work of the USDA bureaus is discussed in A. Hunter Dupree, *Science in the Federal Government: A History of Policies to 1940* (Cambridge: Harvard University Press, 1957), chap. 8, and T. Swann Harding, *Two Blades of Grass* (Norman: University of Oklahoma Press, 1947). See also Fred Wilbur Powell, *The Bureau of Plant Industry* (Baltimore: The Johns Hopkins University Press, 1927); "Reports of the Chief of the Bureau of Plant Industry" in the *Annual Reports of the Department of Agriculture,* 1925-29; and portions of Paul de Kruif's *Hunger Fighters* (New York: Harcourt, Brace, 1928).

15. The Bureau of Home Economics merits further work by historians since the only study of it is narrow and outdated (Paul V. Betters, *The Bureau of Home Economics* [Washington, D.C.: Brookings Institution, 1930]). Similarly, the only studies of Louise Stanley are in *NAW:MP* and Helen T. Finneran, "Louise Stanley: A Study of the Career of a Home Economist, Scientist, and Administrator, 1923-1953" (Master's thesis, American University, 1965). (Stanley was one of the few prominent home economists of her generation who did not serve in the Food Administration during World War I.) See also Ruth O'Brien, "The Program of Textile Research in the Bureau of Home Economics," *Journal of Home Economics* 22(1930): 281-87; Harding, *Two Blades of Grass,* chap. 12, "Food and Raiment"; Virginia Dean Rose, "Trends in Home Economics in Public Service," *Women's Work and Education* 12, no. 4 (Dec. 1941): 1-7. Helen Atwater and Marjorie Hesseltine, "The Home Economist in the Public Service," *Women's Work and Education* 11, no. 1(Feb. 1940): 1-5, claims the BHE had seventy scientists in 1940 (including forty home economists, thirteen chemists, seven physicists, two physiologists, and one bacteriologist). Home economists also did extension work for numerous other depression agencies, such as the Farm Credit Administration, the Rural

Electrification Administration, and the Works Project Administration. See also Gladys L. Baker, "Women in the U.S. Department of Agriculture," *Agricultural History* 50(1976): 190–201.

16. In 1917, in response to the war, the USPHS began to seek out women bacteriologists (W. H. Frost to E. O. Jordan, 16 July 1917, E. O. Jordan Papers, Special Collections, University of Chicago Library). Evans: note 2 to this chapter and Elizabeth M. O'Hern, "Alice Evans: Pioneer Microbiologist," *ASM [American Society for Microbiology] News* 39(1973): 573–78; and idem, "Alice Evans and the Brucellosis Story," *Annali Sclavo* 19(1977): 12–19; G. W. McCoy to Alice Evans, 31 Dec. 1927, in Alice Evans Papers, Cornell University Archives and Regional History Center. In 1930 Charles Herty wrote chemist Mary Sherrill of Mount Holyoke that the passage of the Ransdell bill creating the National Institute of Health would lead to many research jobs for women. (Charles Herty to Mary Sherrill, 9 June, and her reply, 11 June 1930, both in the Charles Holmes Herty Papers, box 117, folder 35, Special Collections, Robert W. Woodruff Library, Emory University. I thank Jeffrey Sturchio for telling me about this collection and archivist Diane Windham for sending copies of these letters.)

17. Charles Armstrong, "George Walker McCoy, 1876–1952," *Science* 116(1952): 468. See also his obituary under "Deaths," *Journal of the American Medical Association* 148(1952): 1519.

18. Branham: Mary Greiner Kelly, "Women Scientists Resemble Housewives in Making Things Do [interview with Branham]," *Washington Post,* 18 May 1934 (clipping in Branham File, National Library of Medicine); J. D. Ratcliff, "Lady of the Lab," *McCall's* 66, no. 7 (Apr. 1939): 7–8; obituaries in *NYT,* 19 Nov. 1962, p. 31, and *ASM [American Society for Microbiology] News* 42(1976): 420–22.

Stewart: transcript of interview with Wyndham D. Miles, 10 Feb. 1964, Historical Library, National Library of Medicine, Bethesda, Md.; Gwen Dobson, "Luncheon with . . . Dr. Sarah Stewart," *Evening Star* (Washington, D.C.), 8 Jan. 1972 (clipping in Stewart File, National Library of Medicine); Bernice E. Eddy, "Sarah Elizabeth Stewart, 1906–1976," *Journal of the National Cancer Institute* 59(1977): 1039–40. Cram: *AMS,* 9th ed. (1955), and Benjamin Schwartz, "Eloise Blaine Cram, 1896–1957," *Proceedings of the Helminthological Society of Washington* 24(1957): 146–47; Pittman: *AMS,* 12th ed. (1972), and *Who's Who in America, 1976–77.* See also "Women in Science," *Medical Woman's Journal* 44(Jan. 1937): 19–20; Donald Swain, "The Rise of a Research Empire: NIH, 1930–1950," *Science* 138(1962): 1233–37; Arthur M. Stimson, *A Brief History of Bacteriological Investigations of the United States Public Health Service,* published as supplement no. 141 to the Public Health Reports (Washington, D.C.: GPO, 1938); and Jeanette Barry, *Notable Contributions to Medical Research by Public Health Service Scientists: A Biobibliography to 1940,* Public Health Service Publication no. 752 (1960).

In 1932 Estella Ford Warner, M.D., a public health official from Oregon, became the first woman commissioned officer in the USPHS elite uniformed service corps. ("Doctor to India: A Woman Directs Our Biggest Health Mission Abroad," *NYT Magazine*, 16 May 1954, p. 44, and Leonard A. Schelle, "Estella Ford Warner," *Public Health Reports* 71[1956]: 375–6.) This service had been opened to women during World War I (Emma Gilmore, M.D., "United States Public Health Service Opened to Women Physicians," *Woman's Medical Journal* 28(1918): 144–45.)

19. *Statistical Work: A Study of Opportunities for Women,* Studies in Occupations, no. 2 (New York: Bureau of Vocational Information, 1921), pp. 42 and 113. The BVI's advisory council included the presidents of eight women's colleges and of five coeducational colleges and universities as well as a variety of other notable educational officials (p. 3). In 1919, 639 women took the Civil Service examination for "statistical clerk"; 245 passed, and 91 were appointed to positions. Several others were appointed in 1920.

20. Lewis B. Sims, "Social Scientists in the Federal Service," in *Public Policy: A Yearbook of the Graduate School of Public Administration, Harvard University, 1940,* ed. Carl J. Friedrich and Edward S. Mason (Cambridge: Harvard University Press, 1940), pp. 280–96; Avis Marion Saint, "Women in the Public Service: no. 4: The Federal Service of the United States," *Public Personnel Studies* 9(1931): 18, citing U.S. Civil Service Commission, *Annual Reports,* 1920–1930, table 1.

Jeanette Studley, "The Challenge of Statistics," *Independent Woman* 19(1931): 6–8, 38; Mary V. Dempsey, "If You Like Figures," *Independent Woman* 18(1939): 336–37, 406–7; "The Public

Health Statistician," *Women's Work and Education* 10, no. 2 (Apr. 1939): 10–11; Aryness Joy Wickens in *CBY, 1962*, pp. 462–64; George Martin, *Madame Secretary: Frances Perkins* (Boston: Houghton Mifflin, 1976), pp. 453–55; and "The Reminiscences of Mrs. A. J. Wickens," Oral History Research Office, Columbia University, 1957; see also Joseph W. Duncan and William C. Shelton, *Revolution in United States Government Statistics, 1926–1976* (Washington, D.C.: U.S. Department of Commerce, 1978).

Some trends evident in government employment also took place in private agencies, as in 1922, when Jessamine Whitney's title at the National Tuberculosis Association was upgraded from "research secretary" to "statistician," and she became supervisor of its (probably all-female) statistical service (Richard Shryock, *National Tuberculosis Association, 1904–1954* [New York: National Tuberculosis Association, 1957], pp. 144, 203, and 229–30).

21. "Knopf" in *NAW:MP*; John Rodgers, "Memorial to Eleanora Bliss Knopf, 1883-1974," Geological Society of America, *Memorials* 6(1977); "Gardner" in *NAW:MP*; Harry S. Ladd, "Memorial to Julia Anna Gardner (1882-1960)," *Proceedings Volume of the Geological Society of America for 1960* (1962): 87–89; R. V. Dietrich, "Memorial to Anna I. Jonas Stose, 1881-1974," Geological Society of America, *Memorials* 6(1977); Joseph J. Fahey, "Memorial of Margaret D. Foster," *American Mineralogist* 56(1971): 687–90; and Margaret D. Foster, "The Chemist in the Water Resources Laboratory," in *The Chemist at Work*, ed. Roy I. Grady and John W. Chittum (Easton, Pa.: *Journal of Chemical Education*, 1940), pp. 49–52. Stadnichenko: obituaries in *Journal of the Washington Academy of Sciences* 48(1958): 407, and Washington D.C., *Evening Star*, 27 Nov. 1958; and an interview with her, "Advice from a Woman in Science," by Virginia Dean Rose in *Women's Work and Education* 10, no. 4(Dec. 1939): 1–4.

22. Underhill: *CBY, 1954*, pp. 617–19; see also Graham D. Taylor, "Anthropologists, Reformers and the Indian New Deal," *Prologue* 7(1975): 151–62; "Carson" in *NAW:MP;* and Paul Brooks, *The House of Life: Rachel Carson at Work* (Greenwich, Conn.: Fawcett Publications, 1972), chap. 2.

23. [Nyswander and Hooks], *Employment of Women*, p. 53.

24. Stanley: see note 15 to this chapter. Martha Eliot: *CBY, 1948*, pp. 184–86, and "Martha May Eliot, M.D., 1891–1978," *American Journal of Public Health* 68(1978): 696–700.

25. Richards: *AMS*, 7th ed. (1944); and John K. McDonald, "We Present Dr. C. Audrey Richards," *Journal of Forestry* 49(1951): 918–19.

26. Cram: see note 18 to this chapter. Dyer: "Interview with Dr. Wyndham D. Miles, March 18, 1968," Historical Library, National Library of Medicine, Bethesda, Md. McCoy also refused to spend laboratory funds on office rugs, fine desks or furniture, or paintings (Armstrong, "George Walker McCoy," p. 468). Bussey: Rexmond C. Cochrane, *Measures for Progress: A History of the National Bureau of Standards* (Washington, D.C.: National Bureau of Standards, 1966), pp. 170 and 443 n.

27. Ruby K. Worner, "Opportunities for Women Chemists in Washington," *Journal of Chemical Education* 16(1939): 584. A later study of the Women's Bureau disagreed, saying that the top women in the Civil Service jobs had not been appointed to those posts but had risen through the ranks (*Women in the Federal Service, Part II: Occupational Information*, WB*B* no. 230–11[1950]; p. 31).

28. [Nyswander and Hooks], *Employment of Women*, pp. 33–41, and Smith and Wright, "Occupations and Salaries," table 8, p. 83 (the average overall salary was $3,137, and the women, with an average salary of $2,299, made up 8.3 percent of the scientists); Worner, "Opportunities for Women Chemists," p. 585.

29. Economy Act of 1932: *Statutes at Large*, 72d Congress, 1st sess. 1932, 47, pt. 1, ch. 314, p. 406; Sarah Slavin Schramm, "Section 213: Woman Overboard" (Paper delivered at Second Berkshire Conference, Radcliffe College, 1974) (on file at SLRC); and Lois Scharf, *To Work and To Wed: Female Employment, Feminism, and the Great Depression* (Westport, Conn.: Greenwood Press, 1980).

30. Allen: C. E. Yarwood and Mabel Nebel, "Ruth Florence Allen, 1879–1963," *Phytopathology* 54(1964): 885; "Densmore" in *NAW:MP;* [Frances Densmore], "Chronology of the Study and Presentation of Indian Music from 1893 to 1944," Frances Densmore Papers, box 1, National Anthropological Archives, Smithsonian Institution. See also Nina Marchetti Archabal, "Frances

Densmore: Pioneer in the Study of American Indian Music," in *Women of Minnesota: Selected Biographical Essays*, ed. Barbara Stuhler and Gretchen Kreuter (St. Paul: Minnesota Historical Society Press, 1977), pp. 94–115 and 354–56.

31. G. Winston Sinclair, "Memorial to Alice E. Wilson (1881-1964)," Geological Society of America, *Proceedings, 1966*, p. 404.

32. Mary Q. Innis, ed., *The Clear Spirit: Twenty Canadian Women and Their Times* (Toronto: University of Toronto, 1966), chap. 14; G. Winston Sinclair to Winifred Goldring, 8 July 1941, Winifred Goldring Papers, New York State Museum and Library. Morris Zaslow, *Reading the Rocks: The Story of the Geological Survey of Canada, 1842–1972* (Ottawa: Macmillan, 1975), mentions Wilson in passing; F. J. Alcock, *A Century in the History of the Geological Survey of Canada,* National Museum of Canada Special Contribution no. 47-1 (Ottawa: King's Printer, 1947), does not. Edmund M. Spieker, "Memorial to Grace Anne Stewart, 1893–1970," Geological Society of America, *Memorials,* 2(1973): 110–14.

Because the *AMS* includes several women employed in Canada, table 8.4 (p. 241) has data on six government agencies there in 1938. They were scattered among the Canadian Department of Agriculture, the Canadian Biological Board, the Dominion Observatory, the National Research Council, the Royal Ontario Museum, and the Canadian Geological Survey. Some of these women seem to have had rather successful careers in the government, as, for example, Margaret Newton of the Dominion Rust Research Laboratory in Winnipeg, a branch of the Department of Agriculture. There, with the help of several male assistants, she did important work on the plant diseases that followed the introduction of the new wonder wheats into western Canada at the turn of the century. Although her researches on the rusts were acclaimed as outstanding and she was elected to the Royal Society of Canada, she seems not to have held offices in the Canadian Phytopathological Society (established in 1930) and to have been twice passed over for promotion to the position of director of the laboratory before her early retirement because of ill health in 1945 ("Flavelle Medal: Margaret Newton," *Transactions of the Royal Society of Canada* 42[1948]: 47–48; I. L. Conners, ed., *Plant Pathology in Canada* [Winnipeg: Canadian Phytopathological Society, 1972], pp. 128–38; James V. Mink, "Chequered Career: Scotland to UCLA, Flora M. Scott," Oral History Transcript, UCLA Oral History Program, 1973, pp. 72–73).

33. May B. Upshaw, *Woman's Place in Civil Service* (New York: Federation of Women's Civil Service Organizations, 1919), surveyed some state and municipal examination and appointment practices.

34. "Murtfeldt" in *AMS,* 2d ed. (1910). None of her obituaries mentions these official posts, although one appends her entry in *AMS* (Herman Schwarz, "Miss Mary E. Murtfeldt," *Entomological News* 24[1913]: 241–42). Office of Experiment Stations, USDA, *List of Botanists of the Agricultural Experiment Stations in the United States,* Experiment Station Bulletin no. 6 (1890), and no. 13(1893); "Bitting" and "Detmers" in *AMS,* 3d ed. (1921). "Smith" in *AMS,* 5th ed. (1933). "Kennedy" in *AMS,* 8th ed (1949). Patch: Arnold Mallis, *American Entomologists* (New Brunswick, N.J.: Rutgers University Press, 1971), p. 230. "Edna Louise Ferry, M.S. 1913," in Yale University, *Obituary Record of Graduates Deceased during the Year Ending July 1, 1920* (New Haven: Yale University, 1921), pp. 1586–87; *Forty-third Annual Report of the Connecticut Agricultural Experiment Station, October 31, 1919* (New Haven: State Government, 1920), p. v.

35. "Maury," "Knopf," and "Jonas," *AMS*, 3d ed. (1921); Chester A. Reeds, "Memorial to Carlotta Joaquina Maury," *Proceedings of the Geological Society of America for 1938,* pp. 157–68; Dietrich, "Memorial to Anna I. Jonas Stose."

36. Barbara E. Brand, "The Influence of Higher Education on Sex-Typing in Three Professions, 1870-1920: Librarianship, Social Work, and Public Health" (Ph.D. diss., University of Washington, 1978), esp. chs. 11-13 (I thank Rosalind Rosenberg for telling me of it.). *Bacteriological Work as a Vocation for Women,* published as Women's Educational and Industrial Union Bulletin no. 9(1911). See also "Richards" in *AMS,* 2d ed. (1910), and *NAW;* other early accounts of the field are Ella V. Mark, "Women as Guardians of the Public Health," *Papers Read at the Association for the Advancement of Women . . . 16th Congress, Detroit, Michigan, November 1888* (Fall River, Mass.: J. H. Franklin, 1889), pp. 45–53, and Adelaide Brown, "Public Health: A Normal Field of Interest and Work for College Women," *JACA* 9(May 1916): 76–78; E. O. Jordan,

G. C. Whipple, and C.-E. A. Winslow, *A Pioneer of Public Health: William Thompson Sedgwick* (New Haven: Yale University Press, 1924), pp. 37, 53, 61, 89, 110, 125, and 138.

Martha Eliot (*CBY, 1948*, pp. 184–86), the eminent authority on rickets and other child health matters in the 1920s and after, recalled in a 1964 interview that Professor Sedgwick had recommended about 1913 that she prepare herself to become a laboratory technician. It reveals a lot about both of them that she pressed him with a second question.

> Then, in my innocence—I suppose it was—I said, "Professor Sedgwick, suppose I was a man sitting here asking you what my education should be if I were to go into public health as a life interest?" He thought for a moment and then said, "Well, I would advise you to go to medical school and become a physician first and then, when you have had some hospital work, to enroll in a school of public health." Miss Eliot replied, "Well I believe you have answered my question very quickly and easily. I want to study medicine myself."

(Interview with Martha Eliot, 28 May 1964, cited in Jean Alonzo Curran, *Founders of the Harvard School of Public Health, 1909–1946* (New York: Josiah Macy Jr. Foundation, 1970), p. 132, and Barbara E. Brand, "Influence of Higher Education," p. 337.)

37. W[illiam] T. S[edgwick], "A Welcome to Women in Public Health Work," *American Journal of Public Health* 8(1918): 164; "Women's New Fields of Work," *American Journal of Public Health* 9(1919): 371–72.

38. "Women in Public Health Work," *NYT*, 16 Sept. 1921, p. 16; "Editorial," *JAAUW* 15, no. 1 (Oct. 1921): 1–2.

39. Anna M. Sexton, *A Chronicle of the Division of Laboratories and Research, New York State Department of Health, The First Fifty Years: 1914–1964* (Lunenburg, Vt.: Stinehour Press, 1967); "Kirkbride" in *AMS*, 3d ed. (1921); *NYT*, 29 Mar. 1967, p. 45, and Elizabeth D. Robinton, "A Tribute to Women Leaders in the Laboratory Section of the American Public Health Association," *American Journal of Public Health* 64(1974): 1006–7; "Gilbert" in *AMS*, 8th ed. (1949); Silverman: James E. Perkins, M.D., "Guessing and Knowing: What Makes the Difference? A Tribute to a Statistician: Hilda Freeman Silverman," *American Journal of Public Health* 47(1957): 1053–64.

40. Kendrick: Robinton, "Tribute to Women Leaders," pp. 1006–7; Connie Tegge, "Pearl Kendrick: A Woman Who Fought Whooping Cough," *Ann Arbor News*, 7 March 1971; and "Whooping Cough Pioneer to Speak," *Ann Arbor News*, 13 June 1975; both clippings in Kendrick File, Bentley Historical Library, University of Michigan.

41. *NYT*, 14 Sept. 1963, p. 25; Elwood A. Seamans, "Emmeline Moore (1872–1963)," in *Leaders of American Conservation*, ed. Henry Clepper (New York: Ronald Press, 1971), p. 229; Faculty File, Vassar College Archives.

42. Quotations from Donald W. Fisher, "Memorial to Winifred Goldring (1888–1971)," Geological Society of America, *Memorials* 3(1974), pp. 96–102; biographical sketch by Sally Gregory Kohlstedt in *NAW:MP;* her papers are at the New York State Museum in Albany; about one hundred other letters are in the Charles Schuchert Papers, Yale University Archives.

43. Harriet Zuckerman has found that a high proportion of American Nobel laureates (15 of 71, or 21 percent) came from New York City at a time when only 6 percent of the U.S. population lived there, and that all but one of them were Jewish. Perhaps this factor is of some importance in New York City's preponderance of government scientists (*Scientific Elite: Nobel Laureates in the United States* [New York: Free Press, 1977], pp. 81–82).

44. "Williams" and "Baker" in *AMS*, 3d ed. (1921); "Williams" in *NAW:MP;* "Baker" in *NAW;* and "President, Dr. S. Josephine Baker," *Women in Medicine* 49(July 1935): 17. S. Josephine Baker, M.D., *Fighting for Life* (New York: Macmillan, 1939); see also "Katherine Collins," in *AMS*, 3d ed. (1921); David A. Blancher, "Workshops of the Bacteriological Revolution: A History of the Laboratories of the New York City Department of Health, 1892–1912" (Ph.D. diss., City University of New York, 1979); Avis Marion Saint, "Women in the Public Service: no. 2: The City of Berkeley," *Public Personnel Studies* 8(1930): 104–7, and idem, "Women in the Public Service: no. 3: The City of Oakland," *Public Personnel Studies* 8 (1930): 119–22. See also Mary Ritter Beard, *Woman's Work in Municipalities* (New York: D. Appleton, 1915), chap. 2 ("Public Health").

45. "Woolley" in *NAW;* "Augusta Bronner" in *NAW:MP;* in *Who's Who in America* 27(1952-53), p. 303; *Harvey Humphrey Baker: Upbuilder of the Juvenile Court* (Boston: Judge Baker Foundation, 1920); and William Healy and Augusta Fox Bronner, "The Child Guidance Clinic: Birth and Growth of an Idea," in *Orthopsychiatry, 1923-1948, Retrospect and Prospect,* ed. Lawson G. Lowrey and Victoria Sloane (New York: American Orthopsychiatric Association, 1948), pp. 14-49; Leta Stetter Hollingworth had been the first Civil Service psychologist in New York City in 1916 and had worked briefly at Bellevue Hospital before getting a job at Teachers College (*NAW*). George S. Stevenson, M.D., *Child Guidance Clinics: A Quarter Century of Development* (New York: Commonwealth Fund, 1934), and Murray Levine and Adeline Levine, *A Social History of Helping Services: Clinic, Court, School and Community* (New York: Appleton-Century-Crofts, Meredith Corporation, 1970).

46. Professor John B. Morgan of Northwestern University quoted in Andrew W. Brown, "The Meeting of the Clinical Section of the APA," *Psychological Exchange* 2 (1933): 176, cited in Donald S. Napoli, "The Architects of Adjustment: The Practice and Professionalization of American Psychology, 1920-1945" (Ph.D. diss., University of California, Davis, 1975), p. 106. Chapter 3 in Napoli discusses the separate labor markets for academic and clinical psychologists. See also Robert I. Watson, "A Brief History of Clinical Psychology," *Psychological Bulletin* 59(1953): 321-46, and idem, "A Brief History of Educational Psychology," *Psychological Record* 11(1961): 209-42; F. H. Finch and M. E. Odoroff, "Employment Trends in Applied Psychology," *Journal of Consulting Psychology* 3(1938): 118-22; Gladys D. Frith, "Psychology as a Profession," *Women's Work and Education* 10, no. 3(Oct. 1939): 1-3, and David B. Tyack, *The One Best System: A History of American Urban Education* (Cambridge: Harvard University Press, 1974), pp. 198-216.

47. David Shakow, "Grace Helen Kent," *Journal of the History of the Behavioral Sciences* 10(1974): 275-80.

CHAPTER 9

1. After decades of expansion, the membership of the American Chemical Society remained almost constant between 1920 (15,582 members) and 1928 (16,240 members) (Charles A. Browne and Mary Elvira Weeks, *A History of the American Chemical Society: Seventy-Five Grateful Years* [Washington, D.C.: American Chemical Society, 1952], p. 127).

2. *Women in Chemistry: A Study of Professional Opportunities* (New York: Bureau of Vocational Information, 1922), p. 48. The bureau's correspondence and questionnaires are in Bureau of Vocational Information Papers, SLRC. See also "The Training of Women Chemists," *JAAUW* 15 (Apr. 1922): 73-75; Marie Riemer, "Women in Chemistry," *Barnard Alumnae Bulletin* 12 (Dec. 1922): 5-7; and "Women in Chemistry: A Study of Professional Opportunities," *Chemical Age* 30 (1922): 268-70. The bureau had hoped for financial support in publishing *Women in Chemistry* from the National Research Council, whose permanent secretary, Vernon Kellogg, was on its advisory council, but he informed them in 1922 that the council was overcommitted at present and unable to help them (Vernon Kellogg to Emma Hirth, 16 and 17 May 1922, BVI Papers, box 21, SLRC).

3. *Women in Chemistry,* pp. 5 and 226-28.

4. Lists of subsequent careers of chemistry majors in Chemistry Department Files, Mount Holyoke College Archives. Carr maintained a strong interest in the professional careers of her students long after graduation. One commented in a remembrance on her retirement that she was just the person to ask questions such as "whether or not to spend her last fifty dollars on a new suit in order to make a good impression on the president of the college to whom she was applying for a position." One suspects the answer was yes! and that this was just the type of question Carr loved to answer (Clara Berissa Tillinghast, "Emma Perry Carr," *Mount Holyoke Alumnae Quarterly* 30 [Aug. 1946]: 1-2).

5. Although women's percentage of the whole was relatively constant, estimates of the numbers of women "chemists" in the 1920s varied widely (from 42 to 1,714): women were 3.2 percent of the members of the American Chemical Society in 1925 (481 of "about 15,000") and 2.3 percent (42 of 1852) in the 1921 edition of the *AMS* (Marie Farnsworth, "Women in Chemistry: A Statistical

Study," *Industrial and Engineering Chemistry* 3[10 Sept. 1925]: 4). They were 5.2 percent (1,714 of 32,941) of the "chemists, assayers and metallugists" in the 1920 federal census and 3.2 percent (70 of 2,172) research chemists listed by the National Research Council ("Women in Chemistry," *Chemical Age*, p. 269). Of the 28 women chemists I counted in the 1921 *AMS*, only 4 (14.3 percent) were employed in industry. (Five others had been previously.) Of the 163 women chemists in the 1938 edition, 20 (12.3 percent; 18 had doctorates) were working in nineteen firms (including 4 as "technical librarians," "chemical secretaries," or "with legal department"—categories not listed in the 1921 edition). Since the 1938 *AMS* contained an estimated 1,850 male chemists in industry, even this fivefold growth led to barely more than 1 percent of the total. Better definitions and numbers are greatly needed here, as for all measures of industrial science. It is to be hoped that Arnold Thackray's "chemical indicators" project will introduce more quantitative rigor into discussions of chemists' employment.

6. "Women Chemists," *Women's Work and Education* 4, no. 2 (Apr. 1933): 1; "Women in Science," *Women's Work and Education* 6, no. 4(Dec. 1935): 1; "Proceedings of the Conference on Women's Work and Their Stake in Public Affairs," mimeographed, sponsored by the Institute of Women's Professional Relations, New York City, March 1935, pp. 137–53; "Jobs in Chemistry," *Women's Work and Education* 7, no. 3(Oct. 1936): 1; "Women in Chemistry," *Women's Work and Education* 9, no. 2(Apr. 1938): 12; *Journal of Chemical Education* 16(Dec. 1939): 577–90; "The Chemist," *Women's Work and Education* 11, no. 1(Feb. 1940): 12; Lucile Nelson, "Success from Test Tubes," *Independent Woman* 19(June 1940): 170–72, 187.

7. Betty Joy Cole, "The Special Library Profession and What It Offers: no. 4: Chemical Libraries," *Special Libraries* 25(1934): 271–77; *Special Libraries* 19(May 1919), no. 4, had also been devoted to chemical libraries; "The Chemical Secretary," *Women's Work and Education* 10, no. 1(Feb. 1939): 1–4; Lois Woodford's comments in "Proceedings of the Conference on Women's Work," pp. 144–46, and her questionnaire, 1920, Bureau of Vocational Information Papers; some of her correspondence of the 1920s and 1930s is included in the Charles Herty Papers (Special Collections, Woodruff Library, Emory University, esp. box 117, folder 35), since she worked for Herty for many years. Female college graduates were in great demand for secretarial work during the depression ("Trends in Occupations for Women," *Occupations* 16[1937]: 175); F. W. Adams, "Opportunities for Women as Research Bibliographers," *Journal of Chemical Education* 16(1939): 581–83, reprinted in *The Chemist at Work*, ed. Roy I. Grady and John W. Chittum (Easton, Pa.: *Journal of Chemical Education*, 1940), pp. 373–79; Janet D. Scott, "My Work with Chemical Abstracts," in Grady and Chittum, eds., *Chemist at Work*, pp. 79–91; Browne and Weeks, *History of the American Chemical Society*, pp. 336–67. See also John Troan, *An Adventure in Knowledge: The Story of the Chemical Abstracts Service* (Washington, D.C.: American Chemical Society [1963]). The history of scientific information deserves a critical assessment.

8. Florence E. Wall, "Vocations in Chemistry," *Journal of Chemical Education* 11 (1934): 96–100, and idem, "Training in Chemistry for the Cosmetic Industry," *Journal of Chemical Education* 13(1936): 432–36; "The Beauty Industry," *Women's Work and Education* 6, no. 3(Oct. 1935): 6, and "Notes on Jobs," *Women's Work and Education* 9, no. 3 (Oct. 1939): 7; Florence E. Wall, "Rebel into Pioneer" (Paper presented to the Meeting of the American Chemical Society, Washington, D.C., September 1979), is autobiographical. See also "Historical Sketch of the Founding of the Society of Cosmetic Chemists," *Journal of the Society of Cosmetic Chemists* 1(1947): i–ii; Dan Dahle, "Opportunities for Chemists in the Cosmetic Industry," *Journal of the Society of Cosmetic Chemists* 1(1947): 67–71; "Directory of Members," *Journal of the Society of Cosmetic Chemists* 2(1951): 198–207 (11 women of 225 members, or 4.9 percent).

9. Florence E. Wall, "The Status of Women Chemists," *Chemist* 15(1938): 174–92; salary data: *Women in Chemistry*, p. 233; Helen L. Wikoff, "Occupations and Earnings of Women in Chemistry," *Industrial and Engineering Chemistry* 25(1933): 471; Ethel L. French, "A Survey of the Training and Placement of Women Chemistry Majors in Women's and Co-educational Colleges," *Journal of Chemical Education* 16(1939): 576 (reprinted in Grady and Chittum, *Chemist at Work*, pp. 351–56); male salaries are in Robert W. French, "The Changing Economic Status of Chemists, 1926–1943," *Chemistry and Engineering News* 24(1946): 1649–55.

10. Data collected on turnover of male and female employees in a large financial house in New

York City (N=400) and in an insurance company (N=635) indicated that persons of both sexes stayed longer when better personnel methods were used and that women stayed longer than men ("Labor Turnover, Comparative Stability of Male and Female Employees," *Monthly Labor Review* 24[1927]: 37-39, reporting on recent research in the *Journal of Personnel Research*).

11. French, "Survey of the Training and Placement," pp. 575-76; W. S. Landis, "Women Chemists in Industry," *Journal of Chemical Education* 16(1939): 577-79; and Walter Savage Landis, "Women Chemists in the Chemical Industries," *Chemical Industries* 44(1939): 502-4; Ruby K. Worner, "Opportunities for Women Chemists in Washington," *Journal of Chemical Education* 16(1939): 584; *Women in Chemistry,* pp. 48, 227-28. As large a percentage as 90 of the women in some WAC companies had "only favorable attitudes toward their WAC commanders" (Mattie E. Treadwell, *The Women's Army Corps,* U.S. Army in World War II, Special Studies [Washington, D.C.: Department of the Army, 1954], p. 677).

12. Forrest A. Anderson, "Industry's Challenge to Chemical Education," *Journal of Chemical Education* 18(1941): 472.

13. "Chemical Secretary," *Women's Work and Education.* Rosabeth Moss Kanter discusses many of these themes of occupational segregation and dead-end jobs in *Men and Women of the Corporation* (New York: Basic Books, 1977), especially pp. 71, 74, and 103. By the 1970s, however, even secretaries' jobs could be seen as channels to nowhere that were in need of restructuring, perhaps as management traineeships.

14. See, for example, F. B. Jewett, "Finding and Encouragement of Competent Men," *Science* 69 (1929): 309-11, and Willis R. Whitney, "Encouraging Competent Men to Continue in Research," *Science* 69 (1929): 3-12; John H. Perry [of E. I. Du Pont de Nemours & Co.], "Man Location," *Chemical and Metallurgical Engineering* 43 (1936): 68-71; and W. A. Gibbons [of U.S. Rubber Co.], "Careers in Research," *Research—A National Resource: Report of the National Resources Committee to the National Resources Planning Board* (Washington, D.C.: GPO, 1941), pp. 108-17. For more on Whitney, see George Wise, "A New Role for Professional Scientists in Industry: Industrial Research at General Electric, 1900-1916," *Technology and Culture* 21(1980): 408-29, and Willis R. Whitney to Frances Cummings, 16 Dec. 1912, which deplores the frequent loss of women chemists through marriage, and Whitney to Emma P. Hirth, 10 Jan. 1918, which reflects the wartime feeling that women chemists' opportunities would inevitably increase (both in BVI Papers, box 23, SLRC). Ronald C. Tobey, *The American Ideology of National Science, 1919-1930* (Pittsburgh: University of Pittsburgh Press, 1971) also discusses this elitist rhetoric of the 1920s. Stanley Cohen reports that in the 1920s about 20 percent of the National Research Council Fellows (for more about them see chapter 10) went to work for industry within a few years of receiving their fellowship ("American Foundations as Patrons of Science: The Commitment to Individual Research," in *The Sciences in American Context: New Perspectives,* ed. Nathan Reingold [Washington, D.C.: Smithsonian Institution Press, 1979], p. 231). Although job advertisements in the *Chemical and Engineering News* in the 1930s did not specify who might or might not apply, almost all asked for photographs. Advertisements posted on university bulletin boards at Columbia University reportedly did specify "Male Christians Only" (Mildred Cohen, commentator, session on "From Amateur to Professional: Women in the Sciences," Third Berkshire Conference on the History of Women, Bryn Mawr College, June 1976). Jews knew they were not welcome in the chemical industry in the 1920s and 1930s (Harriet Zuckerman, *Scientific Elite: Nobel Laureates in the United States* [New York: Free Press, 1977], p. 76); nor could the best black chemists of the time (Percy Julian and L. A. Hall) get jobs there (Louis Haber, *Black Pioneers of Science and Invention* [New York: Harcourt, Brace and World, 1970]).

15. Langmuir was an atypical industrial scientist, even for G. E. Labs (Wise, "New Role," pp. 420-25); Lawrence A. Hawkins, *Adventure into the Unknown: The First Fifty Years of the General Electric Research Laboratory* (New York: William Morrow, 1950), pp. 96-97 (I thank Arthur Norberg for this reference); Blodgett: *CBY, 1952,* pp. 55-57; *Chemistry and Engineering News* 29(9 Apr. 1951): cover and 1408 and a short obituary recently, "Katharine Burr Blodgett," *Physics Today* 33(March 1980): 107; Hayner: *AMS*, 6th ed. (1938), and *NYT*, 25 Sept. 1971, p. 34; Brophy: *AMS,* 6th ed. (1938); Fenwick: *AMS,* 5th ed. (1933); and Armbruster: *AMS,* 8th ed. (1949).

16. Sullivan: *Chemical and Engineering News* 26(12 July 1948): cover, 1767, and 2055; *Chemical and Engineering News* 32(22 Mar. 1954): cover and 1138; Lois Hans, "Women Chemists in the Paper Industry," *Paper Industry and Paper World* 25 (Apr. 1943): 33–35, and "Helen Kiely Honored on 25th Anniversary with A.W.P. Corp.," *Paper Industry and Paper World* 25 (Apr. 1943): 56; Minor: *AMS*, 6th ed. (1938), and 1920 questionnaire in BVI Papers, box 22, SLRC; Cobb: *AMS*, 13th ed. (1976).

17. Wassell: "Expert on Emulsions," *Chemical Week* 71(27 Dec. 1952): 40.

18. Hans, "Women Chemists," p. 33.

19. Eloise Davison, "Home Economics Invades Business," *Independent Woman* 10 (Mar. 1931): 106–7, 127–28; Edith Barber, "Home Economics: A Widening Field," *Independent Woman* 11(Aug. 1932): 282–83; "Proceedings of the Conference on Women's Work," pp. 155–66; see also Institute for Women's Professional Relations, *Business Opportunities for the Home Economist, New Jobs in Consumer Service, A Study Made by the Institute for Women's Professional Relations under the Direction of Chase Going Woodhouse* (New York: McGraw-Hill Book Co., 1938). Kelly: *NYT*, 31 May 1958, p. 15; Husted: *CBY, 1949*, pp. 286–87; Katherine A. Fisher: *NYT*, 17 Mar. 1958, p. 29; Ruth Bien, "She's a Good Cook, Too!" *Good Housekeeping* 143(Aug. 1956): 4; Grace M. White: *NYT*, 22 Dec. 1965, p. 31; Frederick: *NAW:MP* and *NYT*, 8 Apr. 1970, p. 43.

20. Ruth Sheldon, "The Ladies Find Oil," *Scribner's Commentator* 10(May 1941): 28–32; Edgar Wesley Owen, *Trek of the Oil Finders: A History of Exploration for Petroleum*, Memoir 6(Tulsa, Okla.: American Association of Petroleum Geologists, 1975), pp. 522–27; J. Clarence Karcher, "The Reflection Seismograph: Its Invention and Use in the Discovery of Oil and Gas Fields," typescript at American Institute of Physics, New York City, cited in Spencer Weart, "The Physics Business in America, 1919–1940," in *The Sciences*, ed. Reingold, p. 352 n. 109. "Memorials: Esther Richards Applin (1895–1972)," *American Association of Petroleum Geologists Bulletin* 57(1973): 596–97; in the same bulletin find also "Memorial: Fanny Carter Edson (1887–1952)," 37 (1953): 1182–86; "Alva Christine Ellisor (1892–1964)," 49(1965). 467–71; "Dollie Radler Hall," 47(1963): 1484; "Louise Jordan (1908–1966)," 52(1968): 2050–60; "Helen Jeanne Plummer (1891–1951)," 38(1954): 1854–57. "Memorial to Virginia Harriet Kline (1910–1959)," *Proceedings of Geological Society of America, 1960*, pp. 115–17; Jean M. Berdan, "Memorial to Esther Richards Applin, 1895–1972," Geological Society of America, *Memorials* 4(1975): 14–18; E. R. Applin, "Memorial to Laura Lane Weinzierl," *Journal of Paleontology* 2(1928): 383; "Dorothy K. Palmer, 1897–1947," *Journal of Paleontology* 22(1948): 518–19. On occasion Florence Bascom's advice was sought about careers in geology for women. See Frank D. Adams to Florence Bascom, 8 Feb. 1921, and reply (undated fragment); Florence Bascom to Dr. N. M. Fenneman, 31 May 1923; and Florence Bascom to Rose J. Segal, 29 Jan. 1934, all in Florence Bascom Papers, Geology Department, Bryn Mawr College. Ida Ogilvie reportedly discussed careers for women in geology in a paper to the GSA in 1938 (*NYT*, 15 Oct. 1963, p. 39), but no copy of this has been found.

21. Richard U. Light, "Gladys Mary Wrigley," *Geographical Review* 40(1950): 4–6; in the same journal find Gladys Wrigley, "Adventures in Serendipity: Thirty Years of the 'Geographic Review,'" 42(1952): 511–42; Wilma B. Fairchild [Wrigley's successor in the 1950s], "Gladys Mary Wrigley, 1885–1975," 66(1976): 331–33. There are also a great many Gladys Wrigley letters in the Isaiah Bowman Papers, Special Collections, Eisenhower Library, The Johns Hopkins University, Baltimore.

The *Review*'s copyeditor for thirty-five years was Marion Eckert, a Barnard graduate, who was known to some for her "nearly infallible" work but to others for her blunt words and "astringent" (a word not usually found in an obituary) personality (Wilma B. Fairchild, "Marion Eckert, 1899–1969," *Geographical Review* 60[1970]: 120–22).

22. "Elizabeth Towar Platt: Librarian," 33(1943): 658–59; Ralph H. Brown, "Elizabeth Towar Platt," *Science* 98(1943): 210–11; Nordis Felland, "Ena Yonge, 1895–1971," *Geographical Review* 62(1972): 414–17. See also "American Geographical Society Library: Nordis Felland, Librarian," *Library Journal* 81 (1956): 1209.

23. Bregman: *AMS*, 6th ed. (1938), and "Making the Misfits Fit," *New York World*, magazine section, 16 July 1922, pp. 1 and 13, clipping and unpublished biographical sketch in Elsie O.

Bregman Papers, Archives of the History of American Psychology, University of Akron. (I thank Marion McPherson for bringing Bregman and her papers to my attention.) Bartlett: Karl Schwartzwalder, "Memorial of Helen Blair Barlett," *American Mineralogist* 56(1977): 669–70.

24. Georgia Leffingwell, "A Woman Engineers It," *Independent Woman* 9(1930): 271–72 and 304; "Olive Wetzel Dennis," *CBY, 1941*, pp. 220-21; Clarke: *NAW-MP*; "Edith Clarke," *Electrical Engineering* 79(Jan. 1960): 108; Edith Clarke, "Women in Electrical Engineering," *Vassar Alumnae Magazine* 45, no. 6, pt. 2 (June 1940): 11-14; Eaves: "Elsie Eaves Retires from ENR," *Engineering News-Record* 170(21 Mar. 1963): 195–96; "Mary Sink," *CBY, 1964*, pp. 417–19; see also Ethel H. Bailey, "Women as Engineers," *Independent Woman* 11(1932): 316–17 and 334; Elsie Eaves, "Wanted: Women Engineers," *Independent Woman* 21(1942); 132–33 and 158–59; and Alice C. Goff, *Women CAN Be Engineers* (Youngstown Ohio: private, 1946).

25. Isabel Loughlin, "Woman Plods 1000 Miles a Season, Inspects 1,000,000 Potato Plants," *Boston Sunday Globe,* 22 July 1923, clipping in Elizabeth Lawrence Clarke Papers, SLRC.

26. "Quimby," *CBY, 1949,* pp. 492-93, and Edna Yost, *Women of Modern Science* (New York: Dodd, Mead, 1959), pp. 94–107.

27. Hoke: her 1920 questionnaire in the BVI Papers, box 23, SLRC, has many comments; "Women in Science," *Women's Work and Education,* p. 1; Hoke also spoke autobiographically at the session on "General Science," in the "Proceedings of the Conference on Women's Work," pp. 137–41. Florence D. Wall, "Calm Morrison Hoke, 1887–1952," in *American Chemists and Chemical Engineers,* ed. Wyndham D. Miles (Washington, D.C.: American Chemical Society, 1976), pp. 221-22. Pennington: *NAW:MP; American Chemists*, ed. Miles, pp. 386-87; and Anne Hartwell, "Mary E. Pennington, Who Keeps Cold Storage Cold," *Woman's Journal* 15(Nov. 1930): 11 and 42–43.

28. Cynthia Westcott, *Plant Doctoring Is Fun* (Princeton, N.J.: D. Van Nostrand, 1957), pp. 39–69; see also *Phytopathology* 64(1974): 8; Gilbreth: *NAW:MP;* "Dr. Lillian Gilbreth, 1878–1972," *Industrial Engineer* 4(Feb. 1972): 28–33; "Lillian Moeller Gilbreth, 1878–1972," *Journal of Home Economics* 64(Apr. 1972): 13-16; "Lillian Gilbreth," *CBY, 1951,* pp. 233-35; Frank B. Gilbreth, Jr., and Ernestine Gilbreth Gray, *Belles on Their Toes* (New York: Thomas Y. Crowell, 1950), passim, and Frank B. Gilbreth, Jr., *Time Out for Happiness: Lillian Gilbreth* (New York: Thomas Y. Crowell, 1970).

29. Cornelia J. Cannon, *The History of the Women's Educational and Industrial Union: A Civic Laboratory, 1877–1927* (Boston, Mass.: private, 1927), p. 16; [WEIU], *Vocations for the Trained Woman* (New York: Longmans, Green, 1910); *Industrial Chemistry as a Vocation for Women*, WEIU Bulletin, Vocational Series no. 8 (1911), and *Bacteriological Work as a Vocation for Women,* WEIU Bulletin, Vocational Series no. 9 (1911). The papers of the WEIU are at SLRC. WEIU also sponsored a later work, Elizabeth Kemper Adams, *Women Professional Workers: A Study Made for the Women's Educational and Industrial Union* (1921; reprint ed., New York: Macmillan, 1930)—chap. 17 is on the sciences.

30. "Introducing the News-Bulletin," *News-Bulletin of the Bureau of Vocational Information* 1, no. 1 (1 Oct. 1922): 1 and 4.

31. Catherine Filene: interviewed in *NYT,* 26 Sept. 1971, p. 80; "Lincoln Filene," *NCAB* 45(1962): 421–22; Catherine Filene, ed., *Careers for Women* (Boston: Houghton Mifflin, 1920); an unnamed reviewer (in *Woman Citizen* 5[12 Feb. 1921]: 988) commented on Filene's title that women formerly had "occupations" but now had "careers." Filene prepared a second edition in 1933, but times had so changed that the later entries are much more somber and the coverage is somewhat changed. See, for example, Ruth Benedict's refusal to write on anthropology, quoted in note 19 to chapter 6 herein. Catherine Filene, *Careers for Women: New Ideas, New Methods, New Opportunities—to Fit a New World,* rev. ed. (Boston: Houghton, Mifflin, 1934).

32. Filene, *Careers for Women,* p. 432. Maltby: see chapter 1. The Girl Scouts also put out a book on careers in the 1920s: Helen Ferris and Virginia Moore, *Girls Who Did: Stories of Real Girls and Their Careers* (New York: E. P. Dutton, 1926), which included an interview with Margaret Maltby, physicist, pp. 213–26. Doris E. Fleischman (Bernays), comp. and ed., *An Outline of Careers for Women: A Practical Guide to Achievement* (Garden City, N.Y.: Doubleday, Doran,

1931), had one chapter on all of "science," but, curiously, one each on civil engineering and industrial engineering.

33. Isabel Turlington, "Careers for College Girls," *Woman's Journal* 14(Mar. 1929): 15, 43-44; "Institute of Women's Professional Relations," *Journal of Home Economics* 21(1929): 122-23; *Women's Work and Education*, passim, especially 9, no. 1(Feb. 1938): 6, and 9, no. 3(Oct. 1938): 6. Florence Wall, "Status of Women Chemists," pp. 190-92, was eager to cooperate with the institute on a major study of women chemists, but the study apparently never appeared.

As World War II approached, the institute, which had moved to Katharine Blunt's Connecticut College in New London, Connecticut, in the 1930s, could hardly keep up with all the news of defense-related positions that were opening to women. As the war progressed and women's opportunities were at their greatest, the indefatigable director was almost too busy to edit the newsletter. In fact Mrs. Woodhouse not only became a consultant to the federal government's new National Roster of Scientific and Specialized Personnel, but in the early 1940s she also entered Connecticut state politics, holding several city and state offices. In 1944 she was elected to the U.S. House of Representatives as a Republican and served two terms (1945-1947 and 1949-1951). By then the institute and its newsletter, so helpful in their day, had ceased to exist, and the problem of careers for women had entered an entirely new phase. The institute's papers are at SLRC. Some grant proposals of 1934 and 1935 are at the Rockefeller Foundation Archives Center, Tarrytown, New York. "Woodhouse" in *Biographical Directory of the American Congress, 1774-1971* (Washington, D.C.: GPO, 1971), p. 1954; "Mrs. Chase Going Woodhouse," *Omicron Nu* 28, no. 3(Spring 1953): 13-14; and Hope Chamberlain, *A Minority of Members: Women in the U.S. Congress* (New York: Praeger, 1973), pp. 188-91.

34. Dorothy Thomas, "Exploring the Museum Field," *Independent Woman* 12(July 1933): 238-39 and 260; "City's Aquarist, Retired, Added to Fish Lore," *New York Herald Tribune*, 5 May 1929, and Frances Hershfield, "Lady Fish Doctor," *Independent Woman* 9(1930): 103 and 142, are on Ida Mellen, a librarian and expert on fish diseases at the New York Aquarium; see also discussion of museum jobs in "Proceedings of the Conference on Women's Work," pp. 141-44, and of scientific art, pp. 271-77.

CHAPTER 10

1. A generation of research on the sociology of recognition is summarized in M. J. Mulkay, "Sociology of the Scientific Research Community," in *Science, Technology and Society: A Cross-Disciplinary Perspective*, ed. Ina Spiegel-Rosing and Derek J. De Solla Price (Beverly Hills, Calif.: Sage Publications, 1977), pp. 102-3.

2. Margaret A. Firth, ed., *Handbook of Scientific and Technical Awards in the United States and Canada, 1900-1952* (New York: Special Libraries Association, 1956).

3. For example, the many professional honors that came to Annie Jump Cannon after 1920 were the result of the initiative taken by her director, Harlow Shapley (and by Henry Norris Russell), to get her the recognition outside Harvard that was denied her at the observatory. Thus her four American and two foreign (Groningen and Oxford) honorary degrees, Draper Medal of the National Academy of Science, and several women's awards were devised to create such an embarrassing status inconsistency that Harvard would have to recognize her in some way. In 1938, when she was seventy-five years of age, President James Bryant Conant acceded and made her the William Cranch Bond Astronomer at the observatory, a nonfaculty but coveted "corporation" appointment (*NAW*). She was never elected to the National Academy of Science, however. See also later in this chapter.

4. Margaret E. Maltby, comp., *History of the Fellowships Awarded by the American Association of University Women, 1888-1929, with the Vitas of the Fellows* (Washington, D.C.: AAUW, [1929]); Ruth W. Tryon, *Investment in Creative Scholarship: A History of the Fellowship Program of the American Association of University Women, 1890-1956* (Washington, D.C.: AAUW, 1957); updated in AAUW, *Idealism at Work: Eighty Years of AAUW Fellowships* (Washington, D.C.: AAUW, 1967); annual reports of the activities of AAUW fellows appear in the *JAAUW;* Ruth W.

Tryon, comp., *Names Remembered through AAUW Fellowships* (Washington, D.C.: AAUW, 1958); Katherine J. Gallagher ("Thoughts on the Applications for the AAUW Fellowships," *Goucher Alumnae Quarterly* 17 [May 1939]: 18-19) was so impressed with the number and quality of the applicants in the sciences that she feared they might win all the fellowships.

5. See Ronald Tobey, *The American Ideology of National Science, 1919-1930* (Pittsburgh: University of Pittsburgh Press, 1971); Stanley Coben, "Foundation Officials and Fellowships: Innovation in the Patronage of Science," *Minerva* 14(1976): 225-40; *BAAUP*, passim; salary studies by Jessica Peixotto, *Getting and Spending at the Professional Standard of Living* [University of California faculty] (New York: Macmillan, 1927); and Yandell Henderson and Maurice R. Davie, eds., *Incomes and Living Costs of a University Faculty* [Yale] (New Haven: Yale University Press, 1928); "Development of Young Biologists," in Marine Biological Laboratory, Woods Hole File (1927?), Rockefeller Foundation Archive Center; see also chapter 7 herein.

6. *National Research Council Fellowships, 1919-1938* (Washington, D.C., 1938). See also Nathan Reingold, "The Case of the Disappearing Laboratory," *American Quarterly* 29(1977): 79-100, for a detailed account of the founding of the NRC fellowships. Myron J. Rand, "The National Research Fellowships," *Scientific Monthly* 73(Aug. 1951): 71-80, has data on the number of applicants for fellowships (not broken down by sex), as does the *Consolidated Report upon the Activities of the National Research Council, 1919 to 1932* (Washington, D.C., 1932), table 2, p. 20.

7. "Report of the Medical Fellowship Board of the National Research Council, 1922-1926," mimeographed, National Research Council File, Rockefeller Foundation Archives Center, pp. 7 and 9 (copy in Joseph Erlanger Papers, Washington University Medical Center Archives, St. Louis, Missouri).

8. Frank Blair Hanson, diary entry of interview with Professor Mary J. Guthrie, 16 Mar. 1934, University of Missouri File, Rockefeller Foundation Archives. Guthrie never got an NRC fellowship.

9. Biographies of these and later fellows are available in Neva E. Reynolds, comp., *National Research Council Fellowships, 1919-1944* (Washington, D.C., 1944). Payne-Gaposchkin did become chairman of Harvard's astronomy department in the 1950s. Robert Millikan apparently did not stop McClintock from coming to Cal Tech to do genetics in 1931, as he had prevented Fellow Hertha Sponer from coming to do physics there in 1925.

10. Raymond B. Fosdick, *The Story of the Rockefeller Foundation* (New York: Harper & Bros., 1952), chap. 21; Elizabeth Lomax, "The Laura Spelman Rockefeller Memorial: Some of Its Contributions to Early Research in Child Development," *Journal of the History of the Behavioral Sciences* 13(1977): 283-93; *Fellows of the Social Science Research Council, 1925-1951* (New York: SSRC, 1951); Elbridge Sibley, *Social Science Research Council: The First Fifty Years* (New York: SSRC, 1974), chap. 5.

11. Hortense Powdermaker, *Stranger and Friend: The Way of an Anthropologist* (New York: W. W. Norton, 1966), pp. 124 and 133, and *NAW:MP;* Margaret Mead, *Blackberry Winter: My Earlier Years* (New York: Simon and Schuster, 1972), p. 129; idem, "My First Job," *Ladies' Home Journal* 74(Apr. 1957): 199. Rosalind Rosenberg reports, however, that Boas was "appalled" that Mead wanted to go to Polynesia and only relented when she settled upon Samoa, which had a U.S. Navy base (Rosalind Rosenberg, *Beyond Separate Spheres: Intellectual Roots of Modern Feminism* [New Haven: Yale University Press, 1982], chap. 8). Betty Furness also reported that Mead had once told her that she had to get married in order to be allowed to go into the field ("*Ladies' Home Journal* Woman of the Year Award Program," CBS Television Network, 13 Jan. 1980 [KPIX-TV, San Francisco]).

12. [Henry Moe], "Reports of the Secretary, A Review, 1925-1936," in John Simon Guggenheim Memorial Foundation *Annual Report for 1936*, pp. 14, 15, and 18. Other *Annual Reports*, passim. "Hurston" in *NAW:MP*.

13. The Frank R. Lillie Papers in Special Collections, University of Chicago Library, and Ross G. Harrison Papers in Yale University Archives both contain many enthusiastic letters of recommendation for women candidates for such fellowships. The Lafayette B. Mendel Papers have only recently been found and are not yet ready for use at the Yale University Archives. Florence B. Seibert, *Pebbles of the Hill of a Scientist* (St. Petersburg, Fla.: private, 1968), 53-54; Florence

Seibert to Elinor Bluemel, 19 Sept. 1955, copy in Florence Seibert Papers, APS; John Simon Guggenheim Memorial Foundation, *Annual Reports,* passim.

14. Louis B. Wright, "An Education in Twenty-five Years," manuscript [1970?], Henry Moe Papers, American Philosophical Society, Philadelphia.

15. Lagging far behind most areas of science in the 1920s and 1930s was engineering, whose many professional societies were still at the stage of electing their first women members. Part of the women's difficulty arose from the rather strict and formal set of professional requirements for admission to these societies. One had to have not only an engineering degree but also several years of suitable job experience. Since the women were still admitted to few schools of engineering (Pennsylvania State University, MIT, University of Tennessee, University of Colorado, and Newark School of Engineering) and had great difficulty finding professional employment in the 1920s and 1930s, few were eligible for even the lowest form of membership, the "junior member." Hardly any women achieved the higher levels, such as "associate member" or "member." Those women who were admitted usually had some form of family employment and influence, such as husband-wife or father-daughter teams. Lillian Gilbreth was elected an "honorary member" of the American Institute of Industrial Engineering in 1921 (as a personal favor to her husband, engineer Frank Gilbreth, she thought) and only much later to the American Institute of Mechanical Engineers when, as her son put it, "the opposition to her membership had all but disappeared" ("A Tip-Top Engineer," *Women's Work and Education* 2 [Apr. 1931]: 2; *NAW:MP; Civil Engineer* 41[Apr. 1971]: 87; Frank B. Gilbreth, Jr., *Time Out for Happiness: Lillian Gilbreth* [New York: Thomas Y. Crowell, 1970], p. 201). Elsie Eaves of the *Engineering News-Record* was the first woman elected an "associate member" of the American Society of Civil Engineers in 1927 (eighteen years after Nora Barney had lost her suit for membership) ("Tip-Top Engineer," p. 2), and Kate Gleason, the first woman member of the American Institute of Mechanical Engineers, was elected in the mid-1920s (*NAW*, and Eve Chappell, "Kate Gleason's Careers," *Woman Citizen* 10[Jan. 1926]: 19–20; *Transactions of ASME* 56[1934]: RI-19). As late as 1942 there were still only three women in the American Institute of Electrical Engineers among thousands of men. Other groups, like those of chemical, mining, safety, and illuminating engineers, apparently had no women members until after World War II. Thus even though women were beginning to find some employment in engineering in the 1920s and 1930s, as Alice C. Goff's *Women CAN Be Engineers* (Youngstown, Ohio: private, 1946) shows, few were able to obtain even the lowest form of professional recognition—admission to the society. *The Outlook for Women in Architecture and Engineering*, WB *Bulletin* no. 223-5 (1947), has a table of the "Minimum Requirements for Membership in the Principal Engineering Organizations" (p. 5.75).

16. *Proceedings of the American Association for the Advancement of Science, 1900,* pp. xxxvii–xciii.

17. Herman L. Fairchild, *The Geological Society of America, 1888–1930* (New York: GSA, 1932), pp. 109–12 and 239 (By-laws, article 2, section 9), Kim Van Dusen (of GSA) to author, 3 Aug. 1976, enclosing an unpublished list of "Women Fellows of the Geological Society of America." Paleontologist Winifred Goldring was elected a fellow in 1921 but was still hurt years later at rumors that Dr. John Clarke had opposed her election because she was a woman. Her confidant, Charles Schuchert, assured her Clarke had not been on the council then (Winifred Goldring to Charles Schuchert, 12 Aug. 1925, and 16 July 1934, and his reply, 19 July 1934, all in Charles Schuchert Papers, Yale University Archives).

18. T. S. Palmer, "A Brief History of the American Ornithologists' Union," in *Fifty Years' Progress of American Ornithology, 1883–1933,* ed. Frank M. Chapman and T. S. Palmer (Lancaster, Pa., 1933), p. 10; see also Paul H. Oehser, "In Memoriam: Florence Merriam Bailey," *Auk* 69(1952): 19–26, and *NAW.*

19. Althea R. Sherman to Margaret Morse Nice, 25 June and 7 Apr. 1925, Margaret Morse Nice Papers, Cornell University Archives and Regional History Office; see also Margaret Morse Nice, "Some Letters of Althea Sherman," *Iowa Bird Life* 22(1952): 51–55 (I thank Hamilton Cravens for a copy); and Mrs. H. J. Taylor, "Iowa's Woman Ornithologist, Althea Rosina Sherman, 1853–1943," *Iowa Bird Life* 13(1943): 19–35.

20. Margaret Morse Nice, *Research Is a Passion with Me* (Toronto: Consolidated Amethyst

Communications, 1979), is an autobiography. See also *NAW:MP;* Milton B. Trautman, "In Memoriam: Margaret Morse Nice," *Auk* 94(1977): 431-41; and "The President's Page, Margaret Morse Nice, 1883-1974," *Wilson Bulletin* 86(1974): 301-2. See also R. M. Strong, "A History of the Wilson Ornithological Club," *Wilson Bulletin* 51(1939): 3-10, and Olin Sewall Pettingill, Jr., "The Wilson Ornithological Club of Today," *Wilson Bulletin* 51(1939): 11-16; Harold B. Wood, "The History of Bird Banding," *Auk* 62(1945): 256-65, and Robert Welker, *Birds and Men* (Cambridge: Harvard University Press, 1955), pp. 186-91 and 196-204.

21. *Journal of Home Economics* 38(1946): 525.

22. Quotation and membership list in *Journal of Nutrition* 7(1934): 3, and supplement, 1-8; Harold H. Williams, "The Founding of the American Institute of Nutrition, Including Commentaries on the Founders," *Federation* [of American Societies of Experimental Biology] *Proceedings* 36(1977): 1915-18; Henry C. Sherman, "Mary Swartz Rose," *Journal of Nutrition* 21(1941): 209-11; Grace MacLeod, "Mary Swartz Rose, 1874-1941," *Journal of Home Economics* 33(1941): 221-24; *NAW;* Lawrence Cremin et al., eds., *A History of Teachers College, Columbia University* (New York: Columbia University Press, 1954), pp. 56-57; Trulson: *Journal of the American Dietetic Association* 46(1965): 250.

23. Alice I. Bryan and Edwin G. Boring, "Women in American Psychology: Prolegomenon," *Psychological Bulletin* 41(1944): 449. See also Samuel W. Fernberger, "Academic Psychology as a Career for Women," *Psychological Bulletin* 36(1939): 390-94, which is based on APA data.

24. "Abstracts of Letters to Dunlap Concerning Fellows in APA," n.d., Lewis Terman Papers, Stanford University Archives. John M. O'Donnell, in "The Crisis of Experimentalism in the 1920s: E. G. Boring and His Uses of History," *American Psychologist* 34(1979): 289-95, suggests that another, more direct, reason for introducing the two tiers of membership was to mute the rising influence of the applied psychologists (many of whom were women). Donald S. Napoli, "The Architects of Adjustment: The Practice and Professionalization of American Psychology, 1920-1945" (Ph.D. diss., University of California, Davis, 1975), chaps. 3 and 4, discusses the cross-currents within the APA during the depression.

25. Edwin G. Boring, "The Society of Experimental Psychologists, 1904-1938," *American Journal of Psychology* 51(1938): 411.

26. Christine Ladd-Franklin to E. B. Titchener, n.d., and 21 Mar. 1914; Mary Whiton Calkins to Christine Ladd-Franklin, Aug. 14 (1913?) all in Christine Ladd-Franklin Papers, Special Collections, Columbia University Library; Christine Ladd-Franklin to James McKeen Cattell, 28 Apr. 1914, in J. M. Cattell Papers, Manuscript Division, Library of Congress. (Other relevant letters in the Ladd-Franklin correspondence are: Carolyn Fisher to CLF, n.d. [1914?]; John B. Watson to CLF, 14 and 18 Apr. 1916; and CLF to E. B. Titchener, n.d. [1914?].) Ladd-Franklin's tactics so alarmed Edwin G. Boring, a professor at Harvard, that he wrote a second, rather bizarre, article on the Experimentalists in the 1960s, in which he outlined (apparently in all seriousness) the men's fears of to what her presence would ultimately reduce them:

For many years Mrs. Ladd-Franklin, armed with her color-theory, invaded laboratories, took over the director's desk, had the women graduate students manicure her finger-nails, and insisted that everyone meet her arguments for her genetic theory of color. Directors escaped, cowering in their homes until they heard that the coast was clear again, but at Columbia she did get into one session of the Experimentalists and the story is that she was kept out of the meeting on another occasion by being thoughtfully locked into a different room along with Poffenberger, her captive audience for the nonce.

(Edwin G. Boring, "Titchener's Experimentalists," *Journal of History of the Behavioral Sciences* 3[1967]: 322.) Cora Friedline, another Titchener student, later reported that Lucy Day Boring, a Cornell Ph.D. and wife of "Garry," had managed to attend a meeting of the Experimentalists by sitting under a table the entire time (Cora Friedline, "Lectures before Class in History of Psychology at Randolph-Macon College, April 11, 1960," p. 10, and "Lectures . . . , May 3, 1960," p. 9 [both in Archives of the History of American Psychology, University of Akron, Akron, Ohio]). Ladd-Franklin later protested the exclusion of women, especially Edith Wharton, from the American

Academy of Arts and Letters in a letter to the editor, "Women and Letters," *NYT*, 13 Dec. 1921, p. 18. Her article for the *Nation*, under way in 1914, never appeared.

27. "Downey" and "Washburn" in *NAW;* Boring, "Society of Experimental Psychologists," p. 411; Margaret F. Washburn to Karl Dallenbach, 9 Mar. 1930, and reply, 12 Mar. 1930, in Karl Dallenbach Papers, Cornell University Archives and Regional History Office.

28. Boring also revealed that Harry L. Hollingworth, professor at Columbia and husband of Leta Stetter Hollingworth, refused to attend Experimentalists' meetings after a dispute with Titchener in 1910 ("Titchener's Experimentalists," pp. 322-23). Although the dispute was reportedly about revisions in a textbook, one suspects that a second reason for Hollingworth's withdrawal may have been Titchener's probable refusal to invite his wife, Leta Stetter Hollingworth, an experimentalist of note, whom he had married in 1908.

29. Dorothy Stimson, informal remarks, Fiftieth Anniversary Celebration of the History of Science Society, Norwalk, Connecticut, 25 Oct. 1974.

30. "Society of Economic Paleontologists and Mineralogists," *Journal of Paleontology* 1(1927-28): 3-8; "Helen Hart, 1900-1971," *Phytopathology* 61(1971): 1151; Federation of American Societies of Experimental Biology, *1975-76 Directory of Members* (Bethesda, Md.: FASEB, 1975), pp. 28-29; Donald Fisher, "Memorial to Winifred Goldring, 1888-1971," Geological Society of America *Memorials* 3(1974): 96-102; A. T. Poffenberger, "Leta Stetter Hollingworth, 1886-1939," *American Journal of Psychology* 53(1940): 301. Gertrude Hildreth was president of the New York-based Association of Consulting Psychologists (and five other women were officers) in 1936 and 1937, when the organization launched a journal and made other plans to become a national organization. Anxious to forestall this competition, the APA appointed a committee to suggest a merger; the committee finally established the American Association of Applied Psychology in 1937. Its chief difference from the ACP seems to have been its male leaders (*Journal of Consulting Psychology* 1[1937] and 2[1938], passim, especially the elliptical "Organization Reports, ACP to AAAP: Progress," *Journal of Consulting Psychology* 1[1937]: 93-94).

31. Bryan and Boring, "Women in American Psychology," p. 451; Lewis Terman thought that both Leta Stetter Hollingworth and Florence Goodenough should have been elected president of the APA. In a review of Harry Hollingworth's biography of his wife, Terman remarked generously: "Comparable productivity by a man would probably have been revealed by election to the presidency of the American Psychological Association or even to membership in the National Academy of Sciences. This opinion of the reviewer is primarily a reflection on the voting habits of male psychologists" (Lewis Terman, review of Harry L. Hollingworth's *Leta Stetter Hollingworth: A Biography* [1943] in *Journal of Applied Psychology* 28[1944]: 358). Terman wrote Goodenough, his former assistant, of his high regard directly: "I have been hoping very much that some day you would be honored with the presidency of the APA. In my opinion you deserve it more than some who have been elected" (Terman to Florence Goodenough, 22 Jan. 1945). Later he attributed her failure to be elected to her early retirement (Lewis Terman to Florence Goodenough, 13 Sept. 1949, both in Lewis Terman Papers, Stanford University Archives). Edwin G. Boring later theorized differently about the voting habits of male psychologists. He thought they would prefer to vote for women candidates for office in the APA if given the chance (Edwin G. Boring, "The Woman Problem," *American Psychologist* 6[1951]: 681).

32. Similarly I suspect, though I have not pursued the topic, that the women were more active and held more offices in regional and local societies and academies than they did on the national level. It is probably pertinent that these groups seem to have been almost invisible professionally, and thus the women's presence was less threatening than in the major national groups that "mattered" more.

33. Catherine Borrass (at AAAS) to author, 29 July 1976; Anita Newcomb McGee, "Women Officers of the Association for the Advancement of Science," *Science* 59(1924): 577.

34. "Women as Officers of the Association," *BAAUP* 20(1934): 500-501, plus individual issues thereafter.

35. Oscar M. Voorhees, *The History of Phi Beta Kappa* (New York: Crown Publishers, 1945), pp. 357-61; Nicolson obituary: *NYT*, 10 Mar. 1981, p. 14; John Wieler, "Interview with Professor

Marjorie Nicolson," 1975, Oral History Memoir, Columbia University, pp. 354–57 and 380–84.

36. Henry Baldwin Ward and Edward Ellery, comps., *Sigma Xi: Half-Century Record and History, 1886–1936* (Schenectady, N.Y.: Union College, 1937), pp. 8, 18, and 908–10; Raymond J. Seeger (historian of Sigma Xi) to author, 12 Apr. 1976; Rosa Smith Eigenmann or her descendants thought she was an early president of the association, but she was not (*NYT,* 14 Jan. 1947, p. 25).

37. Raymond Pearl to George Ellery Hale, 9 Mar. 1920, G. E. Hale Papers, California Institute of Technology Archives.

38. C. G. Abbott to G. E. Hale, 28 Feb. 1921, and reply, 7 Mar. 1921; telegram from G. E. Hale to C. G. Abbott, 12 Mar. 1921, and reply, 14 Mar. 1921, all in NAS Archives. (I thank Paul McClure, formerly Deputy Archivist, for locating these letters.) The admission of women to the National Academy is also discussed in my "Florence Sabin: Election to the NAS," *American Biology Teacher* 39(1977): 484–86 and 494.

39. Raymond Pearl to E. B. Wilson, 12 Mar. 1923, NAS Archives. Edwin Bidwell Wilson to Harlow Shapley, 31 October 1925 (Harvard College Observatory Papers, HUA), wondered why the astronomers had not nominated Miss Cannon. She was at least as well qualified as Florence Sabin, she already had numerous foreign honors, and several men in the academy had been elected for work based on her *Henry Draper Catalog*. Wilson thought she was the most qualified astronomer not yet elected. (I thank Michael Sokal for bringing this item to my attention.) The only discussion of the politics behind the National Academy's sections is A. Hunter Dupree's "The National Academy of Sciences and the American Definition of Science," in *The Organization of Knowledge in Modern America, 1860–1920*, ed. Alexandra Oleson and John Voss (Baltimore: The Johns Hopkins University Press, 1979), pp. 342–63.

40. Edwin G. Conklin to Ross G. Harrison, 30 Apr. 1925, Edwin G. Conklin Papers, Princeton University Library. There are also several relevant letters in the Ross G. Harrison Papers in the Yale University Archives. Although Sabin was a friend of his, he did not think in 1923 that she deserved membership in the academy. It was not his zoology and animal morphology section at the academy that nominated her, but that for physiology and pathology.

41. G.R.R. (?) to Florence Sabin, n.d., Florence Sabin Papers, APS.

42. S. S. Visher lends considerable support for this rather cynical view of election to the National Academy in the 1920s and 1930s in his *Scientists Starred, 1903–1943, in the "American Men of Science"* (Baltimore: The Johns Hopkins University Press, 1947), p. 4 n. He cited an unpublished survey of members of the academy who ranked merit fourth among the criteria needed for election. "Proper connections" came first, followed by at least two influential friends already in the academy, and thirdly an age of "about fifty." Visher thought stars in the *AMS* were "comparatively democratic and effective" by comparison.

43. *AMS,* 6th ed. (1938), and W. W. Campbell to Dr. Florence Bascom, 24 June 1935, Florence Bascom Papers, Sophia Smith Collection, Smith College (carbon in the NAS Archives). Again I thank Paul McClure for bringing this item to my attention. Similarly, some of Alice Fletcher's friends apparently thought she was a member of the National Academy of Sciences (although she never was) and listed her as a probable one in "[Report of the] Committee on Honorary Membership, Proceedings of the Thirty-Sixth Meeting of the ACA," *JACA* 14(1921): 232 n.

44. American Academy of Arts and Sciences, "Excerpts from Its Archives on The Academy and the Female" [1963]. (I wish to thank Alexandra Oleson for sending me a copy of this list.)

45. Jonathan Cole and Stephen Cole, *Social Stratification in Science* (Chicago: University of Chicago Press, 1973); Gertrude Rand Ferree, "Gold Medal Acceptance Address," *Illuminating Engineering* 58(Nov. 1963): 11A.

46. Selman Waksman, *My Life with the Microbes* (New York: Simon and Schuster, 1954), p. 195.

47. Stephen Visher, *Scientists Starred,* pp. 121, 125, 133, 134, and 140.

48. Carol Green Wilson, *Alice Eastwood's Wonderland: The Adventures of a Botanist* (San Francisco: California Academy of Sciences, 1955), and Susanna Bryant Dakin, *The Perennial Adventure: A Tribute to Alice Eastwood, 1859–1953* (San Francisco: California Academy of Sciences, 1954); *NAW:MP;* James McKeen Cattell, "A Further Statistical Study of American Men of

Science," *Science* 32(1910): 637, 639, and 676; "No Longer Starred in 3rd Edition (1921)," James McKeen Cattell Papers, Manuscript Division, Library of Congress, box 62.

49. Jaques [*sic*] Cattell, "American Men of Science, Scientific Men Receiving Stars in the Seventh Edition," *Science* 100(1944): 127; inequities: *Science* 90(1939): 252; *Science* 97(1943): 465–66, 487; *Science* 98(1943): 281–82, 473–74; *Science* 99(1944): 221–22, 386, and 533–34; *Science* 101(1945): 222–23, 272–73, and 639; *Science* 106(1947): 359–61 and 617; *AMS,* 8th ed. (1948), preface. Ironically, the purpose of Visher's counts and articles had been to examine the roots of American genius (rather than to expose the inadequacies of one of the main reward systems of the time.)

50. James McKeen Cattell to Christine Ladd-Franklin, 24 June 1922, Cattell Papers, Manuscript Division, Library of Congress; Stephen Sargent Visher, *Scientists Starred, 1903–1943, in "American Men of Science,"* pp. 113 and 148–49 (beware of numerous errors in addition); see also "Best Women," *Time* 21(20 Mar. 1933): 38, on the three women starred in the fifth edition, and Ruth Benedict's correspondence with *Time* magazine about being misquoted (Ruth Benedict to *Time*, 16 Mar. 1933, and I. Van Meter to Ruth Benedict, 27 Mar. 1933, promising a correction—though none ever appeared. Both in Ruth Benedict Papers, Vassar College Archives). Alice C. Evans, "Opportunities for Women in Science" (1941) manuscript in Alice C. Evans Papers, Cornell University Archives and Regional History Office.

51. A similar pattern had also occurred in the first two editions of the *American Men of Science* when eight of the twenty-five starred women had been married (32 percent). Five of these were married to scientists who were also starred: Ellen Richards (Mrs. Robert H.), Susanna Phelps Gage (Mrs. Simon), Elizabeth K. Britton (Mrs. Nathaniel), Christine Ladd-Franklin (Mrs. Fabian), and Margaret L. Nickerson (Mrs. Winfield). However, the two other married starred women had not eclipsed their husbands, since they were longtime widows—Mrs. Williamina P. Fleming and Mrs. Lydia DeWitt—who embarked on their successful careers after the deaths of their husbands.

In further regard to marriage and stars in the *AMS,* there is also evidence that marriage to a fellow zoologist had a significant effect on a man's career. Male zoologists are already well known for marrying a disproportionately large number of distinguished women, many of them fellow zoologists (Dean R. Brimhall, "Family Resemblances Among American Men of Science," *American Naturalist* 56 [1922]: 546–47, and *American Naturalist* 57 [1923]: 342–43). Moreover, of the twenty-six married women zoologists in the first three editions of the *AMS,* fifteen were married to men also in the AMS. Of these, fourteen were in zoology or medical sciences, and eight of them were "starred" or considered eminent by their colleagues, a far higher percentage (57.1 percent) than was true of zoologists in general (roughly 33.3 percent in the first edition of 1906, and 11.1 percent in the third of 1921). In other words, marrying a woman in his field had a great impact on a zoologist's chance of winning a star in the *AMS*. She would probably be unemployed and thus able to assist his researches.

52. Bessie Z. Jones and Lyle G. Boyd, *The Harvard College Observatory: The First Four Directorships, 1839–1919* (Cambridge: Harvard University Press, 1971), p. 479 n. 15.

53. Dick: *NAW:MP;* Harriet Zuckerman, *Scientific Elite: Nobel Laureates in the United States* (New York: Free Press, 1977), p. 300; and *NCAB* 51(1969): 107.

54. Cori: *NAW:MP*.

55. It is not clear why Alfred Nobel created the prizes in the fields that he did, although as an inventor and dynamite manufacturer, a self-taught chemist, and a hypochondriac who had earlier supported medical research, there are plausible reasons for at least two of the three science fields. His creation of the peace prize grew out of a long friendship with pacifist Bertha von Suttner. He never married. (Erik Bergengren, *Alfred Nobel: The Man and His Work,* trans. Alan Blair [London: Thomas Nelson and Sons, 1960], chap. 23, and Goran Liljestrand, "The Prize in Physiology or Medicine," in Nobel Foundation, *Nobel, The Man and His Prizes* [Amsterdam: Elsevier Publishing, 1962], pp. 135–38; Olga S. Opfell, *The Lady Laureates: Women Who Have Won the Nobel Prize* [Metuchen, N.J.: Scarecrow Press, 1978].)

56. Margaret Mead, *Male and Female: A Study of the Sexes in a Changing World* (New York: William Morrow, 1949), p. 159. See also her "Sex and Achievement," *Forum and Century* 94(1935): 301–3, and *Sex and Temperament in Three Primitive Societies* (New York: William

Morrow, 1935). In her autobiography, *Blackberry Winter*, pp. 216–22, she describes how she and two of her husbands came to these ideas while doing field work in New Guinea in 1933.

CHAPTER 11

1. For example, in 1934 pathologist Frieda Robscheit-Robbin insisted that prizes were unnecessary in science, since the satisfactions came from good work well done, whether or not one was ever rewarded for it by others. (Frieda Robscheit-Robbins interviewed and quoted in Mildred Bond, "Along the Promenade: No Desire for Honors in Dr. Robbins' Life of Long Research," *Rochester, New York, Democrat & Chronicle,* 18 Nov. 1934, clipping in George H. Whipple Scrapbooks, Historical Library, University of Rochester Medical School.) In 1944 Helen T. Parsons, professor of home economics at the University of Wisconsin, declined to serve on the Borden prize committee of the American Institute of Nutrition on the grounds that "I have served several times on such committees and it becomes increasingly a less congenial service, as I find myself out of sympathy with the principles of such awards. Perhaps I have seen less of the favorable results and have emphasized the jealousies, frustrations and embarassments which I know to have resulted in some cases" (Helen Tracy Parsons to Arthur H. Smith, 2 June 1944, copy in Icie Macy Hoobler Papers, Bentley Historical Library, University of Michigan).

2. J. Stanley Lemons, *The Woman Citizen: Social Feminism in the 1920s* (Urbana: University of Illinois Press, 1973), especially chap. 2.

3. Elizabeth Kemper Adams, *Women Professional Workers: A Study Made for the Women's Educational and Industrial Union* (1921; reprint ed., New York: Macmillan, 1930), p. 29. Ellen Anderson, *Guide to Women's Organizations: A Handbook about National and International Groups* (Washington, D.C.: Public Affairs Press, 1949–50), identifies about one thousand such associations.

4. John Robson, ed., *Baird's Manual of American College Fraternities* (Menasha, Wis.: Baird's Manual Foundation, 1977), contains a brief history of fraternities in the United States (pp. 1–43). Their importance for the history of science education remains to be explored.

5. Phi Upsilon Omicron: Robson, ed., *Baird's Manual,* pp. 580–81.

6. Omicron Nu: ibid., p. 638. Issues of *Omicron Nu,* its journal, are difficult to locate, but a few more recent copies are in the Icie Macy Hoobler Papers, Bentley Historical Library, University of Michigan.

7. Alpha Nu: Agnes Fay Morgan, *Directory and History of the Alpha Nu Honor Society* (Berkeley, Calif.: private, 1927), and unpublished supplement to 1940, both in Icie Macy Hoobler Papers. See also Agnes Fay Morgan, "The History of Nutrition and Home Economics in the University of California, Berkeley, 1914–1962," mimeographed [1962], chap. 2, on student organizations, 1915–1962; and idem, "A Household Science Honor Society," notice sent to *Journal of Home Economics* in 1920 but never published (in Agnes Fay Morgan Papers, Bancroft Library, University of California, Berkeley).

A fourth home economics fraternity was Kappa Omicron Phi, which was founded at Northwest Missouri State Teachers College in 1922. It had nineteen chapters by 1940, including one at Santa Barbara State (Teachers) College (now the University of California at Santa Barbara), which has a scrapbook of photos, clippings, and invitations of the fraternity's activities from 1928 to 1940 in the Special Collections department of the library. Since 1940 Kappa Omicron Phi has grown more rapidly than any of the other home economics fraternities and has many chapters at black as well as white colleges. (Robson, ed., *Baird's Manual,* pp. 625–26.)

8. Agnes Fay Morgan, ed., *A History of Iota Sigma Pi* (Berkeley, Calif.: private, 1963), p. 70. (I wish to thank past president Hoylande Young for loaning this to me.) See also "Iota Sigma Pi: 50 Years of Service," *Chemical and Engineering News* 30(1952): 4644–45.

9. Morgan, ed., *History of Iota Sigma Pi*, pp. 19 and 22.

10. Mary Louise Robbins, ed., *A History of Sigma Delta Epsilon, 1921–1971: Graduate Women in Science* (n.p., 1971), p. 4. (I thank Professor Robbins for sending me this history.) Alvan E. Duerr, ed., *Baird's Manual* [*of*] *American College Fraternities*, 14th ed. (Menasha, Wis.: George Banta Publishing, 1940), pp. 684–85. Charlotte Williams Conable, *Women at Cornell: The Myth of*

Equal Education (Ithaca: Cornell University Press, 1977) does not mention SDE, but does describe the women graduate students' isolation.

11. Conable, *Women at Cornell,* passim. Member Cynthia Westcott has described the place of Sigma Delta Epsilon in her young professional life in her autobiography, *Plant Doctoring Is Fun* (Princeton, N.J.: D. Van Nostrand, 1957), pp. 21 and 234. The extensive administrative correspondence of Sigma Delta Epsilon is located at the Cornell Regional History and Archives Office, Cornell University, Ithaca, New York.

12. *Kappa Mu Sigma, National Directory and Constitution, 1920–27,* pamphlet in Florence Seibert Papers, APS.

13. Louise A. Giblin to Miss Woodford, 19 Feb. 1927, and reply, 25 Feb. 1927, and Louise A. Giblin to Dr. A. W. Thomas, 10 Mar. 1927, all in Charles H. Herty Papers, box 20, folder 4, Special Collections, Woodruff Library, Emory University. Woodford was on the section's executive committee and claimed to have discussed the exclusion several times with members, but to no avail.

14. *Kappa Mu Sigma, National Directory and Constitution, 1920–27,* pamphlet in Florence Seibert Papers, APS.

15. "A New Association," *JACA* 14, no. 1 (Oct. 1920): p. 25.

16. Mary R. Lakeman, "Association of Women in Public Health," *Medical Woman's Journal* 31(1924): 212. Some early correspondence of the association is in the Martha Tracy Papers, Medical College of Pennsylvania Archives. (I thank Sandra Chaff for bringing these items to my attention.)

17. "A New Association," *JACA;* Mary R. Lakeman, "Annual Meeting of Association of Women in Public Health," *Medical Woman's Journal* 42 (1935): 311–12; Meta R. Pennock Newman, "Twenty-First Annual Meeting of the Association of Women in Public Health," *Medical Woman's Journal* 50(1943): 8 and 12; Anderson, *Guide to Women's Organizations,* p. 20.

Harvard's School of Public Health, established in 1921, refused to grant women the D.P.H. degree until 1936. When the president of the Massachusetts League of Voters heard in 1921 that the Rockefeller Foundation was about to underwrite the new Harvard program, she wrote the foundation's president, George Vincent, to urge him to require that women be accepted on the same basis of men. Vincent refused on the grounds that he did not wish to interfere with Harvard's autonomy. As earlier (see chapter 10), with the NRC fellowships, the Rockefeller Foundation did not wish to engage in any "coercive philanthropy." (Jean Alonzo Curran, *Founders of the Harvard School of Public Health,* with Biographical Notes [New York: Josiah Macy Jr. Foundation, 1970], pp. 36 and 131, cited in Barbara E. Brand, "The Influence of Higher Education on Sex-Typing in Three Professions, 1870–1920: Librarianship, Social Work, and Public Health" [Ph.D. diss., University of Washington, 1978], pp. 369 and 371.)

18. Mary A. Nourse, "History of the Society," *Bulletin of the Society of Women Geographers, 25th Anniversary Number, December 1950;* p. 5. (I wish to thank Helen Loerke of the SWG for sending me this.) See also "Globetrotting Artists," *Independent Woman* 20(Mar. 1941): 67–68; Ann Hark, "Ladies of the Roving Foot," *Christian Science Monitor Magazine,* 31 Dec. 1937, p. 6; Blair Niles, "Over the World and Back," *Woman Citizen* 11(Dec. 1926): 25–26 and 43; and Gertrude E. Dole, *Vignettes of Some Early Members of the Society of Woman Geographers in New York* (New York: New York Group of the Society of Woman Geographers, 1978). (I thank Betty Guyot for a copy of this booklet.)

19. Marie K. Farnsworth, "Women in Chemistry: A Statistical Study," *Industrial and Engineering Chemistry* 3(10 Sept. 1925): 4.

20. Lois Woodford to "Lady Rose" [Glenola B. Rose], 20 July 1928, in the Charles Holmes Herty Papers, box 21, folder 8, Special Collections, Robert W. Woodruff Library, Emory University. There is much of Woodford's correspondence about the committee in this collection. I thank Jeffrey Sturchio for telling me of it and Diane Windham, associate archivist, for sending copies. The only biographical information I have found about Rose was her response to a 1920 "Questionnaire for Chemists" in box 21 of the Bureau of Vocational Information Papers, SLRC. Lois Woodford was a graduate of Mount Holyoke College (1913). (I thank college historian Elaine Trehub for sending an obituary and excerpts from the class notes in the *Mount Holyoke Alumnae Quarterly*.)

21. Society of Women Geographers: Helen Fitzgerald to Florence Bascom, 19 Feb., and her reply, 6 Mar. 1930, both in Florence Bascom Papers, Sophia Smith Collection, Smith College

Library. Jessie Ash Arndt, "Women Here for Geology Parley Feted," *Washington Post,* 29 Dec. 1937; "Eminent Woman Geologist Honored at Luncheon Given in Washington," *New York Sun,* 29 Dec. 1937; "Geological Group Pays Florence Bascom Honor" and "Veteran Woman Geologist Feted," unidentified clippings, all in Florence Bascom Papers. Alice S. Allen to Isabel F. Smith, 23 May 1963, also in the Bascom Papers, recalls how awkward the whole affair had been. An undated typescript "Research" may have been Bascom's notes for her brief talk at the luncheon (also in Bascom Papers).

22. The association had considerable difficulty over the years with its prize. It is hard to know whether this was just its own indecision or something inherent in giving a women's prize. Later groups seem not to have had the same trouble finding a winner. The association's procedure was to send all submissions to an (all-male) panel of judges, which included chemist T. W. Richards of Harvard and physiologist William H. Howell of The Johns Hopkins Medical School. Frequently the panel's appraisals were harsh, and the association's members did not know how to evaluate such criticism. (Perhaps they were more used to the enthusiastic letters of recommendation that candidates for the Naples Table award were able to muster.) Florence Sabin, an active member of the association, interpreted one such negative evaluation to another member in 1925 as "He only meant that he would have done it differently himself" (Florence Sabin to Florence Cushing, 27 Mar. 1925, Florence Sabin Papers, APS). Nevertheless, under these circumstances the association often voted not to give the prize to any of its candidates.

Some years, however, after the members had voted to deny the prize to all the candidates, they then agreed to give the money to one as a "grant-in-aid." Some recipients were confused and insulted at being denied the prize if they were worth the money (Mrs. Samuel F. Clarke to Miss Ann C. Davies, 25 Apr. and 30 May 1922, Ida Hyde Papers, AAUW Headquarters). Others were amused at such indecision but glad to get the money after all. (Frances Wick to Percy Bridgman, 4 Mar. 1923, Percy Bridgman Papers, HUA. I thank Clark Elliott for bringing these letters to my attention.) Hindsight and the group's own records reveal that at least two candidates may not have been as unworthy as the judges apparently thought. Alice Evans of the U.S. Public Health Service submitted her important work on brucellosis three times around 1918 to 1921 to no avail (it was controversial but later was vindicated and won her the presidency of the Society of American Bacteriologists), and Helen Dean King lost twice before winning one-half of the association's final prize in 1932.

23. File on "Association to Aid Scientific Research by Women," 1921-22, Mary Thaw Thompson Papers, Vassar College Library; "Data Collected from Former Prize Committee Reports and Records" [1925], Ida Hyde Papers, AAUW Headquarters.

24. Florence Sabin to Stella M. Hague, Sigma Delta Epsilon, 29 Oct. 1929, Florence Sabin Papers, APS. See also Florence Sabin, "Women in Science," *Science* 83(1936): 24-26, which lavishes praise on Bryn Mawr College for its high standards (in contrast with William Welch's feminist address there in 1922 [see note 59 to chapter 5]) but which claims that the (only?) women scientists of all time who had done work equal to that by men were the three Europeans Madame Curie, the obscure Agnes Pockels, and Emmy Noether.

25. H. Jean Crawford, "The Association to Aid Research by Women," *Science* 76 (1932): 492-93. There is apparently no other history of the association, but annual reports appear in the *Journal of the Association of Collegiate Alumnae* (later AAUW), and much correspondence relating to both the Naples Table Association and the Ellen Richards Prize exists in the Florence Sabin Papers at the American Philosophical Society in Philadelphia and the Ida Hyde Papers at the American Association of University Women Archives at their national headquarters in Washington, D.C. The association's longtime secretary-treasurer, Elizabeth Lawrence Clarke (Smith 1883), of Williamstown, Massachusetts, left a small collection of papers to the Smith College Archives. See also E.N.H., "In Memoriam, Elizabeth Lawrence Clarke, 1883," *Smith Alumnae Quarterly* 42(Aug. 1951): 226.

The association had been damaged in 1930 when it refused to let Matilda Moldenhauer Brooks, a research associate at the University of California at Berkeley, use its Naples Table (which was vacant at the time) on the grounds that her previous work was inadequate. This had never before been a criterion, and, as several male scientists wrote members of the association, Brooks was far better than some others who had been chosen in earlier years when the group had been more positive and supportive of women scientists. A few months later the association reversed itself and sent Brooks the money (correspondence in Ida Hyde Papers). The unpleasantness of this whole affair may have been an unstated reason behind the group's desire to disband in 1932.

26. "American Astronomical Society, Annie J. Cannon Prize in Astronomy," in Margaret A. Firth, ed., *Handbook of Scientific and Technical Awards in the United States and Canada, 1900–1952* (New York: Special Libraries Association, 1956), pp. 25–26; "Rules for the Annie J. Cannon Prize in Astronomy," *Publications of American Astronomical Society* 8(1934–36): 317; "The Star-gazer," *Time* 99(20 Mar. 1972): 38; there is also some evidence that there may have been an informal women's group among astronomers in the 1930s (obituary of Helen Bigelow, *Northampton [Mass.] Gazette,* 7 Nov. 1934, in Faculty File, Smith College Archives).

27. Florence E. Wall, "The Status of Women Chemists," *Chemist* 15(1938): 191; Glenola Behling Rose to Icie Macy, n.d. [Apr. 1935], and reply, 30 Apr. 1935, and subsequent correspondence with Rose and Emma Perry Carr, all in Icie Macy Hoobler Papers, Bentley Historical Library, University of Michigan; there is also correspondence about the Garvan Medal, 1936 to 1938, in the Charles Holmes Herty Papers (box 21, folder 3, Special Collections, Robert W. Woodruff Library, Emory University), as he was on the selection committee.

See also "The 94th Meeting of the A.C.S.," *Chemical and Engineering News* 15(1937): 401, on the establishment of the prize, and "Francis P. Garvan, 1875–1937," *Chemical and Engineering News* 15(1937): 539–60, for biography of the founder. A list of winners of the Garvan Medal appears in Herman Skolnik and Kenneth M. Reese, eds., *A Century of Chemistry: The Role of Chemists and the American Chemical Society* (Washington, D.C.: American Chemical Society, 1976), p. 444, which doesn't even mention the Women's Service Committee. Winners are also announced annually in the *Chemical and Engineering News*.

28. Quoted in "Tentative Report," attached to Glenola Behling Rose memo to Dr. Mary E. Pennington et al., 3 Feb. 1927, Charles H. Herty Papers, box 20, folder 4.

29. For difficulties women long faced in getting on an ACS program, see Martha Morse, "The First Fifty Years [of A.C.S. in California]," *Vortex* 12(Dec. 1951): 469 and 482.

30. Virgilia Sapiena (Patterson), Ruth Neely, and Mary Love Collins, *Eminent Women: Recipients of the National Achievement Award* (n.p., George Banta Publishing, 1948); Hinkle in *Who's Who in America* 20(1938–39), p. 1219; Anne O'Hagan, "Beatrice Hinkle, Mind Explorer," *Woman Citizen* 12(July 1927): 13, 46–48.

31. AAUW, *Idealism at Work* (Washington, D.C.: AAUW, 1967), chap. 5, "The Achievement Award."

32. Margaret Ferguson to President Mary E. Woolley, 16 Feb. 1937, Margaret Ferguson Papers, Wellesley College Archives. For list of 1937 honorary degree recipients at Mount Holyoke College, see Jeannette Marks, *Life and Letters of Mary Emma Woolley* (Washington, D.C.: Public Affairs Press, 1955), p. 289 n. 1. Winifred Goldring to Charles Schuchert, 19 Apr. 1937, Charles Schuchert Papers, Yale University Archives.

33. Genevieve Parkhurst, "Dr. Sabin, Scientist, Winner of *Pictorial Review's* Achievement Award," *Pictorial Review* 31(Jan. 1930): 2 and 70–71; "Women Have Helped Her All through Life, Asserts Our Most Noted Woman Scientist," *New York Sun,* n.d., clipping in Florence Sabin Papers, APS; "Women Helped Her To Succeed, Asserts Noted Woman Scientist," unidentified clipping [29 Dec. 1929], Florence Sabin Papers, Sophia Smith Collection, Smith College; "A Feminine 'First,'" *NYT,* 19 Nov. 1929, p. 28.

34. Florence Seibert Papers, APS, boxes 4, 7, and 8 contain much correspondence about her numerous awards in the 1930s and 1940s.

35. Edward L. Bernays, *Biography of an Idea: Memoirs of a Public Relations Counsel* (New York: Simon and Schuster, 1965), chap. 26, pp. 196, 329, and 672; Edward L. Bernays, *Public Relations* (Norman: University of Oklahoma Press, 1952), passim. Keith A. Larson, *Public Relations, the Edward L. Bernayses, and the American Scene: A Bibliography*, Useful Reference Series no. 114 (Westwood, Mass.: F. W. Faxon, 1978), summarizes as well as lists the Bernayses' works. Daniel Boorstin has used the term *pseudo-event* to describe such media manipulation (*The Image: A Guide to Pseudo-events in America* (New York: Harper & Row, 1964).

36. A later generation thought such interaction essential to full participation in science (Martha S. White, "Psychological and Social Barriers to Women in Science," *Science* 170[1970]: 413–16).

BIBLIOGRAPHY

Preparing a bibliography at the end of a long project gives one some distance and perspective on the path just traveled. It is also a process of blatant distortion, since it puts into logical order those sources that one often found quite serendipitously. As recently as 1973 the "literature" on the history of American women scientists consisted of Nancy O. Lurie's pioneering essay on early women anthropologists (1966) and three biographies of major women (Caroline Hunt on Ellen Richards [1912], Helen Wright on Maria Mitchell [1949], and Elinor Bluemel on Florence Sabin [1959]). The only attempts at a unified history were a sketchy chapter in Mozans (1913) and four mimeographed essays around 1940 by members of Sigma Delta Epsilon (Davey, Haber, Owens, and Willard). Even these were mostly on premodern times and merely listed American women at the end.

Under such conditions I necessarily started this project with biographical dictionaries, other reference works, and long runs of certain journals, then moved on to manuscript collections and oral histories, and took advantage of the flood of secondary sources on women's history that began to appear after 1973. Altogether these tools and sources provided the historical equivalent of a higher magnification on a microscope: that which had seemed barely a speck or a trace gave way to a whole new realm teeming with activity and individuality.

I. REFERENCE WORKS

Because reference works are often unsung and are taken for granted—footnotes usually refer only to the final source once it has been located—it is necessary to recall their usefulness. The chief biographical dictionaries used were the first six editions of the *American Men of Science* (1906–38), not because they are "complete," but because it would have been close to impossible to find elsewhere such full information on almost 2,400 women scientists before 1940. Also indispensable were the *Notable American Women* (3 vols., 1971) and its recent supplement, *Notable American Women: The Modern Period* (1980), which have full scholarly biographical essays on over one hundred women scientists as well as the supporting cast of college presidents, government officials, and political activists. Supplementing these were the *National Cyclopedia of American Biography,* Margaret Maltby's directory of AAUW Fellows (1929), and Paul H. Oehser, ed., *Biographies of Members of the American Ornithologists' Union by T. S. Palmer and Others* (1954). Invaluable was the *Biography Index* (10 vols., 1946–76), which is arranged alphabetically but is indexed by occupation and thus leads one quickly to the obituaries (and other biographical material) on scientists in a host of professional journals. One of its sources is *Current Biography,* 1940– (and its *Current Biography*

Yearbook), which have full essays on persons in the news. For the years before the *Biography Index,* it was necessary to fall back on a government mimeograph that is well indexed: U.S. Library of Congress, Division of Bibliography, Florence Hellman, comp., "List of References Relating to Notable American Women," 1931, with supplements in 1937 and 1941.

There are now two bibliographies specifically on women scientists: Audrey B. Davis (1973), which is based on holdings of the Library of Congress and the National Library of Medicine, and Michele Aldrich's interpretive essay in *Signs* in 1978. Supplementing these and introducing one to the wealth of material on vocational guidance and scientific careers are two post–World War II bibliographies. One appears in each of the eight portions of Women's Bureau Bulletin no. 223, *The Outlook for Women in Science* (1947). The other is U.S. Library of Congress, Navy Research Section, Mabel H. Eller and Jack Weiner, comps., Barton Bledsoe, ed., "Scientific Personnel: A Bibliography," mimeographed, 1950, which indexes articles in the *Journal of Chemical Education, Journal of Engineering Education, Education Index, Psychological Abstracts,* and others for the years 1930 to 1948. For similar material in earlier years, the best introduction is the *Industrial Arts Index* (1913–37), continued as the *Applied Science Index,* 1938– , and the *Reader's Guide to Periodical Literature*, 1890–.

In addition, Mark Beach's *A Bibliographic Guide to American Colleges and Universities From Colonial Times to the Present* (1975) is helpful for the history of women's colleges and other institutions. Joseph Kiger's compendium *American Learned Societies* (1963) is such an invaluable introduction to an erratic literature that it ought to be updated to 1980, as should Margaret Firth's unique *Handbook of Scientific and Technical Awards* (1956). Anyone who has tried to find a list of past presidents of a learned society, for example, will understand how helpful these works have been.

Lastly, one wonders what specialized multi-library research must have been like before the Library of Congress's immense and now nearly complete in more than 700 massive volumes *Union Catalog of Pre-1956 Imprints* (1968–) (with supplements for later works) consolidated bibliographical entries so thoroughly that it is now quite easy to verify references to obscure books and to locate copies across the nation.

II. BASIC JOURNALS

Examining long runs of certain basic journals proved profitable as much for the flavor of the times as for the actual articles cited. Among these sources were the following:

Annual Report of the Association for the Advancement of Women, 1873–93.
Papers Presented at the Association for the Advancement of Women, 1873–91.
Publications (later *Journal*) *of the Association of Collegiate Alumnae* (later *American Association of University Women*), 1889–1940.
Women's Bureau (U.S. Department of Labor) Bulletins, 1918–40.
Independent Woman (published by the National Federation of Business and Professional Womens Clubs), 1919–40.
Bulletin of the American Association of University Professors, 1915–40.
Women's Work and Education, 1930–46.

In addition, the alumnae bulletins of the major women's colleges were helpful for both specific items and general atmosphere. Their usefulness would be greatly increased, however, if they were indexed either by themselves or by the major indexing publications, such as the *Biography Index*, the *Education Index*, or the *Reader's Guide*.

III. MANUSCRIPT COLLECTIONS AND ORAL HISTORIES

It is quite remarkable that so many of the major figures in this story have left large, accessible manuscript collections. Although rarely used before by historians of science, these collections were for the most part not hard to locate. Despite some delightful surprises, such as the Margaret Morse Nice Papers at Cornell or the Ida Hyde Papers at the American Association of University Women in Washington, D.C., most collections were in the obvious places: the archives of the women's colleges, those of the major graduate schools, and several other important repositories, such as the American Philosophical Society in Philadelphia, the Manuscript Division of the Library of Congress, and the Schlesinger Library at Radcliffe College. Not only have American archives grown magnificently in the 1970s, but several archival tools are now available to make locating collections easier. Andrea Hinding's two-volume *Women's History Sources* (1980) will expedite future searches (though its index cannot be relied upon) and may in a sense have superseded the older volumes of the *National Union Catalog of Manuscript Collections*. In addition, David Bearman's "Newsletter of the Survey of Sources for the History of Biochemistry and Molecular Biology," which appeared between 1975 and 1979, had many useful suggestions about collections to pursue. This project has since been described in John T. Edsall and David Bearman, "Historical Records of Scientific Activities: The Survey of Sources for the History of Biochemistry and Molecular Biology," *Proceedings of the American Philosophical Society* 123 (1979): 279–92, and its findings presented in their *Archival Sources for the History of Biochemistry and Molecular Biology: A Reference Guide and Report* (Boston: American Academy of Arts and Sciences, and Philadelphia: American Philosophical Society, 1980). It seems to have been a model project that other fields would do well to emulate.

Locating the collections was far easier than finding appropriate items within them. The collections were often vast and arranged either chronologically or alphabetically by correspondent, and I could only follow my hunches as to where to begin. It was here that serendipity and archivists played a large role, since some of the choicest letters, as those between two men about a particular woman, were not indexed in any initially telling way.

The most useful collections are arranged geographically here. Collections of particular importance are preceded by an asterisk (*).

CALIFORNIA

Berkeley

Bancroft Library

Nancy Bayley Oral History
H. E. Bolton Papers
Nina Floy Bracelin Papers
Herbert Evans Papers
Phoebe A. Hearst Papers
Mary Cover Jones Oral History
E. O. Lawrence Papers
A. O. Leuschner Papers

Robert H. Lowie Papers
Jean Walker MacFarlane Oral History
Ynes Mexia Papers
Agnes Fay Morgan Papers
Flora M. Scott Oral History (UCLA copy)
Millicent Shinn Papers
Women's Faculty Club Papers

Office for the History of Science and Technology (Archive for the History of Quantum Physics)
Hertha Sponer Oral History
Ernest Rutherford Microfilms

PASADENA

California Institute of Technology Archives
Biology Division Papers
George Ellery Hale Papers
Robert A. Millikan Papers
H. P. Robertson Papers
A. H. Sturtevant Papers

SAN DIEGO

University of California, San Diego, Mandeville Special Collections
*Maria Goeppart-Mayer Papers

SAN FRANCISCO

California Academy of Sciences
Alice Eastwood Papers

California Historical Society
Millicent Shinn Papers

SAN MARINO

Henry E. Huntington Library
John Bouvier Collection

SANTA BARBARA

University of California, Santa Barbara, Special Collections
Kappa Omicron Phi Scrapbook

SANTA CRUZ

Lick Observatory Archives (at University of California, Santa Cruz)
Outgoing Correspondence

STANFORD

Stanford University Archives
Faculty Files
David Starr Jordan Papers
*Lewis Terman Papers
Women's Faculty Club Papers

CONNECTICUT

NEW HAVEN

Yale University Library
Russell Chittenden Papers
*Ross G. Harrison Papers
Lafayette B. Mendel Papers
Presidential Papers
Charles Schuchert Papers
Robert A. Yerkes Papers

DISTRICT OF COLUMBIA

American Association of University Women Headquarters
Association Archives
Ida Hyde Papers

American Home Economics Association
Ellen Richards Papers

Library of Congress, Manuscript Division
Tasker Bliss Papers
*James McKeen Cattell Papers
Mira Lloyd Dock Papers
Anita Newcomb McGee Papers

Library of Congress, Rare Book Room
*National Science Club Papers

National Academy of Sciences Archives
Cori, Sabin, and Washburn Files
Correspondence regarding Curie visit and Sabin election

National Anthropological Archives (in U.S. National Museum)
Bureau of American Ethnology, Incoming Correspondence
Frances Densmore Papers
Alice C. Fletcher Papers
Matilda C. Stevenson Papers

National Library of Medicine, Historical Library
Ida Bengston File
Sara Branham File
Helen Dyer Oral History
Sarah Stewart Oral History and File

Smithsonian Institution Archives
Assistant Secretary's Papers
Mary Jane Rathbun Papers

GEORGIA

ATLANTA

Emory University, Robert W. Woodruff Library, Special Collections
*Charles Holmes Herty Papers

ILLINOIS

CHICAGO

Chicago Historical Society
World's Columbian Exposition Papers

University of Chicago, Regenstein Library, Special Collections
T. C. Chamberlain Papers
James Franck Papers
*Edwin O. Jordan Papers
Frank R. Lillie Papers
E. H. Moore Papers
*Marion Talbot Papers
Charles O. Whitman Papers

URBANA

University of Illinois Archives
Roger Adams Papers
Louisa Allen Papers
Lita Bane Papers
*Isabel Bevier Papers
Filbey Family Papers
Nellie Goldthwaite Papers
Charles Zeleny Papers

University of Illinois Mathematics Department
*Olive Hazlett Personnel File

INDIANA

BLOOMINGTON

Indiana University Archives
Mary Bidwell Breed Letters
Kate Heuvner Mueller File
Agnes Wells Biographical File

Lilly Library
Ralph Cleland Papers
Paul Weatherwax Papers

KANSAS

LAWRENCE

University of Kansas Archives
Ida Hyde File
Kate Stephens Collection

MANHATTAN

Kansas State University Archives
Margaret Justin File

KENTUCKY

LEXINGTON

University of Kentucky Archives
Ellen Semple Papers

MARYLAND

BALTIMORE

The Johns Hopkins University Archives
Janet Howell Clark and Christine Ladd-Franklin Items

The Johns Hopkins University Special Collections
"Admission of Women" File
Isaiah Bowman Papers
Rachel Carson File

The Alan Mason Chesney Medical Archives of the Johns Hopkins Medical Institutions
Florence Sabin Letters
*William H. Welch Papers

TOWSON

Goucher College Archives
Alumnae bulletins

MASSACHUSETTS

BOSTON

Boston Public Library
Hugo Münsterberg Collection

Massachusetts Historical Society
Edward Atkinson Papers

CAMBRIDGE

Harvard University Archives
*Edwin G. Boring Papers
Percy Bridgman Papers
Annie Jump Cannon Obituaries
Williamina P. Fleming Journal
A. Lawrence Lowell Papers
Frederic W. Putnam Papers

Farlow Herbarium
*William G. Farlow Letterbooks

Gray Herbarium
Historic Letter File

Houghton Library
William James Papers

Museum of Comparative Zoology Archives
Museum Collection

Massachusetts Institute of Technology Archives
Ellen Richards Papers
William Barton Rogers Papers

Massachusetts Institute of Technology Special Collections
Ellen Richards Letters

Radcliffe College—Schlesinger Library
Elizabeth Cary Agassiz Papers
*Bureau of Vocational Information Papers
Elizabeth Clarke Papers
Caroline Dall Papers
Ruth Holden Papers
Institute of Women's Professional Relations Papers
Morgan-Puffer Family Papers
Helen Brewster Owens Papers

NORTHAMPTON

Smith College Archives
Elizabeth Lawrence Clarke ('83) Papers
Faculty Files
Marjorie Hope Nicholson Papers

Sophia Smith Collection
Association of Collegiate Alumnae Papers
*Florence Bascom Papers
Ellen Richards Papers
Florence Sabin Papers

SOUTH HADLEY

Mount Holyoke College History Collection
Cornelia Clapp Papers
Department and Faculty Files
Lydia Shattuck Papers

WELLESLEY

Wellesley College Archives
Faculty Files
Margaret C. Ferguson Papers
Sarah Whiting Papers

MICHIGAN

ANN ARBOR

Bentley Historical Library
Elizabeth Crosby File
*Icie Macy Hoobler Papers
Pearl Kendrick File

Howard Lewis Papers
University of Michigan Faculty Women's Club Papers
Women's Research Club Papers

MISSOURI

COLUMBIA

University of Missouri—Western Historical Manuscripts Collection
Laws Observatory Papers
Lewis J. Stadtler Papers

SAINT LOUIS

Washington University Medical Center Archives
Gerty Cori File
Helen Tredway Graham Papers

NEW JERSEY

PRINCETON

Princeton University Special Collections
Edwin G. Conklin Papers
Henry Norris Russell Papers

NEW YORK

ALBANY

New York State Museum and Library
*Winifred Goldring Papers

ITHACA

Cornell University Archives and Regional History Office
Liberty Hyde Bailey Papers
Anna B. Comstock Papers
John Comstock Papers
Karl Dallenbach Papers
*Alice Evans Papers
Anna E. Jenkins Papers
*Margaret Morse Nice Papers
*Sigma Delta Epsilon Papers
Edward B. Titchener Papers

NEW YORK CITY

American Institute of Physics
Elizabeth Laird Autobiography
Melba Phillips Oral History

Barnard College Archives
Faculty and Department Files

Columbia University Oral History Office
Marjorie Hope Nicolson Oral History
Aryness Joy Wickens Oral History

Columbia University Library, Rare Books and Manuscripts Library (formerly Special Collections, Butler Library)
*Christine Ladd-Franklin Papers
Meloney-Curie Correspondence

New York Botanic Garden Archives
*Elizabeth G. Britton Papers
John Torrey Papers
Anna Vail Papers

New York Public Library (Annex)
Maria Trumbull Silliman Church Papers

NORTH TARRYTOWN

Rockefeller Foundation Archives Center
Files for Bryn Mawr College, University of California, Duke University, The Johns Hopkins University, Marine Biological Laboratory, University of Missouri, Mount Holyoke College, and National Research Council

POUGHKEEPSIE

Vassar College Archives
*Ruth Benedict Papers
Faculty Files
Caroline Furness Papers
Christine Ladd-Franklin Diaries
*Maria Mitchell Papers
Ellen Richards Papers
Mary Thaw Thompson Papers

ROCHESTER

University of Rochester, Rush Rhees Library, Special Collections
Alan Valentine Papers

University of Rochester Medical School Archives
George Whipple Scrapbooks

NORTH CAROLINA

DURHAM

Duke University Archives
William Hane Wannamaker Papers
William Preston Few Papers

OHIO

AKRON

University of Akron, Archives of the History of American Psychology
Elsie O. Bregman Papers
Cora Friedline Oral History

PENNSYLVANIA

BRYN MAWR

Bryn Mawr College Archives
Geology Department History

Bryn Mawr College, Geology Department
Florence Bascom Papers

PHILADELPHIA

American Philosophical Society
Simon Flexner Papers
Maria Mitchell Memorabilia (microfilm)
Henry A. Moe Papers
Peter Olitsky Papers
W. Osterhout Papers
Elsie Clews Parsons Papers
Raymond Pearl Papers
*Florence Sabin Papers
*Florence Seibert Papers

Medical College of Pennsylvania Archives and Special Collections
Rachel Bodley Papers
Louise Pearce Papers
Martha Tracy Papers

University of Pennsylvania Archives
Files on women

TEXAS

AUSTIN

University of Texas Archives, Barker Texas History Center
*A. Caswell Ellis Papers
University of Texas Faculty Women's Club Papers

University of Texas, Humanities Research Library
Botanical Society of America Archives

WISCONSIN

MADISON

University of Wisconsin Archives
Conrad Elvehejm Papers
E. B. Fred Oral History
E. B. Fred Papers
Ruth Henderson Oral History
Elizabeth McCoy File
*Helen Tracy Parsons Oral History
May S. Reynolds Oral History

WYOMING

LARAMIE

University of Wyoming, Western History Collections
Vernon and Florence Merriam Bailey Papers
Frederic and Edith Clements Papers
June Downey File

IV. PRINTED SOURCES

Although most of the work on the history of women scientists has been published in the last decade, many older printed items, such as biographies, government reports, statistical studies, and works of vocational guidance, document well particular aspects of this history. Only the more general of these works are listed here, because all are cited fully in the relevant notes.

Adams, Elizabeth Kemper. *Women Professional Workers: A Study Made For the Women's Educational and Industrial Union*. 1921. Reprint. New York: Macmillan, 1930.

Adams, J.F.A. "Is Botany a Suitable Study for Young Men?" *Science* 9 (1887): 117–18.

Aldrich, Michele L. "Review Essay: Women in Science." *Signs* 4 (1978): 126–35.

American Association of University Professors, Committee W. "Preliminary Report of Committee W on Status of Women in College and University Faculties." *Bulletin of the American Association of University Professors* 7 (1921): 21–32.

———. "Second Report of Committee W on the Status of Women in College and University Faculties." *Bulletin of the American Association of University Professors* 19 (1924): 65–73.

American Association of University Professors, Committee Y. *Depression, Recovery, and Higher Education*. New York: McGraw-Hill, 1937.

Baker, Gladys. "Women in the United States Department of Agriculture." *Agricultural History* 50 (1976): 190–201.

Berkin, Carol Ruth, and Norton, Mary Beth, eds. *Women of America: A History*. Boston: Houghton Mifflin, 1979.

Bernard, Jessie. *Academic Women*. Cleveland: World Publishing, Meridian Books, 1966.

Bernays, Edward L. *Biography of an Idea: Memoirs of a Public Relations Counsel*. New York: Simon & Schuster, 1965.

Bevier, Isabel. *Home Economics in Education*. Philadelphia: J. B. Lippincott, 1924.

Bevier, Isabel, and Usher, Susannah. *The Home Economics Movement*. Boston: Whitcomb & Barrows, 1912.

Bluemel, Elinor. *Florence Sabin: Colorado Woman of the Century*. Boulder: University of Colorado Press, 1959.

Bolzau, Emma L. *Almira Hart Lincoln Phelps: Her Life and Work*. Philadelphia: University of Pennsylvania Press, 1936.

Boring, Edwin G. "The Society of Experimental Psychologists, 1904–1938." *American Journal of Psychology* 51 (1938): 410–23.

———. "Titchener's Experimentalists." *Journal of the History of the Behavioral Sciences* 3 (1967): 315–25.

Brimhall, Dean R. "Family Resemblances among American Men of Science." *American*

Naturalist 56 (1922): 504–47 and *American Naturalist* 57 (1923): 74–88, 137–52, and 326–44.

Bryan, Alice I., and Boring, Edwin G. "Women in American Psychology: Prolegomenon." *Psychological Bulletin* 41 (1944): 447–54.

Burstyn, Joan. "Early Women in Education: The Role of the Anderson School of Natural History." *Boston University Journal of Education* 159 (1977): 50–64.

Chafe, William H. *The American Woman: Her Changing Social, Economic, and Political Roles, 1920–1970*. New York: Oxford University Press, 1972.

Clarke, Robert. *Ellen Swallow: The Woman Who Founded Ecology*. Chicago: Follett Publishing, 1973.

Clements, Edith S. *Adventures in Ecology*. New York: Hafner Publishing, 1960.

Coben, Stanley. "Foundation Officials and Fellowships: Innovation in the Patronage of Science." *Minerva* 14 (1976): 225–40.

Cole, Jonathan, and Cole, Stephen. *Social Stratification in Science*. Chicago: University of Chicago Press, 1973.

Conable, Charlotte Williams. *Women at Cornell: The Myth of Equal Education*. Ithaca: Cornell University Press, 1977.

"Confessions of a Woman Professor." *Independent* 55 (1903): 954–58.

Crawford, H. Jean. "The Association to Aid Research by Women." *Science* 76 (1932): 492–93.

Dash, Joan. *A Life of One's Own: Three Gifted Women and the Men They Married*. New York: Harper & Row, 1973.

Davey, Laura Gunn. "The History of Women in Physics." Mimeographed. State College: Pennsylvania State University, 1940.

Davis, Audrey B. *Bibliography on Women: With Special Emphasis on Their Roles in Science and Society*. New York: Science History Publications, 1974.

Deegan, Mary Jo. "Women and Sociology." *Journal of the History of Sociology* 1 (1978): 11–32.

Dexter, Ralph S. "The Annisquam Sea-Side Laboratory of Alpheus Hyatt, Predecessor of the Marine Biological Laboratory at Woods Hole, 1880–1886." In *Oceanography: The Past*, edited by Mary Sears and Daniel Merriman, pp. 94–100. New York: Springer-Verlag, 1980.

———. "Guess Who's Not Coming to Dinner: Frederic Ward Putnam and the Support of Women in Anthropology." *History of Anthropology Newsletter* 5, no. 2 (1978): 5–6.

Doolittle, Dortha Bailey. "Women in Science." *Journal of Chemical Education* 22 (1945): 171–74.

Drake, Thomas E. *A Scientific Outpost: The First Half-Century of the Nantucket Maria Mitchell Association*. Nantucket, Mass.: Nantucket Maria Mitchell Association, 1968.

Eagles, Juanita Archibald; Pye, Orrea Florence; and Taylor, Clara Mae. *Mary Swartz Rose, 1874–1941: Pioneer in Nutrition*. New York: Teachers College Press, 1979.

Eells, Walter Crosby. "Earned Doctorates for Women in the Nineteenth Century." *Bulletin of the American Association of University Professors* 42 (1956): 644–51.

Farnsworth, Marie K. "Women in Chemistry: A Statistical Study." *Industrial and Engineering Chemistry* 3 (1925): 4.

Filene, Catherine. *Careers for Women*. 1920. Revised. Boston: Houghton Mifflin, 1934.

Finch, Edith. *Carey Thomas of Bryn Mawr*. New York: Harper & Bros., 1947.

Fleischman [Bernays], Doris E., comp. and ed. *An Outline of Careers for Women: A Practical Guide to Achievement*. Garden City, N.Y.: Doubleday, Doran, 1931.

Fleming, Mrs. M. (Williamina P.) "A Field for Woman's Work in Astronomy." *Astronomy and Astrophysics* 12 (1893): 683-89.

Flexner, Eleanor. *Century of Struggle: The Woman's Rights Movement in the United States*. Cambridge: Harvard University Press, 1959.

Flexner, Simon. "The Scientific Career for Women." *Scientific Monthly* 13 (1921): 97-105.

Frankfort, Roberta. *Collegiate Women: Domesticity and Career in Turn-of-the-Century America*. New York: New York University Press, 1977.

Goff, Alice C. *Women CAN Be Engineers*. Youngstown, Ohio: private, 1946.

Goldberg, Morris. "[Jewish] Women in the Realm of Science." *American Hebrew* 126 (1929-30): 312.

A Growing College: Home Economics at Cornell University. Ithaca: Cornell University Press, 1969.

Haber, Julia Moesel. "Women in the Biological Sciences." Mimeographed. State College: Pennsylvania State University, 1939.

Hamilton, Alice C. *Exploring the Dangerous Trades: The Autobiography of Alice Hamilton, M.D.* Boston: Little, Brown, 1943.

Harmon, Lindsey, and Soldz, Herbert, comps. *Doctorate Production in United States Universities, 1920-1962*. Washington, D.C.: National Academy of Sciences-National Research Council Publication no. 1142 (1963).

Harshbarger, John W. *The Botanists of Philadelphia and Their Work*. Philadelphia: T. C. Davis & Son, 1899.

Hawthorne, Marion O. "Women as College Teachers." *Annals of the American Academy of Political and Social Sciences* 143 (1929): 146-53.

Ho, Ching-Ju. *Personnel Studies of Scientists in the United States*. Teachers College, Columbia University, Contributions to Education no. 298 (1928).

Hollingworth, Harry L. *Leta Stetter Hollingworth: A Biography*. Lincoln: University of Nebraska Press, 1943.

Hummer, Patricia. *Decade of Elusive Promise: Professional Women in the United States, 1920-1930*. Studies in American History and Culture no. 5. [Ann Arbor]: UMI-Research Press, 1979.

Hunt, Caroline L. *The Life of Ellen H. Richards*. Boston: Whitcomb & Barrows, 1912.

Hutchinson, Emilie. *Women and the Ph.D.* Institute of Women's Professional Relations, Greensboro, N.C., Bulletin no. 2 (1929).

Hyde, Ida. "Before Women Were Human Beings: Adventures of an American Fellow in German Universities of the '90s." *Journal of the American Association of University Women* 31 (1938): 226-36.

Ingels, Margaret. "Petticoats and Slide Rules." *Midwest Engineer* 5 (1952): 2-4 and 10-16.

Jones, Bessie Z., and Boyd, Lyle. *The Harvard College Observatory: The First Four Directorships, 1839-1919*. Cambridge: Harvard University Press, 1971.

Kanter, Rosbeth Moss. *Men and Women of the Corporation*. New York: Basic Books, 1977.

Kendall, Elaine. *"Peculiar Institutions": An Informal History of the Seven Sister Colleges*. New York: Putnam, 1976.

Kendall, Phebe Mitchell, comp. *Maria Mitchell: Life, Letters, and Journals*. Boston: Lee & Shepard Publishers, 1896.

Kerber, Linda. *Women of the Republic: Intellect and Ideology in Revolutionary America*. Chapel Hill: University of North Carolina Press, 1980.

Klein, Viola. *The Feminine Character: History of an Ideology*. 2d ed. 1971. Reprint. Urbana: University of Illinois Press, 1972.

Kohlstedt, Sally Gregory. "In from the Periphery: American Women in Science, 1830–1880." *Signs* 4 (1978): 81–96.

______. "Maria Mitchell: The Advancement of Women in Science." *New England Quarterly* 51(1978): 39–63.

______. "The Nineteenth-Century Amateur Tradition: The Case of the Boston Society of Natural History." In *Science and Its Public: The Changing Relationship*, edited by Gerald Holton and William A. Blanpied, pp. 173–90. Dordrecht, Holland: D. Reidel, 1976.

______. "Single-Sex Education and Leadership: The Early Years of Simmons College." In *Women and Educational Leadership: A Reader*, edited by Sari Knopp Biklen and Marilyn B. Brannigan, pp. 93–112. Lexington: Lexington Press, 1980.

Landis, W. S. "Women Chemists in Industry." *Journal of Chemical Education* 16 (1939): 577–79.

Lemons, J. Stanley. *The Woman Citizen: Social Feminism in the 1920s*. Urbana: University of Illinois Press, 1973.

Lomax, Elizabeth. "The Laura Spelman Rockefeller Memorial: Some of Its Contributions to Early Research in Child Development." *Journal of the History of the Behavioral Sciences* 13 (1977): 283–93.

Lurie, Nancy Oestreich. "Women in Early Anthropology." In *Pioneers of American Anthropology*, edited by June Helm [MacNeish], pp. 29–81. Seattle: University of Washington Press, 1966.

Lutz, Alma. *Emma Willard: Pioneer Educator of American Women*. Boston: Beacon Press, 1964.

[MacMillan, Lucille Foster.] *Women in the Federal Service*. Washington, D.C.: U.S. Civil Service Commission, 1938.

Maltby, Margaret E., comp. *History of the Fellowships Awarded by the American Association of University Women, 1888–1929, With the Vitas of the Fellows*. Washington, D.C.: American Association of University Women [1929].

Mead, Margaret. *Blackberry Winter: My Earlier Years*. New York: Simon and Schuster, 1972.

______. *Male and Female: A Study of the Sexes in a Changing World*. New York: William Morrow, 1949.

______. "Sex and Achievement." *Forum and Century* 94 (1935): 301–3.

Menard, Henry. *Science: Growth and Change*. Cambridge: Harvard University Press, 1971.

Mozans, H. J. [pseud.] *Woman in Science*. 1913. Reprint. Cambridge: MIT Press, 1974.

Newcomer, Mabel. *A Century of Higher Education for American Women*. New York: Harper & Bros., 1959.

Nice, Margaret Morse. *Research Is a Passion with Me*. Toronto: Consolidated Amethyst Communications, 1979.

Nienburg, Bertha M. *Women in the Government Service*. Women's Bureau Bulletin no. 8 (1920).

Norton, Mary Beth. *Liberty's Daughters: The Revolutionary Experience of American Women, 1750–1800*. Boston: Little, Brown, 1980.

[Nyswander, Rachel Fesler, and Hooks, Janet M.] *Employment of Women in the Federal Government, 1923 to 1939*. Women's Bureau Bulletin no. 182 (1941).

Opfell, Olga S. *The Lady Laureates: Women Who Have Won the Nobel Prize*. Metuchen, N.J.: Scarecrow Press, 1978.

Organization and Historical Sketch of the Women's Anthropological Society of America. Washington, D.C.: The Society, 1889.

Owens, Helen Brewster. "Early Scientific Work of Women and Women in Mathematics." Mimeographed. State College: Pennsylvania State University [1940].

Payne, Alma Smith. *Partners in Science*. Cleveland: World Publishing, 1968.

[Phelps], Almira H. Lincoln. *Familiar Lectures on Botany*. Hartford: H. and F. J. Huntington, 1839.

Pollard, Lucille Addison. *Women on College and University Faculties: A Historical Survey and a Study of Their Present Academic Status*. New York: Arno Press, 1977.

Powdermaker, Hortense. *Stranger and Friend: The Way of an Anthropologist*. New York: W. W. Norton, 1966.

Reid, Robert. *Marie Curie*. New York: Saturday Review Press, 1974.

Remington, Jeanne E. "Katharine Jeannette Bush: Peabody's Mysterious Zoologist." *Discovery* 12, no. 3 (1977): 3-8.

Robinson, Mabel L. *The Curriculum of the Woman's College*. U.S. Bureau of Education Bulletin 1918, no. 6.

Rose, Mary Swartz. "University Teaching of Nutrition and Dietetics in the United States." *Nutrition Abstracts and Reviews* 4 (1935): 439-46.

Rosenberg, Rosalind. *Beyond Separate Spheres: Intellectual Roots of Modern Feminism*. New Haven: Yale University Press, 1982.

Rossiter, Margaret W. "Doctorates for American Women." *History of Education Quarterly*, in press.

———. "Florence Sabin: First Woman in the National Academy of Sciences." *American Biology Teacher* (1977): 484-86 and 494.

———. "Sexual Segregation in the Sciences: Some Data and a Model." *Signs* 4 (1978): 146-51.

———. "Women Scientists in America before 1920." *American Scientist* 62 (1974): 312-23. Reprinted in *Dynamos and Virgins Revisited: Women and Technological Change in History, An Anthology*, edited by Martha Moore Trescott, pp. 120-48. Metuchen, N.J.: Scarecrow Press, 1979.

———. " 'Women's Work' in Science, 1880-1910." *Isis* 71 (1980): 381-98.

Rudolph, Emmanuel D. "How It Developed that Botany Was the Science Thought Most Suitable for Victorian Young Ladies." *Children's Literature* 2 (1973): 92-97.

Saint, Avis Marion. "Women in the Public Service: no. 2: The City of Berkeley." *Public Personnel Studies* 8 (1930): 104-7.

———. "Women in the Public Service: no. 3: The City of Oakland." *Public Personnel Studies* 8 (1930): 119-22.

———. "Women in the Public Service: no. 4: The Federal Service of the United States." *Public Personnel Studies* 9 (1931): 14-19.

Scharf, Lois. *To Work and to Wed: Female Employment, Feminism, and the Great Depression*. Westport, Conn.: Greenwood Press, 1980.

Scott, Anne Firor. "The Ever Widening Circle: The Diffusion of Feminist Values from the Troy Female Seminary, 1822-1872." *History of Education Quarterly* 19 (1979): 3-25.

Seibert, Florence. *Pebbles on the Hill of a Scientist*. St. Petersburg, Fla.: private, 1968.

Seller, Maxine. "G. Stanley Hall and Edward Thorndike on the Education of Women: Theory and Policy in the Progressive Era." *Educational Studies*, in press.

Sexton, Anna M. *A Chronicle of the Division of Laboratories and Research: New York State Department of Health, The First Fifty Years*. Lunenburg, Vt.: Stinehour Press, 1967.

Sheldon, Ruth. "The Ladies Find Oil." *Scribner's Commentator* 10 (1941): 28-32.

Sims, Lewis B. "Social Scientists in the Federal Service." In *Public Policy: A Yearbook of the Graduate School of Public Administration, Harvard University, 1940*, edited by Carl J. Friedrich and Edward S. Mason, pp. 280-96. Cambridge: Harvard University Press, 1940.

Sloan, Jan Butin. "The Founding of the Naples Table Association for Promoting Scientific Research by Women, 1897." *Signs* 4 (1978): 208-16.

Smith, Isabel Fothergill. *The Stone Lady: A Memoir of Florence Bascom*. Bryn Mawr, Pa.: Bryn Mawr College Library, 1981.

Smith, Malcolm L., and Wright, Kathryn R. "Occupations and Salaries in Federal Employment." *Monthly Labor Review* 52 (1941): 66-85.

Statistical Work: A Study of Opportunities for Women. Studies in Occupations no. 2. New York City: Bureau of Vocational Information, 1921.

The Status of Women in the Government Service in 1925. Women's Bureau Bulletin no. 53 (1926).

Talbot, Marion. *The Education of Women*. Chicago: University of Chicago Press, 1910.

______. *More Than Lore: Reminiscences of Marion Talbot*. Chicago: University of Chicago Press, 1936.

Talbot, Marion, and Rosenberry, Lois K. M. *The History of the American Association of University Women, 1881-1931*. Boston: Houghton Mifflin, 1931.

Thomas, M. Carey. *Education of Women*. Volume 7 of *Monographs on Education in the United States*, edited by Nicholas Murray Butler. Department of Education for the United States Commission to the Paris Exposition of 1900. Albany: J. B. Lyon, 1900.

Tobey, Ronald C. *The American Ideology of National Science, 1919 1930*. Pittsburgh: University of Pittsburgh Press, 1971.

Trecker, Janice Law. "Sex, Science, and Education." *American Quarterly* 26 (1974): 353-66.

Turner, Thomas B. *Heritage of Excellence: A History of The Johns Hopkins Medical Institutions, 1919-1947*. Baltimore: The Johns Hopkins University Press, 1974.

Visher, Stephen Sargent. *Scientists Starred, 1903-1943, in "American Men of Science."* Baltimore: The Johns Hopkins University Press, 1947.

Wall, Florence E. "The Status of Women Chemists." *Chemist* 15 (1938): 174-92.

Walsh, Mary Roth. *"Doctors Wanted, No Women Need Apply": Sexual Barriers in the Medical Profession, 1835-1975*. New Haven: Yale University Press, 1977.

Warner, Deborah. *Graceanna Lewis: Scientist and Humanitarian*. Washington, D.C.: Smithsonian Institution Press, 1979.

______. "Science Education for Women in Antebellum America." *Isis* 69 (1978): 58-67.

Weigley, Emma Seifrit. "It Might Have Been Euthenics: The Lake Placid Conferences and the Home Economics Movement." *American Quarterly* 26 (1974): 79-96.

Welch, William H. "Contribution of Bryn Mawr College to the Higher Education of Women." *Science* 56 (1922): 1-8.

Welt, Ida. "The Jewish Woman in Science." *Hebrew Standard* 50, no. 11 (1907): 4.

Welter, Barbara. "Anti-Intellectualism and the American Woman, 1800-1860." *Mid-America* 48 (1966): 258-70.

Westcott, Cynthia. *Plant Doctoring Is Fun*. Princeton, N.J.: Van Nostrand, 1957.

Whitney, Mary W. "Scientific Study and Work for Women." *Education* 3 (1882): 58-69.

Willard, Mary L. "Pioneer Women in Chemistry." Mimeographed. State College: Pennsylvania State University [1940].

Wilson, Joan Hoff. "Dancing Dogs of the Colonial Period: Women Scientists." *Early American Literature* 7 (1973): 225-35.

Women in Chemistry: A Study of Professional Opportunities. New York: Bureau of Vocational Information, 1922.

Woody, Thomas. *A History of Women's Education in the United States*. 2 vols. New York: Science Press, 1929.

Worner, Ruby K. "Opportunities for Women Chemists in Washington." *Journal of Chemical Education* 16 (1939): 583-85.

Wright, Helen. *Sweeper in the Sky: The Life of Maria Mitchell, First Woman Astronomer in America*. New York: Macmillan, 1949.

Yost, Edna. *American Women of Science*. 1943. Revised. Philadelphia: Lippincott, 1955.

———. *Women of Modern Science*. New York: Dodd, Mead, 1959.

V. SELECTED UNPUBLISHED MATERIAL

Amatniek, Joan Cindy. "The Women's Anthropological Society of America: A Dual Role—Scientific Society and Woman's Club." Undergraduate honors thesis, Harvard University, 1979.

Bever, Marilyn A. "The Women of MIT, 1871-1941: Who They Were, What They Achieved." Undergraduate thesis, MIT, 1976.

Brand, Barbara E. "The Influence of Higher Education on Sex-Typing in Three Professions, 1870-1920: Librarianship, Social Work, and Public Health." Ph.D. dissertation, University of Washington, 1978.

Finneran, Helen T. "Louise Stanley: A Study of the Career of a Home Economist, Scientist, and Administrator, 1923-1953." Master's thesis, American University, 1965.

Mack, Pamela E. "Women in Astronomy in the United States, 1875-1920." Undergraduate honors thesis, Harvard University, 1977.

Napoli, Donald S. "The Architects of Adjustment: The Practice and Professionalization of American Psychology, 1920-1945." Ph.D. dissertation, University of California, Davis, 1975.

INDEX